STUDENT SOLUTIONS MANUAL FOR

ALGEBRA & TRIGONOMETRY

Second Edition

BY STANLEY I. GROSSMAN

GEORGE BRADLEY & DANIEL BARBUSH

Duquesne University

SAUNDERS COLLEGE PUBLISHING
Harcourt Brace Jovanovich College Publishers
Fort Worth Philadelphia San Diego New York Orlando Austin
San Antonio Toronto Montreal London Sydney Tokyo

Printed in the United States of America.

Grossman: Student Solutions Manual to accompany ALGEBRA & TRIGONOMETRY, 2/E

ISBN 0-03-053162-4

234 095 987654321

Preface

This Student Solutions Manual accompanies <u>Algebra and Trigonometry</u>, second edition by Stanley I. Grossman. It contains complete solutions to the odd-numbered problems in the book. Also included are chapter objectives, hints on how to solve problems, and lists of important equations and definitions summarized from each chapter. While solving problems from the book you should try each problem first on your own before consulting this manual. It is only by practicing mathematics that you can truly learn the subject.

George Bradley
Daniel Barbush

TABLE OF CONTENTS

CHAPTER ONE
Basic Concepts of Algebra

Chapter <u>Objectives</u>

By working through this chapter you will become familiar with many of the basic "nuts and bolts" type operations of algebra (arithmetic with letters). The ability to smoothly manipulate expressions into nicer forms, often called simplifying, and the understanding of <u>WHY</u> the simplifications are valid in terms of the properties of real numbers are the major objectives. Fluency in working with exponents (positive, negative, fractional), radicals, polynomials and rational expressions (fractions!) will only result from doing <u>MANY</u> problems.

Summary

Numbers

- The natural numbers, denoted by N, are 1, 2, 3, . . .
- The integers, denoted by Z, are 0, ±1, ±2, ±3, . . .
- The rational numbers, denoted by Q, are all numbers that can be written r=m/n where m and n are integers and $n \neq 0$
- The irrational numbers consist of all real numbers that are not rational.
- The real numbers are the rationals combined with the irrationals.

<u>Rules</u> <u>of</u> <u>Signs</u>: Let a and b be real numbers

- $-a = (-1)a$
- $-(-a) = a$
- $(-a)(-b) = ab$
- $-a + (-b) = -(a+b)$

<u>Rules</u> <u>of</u> <u>Fractions</u>

- $\dfrac{a}{b} + \dfrac{c}{b} = \dfrac{a+c}{b}$ $\dfrac{a}{b} + \dfrac{c}{d} = \dfrac{ad+bc}{bd}$

- $\dfrac{ad}{bd} = \dfrac{a}{b}$ $\dfrac{a}{b} \cdot \dfrac{c}{d} = \dfrac{ac}{bd}$

Properties of Inequalities

- If $a < b$, then $a + c < b + c$
- If $a < b$ and $b < c$, then $a < c$
- If $a < b$ and $c > 0$, then $ac < bc$
- If $a < b$ and $c < 0$, then $ac > bc$
- If $ab > 0$ and $a < b$, then $\frac{1}{a} > \frac{1}{b}$

Properties of Absolute Value

- $|a| = a$ if $a \geq 0$
- $|a| = -a$ if $a < 0$ (note that in this case $-a$ is a positive number)
- $|-a| = |a|$
- $|ab| = |a| \cdot |b|$
- $|a + b| \leq |a| + |b|$ $\bullet$ $|a|^2 = a^2$

Properties of Exponents: assume $a \neq 0$ and m and n are integers

- $a^n = a \cdot a \cdot a \cdot \ldots \cdot a$ n times if $n > 0$ $a^{-n} = \frac{1}{a^n}$

- $a^{1/n} =$ the nth root of a $a^{m/n} = (a^{1/n})^m$

- $a^0 = 1$ $a^1 = a$ $\frac{a^m}{a^n} = a^{m-n} = \frac{1}{a^{n-m}}$ $a^m \cdot a^n = a^{m+n}$

- $(ab)^n = a^n b^n$ $(a/b)^n = a^n / b^n$ $(a^m)^n = a^{mn}$

- $a^{-1} = \frac{1}{a}$ $\left(\frac{a}{b}\right)^{-1} = \frac{b}{a}$ $\left(\frac{a}{b}\right)^{-n} = \left(\frac{b}{a}\right)^n$

- For $a > 0$, $a^{1/2} = \sqrt{a}$ (which is positive) and $-a^{1/2} = -\sqrt{a}$ (which is negative)

- If $a, b > 0$, $\sqrt{ab} = \sqrt{a}\sqrt{b}$ and $\frac{\sqrt{a}}{\sqrt{b}} = \sqrt{\frac{a}{b}}$

Properties of Polynomials

- A polynomial in x is an expression of the form

$$p(x) = a_n x^n + a_{n-1} x^{n-1} + \ldots + a_2 x^2 + a_1 x + a_0$$

where $a_0, a_1, \ldots a_n$ are real numbers called the coefficients of the polynomial p.

- To add (or subtract) two polynomials, add (or subtract) their corresponding coefficients.

- If a, b, and c are polynomials then $a(b+c) = ab + ac$ and $(a+b)c = ac + bc$

- $(x + y)^2 = x^2 + 2xy + y^2$ $(x - y)^2 = x^2 - 2xy + y^2$ $(x + y)(x - y) = x^2 - y^2$

- $(x + y)^3 = x^3 + 3x^2y + 3xy^2 + y^3$ $(x - y)^3 = x^3 - 3x^2y + 3xy^2 - y^3$

- $x^3 + y^3 = (x + y)(x^2 - xy + y^2)$ $x^3 - y^3 = (x - y)(x^2 + xy + y^2)$

Rules for Simplifying Rational Expressions

- $\dfrac{a(x)\,d(x)}{b(x)\,d(x)} = \dfrac{a(x)}{b(x)}$ (you can divide out common factors)

- $\dfrac{a(x)}{b(x)} \pm \dfrac{c(x)}{b(x)} = \dfrac{a(x) \pm c(x)}{b(x)}$

- $\dfrac{a(x)}{b(x)} \pm \dfrac{c(x)}{d(x)} = \dfrac{a(x)\,d(x) \pm b(x)\,c(x)}{b(x)d(x)}$ (use this only if b(x) and d(x) have no common factor)

- $\dfrac{a(x)}{b(x)} \cdot \dfrac{c(x)}{d(x)} = \dfrac{a(x)\,c(x)}{b(x)\,d(x)}$

- $\dfrac{a(x)\,/\,b(x)}{c(x)\,/\,d(x)} = \dfrac{a(x)}{b(x)} \cdot \dfrac{d(x)}{c(x)}$

SOLUTIONS TO CHAPTER ONE PROBLEMS

Problems 1.2

1. Let r = 0.2222222...   since one digit repeats, multiply both sides by 10^1 or 10.

   10r = 2.2222222...   now subtract the top equation from the bottom one.

   9r = 2   so $r = \frac{2}{9}$ by dividing both sides by 9.

3. Let r = 0.147147...   since three digits repeat, multiply both sides by 10^3

   1000r= 147.147147...   subtract top from bottom to get

   999r = 147 so $r = \frac{147}{999} = \frac{49}{333}$

5. Let r = 2.123123123...   three digits repeat (the 123); multiply by 1000

   1000r= 2123.123123123.....

   999r = 2121   so $r = \frac{2121}{999} = \frac{707}{333}$

7. Keystrokes: ⑤ ÷ ⑥ ▭ after which you should see displayed 0.83333333 or $0.8\bar{3}$

9. Display should be 0.41666667 which is the rounded version of 0.416666666.... or $0.41\bar{6}$

11. Display should be 0.1287129, a rounded version of 0.128712871287... or $0.\overline{1287}$

13. 1.78125 (if you want to see something repeating, write it as 1.7812500000... or $1.78125\bar{0}$)

15. Choose irrational opposites like a = $\sqrt{5}$ and b= $-\sqrt{5}$ which, when added, will sum to 0 which is

    a rational number.

17. a) 1/9 done on a calculator yields 0.11111111....

    1/11 similarly gives 0.09090909.....

    b) 1/27 gives 0.037037037.... ;   1/37 gives 0.027027027.....

    c) The denominators of the fractions must produce either 99 or 999 or 9999 etc. when

    multiplied together.

<u>Problems</u> <u>1.3</u>

1. The order of addition has been changed (reversed) which is what $a + b = b + a$
   (additive commutativity) says.

3. The factors have been regrouped but the order has not been changed. This is allowable by the
   associative law of multiplication: $a(bc) = (ab)c$.

5. z has been multiplied by 1 giving an "identical" z; this is justified by the multiplicative identity
   property.

7. Here the addends are regrouped (no change in order) so this is an example of the associative law of
   addition.

9. The number xyz is being multiplied by its reciprocal $\frac{1}{xyz}$ (OK since $xyz \neq 0$) which makes 1. We
   call this the (existence of the) multiplicative inverse.

11. The 3, on the left, is distributed (by multiplication) over the sum $u + w$ and this is called the left
    distributive law.

13. If equals (equal expressions), in this case 2 and 2, are added to equals (x and y), the results are
    equal (Additive substitution).

15. If equals are multiplied by equals, the results are equal. In symbols: if $a = b$ then $ac = bc$
    (Multiplicative substitution).

17. Rewriting x - 3 = y - 3 as $x + (-3) = y + (-3)$ allows us now to state that $x = y$ because of the
    additive reduction law: if $a + c = b + c$ then $a = b$.

19. 3 has been added to both sides of the equality $a = 7$ and this is justified by additive substitution.

21. $\frac{2}{3} + \frac{5}{3} = \frac{2 + 5}{3} = \frac{7}{3}$

23. $\frac{1}{4} + \frac{1}{3} = \frac{1 \cdot 3 + 1 \cdot 4}{4 \cdot 3} = \frac{3 + 4}{12} = \frac{7}{12}$ (denominators had no factors in common)

25. $\frac{5}{7} - \frac{3}{7} = \frac{2}{7}$ 27. $\frac{4}{5} - \frac{3}{4} = \frac{4 \cdot 4 - 3 \cdot 5}{5 \cdot 4} = \frac{16 - 15}{20} = \frac{1}{20}$ 29. $3 \cdot \frac{7}{4} = \frac{3}{1} \cdot \frac{7}{4} = \frac{21}{4}$

31. $\frac{7}{2} \cdot 4 = \frac{7}{1} \cdot 2 = 14$ 33. $\frac{3}{4} \cdot \frac{4}{5} = \frac{3}{1} \cdot \frac{1}{5} = \frac{3}{5}$ 35. $\frac{7}{3} \cdot \frac{6}{7} = \frac{1}{1} \cdot \frac{2}{1} = 2$

37. $\frac{3}{4} \cdot \frac{5}{7} = \frac{3 \cdot 5}{4 \cdot 7} = \frac{15}{28}$ 39. $\frac{3}{4} \cdot \frac{2}{5} = \frac{3}{2} \cdot \frac{1}{5} = \frac{3}{10}$ 41. $\frac{3/5}{4/7} = \frac{3}{5} \div \frac{4}{7} = \frac{3}{5} \cdot \frac{7}{4} = \frac{21}{20}$

43. $\frac{1}{8/15} = 1 \div \frac{8}{15} = 1 \cdot \frac{15}{8} = \frac{15}{8}$ 45. $\frac{1}{2} + \frac{1}{3} + \frac{1}{4} = \frac{?}{12} + \frac{?}{12} + \frac{?}{12} = \frac{6}{12} + \frac{4}{12} + \frac{3}{12} = \frac{13}{12}$

47. $\frac{3}{4} + \frac{2}{5} - \frac{5}{6} =$ LCD is 60 $\frac{45}{60} + \frac{24}{60} - \frac{50}{60} = \frac{45 + 24 - 50}{60} = \frac{19}{60}$

49. $\frac{2/3 + 1/4}{3/5 - 1/6} = \left(\frac{2}{3} + \frac{1}{4}\right) \div \left(\frac{3}{5} - \frac{1}{6}\right) = \left(\frac{8 + 3}{12}\right) \div \left(\frac{18 - 5}{30}\right) = \frac{11}{12} \cdot \frac{30}{13} = \frac{11}{2} \cdot \frac{5}{13} = \frac{55}{26}$

The solutions to many of the theoretical exercises will be presented in the form:

> *Hypothesis: (what's given or known at the start)*
> *Conclusion: (what we're supposed to prove is true)*

> *Proof:*
> > *(series of statements)   (series of reasons)*

The abbreviations H, C, and P will stand for hypothesis, conclusion, and proof.

51. H: $ac = bc$ and $c \neq 0$

C: $a = b$

P: $ac = bc$ given

   $(ac) \cdot \frac{1}{c} = (bc) \cdot \frac{1}{c}$ Multiplicative substitution

   $a(c \cdot \frac{1}{c}) = b(c \cdot \frac{1}{c})$ Associative law for multiplication

   $a(1) = b(1)$ Multiplicative inverse idea

   $a = b$ Multiplicative identity

53. *hint: often, when we're trying to show that someting is unique, our ploy is to assume there are two of the particular item and then show, using the properties, that these two items are really one and the same.*

    H: a is some real number

    C: The additive inverse of a, denoted by -a, is unique

      P: assume that b is another additive inverse for a (in additon to -a)

    so we have a + (-a) = 0 and a + b = 0 both being true.

    therefore, a + (-a) = a + b   (since they're both equal to 0)

    so now we can say that (-a) = b by the additive reduction law and this

    means that the two additive inverses, -a and b, are really the same.

55. H: a is a real number

    C: -(-a) = a

    P: $-(-a) = -1(-1 \cdot a)$ *multiplying a number by -1 produces its additive inverse*

           $= (-1 \cdot -1) \cdot a$ *associative property of multiplication*

           $= (1) \cdot a$ *multiplying a number by -1 produces the additive inverse of the number*

           $= a$ (1) *is the multiplicative identity*

57. *H: a and b are real numbers*

    C: $(-a)(-b) = ab$

    P: $(-a)(-b) = (-1 \cdot a)(-1 \cdot b)$ *multiplication by -1 produces additive inverse*

           $= (-1) \cdot (a \cdot -1) \cdot b$ *associative property of multiplication*

           $= (-1) \cdot (-1 \cdot a) \cdot b$ *commutative property of multiplication*

           $= (-1 \cdot -1) \cdot (a \cdot b)$ *associative property of multiplication*

           $= (1) \cdot (ab)$ *multiplication by -1*

           $= ab$ *multiplicative identity*

59. *H: a, b, and c are real numbers*

    C: $a(b - c) = ab - ac$

    P: $a(b - c) = a(b + (-c))$ *definition of subtraction*

           $= ab + a(-c)$ *left distributive law*

           $= ab + (-ac)$ *one of the rules of signs (property c)*

           $= ab - ac$ *definition of subtraction*

61. *H:* $ab = 0$

C: $a = 0$ or $b = 0$ or both $a = 0$ and $b = 0$

P: (this proof is indirect: we assume the OPPOSITE of what we want to prove and show that this leads to a contradiction of a known fact; arriving at a contradiction tells us that whatever we assumed must not be true.)

ASSUME that neither $a = 0$ nor $b = 0$ is true; then both a and b have multiplicative inverses $\frac{1}{a}$ and $\frac{1}{b}$ respectively.

$ab = 0$ *given*

$\frac{1}{a}(ab)\frac{1}{b} = \frac{1}{a}(0)\frac{1}{b}$ *multiplicative substitution*

$(\frac{1}{a} \cdot a)(b \cdot \frac{1}{b}) = 0$ *associative property of multiplication and multiplicative property of zero (see #34)*

$(1)(1) = 0$ *action of multiplicative inverses*

$1 = 0$ *identity of multiplication*

BUT THIS IS ABSURD!

therefore our assumption must be false and if the statement that neither $a = 0$ nor $b = 0$ is false then the statement that either $a = 0$ or $b = 0$ or both a and b are 0 must be true.

63. To show that a - b = b - a is false is easy; any choices for a and b except those where $a = b$ should show a violation. For example: $a = 5$; $b = 2$ would give us 5 - 2 = 2 - 5.   (does 3 = -3?)

65. Let's try $a = 1$; $b = 2$; $c = 3$; $d = 4$: $\frac{a + b}{c + d} = \frac{1+2}{3+4} = \frac{3}{7}$

but $\frac{a}{c} + \frac{b}{d} = \frac{1}{3} + \frac{2}{4} = \frac{10}{12} = \frac{5}{6}$ and $\frac{3}{7} \neq \frac{5}{6}$

67. Let's subtract -8 from -2:

If the negative numbers are closed under subtraction, the result should still be negative: -2 - (-8) = -2 + 8 = 6 which ISN'T negative so the negative numbers aren't closed under subtraction.

Problems 1.4

1. $-3 < 2$ since -3 is to the left of 2 on a number line.

3. $-3 > -8$; -3 is to the right of -8.

5. $\frac{2}{3} = 0.666666....$ and $0.6666666... > 0.6$ so $\frac{2}{3} > 0.6$

7. $-\frac{1}{3} > -\frac{1}{2}$ because of the fact that $\frac{1}{3} < \frac{1}{2}$ and the order reversing property (multiply by -1).

9. $-\frac{1}{3} < \frac{1}{2}$ since ANY negative number is to the left of ANY positive one.

11. $|\,7\,| - |\,3\,| = 7 - 3 = 4$ (absolute value signs are signs of grouping and must be evaluated before the subtraction.)

13. $|\,4\,| - |\,-9\,| = 4 - 9 = 4 + (-9) = -5$

15. $|\,\sqrt{2} - 4\,|$:   must decide which is larger in absolute value; since $\sqrt{2} \approx 1.414$ so 4 is larger!
    $= 4 - \sqrt{2}$ which is definitely a positive number.

17. No reference numbers are given so all we can conclude is that e.) $x < y$ is true since x is to the left of y.

19. Since x is to the right of 3 and therefore to the right of 0, we can conclude that x is positive so (c) is true; also, since x is to the right of y we know that (f.), $y < x$, is true. We can't make any statement about the sign of y because we don't know the scale or the location of 0.

21. (a), x is negative, is true (x is to the left of 0)
     (d), y is positive, is true (y is to the right of 0)
     (e), $x < y$, is true (x is to the left of y)

23. If $a > b$ then $2a > a + b$ is justified by the property which states that adding the same quantity to both sides of an inequality preserves the inequality. (if $a < b$ then $a + c < b + c$)

25. If $w > 3z$ then $2w > 6z$ because of Inequalities property (c): here both sides have been multiplied by $+2$.

27. If $y < 3x$ then $-9x < -3y$ because of Inequalities property (d): here each side was multiplied by -3 and then the inequality was written in reverse fashion.

29. If $a - b < c - d$ then $b - a > d - c$ because both sides were multiplied by -1; property (d).

31. If $0 < \frac{3}{x} < \frac{y}{4}$ then $\frac{4}{y} < \frac{x}{3}$ because of Inequalities property (e), the reciprocal property.

In problems 27 - 35 we'll use the fact that distance from a to b $= \mid b - a \mid$.

33. distance from 2 to $7 = \mid 7 - 2 \mid = \mid 5 \mid = 5$

35. $\mid -7 - 2 \mid = \mid -9 \mid = 9$

37. $\mid 0 - (-5) \mid = \mid 5 \mid = 5$

39. $\mid 3.8 - 1.6 \mid = \mid 2.2 \mid = 2.2$

41. $\mid -3.8 - 1.6 \mid = \mid -5.4 \mid = 5.4$

43. H: a is a real number

   C: $\mid -a \mid = \mid a \mid$

   P: We'll have to examine the three cases: $a > 0$, $a < 0$, and $a = 0$. In each case, we'll look at the left and right sides of the conclusion separately.

   I. if $a > 0$, then $-a < 0$. (ie, if a is positive, its opposite is negative)

   Working first with the left side of our conclusion, $\mid -a \mid$:

   $\mid -a \mid = -(-a)$ reason: definition of absolute value ($-a$ is negative here)

   $= a$

   Now examining the right side of the conclusion, $\mid a \mid$:

   $\mid a \mid = a$ (since a is definitely positive)

   So the conclusion is true in the first case since the left and

   right sides have "matched" at the value a.

- 10 -

43. (continued)

II.   second case: if a < 0, then - a > 0.

The left side:

$| - a | = - a$ (remember, - a is now POSITIVE)

The right side:

$| a | = - a$ (remember, a itself is negative in this case)

So the conclusion is true in the second case.

III.   if a = 0 then $| - 0 | = | 0 | = 0$ and the proof is now complete as we

have shown in all cases that $| - a | = | a |$ holds true.

45. H: a > 0 and b > 0

C: $| a + b | = | a | + | b |$

P: Working first with the left side of the conclusion:

since both a and b are positive, their sum a + b is also positive.

therefore, $| a + b | = (a + b) = a + b$.

Working now with the right side of the conclusion:

$| a | + | b | = a + b$ since a and b are each positive.

So we see that the left side matches the right side and our proof is complete.

47. H: a < 0 and b > 0

C: $| a + b | < | a | + | b |$

P: general idea: since a and b have opposite signs, when added they will combine "destructively"

on the left before the absolute value of the sum is taken. On the right, the absolute values,

definitely positive numbers, will combine "constructively" always making a larger number than

the number on the left.

case I: magnitude of a > magnitude of b (ie, $| a | > | b |$) (for example: a = - 7 and b = 3)

left side: $| a + b | = | - (| a | - | b |) |$ because the answer to a + b must be negative in this case

$= - [- (| a | - | b |)]$ definition of absolute value

$= | a | - | b |$ rules of signs (elimination of the double negative)

47. (continued)

right side: | a | + | b | cannot be simplified but it is fairly obvious that | a | + | b | is the sum of two positive numbers while the left side is a difference of those same two positive numbers and it should be clear that | a | - | b | < | a | + | b |.

More rigorously, | a | - | b | < | a | + | b | if

$$(| a | + | b |) - (| a | - | b |) \text{ is positive.}$$

And this expression simplifies to 2| b | which is positive.

case II: magnitude of b > magnitude of a (for example a = - 3 and b = 8)

left side: | a + b | = | (| b | - | a |) | = | b | - | a | which will be less than the right side, | a | + | b | for the same reason as above in case I.

case III: magnitude of a = magnitude of b (for example: a = - 5 and b = 5)

left side: | a + b | = | 0 | = 0 which would have to be less than the right side of | a | + | b | which is definitely positive.

49. Let's do the case for a = 0:

H: a = 0, b is any real number

C: | a + b | = | a | + | b |

P: Left side: | a + b | = | 0 + b | = | b |

Right side: | a | + | b | = | 0 | + | b | = 0 + | b | = | b |

And as you can see, the left side has come out to equal the right side!

Problems 1.5

1. $3^2 = 3 \cdot 3 = 9$

3. $4^{-3} = \frac{1}{4^3} = \frac{1}{64}$

5. $(\frac{1}{3})^2 = \frac{1}{3} \cdot \frac{1}{3} = \frac{1}{9}$

7. $(\frac{1}{2})^{-4} = (\frac{2}{1})^4 = 2^4 = 16$

9. $4^{-1} \cdot 2^3 = \frac{1}{4} \cdot 8 = 2$

11. $(-3)^{-3} \cdot 3^3 = \frac{1}{(-3)^3} \cdot 27 = -\frac{1}{27} \cdot 27 = -1$

13. $4^2 \cdot 2^3 = 16 \cdot 8 = 128$ 15. $2^2 \cdot 3^{-2} = 4 \cdot \frac{1}{9} = \frac{4}{9}$ 17. $\left(2^2\right)^{-3} = 2^{-6} = \frac{1}{2^6} = \frac{1}{64}$

19. $23^0 \cdot 5^{-1} = 1 \cdot \frac{1}{5} = \frac{1}{5}$ 21. $(-1)^{-1} = \frac{1}{(-1)} = -1$

23. $(-1)^{101} = -1$ (since exponent is an odd integer)

25. $(-10)^1 = -10$ (any number to the first power is itself)

27. $(-10)^3 = (-10)(-10)(-10) = -1000$

29. $\left(\frac{257}{1049}\right)^{-1} = \frac{1049}{257}$ 31. $\frac{x^3}{x^5} = \frac{1}{x^{5-3}} = \frac{1}{x^2}$

33. $\frac{y^6}{y^4} = y^{6-4} = y^2$ 35. $\left(\frac{2}{x}\right)^2 = \frac{2^2}{x^2} = \frac{4}{x^2}$

37. $\left(\frac{2b}{3}\right)^{-1} = \left(\frac{3}{2b}\right)^1 = \frac{3}{2b}$ 39. $\left(\frac{7d}{5e}\right)^{-2} = \left(\frac{5e}{7d}\right)^2 = \frac{25e^2}{49d^2}$

41. $\left(\frac{5x}{2y}\right)^{-3} = \left(\frac{2y}{5x}\right)^3 = \frac{8y^3}{125x^3}$ 43. $\frac{5a^2b^3}{10ab^4} = \frac{5a^2b^3}{5 \cdot 2ab^4} = \frac{a^{2-1}}{2b^{4-3}} = \frac{a}{2b}$

45. $\left(\frac{u^2v^2}{u^3v}\right)^{-1} = \left(\frac{v}{u}\right)^{-1} = \frac{u}{v}$ 47. $\left(\frac{x^3z^4w^3}{xz^5w^2}\right)^2 = \left(\frac{x^2w}{z}\right)^2 = \frac{x^4w^2}{z^2}$

49. $\frac{a^2a^{-5}b^{-3}b^4}{ab^6b^{-2}} = \frac{a^{-3}b}{ab^4} = \frac{1}{a \cdot a^3 \cdot b^{4-1}} = \frac{1}{a^4b^3}$ 51. $(abc)^2 = a^2b^2c^2$

53. $x^{-1}y(xy)^2 = x^{-1}y \cdot x^2y^2 = xy^3$ 55. $(wz)^4w^{-2}z^{-3} = w^4z^4w^{-2}z^{-3} = (w^4w^{-2})(z^4z^{-3}) = w^2z$

57. $(ab)^{-1}(ab)^2 = (ab)^1 = ab$ 59. $\left(\frac{x}{2y^2}\right)^{-1}\left(\frac{y}{4x}\right)^2 = \left(\frac{2y^2}{x}\right)\left(\frac{y^2}{16x^2}\right) = \frac{y^4}{8x^3}$

61. $\left(\dfrac{2x^{-1}y^{-2}}{3x^2}\right)^{-1}\left(\dfrac{6x}{5y}\right)^{-2} = \left(\dfrac{2}{3x^3y^2}\right)^{-1}\left(\dfrac{5y}{6x}\right)^2 = \left(\dfrac{3x^3y^2}{2}\right)\left(\dfrac{25y^2}{36x^2}\right) = \dfrac{25xy^4}{24}$

63. $\left(\dfrac{4x^2}{5a^3}\right)\left(\dfrac{9}{2x}\right)^3 = \left(\dfrac{4x^2}{5a^3}\right)\left(\dfrac{729}{8x^3}\right) = \dfrac{729}{10a^3x}$

65. $\left(\dfrac{x^3wz^{-5}}{4x^2w^2z^{-8}}\right)^0 = 1$

67. $\left(\dfrac{x^2y^3}{xy^5}\right)^0\left(\dfrac{xy^5}{x^2y^3}\right) = 1 \cdot \left(\dfrac{y^2}{x^1}\right) = \dfrac{y^2}{x}$

69. $\left(\dfrac{w^{-2}v^5}{wv^{-1}}\right)^0\left(\dfrac{v}{w}\right)^{-1} = 1 \cdot \dfrac{w}{v} = \dfrac{w}{v}$

71. $\left(\dfrac{x^2y^{-2}z^{-3}}{z^3x^4y^{-1}}\right)^{-2} = \left(\dfrac{1}{z^6x^2y^1}\right)^{-2} = (z^6x^2y)^2 = z^{12}x^4y^2$

on converting into scientific notation: if your original number is greater than one, the exponent will be at least 0; if your original number is a fraction less than one in value, the exponent on 10 will DEFINITELY be negative.

73. $365 = 3.65 \text{ X } 100 = 3.65 \text{ X } 10^2$

75. $521.236 = 5.21236 \text{ X } 10^2$

77. $0.0000001 = 1 \text{ X } 10^{-7}$ (the decimal point in 1 would have to be moved seven places to the left to obtain the original number.)

79. $29{,}028 \text{ ft} = 2.9028 \text{ X } 10^4 \text{ ft}$

81. $5{,}878{,}499{,}830{,}120 \text{ mi} = 5.87849983012 \text{ X } 10^{12} \text{ mi.}$

83. $0.000000000000000000000000000000911 \text{ kg} = 9.11 \text{ X } 10^{-31} \text{ kg}$

85. $92{,}900{,}000 \text{ mi} = 9.29 \text{ X } 10^7 \text{ mi}$

87. $0.000000555 \text{ m} = 5.55 \text{ X } 10^{-7} \text{ m}$

89. $(-25.4)^3 = (-2.54 \text{ X } 10^1)^3 = (-2.54)^3 \text{ X } 10^3 = -1.6387064 \text{ X } 10^4$

91. $(0.235)^{-4} = (2.35 \text{ X } 10^{-1})^{-4} = (2.35)^{-4} \text{ X } 10^4 = 3.278902863 \text{ X } 10^2$

93. $(2{,}310{,}624)^5 = (2.310624 \text{ X } 10^6)^5 = (2.310624)^5 \text{ X } 10^{30} = 6.586374179 \text{ X } 10^{31}$

95. $(-8.2403 \times 10^{-3}) \times (4.106 \times 10^2) = -3.38346718 \times 10^0$

97. $\dfrac{-6.105 \times 10^{-3}}{9.011 \times 10^4} = \dfrac{-6.105}{9.011} \times 10^{-7} = -6.775052713 \times 10^{-8}$

99. $\dfrac{3.2105 \times 10^{-4}}{-5.6615 \times 10^{-7}} \div 7.1124 \times 10^2 = \dfrac{3.2105}{(-5.6615)(7.1124)} \times 10^{-4+7-2}$

$$= -7.97305921 \times 10^{-1}$$

101. $(-1.12724 \times 10^{-7}) \div [(3.4107 \times 10^{-8}) \times (-8.2993 \times 10^{-6})]$

$$= \dfrac{-1.12724}{3.4107(-8.2993)} \times 10^{-7+8+6} = 4.495092941 \times 10^5$$

103. $(1.01)^{100} = 2.704813829 \times 10^0$

105. by trial and error: $(1/2)^9 = 0.001953$ and $(1/2)^{10} = 0.000977$ so the smallest integer is 10.

107. $(1.2)^{37} = 850.6$ and $(1.2)^{38} = 1020.7$ so it's 38

109. $4^5 = 1024$ is greater than 5^4 which is 625

111. Go with the larger exponent (you get another entire multiplier this way): 1000^{1001} will be the larger.

Problems 1.6

1. $4^{5/2} = (4^{1/2})^5 = 2^5 = 32$

3. $4^{9/2} = (4^{1/2})^9 = 2^9 = 512$

5. $8^{-1/3} = \dfrac{1}{8^{1/3}} = \dfrac{1}{2}$

7. $(-8)^{2/3} = [(-8)^{1/3}]^2 = [-2]^2 = 4$

9. $8^{-2/3} = \dfrac{1}{8^{2/3}} = \dfrac{1}{(8^{1/3})^2} = \dfrac{1}{2^2} = \dfrac{1}{4}$

11. $(-1/64)^{-2/3} = (-64)^{2/3} = ((-64)^{1/3})^2$
$$= (-4)^2 = 16$$

- 15 -

13. $27^{1/3}$ = the cube root of $27 = 3$

15. $(-27)^{-2/3} = \dfrac{1}{(-27)^{2/3}} = \dfrac{1}{[(-27)^{1/3}]^2} = \dfrac{1}{[-3]^2} = \dfrac{1}{9}$

17. $100^{1/2}$ = the principal square root of $100 = 10$

19. $(-100)^{1/2}$ does not exist as a real number (no real number multiplied by itself could yield a negative result).

21. $9^{5/2} = (9^{1/2})^5 = 3^5 = 243$
23. $100^{5/2} = (100^{1/2})^5 = 10^5 = 100{,}000$

25. $(-100)^{5/2}$ is undefined as a real number; $(-100)^{1/2}$ is the problem.

27. $-(1000^{1/3}) = -(10) = -10$
29. $(-1000)^{1/3} = -10$ (since -10 cubed makes -1000)

31. $(128)^{1/7} = 2$ (since $2^7 = 128$)
33. $128^{5/7} = (128^{1/7})^5 = 2^5 = 32$

35. $(-128)^{6/7} = [(-128)^{1/7}]^6 = [-2]^6 = 64$
37. $64^{-1/6} = \dfrac{1}{64^{1/6}} = \dfrac{1}{2}$

39. $4^{1/2} \cdot 4^{3/2} = 4^{4/2} = 4^2 = 16$
41. $10^{1/3} \cdot 10^{5/3} = 10^2 = 100$

43. $\dfrac{(11)^{3/4}}{(11)^{-5/4}} = (11)^{3/4 - (-5/4)} = 11^{8/4} = 11^2 = 121$

45. $\dfrac{8^3}{24^3} = \left(\dfrac{8}{24}\right)^3 = \left(\dfrac{1}{3}\right)^3 = \dfrac{1}{27}$ (first step possible because of same exponents)

47. $\dfrac{(3.15)^3}{(6.3)^3} = \left(\dfrac{3.15}{6.3}\right)^3 = \left(\dfrac{1}{2}\right)^3 = \dfrac{1}{8}$

49. $\dfrac{(-64)^{1/3}(-64)^{-2/3}}{(-64)^{1/3}(-64)^{2/3}} = (-64)^{\frac{1}{3}-\frac{2}{3}-\frac{1}{3}-\frac{2}{3}} = (-64)^{-4/3} = \dfrac{1}{(-64)^{4/3}} = \dfrac{1}{(-4)^4} = \dfrac{1}{256}$

51. $\sqrt{108} = \sqrt{36 \cdot 3} = \sqrt{36} \cdot \sqrt{3} = 6\sqrt{3}$ (didn't use $4 \cdot 27$ as a replacement for 108 because 36 is a larger perfect square than 4)

53. $\sqrt[3]{-5000} = \sqrt[3]{-1000} \cdot \sqrt[3]{5} = -10(\sqrt[3]{5})$

55. $\dfrac{100}{\sqrt{3}} = \dfrac{100}{\sqrt{3}} \cdot \dfrac{\sqrt{3}}{\sqrt{3}} = \dfrac{100\sqrt{3}}{3}$ (in the denoiminator, $\sqrt{3} \cdot \sqrt{3} = \sqrt{9} = 3$)

57. $\dfrac{\sqrt{0.1}}{\sqrt{0.4}} = \sqrt{\dfrac{0.1}{0.4}} = \sqrt{\dfrac{1}{4}} = \dfrac{1}{2}$

59. $\dfrac{-5}{\sqrt[3]{-3}} = \dfrac{-5}{\sqrt[3]{-3}} \cdot \dfrac{\sqrt[3]{-9}}{\sqrt[3]{-9}} = \dfrac{-5\,\sqrt[3]{-9}}{\sqrt[3]{27}} = \dfrac{-5\,\sqrt[3]{-9}}{3}$ or $\dfrac{5 \cdot \sqrt[3]{9}}{3}$

61. $x^{3/2} \cdot x^{-3/2} = x^0 = 1$

63. $\dfrac{y^{3/7}\,y^{8/7}}{y^{1/7}\,y^{10/7}} = \dfrac{y^{11/7}}{y^{11/7}} = 1$

65. $(x^2 y^4)^{1/2} = x^{2 \cdot 1/2}\,y^{4 \cdot 1/2} = xy^2$

67. $\left(\dfrac{u}{2u}\right)^2 = \left(\dfrac{1}{2}\right)^2 = \dfrac{1}{4}$

69. $\left(\dfrac{a^{1/2}}{b^2}\right)^2 \left(\dfrac{b^{3/2}}{a^{2/3}}\right)^3 = \left(\dfrac{a}{b^4}\right)\left(\dfrac{b^{9/2}}{a^2}\right) = \dfrac{b^{9/2-4}}{a^{2-1}} = \dfrac{b^{1/2}}{a}$

71. $\left(\dfrac{y^{2/3}\,z^{1/4}}{z}\right)^{-1} = \dfrac{z}{y^{2/3}\,z^{1/4}} = \dfrac{z^{3/4}}{y^{2/3}}$

73. $\dfrac{y^{1.3}\,y^{2.6}}{y^{-2.4}\,y^3} = \dfrac{y^{1.3}\,y^{2.4}}{y^{2.6}\,y^3} = \dfrac{y^{3.7}}{y^{5.6}} = \dfrac{1}{y^{1.9}}$

75. $\dfrac{\sqrt{x}}{\sqrt{y}} = \dfrac{\sqrt{x}}{\sqrt{y}} \cdot \dfrac{\sqrt{y}}{\sqrt{y}} = \dfrac{\sqrt{xy}}{y}$

77. $\sqrt[6]{x^4 y^{12}} = (x^4 y^{12})^{1/6} = x^{2/3} y^2 = y^2 \cdot \sqrt[3]{x^2}$

79. $\sqrt{xy^2} \cdot \sqrt{yx^2} = \sqrt{x^3y^3} = \sqrt{(x^2y^2)(xy)} = \sqrt{x^2y^2} \cdot \sqrt{xy} = xy\sqrt{xy}$

81. $\sqrt{\sqrt{y^{16}}} = \sqrt{y^8} = y^4$ (in the first step, take care of the inner radical)

83. $\dfrac{\sqrt{w} \cdot \sqrt[3]{w}}{\sqrt[4]{w}} = \dfrac{w^{1/2} \cdot w^{1/3}}{w^{1/4}} = w^{1/2 + 1/3 - 1/4} = w^{\frac{6+4-3}{12}} = w^{7/12} = \sqrt[12]{w^7}$

85. $\sqrt{(x-1)^5 y^2} = \sqrt{(x-1)^4 y^2 \cdot (x-1)} = (x-1)^2 y\sqrt{x-1}$ or $y(x-1)^{5/2}$

87. $\sqrt{x^{12}/y^9} = (x^{12}/y^9)^{1/2} =$ mult. inside by $\dfrac{y}{y}$ to get $(x^{12}y/y^{10})^{1/2} = x^6 y^{1/2}/y^5$ or $\dfrac{x^6\sqrt{y}}{y^5}$

89. $\sqrt{8} + \sqrt{18} = \sqrt{4 \cdot 2} + \sqrt{9 \cdot 2} = 2\sqrt{2} + 3\sqrt{2} = 5\sqrt{2}$ (can only add "like" radicals)

91. $x^n x^{2n} = x^{n+2n} = x^{3n}$

93. $\sqrt{\dfrac{y^{4k}y^{7k}}{y^{8k}}} = \sqrt{y^{3k}} = \sqrt{y^{2k}y^k} = \sqrt{y^{2k}} \cdot \sqrt{y^k} = y^k \cdot \sqrt{y^k}$ or $y^{3k/2}$

95. $2^x = \frac{1}{4}$; x must be - 2 "by inspection"

97. $16^{-x} = \frac{1}{2}$; 16 is the same as 2^4 and $\frac{1}{2}$ is the same as 2^{-1}; making these substitutions yields

$(2^4)^{-x} = 2^{-1}$ which simplifies to $2^{-4x} = 2^{-1}$ and equating the exponents yields $-4x = -1$ so $x = \frac{1}{4}$.

99. $8^{x+3} = 4$: we'll do this one by changing to like bases: $8 = 2^3$ and $4 = 2^2$

making these substitutions yields:

$(2^3)^{x+3} = 2^2$ so $2^{3x+9} = 2^2$ and for this to be true $3x + 9 = 2$; $3x = -7$; $x = -\frac{7}{3}$

101. $(\frac{1}{2})^{3x/2} = 16$ (change of base again: $\frac{1}{2} = 2^{-1}$ and $16 = 2^4$)

$(2^{-1})^{3x/2} = 2^4$

$2^{-3x/2} = 2^4$ and so it seems that -3x/2 = 4 ; -3x = 8 ; x = -8/3

103. $(0.01)^{-\frac{1}{x}} = 1000$ (go to base 10: $0.01 = 10^{-2}$; $1000 = 10^3$)

$(10^{-2})^{-\frac{1}{x}} = 10^3$

$10^{\frac{2}{x}} = 10^3$ so at this point we see that $\frac{2}{x} = 3$; 2 = 3x ; $x = \frac{2}{3}$.

105. same as $10^{1/3}$: take care of exponent first: tap $\boxed{1}$, $\boxed{\div}$, $\boxed{3}$, $\boxed{\text{sto}}$ and 1/3 is in memory; then clear and enter 10 and then tap $\boxed{y^x}$ and $\boxed{\text{rcl}}$ to recall the exponent from memory; depress $\boxed{=}$; you should see 2.15443469

107. same as $(-1000)^{1/5}$ so enter 1000; depress $\boxed{+/-}$; depress $\boxed{y^x}$; enter .2, the decimal equivalent of 1/5; hit $\boxed{=}$ and you'll see -3.981071706

109. enter 3 $\boxed{\div}$ 7 $\boxed{+/-}$ $\boxed{=}$ $\boxed{\text{sto}}$ and then enter 80 $\boxed{y^x}$ $\boxed{\text{mr}}$ $\boxed{=}$ and you'll get 0.152893846

111. enter 100000, depress the "y to the x" key and then enter .1182 and hit the $\boxed{=}$ key and 3.899419867 should result.

113. enter 0.235, depress the $\boxed{y^x}$ key and then 1.46 and $\boxed{=}$ to get 0.120714399

115. a) $A(t) = P(1 + r)^t$ becomes $A(5) = 1000(1 + 0.06)^5 = 1000(1.33823) = \$1,338.23$
    b) $A(7.5) = 1000(1 + 0.06)^{7.5} = 1000(1.54808) = \$1,548.08$
    c) $A(11.75) = 1000(1 + 0.06)^{11.75} = 1000(1.9831) = \$1,983.10$

117. A time period is now $\frac{1}{12}$ of a year so the monthly interest rate is $\frac{12\%}{12} = 1\% = 0.01$ and the number of time periods is 10 yr X 12 mo/yr = 120 mo.
    The formula $A(t) = P(1 + r)^t$ becomes $A(120) = 5000(1 + 0.01)^{120} = 5000(3.30039) = \$16,501.95$

119. If the mass of the given star is three times the mass of the sun, then the ratio $\frac{M}{M_0}$ must be 3. Since 3 is between 2.5 and 5 (fourth row in the table) we can see that the corresponding r value to use is 3.95. The relation $\frac{L}{L_0} = \left(\frac{M}{M_0}\right)^r$ becomes $\frac{L}{L_0} = (3)^{3.95} \approx 76.6706$ and this is the ratio that was asked for.

Problems 1.7

1. degree is 4 because of x^4

3. degree is 0 because $6 = 6(1) = 6x^0$

5. degree is 6 because of x^6

7. $2p(x) = 2(2x^2 - 3x + 4) = 4x^2 - 6x + 8$

9. $p(x) + q(x) = (2x^2 - 3x + 4) + (3x^3 - x^2 + 5x - 3) = 3x^3 + (2-1)x^2 + (-3+5)x + (4-3)$
$$= 3x^3 + x^2 + 2x + 1$$

11. $q(x) - p(x) = (3x^3 - x^2 + 5x - 3) - (2x^2 - 3x + 4) = 3x^3 - x^2 + 5x - 3 - 2x^2 + 3x - 4$
$$= 3x^3 - 3x^2 + 8x - 7$$

13. $3q(x) - 4p(x) = 3(3x^3 - x^2 + 5x - 3) - 4(2x^2 - 3x + 4) = 9x^3 - 3x^2 + 15x - 9 - 8x^2 + 12x - 16$
$$= 9x^3 - 11x^2 + 27x - 25$$

15. $p(x)q(x) = (2x^2 - 3x + 4)(3x^3 - x^2 + 5x - 3) = 6x^5 - 2x^4 + 10x^3 - 6x^2$
$$-9x^4 + 3x^3 - 15x^2 + 9x$$
$$\underline{+12x^3 - 4x^2 + 20x - 12}$$
$$6x^5 - 11x^4 + 25x^3 - 25x^2 + 29x - 12$$

17. $3q(x) = 3(x^7 - 2x + 3) = 3x^7 - 6x + 9$

19. $q(x) - p(x) = (x^7 - 2x + 3) - (x^4 - 3) = x^7 - 2x + 3 - x^4 + 3 = x^7 - x^4 - 2x + 6$

21. pq will contain the term x^{11} (from $x^7 \cdot x^4$) so the degree is 11.

23. $(x + 2)(x + 4) = x^2 + 4x + 2x + 8 = x^2 + 6x + 8$; degree is 2

25. $(x + 5)(x - 5) = x^2 - 5^2 = x^2 - 25$; 2 (use of formula 6 from text)

27. $(3x - 5)(-5x + 2) = -15x^2 + 6x + 25x - 10 = -15x^2 + 31x - 10$; 2

29. $(3x - 2)(4x + 1) = 12x^2 + 3x - 8x - 2 = 12x^2 - 5x - 2$; 2

31. $(7x + 2)(7x + 1) = 49x^2 + 7x + 14x + 2 = 49x^2 + 21x + 2$; 2

33. $(8x + 1)(3x + 4) = 24x^2 + 32x + 3x + 4 = 24x^2 + 35x + 4$; 2

35. $(x - 3)(2x - 9) = 2x^2 - 9x - 6x + 27 = 2x^2 - 15x + 27$; 2

37. $(2 - x)(5 - 2x) = 10 - 4x - 5x + 2x^2 = 10 - 9x + 2x^2$ or $2x^2 - 9x + 10$; 2

39. $(x^4 - 4)(x^4 + 4) = (x^4)^2 - 4^2 = x^8 - 16$; 8   (formula 6 again)

41. $(-3x^2 - 4x + 2)(6x^2 - 3x + 2) = -18x^4 + 9x^3 - 6x^2$
$$-24x^3 + 12x^2 - 8x$$
$$+12x^2 - 6x + 4$$
$$\overline{-18x^4 - 15x^3 + 18x^2 - 14x + 4}$$ (degree is 4)

43. $(x^3 - 1)(x^2 + 1) = x^5 + x^3 - x^2 - 1$ and the degree is 5

45. $(2x^2 - 3x)(2x^3 - 3x^2 + 4) = 4x^5 - 6x^4 \quad\quad + 8x^2$
$$-6x^4 \quad + 9x^3 \quad - 12x$$
$$\overline{4x^5 - 12x^4 + 9x^3 + 8x^2 - 12x}\ ;\ \text{5th degree}$$

47. $(ax^3 + bx)(cx^3 + dx + e) = acx^6 + adx^4 + aex^3$
$$bcx^4 \quad\quad + bdx^2 + bex$$
$$\overline{acx^6 + (ad + bc)\,x^4 + aex^3 + bdx^2 + bex;\ \text{6th degree}}$$

49. $x^{10}(x^{20} - 2) = x^{30} - 2x^{10}$; 30

51. $(x^4 - 2x^2 + 3)(x^4 + 2x^2 - 3) = x^8 + 2x^6 - 3x^4$

$\qquad\qquad\qquad\qquad\qquad -2x^6 - 4x^4 + 6x^2$

$\qquad\qquad\qquad\qquad\qquad\qquad +3x^4 + 6x^2 - 9$

$\qquad\qquad\qquad\qquad\qquad x^8 \qquad -4x^4 + 12x^2 - 9$; 8th degree

53. $(x^2 - 4)(x^2 + 4) = (x^2)^2 - 4^2 = x^4 - 16$; degree 4

55. $(3x - 2)^2 = (3x)^2 - 2(3x)(2) + 2^2 = 9x^2 - 12x + 4$; 2 (property 5 was used)

57. $(x - 1)^3 = x^3 - 3x^2(1) + 3x(1)^2 - 1^3 = x^3 - 3x^2 + 3x - 1$; 3 (prop. 8 with x=x and y=1)

59. $(x + 1)(x - 2)(x + 3) = (x + 1)[x^2 + x - 6] = x^3 + x^2 - 6x$

$\qquad\qquad\qquad\qquad\qquad\qquad\qquad\qquad\qquad 1x^2 + x - 6$

$\qquad\qquad\qquad\qquad\qquad\qquad\qquad\qquad x^3 + 2x^2 - 5x - 6$ (degree 3)

61. $(x+3)^4 = (x+3)^2(x+3)^2 = (x^2+6x+9)(x^2+6x+9) = x^4+6x^3+9x^2$

$\qquad\qquad\qquad\qquad\qquad\qquad\qquad\qquad\qquad\qquad +6x^3+36x^2+54x$

$\qquad\qquad\qquad\qquad\qquad\qquad\qquad\qquad\qquad\qquad\quad +9x^2+ 54x+81$

$\qquad\qquad\qquad\qquad\qquad\qquad\qquad\qquad\qquad x^4+12x^3+54x^2+108x+81$; 4

63. a) $\deg(p) = 3$ and $\deg(q) = 4$ so the degree of pq is $3 + 4 = 7$

    b) $(2x^3)(3x^4) = 6x^7$ so the leading coefficient is 6

    c) $(3)(4) = 12$

65. a) $4 + 5 = 9$

    b) $(x^4)(x^5) = x^9$ so the leading coefficient is 1

    c) $(1)(0) = 0$ (q's constant term is 0)

67. a) $2 + 3 = 5$ (p will begin $4x^2$; q will begin x^3)

    b) $(2x)^2(x^3) = 4x^5$ so the leading coefficient is 4

    c) $(9)(-125) = -1125$

69. a) $4 + 4 = 8$ (p begins with $16x^4$; q begins with x^4)

b) $16 \cdot 1 = 16$

c) p will end in $(-3)^4 = 81$ while q will end in $2^4 = 16$ so pq will end in $(81)(16) = 1296$

71. p begins with x^n and q begins with x^m so pq begins with x^{n+m} so

a) degree of pq $= n+m$

b) the leading coefficient is 1

c) p ends in $(+1)^n = 1$ and q ends with $(-1)^m = +1$ since m is even so pq will end in $1 \cdot 1 = 1$

73. $x^3y^5 - 3x^2y^4 + 8xy^6$ is of degree 3 in x, of degree 6 in y, and of degree 8 (from the sum of the exponents in the first term) overall.

75. $xy + 3x^2y^2 - 5x^3 - y^3$ is of degree 3 in x, 3 in y, and 4 in general (sum of exponents in term 2).

77. $x^3z^5 - 2xy^2z^2 + 3x^5y - 8x^2y^3z^4$ is of degree 5 in x, 3 in y, 5 in z and 9 (last term) overall.

79. $(2a - b)^2 = (2a)^2 - 2(2a)(b) + (b)^2 = 4a^2 - 4ab + b^2$

(followed property 5 with x $= 2a$ and y $= b$)

81. $(3w - 4z)^3 = (3w)^3 - 3(3w)^2(4z) + 3(3w)(4z)^2 - (4z)^3 = 27w^3 - 108w^2z + 144wz^2 - 64z^3$

(followed property 8 with x $= 3w$ and y $= 4z$)

83. $\left(\frac{1}{x} - \frac{1}{y}\right)^2 = \left(\frac{1}{x}\right)^2 - 2\left(\frac{1}{x}\right)\left(\frac{1}{y}\right) + \left(\frac{1}{y}\right)^2 = \frac{1}{x^2} - \frac{2}{xy} + \frac{1}{y^2}$

85. $(x+y-z)^2 = (x+y-z)(x+y-z) = x^2 + xy - xz$

$\qquad\qquad\qquad\qquad\qquad\quad + xy \qquad + y^2 - yz$

$\qquad\qquad\qquad\qquad\qquad\qquad\quad - xz \qquad - yz + z^2$

$\qquad\qquad\qquad\qquad\qquad\overline{\qquad\qquad\qquad\qquad\qquad\qquad}$

$\qquad\qquad\qquad\qquad\qquad x^2 + 2xy - 2xz + y^2 - 2yz + z^2$

same problem now done a little differently: a little creative grouping and property 5:

view problem as $[(x+y) - z]^2 = (x+y)^2 - 2(x+y)z + z^2$

$\qquad\qquad\qquad\qquad\qquad\quad = x^2 + 2xy + y^2 - 2xz - 2yz + z^2$

87. view problem as $[(x-y) + z]^2 = (x-y)^2 + 2(x-y)z + z^2$

$$= x^2 - 2xy + y^2 + 2xz - 2yz + z^2$$

89. view problem as $[(x + y) + z][(x + y) - z]$ which is allowable by the associative property

$$= (x + y)^2 - z^2 \quad \text{by formula (6)}$$
$$= x^2 + 2xy + y^2 - z^2 \quad \text{by formula (4)}$$

91. view as $(x^2 - y^2 + z^2)(x^2 - y^2 + z^2)$: distribute each term in 1st factor to each term in 2nd

$$= x^4 - x^2y^2 + x^2z^2 - y^2x^2 + y^4 - y^2z^2 + z^2x^2 - z^2y^2 + z^4$$
$$= x^4 - 2x^2y^2 + 2x^2z^2 - 2y^2z^2 + y^4 + z^4$$

93. $\left(x + \frac{1}{x} + y\right)\left(x + \frac{1}{x} + y\right) = x^2 + 1 + xy + 1 + \frac{1}{x^2} + \frac{y}{x} + yx + \frac{y}{x} + y^2$

$$= x^2 + 2xy + y^2 + \frac{2y}{x} + \frac{1}{x^2} + 2$$

95. $(x + y + z + w)(x + y + z + w)$: just distribute!

$$= x^2 + xy + xz + xw + yx + y^2 + yz + yw + zx + zy + z^2 + zw + wx + wy + wz + w^2$$
$$= x^2 + y^2 + z^2 + w^2 + 2xy + 2xz + 2xw + 2yz + 2yw + 2zw$$

: each variable got squared and there were two of each kind of cross product!

97. view as $[(x + y) + (z + w)][(x + y) - (z + w)]$: so we can use formula (4): $(a+b)(a-b) = a^2 - b^2$

$$= (x + y)^2 - (z + w)^2$$
$$= (x^2 + 2xy + y^2) - (z^2 + 2zw + w^2)$$
$$= x^2 + 2xy + y^2 - z^2 - 2zw - w^2$$

99. $(\sqrt{x} - \sqrt{y})^2 = (\sqrt{x})^2 - 2(\sqrt{x})(\sqrt{y}) + (\sqrt{y})^2 = x - 2\sqrt{xy} + y$

101. $\left(w^2 - \frac{1}{w^2}\right)^3 = (w^2)^3 - 3(w^2)^2\left(\frac{1}{w^2}\right) + 3(w^2)\left(\frac{1}{w^2}\right)^2 - \left(\frac{1}{w^2}\right)^3$ (property 8 with $x=w^2$ and $y=\frac{1}{w^2}$)

$$= w^6 - 3(w^4)\left(\frac{1}{w^2}\right) + 3(w^2)\left(\frac{1}{w^4}\right) - \frac{1}{w^6}$$

$$= w^6 - 3w^2 + \frac{3}{w^2} - \frac{1}{w^6}$$

103. $(3x^2 - 12y^2)(10x^3 + 4y) = 30x^5 - 120x^3y^2 + 12x^2y - 48y^3$

105. $(a + b)(c + d) = ac + ad + bc + bd$

107. $(z + \frac{1}{z})(z - \frac{1}{z}) = z^2 - (\frac{1}{z})^2 = z^2 - \frac{1}{z^2}$

109. $(x^3 + 2y + 3z^4)(4x - y^3 + 5z^2) = 4x^4 - x^3y^3 + 5x^3z^2 + 8xy - 2y^4 + 10yz^2 + 12xz^4 - 3z^4y^3 + 15z^6$

   (NO like terms!)

111. $\left(\frac{2}{x} - 3y\right)^3$: use formula (8) with $x = \frac{2}{x}$ and $y = 3y$

$$= \left(\frac{2}{x}\right)^3 - 3\left(\frac{2}{x}\right)^2(3y) + 3\left(\frac{2}{x}\right)(3y)^2 - (3y)^3$$

$$= \frac{8}{x^3} - 3\left(\frac{4}{x^2}\right)3y + \frac{6}{x}(9y^2) - 27y^3 = \frac{8}{x^3} - \frac{36y}{x^2} + \frac{54y^2}{x} - 27y^3$$

113. a) 93 X 88   93 = 100 - 7

   88 = 100 - 12

   7 X 12 = 84

   7 + 12 = 19

   100 - 19 = 81   so answer should be 8184

   b) 96 X 85   96 = 100 - 4

   85 = 100 - 15

   4 X 15 = 60

   4 + 15 = 19

   100 - 19 = 81   so answer should be 8160

Problems 1.8

1. $2x + 4 = 2 \cdot x + 2 \cdot 2 = 2(x + 2)$ 3. $10x + 25 = 5 \cdot 2 \cdot x + 5 \cdot 5 = 5(2x + 5)$

5. $8x + 12y = 4(2x + 3y)$ since 4 is the GCD (greatest common divisor) of 8 and 12

7. $ab + 2ac = a(b + 2c)$

9. $8x^3 + 14x^2$: 2 is the GCD of 8 and 14; x^2 is the GCD of x^3 and x^2 }

$\qquad = 2x^2(4x + 7)$

11. $(x + 2)(x - 3) + (x^2 + 1)(x + 2)$: $(x + 2)$ is the GCD of the two **TERMS**

$\qquad = (x + 2) [(x - 3) + (x^2 + 1)]$

note: brackets used because terms in original problem contained ( )'s

$\qquad = (x + 2) [x^2 + x - 2] = (x + 2)[(x + 2)(x - 1)]$

$\qquad = (x + 2)^2(x - 1)$

PATTERN FOR DIFFERENCE OF TWO SQUARES: $A^2 - B^2 = (A + B)(A - B)$

13. $x^2 - 1 = x^2 - 1^2 = (x + 1)(x - 1)$ where we used x in place of A and 1 in place of B

15. $z^2 - w^2 = (z + w)(z - w)$

17. $(x - 2)^2 - 9 = (x - 2)^2 - 3^2 = [(x - 2) + 3] [(x - 2) - 3] = [x + 1] [x - 5]$ or $(x + 1)(x - 5)$

19. $(y + 4)^2 - (z - 2)^2 = [(y + 4) + (z - 2)] [(y + 4) - (z - 2)] = [y + z + 2] [y - z + 6]$

21. $64 - 36b^4$: try to get in the habit of factoring out common factors first!

$\qquad = 4(16 - 9b^4)$

$\qquad = 4(4 + 3b^2)(4 - 3b^2)$

23. $4z^2 - 25 = (2z)^2 - 5^2 = (2z + 5)(2z - 5)$

25. $w^6 - 9u^6 = (w^3)^2 - (3u^3)^2 = (w^3 + 3u^3)(w^3 - 3u^3)$

PATTERNS FOR PERFECT TRINOMIAL SQUARES: $A^2 \pm 2 \cdot A \cdot B + B^2 = (A \pm B)^2$

27. (It's nice to have perfect squares grouped together and pointed out like this!)

$\qquad w^2 - 2w + 1 = w^2 - 2 \cdot w \cdot 1 + 1^2 = (w - 1)^2$ $\qquad$ 29. $q^2 - 6q + 9 = q^2 - 2 \cdot q \cdot 3 + 3^2 = (q - 3)^2$

31. $16w^4 - 24w^2z + 9z^2 = (4w^2)^2 - 2(4w^2)(3z) + (3z)^2 = (4w^2 - 3z)^2$

33. $-x^2 + 4x - 4 = -1(x^2 - 4x + 4) = -(x - 2)^2$

35. $x^4 - 6x^2 + 9 = (x^2)^2 - 2 \cdot (x^2) \cdot 3 + 3^2 = (x^2 - 3)^2$

37. $16w^4 - 8w^2 + 1 = (4w^2)^2 - 2 \cdot (4w^2) \cdot 1 + 1^2 = (4w^2 - 1)^2$

39. $3r^2 - 30r + 75 = 3(r^2 - 10r + 25) = 3(r - 5)^2$

41. $a^6 + 2a^3b^3 + b^6 - 9$: first three terms fit the pattern; group them!
$$= (a^6 + 2a^3b^3 + b^6) - 9$$
$$= (a^3 + b^3)^2 - 3^2 \quad \text{(the difference of two squares!)}$$
$$= [(a^3 + b^3) + 3] \, [(a^3 + b^3) - 3] \quad \text{or} \quad (a^3 + b^3 + 3)(a^3 + b^3 - 3)$$

****

Some hints on factoring trinomials by trial and errror:

1. On signs: a) if the last sign is +'ve in the trinomial, both signs in the factors will have to agree with the middle sign in the trinomial.

    b) if the last sign is -'ve, the signs in the factors will be "mixed".

2. If the leading coefficient is 1, the numbers in the factors (signs included) must add to the coefficient of the middle term in the original trinomial. These are usually the easiest to see.

3. If the leading coefficient is not one, jot down all factor pairs of the leading coefficient as well as of the last term and look for the possibility of the middle coefficient coming about from the sum of products of these numbers taken two at a time, one from each factor pair.

4. If you suspect that $ax^2 + bx + c$ cannot be factored, evaluate $b^2 - 4ac$. If the number you get is NOT a perfect square, your suspicion is correct.

5. Any attempted factorization can be checked by simply multiplying it out and seeing if a match of the original expression is obtained.

6. Never waste time trying a factor with a common factor, eg., $(2x + 4)(\quad)$

   (If the original expression didn't have a common factor, how can one of its factors?)

****

43. $x^2 + 7x + 10$: last sign is +'ve; 5 and 2 are factors of 10 that sum to 7
$$= (x + 5)(x + 2)$$

45. $x^2 + 7x - 10$: last sign is -'ve so signs must be mixed, ie, one + and one -

   : possibilities are 5 and - 2, - 5 and 2, 10 and -1, -10 and 1

   : none of these ADD to +7 so this trinomial is PRIME and can't be factored

   : also note that $b^2 - 4ac = 7^2 - 4 \cdot 1 \cdot (-10) = 49 + 40 = 89$ (NOT a perfect square!)

47. $x^2 + x - 6$: need mixed sign factors of - 6 that add to +1;   how about +3 and -2?
$$= (x + 3)(x - 2)$$

49. $x^2 - 12x - 28$: need mixed sign factors of - 28 that sum to - 12

   : only possibilities are 4 and - 7 or - 4 and 7 or 28 and -1 or - 28 and 1 or 14 and - 2

   or - 14 and 2

   : since the last pair, - 14 and 2 add to - 12, the factors are $(x + 2)(x - 14)$

51. $x^2 - 13x + 42$: need -'ve factors of 42 that add to -13
$$= (x - 7)(x - 6)$$

53. $x^2 + 2x - 4$: something seems rotten in the state of Denmark!

   : $b^2 - 4ac = 2^2 - 4 \cdot 1 \cdot (-4) = 4 + 16 = 20$ (Not a perfect square)

   : CANNOT BE FACTORED

55. $-2x^2 + 6x - 4$: factor out -2 first
$$= -2(x^2 - 3x + 2)$$
$$= -2(x - 2)(x - 1)$$

57. $5x^2 - 3x - 2$: leading coefficient not 1 (rats!)

       : only factor pair for 5 is 5,1

       : mixed sign factor pairs for -2 are 1,-2 and -1,2

       : first try- $(5x + 1)(x - 2)$ which has -9x as its middle term

       : second try- $(5x - 2)(x + 1)$ which has +3x as its middle term

       : since the above guess was only off by the SIGN of the middle

        term, we need only flip-flop the signs in the factors.

     $= (5x + 2)(x - 1)$

59. $12x^2 + 13x - 14$: for 12 we have 4,3 or 6,2 or 12,1

           : for -14 there's -2,7 or 2,-7 or 1,-14 or -1,14

           : try $(4x - 7)(3x + 2)$ ... but this has a middle term of -13x

           : Close but no cigar! (yet!)

        $= (4x + 7)(3x - 2)$

61. $x^2 - 3xy + 2y^2 = (x - 2y)(x - y)$ a relatively easy one!

63. $6x^2 - 13xy - 5y^2$: for 6 we have 3,2 or 6,1

          : for - 5 there's 5,-1 or -5,1

          : if you doubt the problem can be factored check out $b^2 - 4ac = (-13)^2 - 4 \cdot 6 \cdot (-5) =$

         $169 + 120 = 289$ which is a perfect square (17^2) so problem can be factored !

       $= (3x + y)(2x - 5y)$

65. $x^2 + (m-n)x - mn$: need mixed sign factors of -mn that add to (m-n)

          : possibilities are m,-n and -m,n

       $= (x + m)(x - n)$

THE PATTERNS FOR FACTORING THE SUM AND DIFFERENCE OF TWO CUBES:

$A^3 + B^3 = (A + B)(A^2 - AB + B^2)$,   the "s" pattern ("s" for sum)

$A^3 - B^3 = (A - B)(A^2 + AB + B^2)$,   the "d" pattern ("d" for difference)

67. $x^3 - 27 = x^3 - 3^3$

: use "d" with $A = x$ and $B = 3$

$= (x - 3)(x^2 + x \cdot 3 + 3^2)$

$= (x - 3)(x^2 + 3x + 9)$

69. $x^3 + 64 = x^3 + 4^3$

: use "s" with $A = x$ and $B = 4$

$= (x + 4)(x^2 - x \cdot 4 + 4^2)$

$= (x + 4)(x^2 - 4x + 16)$

71. $8x^3 + 1 = (2x)^3 + 1^3$

: use "s" with $A = (2x)$ and $B = 1$

$= (2x + 1)[(2x)^2 - (2x) \cdot 1 + 1^2]$

$= (2x + 1)(4x^2 - 2x + 1)$

73. $x^3 + 27y^3 = x^3 + (3y)^3 = (x + 3y)(x^2 - 3xy + 9y^2)$ (used "s")

75. $2x^3y^3 + 2y^3z^3 = 2y^3(x^3 + z^3) = 2y^3(x + z)(x^2 - xz + z^2)$ (used "s")

77. $xy + y^2 - x - y$: four terms usually imply a grouping technique

$= (xy + y^2) - (x + y)$ (note sign change in second group)

$= y(x + y) - 1(x + y)$ (the 1 seems to help later)

: $(x + y)$ is now a common factor!

$= (x + y)(y - 1)$

79. $2x^2y + 4x + 9xy^3 + 18y^2 = 2x(xy + 2) + 9y^2(xy + 2) = (xy + 2)(2x + 9y^2)$

81. $x^2 - 4xy - 4 + 4y^2$: normal 2X2 grouping doesn't lead anywhere

: try 3X1 grouping with a little rearranging...

$= (x^2 - 4xy + 4y^2) - 4$

: possibility of difference of two squares looms

$= (x - 2y)^2 - 2^2$

$= [(x - 2y) + 2][(x - 2y) - 2]$ $= (x - 2y + 2)(x - 2y - 2)$

83. $2x^2 - 2 = 2(x^2 - 1) = 2(x + 1)(x - 1)$

85. $-6x^2 - 12x - 6 = -6(x^2 + 2x + 1) = -6(x + 1)^2$

87. $-3x^2 + 21x - 30 = -3(x^2 - 7x + 10) = -3(x - 5)(x - 2)$

89. $x^5 - 8x^2 = x^2(x^3 - 8) = x^2(x - 2)(x^2 + 2x + 4)$

91. $x^3y - x^2y^2 - 6xy^3 = xy(x^2 - xy - 6y^2) = xy(x - 3y)(x + 2y)$

93. $(x + y)^2 - 4xy$: this is NOT the difference of two squares

$\qquad$: multiply it out and hope for like terms

$\qquad = x^2 + 2xy + y^2 - 4xy$

$\qquad = x^2 - 2xy + y^2$

$\qquad = (x - y)^2$

95. $(x^2 - y^2)^2 + 4x^2y^2 = x^4 - 2x^2y^2 + y^4 + 4x^2y^2$

$\qquad = x^4 + 2x^2y^2 + y^4$

$\qquad = (x^2 + y^2)^2$

97. $(x + 1)(x^2 + 3x + 2) + (x + 1)(x^2 - 1) = (x + 1)\,[(x^2 + 3x + 2) + (x^2 - 1)\,]$

$\qquad = (x + 1)\,[\,2x^2 + 3x + 1]$

$\qquad = (x + 1)\,(2x + 1)(x + 1)$

$\qquad = (x + 1)^2\,(2x + 1)$

99. $x^{-2} - x^{-5}$: smallest exponent is -5; if -5 is subtracted from -2, ie -2 - (-5), you get +3

$\qquad = x^{-5}(x^3 - 1) = x^{-5}(x - 1)(x^2 + x + 1)$

101. $6x^{-5} + 15x^{-3} = 3x^{-5}(2 + 5x^2)$

103. $3x^2 - 6x^{-2} + 12x^{-1}$: smallest exponent is - 2; right?

$\qquad = 3x^{-2}(x^{2 - (-2)} - 2x^{-2 - (-2)} + 4x^{-1 - (-2)}$

$\qquad = 3x^{-2}(x^4 - 2 + 4x)$

105. $x^2 - x^{-2} = x^{-2}(x^4 - 1) = x^{-2}(x^2 - 1)(x^2 + 1) = x^{-2}(x + 1)(x - 1)(x^2 + 1)$

107. $x^{-4} + 2x + x^6 = x^{-4}(1 + 2x^5 + x^{10}) = x^{-4}(1 + x^5)^2$

109. $25x^2 - 20x^{-2} + 4x^{-6} = x^{-6}(25x^8 - 20x^4 + 4) = x^{-6}(5x^4 - 2)^2$

111. $y^{-2} - 3y^{-1} + 2 = y^{-2} - 3y^{-1} + 2y^0 = y^{-2}(1 - 3y + 2y^2) = y^{-2}(1 - y)(1 - 2y)$

113. $u^{-4} - 4u^{-2} - 5 = u^{-4}(1 - 4u^2 - 5u^4) = u^{-4}(1 - 5u^2)(1 + u^2)$

115. $2z^{-2} - 7z^{-1} - 4 = z^{-2}(2 - 7z - 4z^2) = z^{-2}(2 + z)(1 - 4z)$

Problems 1.9

1. $\dfrac{6x}{3} = \dfrac{3 \cdot 2x}{3} = 2x$

3. $\dfrac{2x + 4}{6x^2 + 8} = \dfrac{2(x + 2)}{2(3x^2 + 4)} = \dfrac{x + 2}{3x^2 + 4}$

5. $\dfrac{12y^4}{24y^7} = \dfrac{12y^4 \cdot 1}{12y^4 \cdot 2y^3} = \dfrac{1}{2y^3}$

7. $\dfrac{x^2y^2}{x^3y^3} = \dfrac{x^2y^2 \cdot 1}{x^2y^2 \cdot xy} = \dfrac{1}{xy}$

9. $\dfrac{3}{x} \cdot \dfrac{x}{6} = \dfrac{3}{x} \cdot \dfrac{x}{3 \cdot 2} = \dfrac{1}{2}$

11. $\dfrac{64}{x^3} \cdot \dfrac{x^5}{12} = \dfrac{64}{12} \cdot \dfrac{x^5}{x^3} = \dfrac{16}{3} \cdot x^2$ or $\dfrac{16x^2}{3}$

13. $\dfrac{x}{x + 1} \cdot \dfrac{(x + 1)^2}{x^4} = \dfrac{x}{x + 1} \cdot \dfrac{(x + 1)(x + 1)}{x \cdot x^3} = \dfrac{x + 1}{x^3}$

15. $\dfrac{x/(x+1)}{x/(x+2)} = \dfrac{x}{x + 1} \cdot \dfrac{x + 2}{x} = \dfrac{x + 2}{x + 1}$

17. $\dfrac{z/(z + 1)}{(z + 1)/z} = \dfrac{z}{z + 1} \cdot \dfrac{z}{z + 1} = \dfrac{z^2}{(z + 1)^2}$

19. $\dfrac{(w^2 + 1)/w}{w + \frac{1}{w}} = \dfrac{(w^2 + 1)/w}{w + \frac{1}{w}} \cdot \dfrac{w}{w} = \dfrac{(w^2 + 1)}{w^2 + 1} = 1$

21. $\dfrac{x^2 + 1}{(x + 1)^2}$ cannot be simplified because $x^2 + 1$ cannot be factored

23. $\dfrac{x^2 - 2x - 3}{x^2 + x - 12} = \dfrac{(x - 3)(x + 1)}{(x - 3)(x + 4)} = \dfrac{x + 1}{x + 4}$

25. $\dfrac{z^2 + 2z - 8}{z^2 - 5z - 36} = \dfrac{(z + 4)(z - 2)}{(z - 9)(z + 4)} = \dfrac{z - 2}{z - 9}$

27. $\dfrac{6x^2 - 5x + 1}{4x^2 - 4x + 1} = \dfrac{(3x - 1)(2x - 1)}{(2x - 1)(2x - 1)} = \dfrac{3x - 1}{2x - 1}$

29. $\dfrac{36z^2 - 24z - 5}{36z^2 - 72z + 35} = \dfrac{(6z - 5)(6z + 1)}{(6z - 7)(6z - 5)} = \dfrac{6z + 1}{6z - 7}$

31. $\dfrac{x^3 - y^3}{(x - y)^3} = \dfrac{(x - y)(x^2 + xy + y^2)}{(x - y)(x - y)^2} = \dfrac{x^2 + xy + y^2}{(x - y)^2}$

33. $\dfrac{\dfrac{x^2 - 9}{x^2 - 16}}{\dfrac{x^2 + 3x - 4}{x^2 + 8x + 16}} = \dfrac{(x + 3)(x - 3)}{(x - 4)(x + 4)} \cdot \dfrac{(x + 4)(x + 4)}{(x + 4)(x - 1)} = \dfrac{(x + 3)(x - 3)}{(x - 4)(x - 1)}$ or $\dfrac{x^2 - 9}{x^2 - 5x + 4}$

35. $\dfrac{(4z - 12)/(z^2 - 4z + 3)}{(5z + 25)/(z^2 + 8z + 15)} = \dfrac{4(z - 3)}{(z - 3)(z - 1)} \cdot \dfrac{(z + 5)(z + 3)}{5(z + 5)} = \dfrac{4(z + 3)}{5(z - 1)}$

37. $2 + \dfrac{y}{2} = \dfrac{2}{1} + \dfrac{y}{2} = \dfrac{2 \cdot 2 + 1 \cdot y}{1 \cdot 2} = \dfrac{4 + y}{2}$ (used fact that $\dfrac{a}{b} + \dfrac{c}{d} = \dfrac{ad + bc}{bd}$)

39. $\dfrac{1}{2} + 2x = \dfrac{1}{2} + \dfrac{2x}{1} = \dfrac{1 + 4x}{2}$

41. $\dfrac{x - 3}{x^2} + \dfrac{2x - 5}{x^2} = \dfrac{(x - 3) + (2x - 5)}{x^2} = \dfrac{3x - 8}{x^2}$

43. $\dfrac{1/2}{1} - \dfrac{1/2}{x - 1} = \dfrac{1}{2} - \dfrac{1}{2} \cdot \dfrac{1}{x - 1} = \dfrac{1}{2} \cdot \dfrac{x - 1}{x - 1} - \dfrac{1}{2(x - 1)} = \dfrac{(x - 1) - 1}{2(x - 1)} = \dfrac{x - 2}{2(x - 1)}$

45. $\dfrac{5}{z} - \dfrac{z}{4} = \dfrac{5 \cdot 4 - z \cdot z}{z \cdot 4} = \dfrac{20 - z^2}{4z}$

47. $\dfrac{2}{3s} - \dfrac{4}{5s} = \dfrac{2}{3s} \cdot \dfrac{5}{5} - \dfrac{4}{5s} \cdot \dfrac{3}{3} = \dfrac{10 - 12}{15s} = \dfrac{-2}{15s}$ since lowest common denominator is 15s

49. $\dfrac{7}{2y^2} + \dfrac{8}{3y^2}$: LCD is $6y^2$

$$= \dfrac{7}{2y^2} \cdot \dfrac{3}{3} + \dfrac{8}{3y^2} \cdot \dfrac{2}{2} = \dfrac{21 + 16}{6y^2} = \dfrac{37}{6y^2}$$

51. $\dfrac{12}{x - b} + \dfrac{3}{x - a}$: LCD is $(x - a)(x - b)$

$$= \dfrac{12}{x - b} \cdot \dfrac{x - a}{x - a} + \dfrac{3}{x - a} \cdot \dfrac{x - b}{x - b} = \dfrac{12x - 12a + 3x - 3b}{(x - a)(x - b)} = \dfrac{15x - 3b - 12a}{(x - a)(x - b)}$$

53. $\dfrac{2}{x - 3} - \dfrac{4}{x + 5} = \dfrac{2(x + 5) - 4(x - 3)}{(x - 3)(x + 5)} = \dfrac{2x + 10 - 4x + 12}{(x - 3)(x + 5)} = \dfrac{-2x + 22}{(x - 3)(x + 5)}$

55. $\dfrac{2x}{x^2 - 4} + \dfrac{5}{x - 2} - \dfrac{3}{x + 2}$: since $x^2 - 4$ factors into $(x+2)(x-2)$, it is the LCD

$$= \dfrac{2x}{(x + 2)(x - 2)} + \dfrac{5}{x - 2} \cdot \dfrac{x + 2}{x + 2} - \dfrac{3}{x + 2} \cdot \dfrac{x - 2}{x - 2}$$

$$= \dfrac{2x + 5x + 10 - 3x + 6}{(x + 2)(x - 2)} = \dfrac{4x + 16}{(x + 2)(x - 2)}$$

57. $\dfrac{4}{x + y} - \dfrac{3}{y - x} + \dfrac{2}{x^2 - y^2}$: do you see that the middle term has a denominator not in "sync" with the other denominators? To remedy this, multiply by middle term by -1/-1

to get $-\dfrac{-3}{-y + x}$ or $+\dfrac{3}{x - y}$

$$= \dfrac{4}{x + y} + \dfrac{3}{x - y} + \dfrac{2}{(x - y)(x + y)}$$

$$= \dfrac{4}{x + y} \cdot \dfrac{x - y}{x - y} + \dfrac{3}{x - y} \cdot \dfrac{x + y}{x + y} + \dfrac{2}{(x - y)(x + y)}$$

57. (continued)

$$= \frac{4x - 4y + 3x + 3y + 2}{(x + y)(x - y)}$$

$$= \frac{7x - y + 2}{(x - y)(x + y)}$$

59. $\frac{3x}{(x - 2)^2} - \frac{4}{x - 2}$: LCD is $(x - 2)^2$

$$= \frac{3x}{(x - 2)^2} \cdot \frac{1}{1} - \frac{4}{(x - 2)} \cdot \frac{x - 2}{x - 2} = \frac{3x - 4(x - 2)}{(x - 2)^2} = \frac{3x - 4x + 8}{(x - 2)^2} = \frac{-x + 8}{(x - 2)^2}$$

61. $\frac{3}{y - 3} - \frac{7}{y + 6} + \frac{2y - 3}{y^2 + 3y - 18}$: since $y^2 + 3y - 18$ factors into $(y + 6)(y - 3)$

that qualifies as our LCD

$$= \frac{3}{y - 3} \cdot \frac{y + 6}{y + 6} - \frac{7}{y + 6} \cdot \frac{y - 3}{y - 3} + \frac{2y - 3}{(y + 6)(y - 3)}$$

$$= \frac{3y + 18 - 7y + 21 + 2y - 3}{(y + 6)(y - 3)}$$

$$= \frac{-2y + 36}{(y + 6)(y - 3)} \text{ or } \frac{-2(y - 18)}{(y + 6)(y - 3)}$$

63. $\frac{2 + \frac{1}{x}}{3 - \frac{1}{x}}$: x is the LCD for all the fractions so multiply by $\frac{x}{x}$

$$= \frac{2 + \frac{1}{x}}{3 - \frac{1}{x}} \cdot \frac{x}{x} = \frac{2x + 1}{3x - 1}$$

65. $\frac{\frac{3}{x} - \frac{5}{y}}{\frac{6}{y} + \frac{2}{x}}$: xy is the LCD so multiply by $\frac{xy}{xy}$ to get : $\frac{\frac{3}{x} - \frac{5}{y}}{\frac{6}{y} + \frac{2}{x}} \cdot \frac{xy}{xy} = \frac{3y - 5x}{6x + 2y}$

67. $\dfrac{\frac{1}{x} - \frac{3}{x^2} + \frac{7x}{x^3}}{-\frac{4}{x} + \frac{3x - 2}{x^2} + \frac{3x^2 - 5x + 2}{x^3}}$: LCD is x^3 so we'll multiply this complex fraction by x^3/x^3

(in actuality, each term in both the numerator and denominator gets multiplied by x^3); the result of this is...

$$= \dfrac{x^2 - 3x + 7x}{-4x^2 + x(3x - 2) + (3x^2 - 5x + 2)}$$

$$= \dfrac{x^2 + 4x}{-4x^2 + 3x^2 - 2x + 3x^2 - 5x + 2} \quad = \dfrac{x^2 + 4x}{2x^2 - 7x + 2}$$

69. $\dfrac{\frac{1}{(x + h)^2} - \frac{1}{x^2}}{h}$: a decent plan is to combine the two fractions in the numerator and then multiply the result by $\frac{1}{h}$, the reciprocal of h

$$= \dfrac{1 \cdot x^2 - 1 \cdot (x + h)^2}{(x + h)^2 x^2} \cdot \dfrac{1}{h} = \dfrac{x^2 - (x^2 + 2xh + h^2)}{(x + h)^2 x^2} \cdot \dfrac{1}{h}$$

$$= \dfrac{x^2 - x^2 - 2xh - h^2}{(x + h)^2 x^2 h} = \dfrac{-2xh - h^2}{(x + h)^2 x^2 h} = \dfrac{h(-2x - h)}{(x + h)^2 x^2 h} = \dfrac{-2x - h}{(x + h)^2 x^2}$$

71. $x^{-1} + x^{-3} = \frac{1}{x} + \frac{1}{x^3}$ (by definition of negative exponents; now the lcd is x^3)

$$= \frac{1}{x} \cdot \frac{x^2}{x^2} + \frac{1}{x^3} = \dfrac{x^2 + 1}{x^3}$$

73. $(4x + 3)(x + 5)^{-1/3} + (2x - 7)(x + 5)^{-4/3} = \dfrac{4x + 3}{(x + 5)^{1/3}} + \dfrac{2x - 7}{(x + 5)^{4/3}}$

now factor out $\dfrac{1}{(x + 5)^{1/3}}$: $\qquad = \dfrac{1}{(x + 5)^{1/3}} \left(\dfrac{4x + 3}{1} + \dfrac{2x - 7}{x + 5} \right)$

- 36 -

73. (continued)

now combine the fractions in $(\quad)$'s: $\qquad = \dfrac{1}{(x+5)^{1/3}} \left(\dfrac{4x^2 + 23x + 15 + 2x - 7}{x + 5} \right)$

multiply back through and simplify: $\qquad = \dfrac{4x^2 + 25x + 8}{(x+5)^{4/3}}$

75. $x^{-3} + x^{-4} - x^{-5} - x^{-6} = \dfrac{1}{x^3} + \dfrac{1}{x^4} - \dfrac{1}{x^5} - \dfrac{1}{x^6}$: the lcd is x^6

$\qquad = \dfrac{1}{x^3} \cdot \dfrac{x^3}{x^3} + \dfrac{1}{x^4} \cdot \dfrac{x^2}{x^2} - \dfrac{1}{x^5} \cdot \dfrac{x^1}{x^1} - \dfrac{1}{x^6} = \dfrac{x^3 + x^2 - x - 1}{x^6}$

77. $\dfrac{3}{\sqrt{x}} = \dfrac{3}{\sqrt{x}} \cdot \dfrac{\sqrt{x}}{\sqrt{x}} = \dfrac{3\sqrt{x}}{x}$ $\qquad$ (in the denominator, $\sqrt{x} \cdot \sqrt{x} = \sqrt{x^2} = x$)

79. $\dfrac{\sqrt{5}}{\sqrt[4]{x+1}}$: must build up quantity in radical to be a perfect fourth power, ie $(x + 1)^4$

it's already to the first power so we need to multiply by $(x + 1)^3$ (under the $\sqrt{}$)

$\qquad = \dfrac{\sqrt{5}}{\sqrt[4]{x+1}} \cdot \dfrac{\sqrt[4]{(x+1)^3}}{\sqrt[4]{(x+1)^3}} = \dfrac{\sqrt{5}\,(x+1)^{3/4}}{x + 1}$

81. $\dfrac{1}{\sqrt{5} + \sqrt{3}}$: this denominator is a binomial; change middle sign only in forming multipliers

$\qquad = \dfrac{1}{\sqrt{5} + \sqrt{3}} \cdot \dfrac{\sqrt{5} - \sqrt{3}}{\sqrt{5} - \sqrt{3}} = \dfrac{\sqrt{5} - \sqrt{3}}{5 - 3} = \dfrac{\sqrt{5} - \sqrt{3}}{2}$

83. $\dfrac{-1}{2 - \sqrt{2x}} = \dfrac{-1}{2 - \sqrt{2x}} \cdot \dfrac{2 + \sqrt{2x}}{2 + \sqrt{2x}} = \dfrac{-2 - \sqrt{2x}}{4 - 2x}$ or $\dfrac{2 + \sqrt{2x}}{2x - 4}$ after multiplying by $\dfrac{(-1)}{(-1)}$

85. $\dfrac{\sqrt{x} - \sqrt{y}}{\sqrt{x} + \sqrt{y}} = \dfrac{\sqrt{x} - \sqrt{y}}{\sqrt{x} + \sqrt{y}} \cdot \dfrac{\sqrt{x} - \sqrt{y}}{\sqrt{x} - \sqrt{y}} = \dfrac{x - 2\sqrt{xy} + y}{x - y}$

87. $\dfrac{2}{1 - \sqrt[3]{x}} \cdot \dfrac{1 + \sqrt[3]{x} + (\sqrt[3]{x})^2}{1 + \sqrt[3]{x} + (\sqrt[3]{x})^2} = \dfrac{2(1 + x^{1/3} + x^{2/3})}{1 - x}$ or $\dfrac{2(1 + \sqrt[3]{x} + \sqrt[3]{x^2})}{1 - x}$

this worked because $(a - b)(a^2 + ab + b^2) = a^3 - b^3$ and $(\sqrt[3]{x})^3 = x$

89. $\dfrac{\sqrt{x + h + 4} - \sqrt{x + 4}}{h} = \dfrac{\sqrt{x + h + 4} - \sqrt{x + 4}}{h} \cdot \dfrac{\sqrt{x + h + 4} + \sqrt{x + 4}}{\sqrt{x + h + 4} + \sqrt{x + 4}}$

$= \dfrac{(x + h + 4) - (x + 4)}{h(\sqrt{x + h + 4} + \sqrt{x + 4}}$

$= \dfrac{h}{h(\sqrt{x + h + 4} + \sqrt{x + 4})} = \dfrac{1}{\sqrt{x + h + 4} + \sqrt{x + 4}}$

91. $\dfrac{\dfrac{1}{\sqrt{x + 2}} - \dfrac{1}{\sqrt{x}}}{2} = \dfrac{1 \cdot \sqrt{x} - 1 \cdot \sqrt{x + 2}}{\sqrt{x + 2} \cdot \sqrt{x}} \cdot \dfrac{1}{2} = \dfrac{\sqrt{x} - \sqrt{x + 2}}{\sqrt{x}\sqrt{x + 2}} \cdot \dfrac{1}{2}$

now multiply by $\dfrac{\sqrt{x} + \sqrt{x + 2}}{\sqrt{x} + \sqrt{x + 2}}$ to get $\dfrac{x - (x + 2)}{\sqrt{x}\sqrt{x + 2}(\sqrt{x} + \sqrt{x + 2})} \cdot \dfrac{1}{2}$ which equals

$\dfrac{-1}{\sqrt{x}\sqrt{x + 2}(\sqrt{x} + \sqrt{x + 2})}$

93. $\dfrac{\dfrac{1}{\sqrt{(x + h)^2}} - \dfrac{1}{x}}{h}$: $\sqrt{a^2} = |a|$ is a commonly used identity not mentioned in the text

$= \dfrac{\dfrac{1}{|x + h|} - \dfrac{1}{x}}{h} = \left(\dfrac{x - |x + h|}{|x + h|x}\right) \cdot \dfrac{1}{h} = \dfrac{x - |x + h|}{xh|x + h|}$

95. $\dfrac{1}{R} = \dfrac{1}{2 \times 10^{-3}} + \dfrac{1}{5 \times 10^{-4}} = 0.5 \times 10^3 + 0.2 \times 10^4 = 0.5 \times 10^3 + 2 \times 10^3 = 2.5 \times 10^3$

so taking the reciprocal of both sides gives $R = \dfrac{1}{2.5 \times 10^3} = 0.4 \times 10^{-3} = 4 \times 10^{-4}$

Review Exercises for Chapter One

1. let r = 0.424242...

   $100r = 42.424242....$

   so $99r = 42$ which means that $r = \frac{42}{99} = \frac{14}{33}$

3. $5 \div 6$ done on a calculator yields 0.83333333....

5. $\frac{1}{5} - \frac{2}{7} = \frac{7 - 10}{35} = -\frac{3}{35}$ 7. $\frac{4}{3} \cdot \frac{15}{16} = \frac{4}{3} \cdot \frac{5 \cdot 3}{4 \cdot 4} = \frac{5}{4}$

9. $\frac{3}{8}\left(\frac{4}{3} - \frac{11}{6}\right) = \frac{1}{2} - \frac{11}{16} = \frac{8}{16} - \frac{11}{16} = -\frac{3}{16}$

11. distance from -2 to 5 $= |\,5 - (-2)\,| = |\,5 + 2\,| = |\,7\,| = 7$

13. $5^2 = 5 \cdot 5 = 25$ 15. $(1/2)^4 = 1^4/2^4 = 1/16$ 17. $3^{-1} \cdot 9^3 = 3^{-1} \cdot (3^2)^3 = 3^{-1} \cdot 3^6 = 3^5 = 243$

19. $15^2 \cdot 15^{-4} = 15^{-2} = \frac{1}{15^2} = \frac{1}{225}$ 21. $\frac{w^4}{w^2} = w^{4-2} = w^2$

23. $\left(\frac{4}{5d^2}\right)^{-1} = \frac{5d^2}{4}$ 25. $\left(\frac{3x}{4y^2}\right)^2 \left(\frac{2x}{y}\right)^{-1} = \left(\frac{9x^2}{16y^4}\right)\left(\frac{y}{2x}\right) = \frac{9x}{32y^3}$

27. $\left(\frac{4xy^{-3}z^2}{3x^2y^{-5}z}\right)^{-1} = \frac{3x^2y^{-5}z}{4xy^{-3}z^2} = \frac{3x^2y^3z}{4xy^5z^2} = \frac{3x}{4y^2z}$

29. $0.0003729 = 3.729 \times 10^{-4}$ 31. $3,666,100,000 = 3.6661 \times 10^9$

33. $(13.203)^{-3} = 4.344923458 \times 10^{-4}$ 35. $\frac{9.256 \times 10^5}{4.213 \times 10^8} = 2.197009257 \times 10^{-3}$

37. $9^{-3/2} = \frac{1}{9^{3/2}} = \frac{1}{(9^{1/2})^3} = \frac{1}{3^3} = \frac{1}{27}$ 39. $(1/16)^{-1/2} = 16^{1/2} = 4$

41. $17^{3/4} 17^{5/4} = 17^{8/4} = 17^2 = 289$

43. $\dfrac{25^{3/2} \, 25^{1/2}}{25^{9/2} \, 25^{-11/2}} = 25^{3/2 + 1/2 - 9/2 + 11/2} = 25^3 = 15{,}625$

45. $\sqrt[3]{-14} = (-14)^{1/3} = -2.410142264$

47. $(14.218)^{-0.237} = 0.533061446$

49. $\sqrt[3]{270} = \sqrt[3]{27 \cdot 10} = \sqrt[3]{27} \cdot \sqrt[3]{10} = 3 \sqrt[3]{10}$

51. $\dfrac{\sqrt{80}}{\sqrt{5}} = \sqrt{\dfrac{80}{5}} = \sqrt{16} = 4$

53. $\dfrac{x^{3/4} \, x^{1/5}}{x^{3/5}} = x^{3/4 + 1/5 - 3/5} = x^{15/20 + 4/20 - 12/20} = x^{7/20}$

55. $\left(\dfrac{2w}{3w}\right)^4 = \left(\dfrac{2}{3}\right)^4 = \dfrac{16}{81}$

57. $\dfrac{x^{-2.3} x^{1.5}}{x^{4.6} x^{-3.2}} = x^{-2.3 + 1.5 - 4.6 + 3.2} = x^{-2.2}$

59. $p(x) - q(x) = (4x^2 - 5x + 3) - (x^3 - 2x^2 + 5x - 1)$
$$= 4x^2 - 5x + 3 - x^3 + 2x^2 - 5x + 1$$
$$= -x^3 + 6x^2 - 10x + 4$$

61. $3p(x) - 4q(x) = 3(4x^2 - 5x + 3) - 4(x^3 - 2x^2 + 5x - 1)$
$$= 12x^2 - 15x + 9 - 4x^3 + 8x^2 - 20x + 4$$
$$= -4x^3 + 20x^2 - 35x + 13$$

63. $p(x)q(x) = (4x^2 - 5x + 3)(x^3 - 2x^2 + 5x - 1)$
$$= 4x^5 - 8x^4 + 20x^3 - 4x^2$$
$$- 5x^4 + 10x^3 - 25x^2 + 5x$$
$$+ 3x^3 - 6x^2 + 15x - 3$$
$$= 4x^5 - 13x^4 + 33x^3 - 35x^2 + 20x - 3$$

65. $(x - 7)(x - 7) = x^2 - 2(x)(7) + 7^2 = x^2 - 14x + 49$ and is of degree 2

67. $(x + 8)(x - 8) = x^2 - 8^2 = x^2 - 64$ and is also of degree 2

69. $(x^2 - 2x + 2)(x - 4) = x^3 - 4x^2$
$$- 2x^2 + 8x$$
$$+ 2x - 8$$
$$= \overline{x^3 - 6x^2 + 10x - 8} \; ; \text{ degree of } 3$$

71. $(3x^2 - 2x + 4)(2x + 3) = 6x^3 + 9x^2$
$$-4x^2 - 6x$$
$$8x + 12$$
$$= \overline{6x^3 + 5x^2 + 2x + 12} \quad \text{of degree 3}$$

73. $(x + 2)^3 = x^3 + 3x^2(2) + 3x(2)^2 + 2^3 = x^3 + 6x^2 + 12x + 8 \quad$ of degree 3

75. $(4x - y)^2 = (4x)^2 - 2(4x)y + y^2 = 16x^2 - 8xy + y^2$

77. $\left(\dfrac{2}{x} - \dfrac{3}{y}\right)^2 = \left(\dfrac{2}{x}\right)^2 - 2\left(\dfrac{2}{x}\right)\left(\dfrac{3}{y}\right) + \left(\dfrac{3}{y}\right)^2 = \dfrac{4}{x^2} - \dfrac{12}{xy} + \dfrac{9}{y^2} = \dfrac{4y^2 - 12xy + 9x^2}{x^2y^2} \quad \text{or} \quad \dfrac{(2y - 3x)^2}{x^2y^2}$

79. $(x + 2y - 3z)^2 = [\, (x + 2y) - 3z\,]^2 = (x + 2y)^2 - 2(x + 2y)(3z) + (3z)^2$
$$= x^2 + 4xy + 4y^2 - 6xz - 12yz + 9z^2$$

81. $(x^{3/4} - y^{1/2})(x^{1/4} + y^{3/2}) = x + x^{3/4}y^{3/2} - x^{1/4}y^{1/2} - y^2$

83. $3x^2 + 12 = 3(x^2 + 4)$ 85. $4x^2 - 16y^2 = 4(x^2 - 4y^2) = 4(x + 2y)(x - 2y)$

87. $4x^2 - 9w^2 = (2x + 3w)(2x - 3w)$ 89. $z^2 + 14z + 49 = (z + 7)(z + 7)$ or $(z + 7)^2$

91. $4r^2 - 12rs + 9s^2 = (2r - 3s)(2r - 3s)$ or $(2r - 3s)^2$ 93. $x^2 + 3x - 10 = (x + 5)(x - 2)$

95. $4z^2 - 3z - 10$: factors of 4 are 2 and 2 or 4 and 1
     : factors of 10 are 5,2 and 10,1; for -10 we have 5,-2 and -5,2 and 10,-1 and -10 and 1
     : first try $(4z - 5)(z + 2)$;   the middle term would be $+3z$ so we're off only by signs
     $= (4z + 5)(z - 2)$

97. $x^3 - 1000 = x^3 - 10^3 = (x - 10)(x^2 + 10x + 100)$

99. $x^3y^3 - 8 = (xy)^3 - 2^3 = (xy - 2)(x^2y^2 + 2xy + 4)$

101. $x^2 + 9y^2 - 16 + 6xy$: grouping (3 X 1) might work

$$= (x^2 + 6xy + 9y^2) - 16$$
$$= (x + 3y)^2 - 4^2 \quad \text{(the difference of two squares)}$$
$$= [(x + 3y) + 4] [(x + 3y) - 4] = (x + 3y + 4)(x + 3y - 4)$$

103. $x^3 + 5x^2 - 14x = x(x^2 + 5x - 14)$

$$= x(x + 7)(x - 2)$$

105. $(x + 2)(x^2 - x - 6) + (x + 2)(x^2 - 4) = (x + 2) [(x^2 - x - 6) + (x^2 - 4)]$

$$= (x + 2) [2x^2 - x - 10]$$
$$= (x + 2) (2x - 5)(x + 2)$$
$$= (x + 2)^2(2x - 5)$$

107. $\dfrac{x^3y^5}{x^2y^7} = \dfrac{x^{3-2}}{y^{7-5}} = \dfrac{x}{y^2}$

109. $\dfrac{(x + 3)/(x + 4)}{(x + 4)/(x + 2)} = \dfrac{x + 3}{x + 4} \cdot \dfrac{x + 2}{x + 4} = \dfrac{(x + 3)(x + 2)}{(x + 4)^2}$

111. $\dfrac{3x^2 - x - 10}{x^2 - 4} = \dfrac{(3x + 5)(x - 2)}{(x + 2)(x - 2)} = \dfrac{3x + 5}{x + 2}$

113. $\dfrac{(x^2 + 3x + 2)/(x^2 + 7x + 12)}{(x^2 - x - 6)/(x^2 + 2x - 3)}$: plan is to factor crazily and invert denominator

$$= \dfrac{(x + 2)(x + 1)}{(x + 4)(x + 3)} \cdot \dfrac{(x + 3)(x - 1)}{(x - 3)(x + 2)} = \dfrac{(x + 1)(x - 1)}{(x + 4)(x - 3)}$$

115. $\dfrac{1}{3} + x = \dfrac{1}{3} + \dfrac{x}{1} = \dfrac{1 \cdot 1 + 3 \cdot x}{3 \cdot 1} = \dfrac{1 + 3x}{3}$

117. $\dfrac{5}{x^2} + \dfrac{x^2}{5} = \dfrac{25 + x^4}{5x^2}$

119. $\dfrac{3x}{x-2} + \dfrac{4}{x+3} = \dfrac{3x(x+3)+4(x-2)}{(x-2)(x+3)} = \dfrac{3x^2+9x+4x-8}{(x-2)(x+3)} = \dfrac{3x^2+13x-8}{(x-2)(x+3)}$

121. $\dfrac{1}{x} + \dfrac{2}{x+1} + \dfrac{3}{x-5}$: LCD is $x(x+1)(x-5)$

$$= \dfrac{1}{x} \cdot \dfrac{x+1}{x+1} \cdot \dfrac{x-5}{x-5} + \dfrac{2}{x+1} \cdot \dfrac{x}{x} \cdot \dfrac{x-5}{x-5} + \dfrac{3}{x-5} \cdot \dfrac{x}{x} \cdot \dfrac{x+1}{x+1}$$

$$= \dfrac{(x^2-4x-5)+(2x^2-10x)+(3x^2+3x)}{x(x+1)(x-5)}$$

$$= \dfrac{6x^2-11x-5}{x(x+1)(x-5)}$$

123. $\dfrac{\dfrac{5}{x+2} - \dfrac{6}{x-2}}{\dfrac{2x+3}{x^2-4}}$: LCD is $(x+2)(x-2)$ so multiply by $\dfrac{(x+2)(x-2)}{(x+2)(x-2)}$ to get

$$= \dfrac{5(x-2)-6(x+2)}{2x+3} = \dfrac{5x-10-6x-12}{2x+3} = \dfrac{-x-22}{2x+3}$$

125. $\dfrac{\sqrt{x}+\sqrt{y}}{2\sqrt{x}-3\sqrt{y}} = \dfrac{\sqrt{x}+\sqrt{y}}{2\sqrt{x}-3\sqrt{y}} \cdot \dfrac{2\sqrt{x}+3\sqrt{y}}{2\sqrt{x}+3\sqrt{y}} = \dfrac{2x+5\sqrt{xy}+3y}{4x-9y}$

CHAPTER TWO
Equations and Inequalities in One Variable

Chapter Objectives

In this chapter you will learn to solve linear and quadratic and other non-linear equations. You will then use these methods to solve applied problems. Algebraic properties of complex numbers are discussed, along with complex numbers as solutions to important equations. Finally, you will learn how to solve linear and non-linear inequalities, and inequalities involving absolute value.

Chapter Summary

- **Linear Equation**

  A linear equation is an equation of the form $ax + b = 0$, where $a \neq 0$.
  Its unique solution is $x = -\frac{b}{a}$.

- **Quadratic Equation**

  An equation of the form $ax^2 + bx + c = 0$, where $a \neq 0$, is a quadratic equation.

- **Solution by Factoring**

  If $ax^2 + bx + c = a(x - r)(x - s)$, then the solutions to $ax^2 + bx + c = 0$ are $x = r$ and $x = s$.

- **Completing the Square**

  $$x^2 + bx + c = \left(x + \frac{b}{2}\right)^2 + c - \frac{b^2}{4}$$

- **The Quadratic Formula**

  $x = \dfrac{-b \pm \sqrt{b^2 - 4ac}}{2a}$ are the solutions to $ax^2 + bx + c = 0$.

- **The sum of the roots to the quadratic equation $ax^2 + bx + c = 0$ is $-\frac{b}{a}$ and the product of the roots is $\frac{c}{a}$.**

- **Discriminant**

  The expression $b^2 - 4ac$ is called the discriminant of $ax^2 + bx + c = 0$. If the discriminant is positive, there are two real solutions; if zero, there is one real solution (called a double root); if negative, there are two complex conjugate solutions.

- **Imaginary Unit**

  The imaginary unit i is defined by $i = \sqrt{-1}$. If $a > 0$, then $\sqrt{-a} = \sqrt{a}\ i$

- **Complex Number**

  A complex number is a number of the form $a + bi$ where a and b are real.

- **Algebra of Complex Numbers**

  $(a + bi) + (c + di) = (a + c) + (b + d)i$

  $(a + bi) - (c + di) = (a - c) + (b - d)i$

  $(a + bi)(c + di) = (ac - bd) + (ad + bc)i$

  $(a + bi)(a - bi) = a^2 + b^2$

  $$\frac{a + bi}{c + di} = \frac{(a + bi)(c - di)}{(c + di)(c - di)} = \frac{ac + bd}{c^2 + d^2} + \frac{bc - ad}{c^2 + d^2}\ i$$

  $\overline{a + bi} = a - bi$ is called the complex conjugate of $a + bi$

- **Intervals**

  open interval $(a, b) = \{x: a < x < b\}$

  closed interval $[a, b] = \{x: a \leq x \leq b\}$

  half-open intervals $[a, b) = \{x: a \leq x < b\}$ and $(a, b] = \{x: a < x \leq b\}$

  $[a, \infty) = \{x: x \geq a\}$

  $(a, \infty) = \{x: x > a\}$

  $(-\infty, a] = \{x : x \leq a\}$

  $(-\infty, a) = \{x: x < a\}$

  $(-\infty, \infty) = $ The set of all real numbers

- **Absolute value inequalities**

  If a > 0, then

  $|x| < a$ is equivalent to $-a < x < a$

  $|x| \leq a$ is equivalent to $-a \leq x \leq a$

  $|x| > a$ is equivalent to $x > a$ or $x < -a$

  $|x| \geq a$ is equivalent to $x \geq a$ or $x \leq -a$

SOLUTIONS TO CHAPTER TWO PROBLEMS

Problems 2.1

1. $4z + 2 = 7$

    $4z + 2 - 2 = 7 - 2$

    $4z = 5$

    $\frac{1}{4} \cdot 4z = \frac{1}{4} \cdot 5$

    $z = \frac{5}{4}$

3. $-2x + 7 = 7$

    $-2x = 0$

    $x = 0$

5. $3(1 - 4y) = -6(7y - 2)$

    $3 - 12y = -42 y + 12$

    $30y = 9$

    $y = \frac{9}{30} = \frac{3}{10}$

7. $(z - 2)^2 = (z - 1)(z - 5)$

    $z^2 - 4z + 4 = z^2 - 6z + 5$

    $2z = 1$

    $z = \frac{1}{2}$

9. $(x - 6)(x - 2) = 0$

    $x^2 - 8x + 12 = x^2 + 6x + 5$

    $7 = 14x$ so $\frac{1}{2} = x$

11. $(4v + 1)(2v - 3) = (8v - 2)(v + 2)$

    $8v^2 - 10v - 3 = 8v^2 + 14v - 4$

    $1 = 24v$

    $\frac{1}{24} = v$

In equations involving fractions, often you should start by multiplying both sides of the equation by the LCD (least common denominator) of all the fractions.

13. $\frac{z-2}{z+1} = 2$

$\frac{z-2}{z+1} \cdot (z+1) = 2(z+1)$

$z - 2 = 2z + 2$

$-4 = z$

15. $\frac{1}{x} + 1 = 2$

$\frac{1}{x} = 1$

$\frac{1}{x} \cdot x = 1 \cdot x$

$1 = x$

17. $\frac{3}{w} + \frac{2}{w} - \frac{8}{w} = 2$ Multiply both sides of the equation by w to obtain

$3 + 2 - 8 = 2w$

$-3 = 2w$

$-\frac{3}{2} = w$

19. $y = 3 - \frac{4}{2/y}$

$y = 3 - 4 \cdot \frac{y}{2}$ Division by a fraction equals multiplication by the reciprocal

$y = 3 - 2y$

$3y = 3$

$y = 1$

21. $\frac{4}{m-1} = \frac{3}{m+2}$

$\frac{4}{m-1} \cdot (m-1)(m+2) = \frac{3}{m+2} \cdot (m-1)(m+2)$

$4(m+2) = 3(m-1)$

$4m + 8 = 3m - 3$

$m = -11$

23. $\dfrac{s-2}{s+4} = \dfrac{s+3}{s-8}$

$\dfrac{s-2}{s+4} \cdot (s+4)(s-8) = \dfrac{s+3}{s-8} \cdot (s+4)(s-8)$

$(s-2)(s-8) = (s+3)(s+4)$

$s^2 - 10s + 16 = s^2 + 7s + 12$

$4 = 17s$

$\dfrac{4}{17} = s$

25. $(x-1)^3 - (x+2)^3 + 9x^2 = 4$

$x^3 - 3x^2 + 3x - 1 - (x^3 + 6x^2 + 12x + 8) + 9x^2 = 4$ (now change sign of all terms in (), add)

$3x - 1 - 12x - 8 = 4$

$-9x - 9 = 4$

$-9x = 13$

$x = \dfrac{13}{-9} = -\dfrac{13}{9}$

27. $\dfrac{1}{x-1} + \dfrac{1}{x+1} = \dfrac{1}{(x-1)(x+1)}$ (Factoring denominator of the right side)

Multiply both sides of the equation by $(x-1)(x+1)$ to obtain

$x + 1 + x - 1 = 1$

$2x = 1$

$x = \dfrac{1}{2}$

29. $\dfrac{2a}{a+5} - \dfrac{a}{a+3} = 1$ Multiply both sides of the equation by $(a+5)(a+3)$

$2a(a+3) - a(a+5) = 1(a+3)(a+5)$

$2a^2 + 6a - a^2 - 5a = a^2 + 8a + 15$

$-7a = 15$

$a = -\dfrac{15}{7}$

31. $1.207x = 4.103$

    $x = \dfrac{4.103}{1.207} \approx 3.399$

33. $12.3156z + 15.2178 = 18.1432$

    $12.3156z = 2.9254$

        $z \approx 0.2375$

35. $8.34 \times 10^{-7}p = 5 \times 10^{-9}$

    $p = \dfrac{5 \times 10^{-9}}{8.34 \times 10^{-7}} \approx 6.0 \times 10^{-3}$

37. $\dfrac{4.106}{x} = -10.321$

    $x = \dfrac{4.106}{-10.321} \approx -0.398$

39. $\dfrac{-5.7}{4w} + \dfrac{8.2}{5w} = 6$ Multiply both sides by 20w to obtain

    $-28.5 + 32.8 = 120w$ or $w = \dfrac{4.3}{120} \approx 0.036$

41. $2(x + 3) = 2x - 1$

    $2x + 6 = 2x - 1$

    $6 = -1$ is not possible, so no solutions.

43. $(x + 5)(x - 3) = (x + 1)^2$

    $x^2 + 2x - 15 = x^2 + 2x + 1$

    $-15 = 1$ is not possible, so no solutions.

45. $(x + 2)(x + 3) = (x + 1)(x + 5) - x$

    $x^2 + 5x + 6 = x^2 + 6x + 5 - x$

        $6 = 5$ is not possible, so no solutions.

47. $\dfrac{5}{y + 2} + 6 = 6$

    $\dfrac{5}{y + 2} = 0$ is not possible, since a fraction $= 0$ only when the numerator $= 0$,

        and $5 \neq 0$. Therefore, no solutions.

49. $\dfrac{2}{q + 1} - \dfrac{6}{q(q + 1)} = \dfrac{2}{q}$ Multiply both sides by q(q+1) and obtain

    $2q - 6 = 2q + 2$

   $-6 = 2$ is not possible, so no solutions.

51. $(x - 1)^2 - 1 = x^2 - 2x + 1 - 1$

        $= x^2 - 2x$

        $= x(x - 2)$

53. $(x + 5)^2 = x^2 + 10x + 25$

        $= x^2 + 10x + 21 + 4$

        $= (x + 3)(x + 7) + 4$

55. $\frac{5}{x} - \frac{x}{7} = \frac{5}{x} \cdot \frac{7}{7} - \frac{x}{7} \cdot \frac{x}{x}$

$= \frac{35}{7x} - \frac{x^2}{7x} = -\frac{(x^2 - 35)}{7x}$

57. $y = 4z^3 w$

$w = \frac{y}{4z^3}$

59. $abc = bcd$ so $d = \frac{abc}{bc} = a$

61. $\frac{z}{4} - xy + \frac{3}{x} = 10$

$\frac{z}{4} + \frac{3}{x} - 10 = xy$ Now multiply through by the common denominator $4x$ to obtain

$xz + 12 - 40x = 4x^2 y$ Finally, divide both sides by $4x^2$

$y = \frac{zx - 40x + 12}{4x^2}$

63. $\frac{az + b}{c} = 1 - b$

$az + b = c - bc$

$az = c - bc - b$

$z = \frac{c - bc - b}{a}$

65. $\frac{3}{x + a} - \frac{2}{x - b} = 3$

Multiply both sides by $(x + a)(x - b)$ to obtain

$3(x - b) - 2(x + a) = 3(x + a)(x - b)$

$3x - 3b - 2x - 2a = 3x^2 + 3ax - 3bx - 3ab$

$3ab - 3ax - 2a = 3x^2 - 3bx + 3b - x$

$a(3b - 3x - 2) = 3x^2 - 3bx + 3b - x$ or $a = \frac{3x^2 - 3bx + 3b - x}{3b - 3x - 2} = \frac{-3x^2 + 3bx - 3b + x}{3x - 3b + 2}$

67. $\frac{1}{b + q} + \frac{1}{q} = 0$ Multiply both sides by $(b + q)q$ to obtain

$q + b + q = 0$

$2q = -b$ or $q = -\frac{b}{2}$

69. $x^2 - 1$ has two solutions, $x = 1$ and $x = -1$.

71. No. It is true that $x = 2$ is the only real root for both, but we will see that they have different complex solutions (the first equation has two and the second equation has four).

73. To go from line 6 to line 7, both sides of the equation are being divided by
    $(x - 5)$. But since $x = 5$, $x - 5 = 0$ and division by 0 is not permitted.

Problems 2.2

1. Divide both sides of the equation by 2π to obtain $r = \dfrac{C}{2\pi}$

3. Divide both sides of the equation by πr^2 to obtain $h = \dfrac{V}{\pi r^2}$

5. Multiply both sides of the equation by $\dfrac{R}{I}$ to obtain $R = \dfrac{E}{I}$

7. Multiply both sides of the equation by $\dfrac{r^2}{Gm_2}$ to obtain $m_1 = \dfrac{Fr^2}{Gm_2}$

9. Multiply both sides by $\left(1 + \dfrac{r}{n}\right)^{-nt}$ to obtain $P = A\left(1 + \dfrac{r}{n}\right)^{-nt}$

For the following problems, you must keep in mind that to convert from percentage to decimal you divide by 100; to convert from decimal to percentage you multiply by 100.

11. Let p = the original price of the bicycle (in dollars).

$$p + 0.23p = 172.20$$
$$1.23p = 172.20$$
$$p = \frac{172.20}{1.23} = \$140 \text{ is the original price of the bicycle.}$$

13. Let p = the original price of the refrigerator (in dollars).

$$p - .15p = 528.70$$
$$p = \$622.00 \text{ is the original price of the refrigerator.}$$

15. Let r = interest rate. Since $I = Prt$,

$$9400 = (20{,}000)(5)r$$
$$9400 = 100{,}000r, \text{ or } r = \frac{9400}{100{,}000} = .094 = 9.4\% \text{ interest rate.}$$

17. Let t = the length of time the money was invested. Solving the simple
    interest formula for t, $\quad t = \dfrac{I}{Pr} = \dfrac{1350}{5000(.06)} = 4.5$ years

19. Let P = amount she invests. Then $P = \dfrac{I}{rt} = \dfrac{55{,}000}{(.105)1} = \$523{,}810$.

21. Let x = the amount deposited in $5\frac{1}{2}\%$ account. Then since the total she invests is $8000, 8000 - x is the amount deposited in the 6% account.

   Using I = Prt, obtain $4(.055x) + 4(.06)(8000 - x) = 1851.60$

$$.22x + 1920 - .24x = 1851.60$$
$$-.02\,x = -68.4$$
$$x = \frac{-68.4}{-.02} = 3420$$

   a) Therefore, the amount invested in the $5\frac{1}{2}\%$ account is \$3420 and the amount invested in the 6% account is 8000 - 3420 = \$4580.

   b) Her interest in one year was $\dfrac{1851.60}{4} = \$462.90$. $\dfrac{462.90}{8000} = 0.05786 = 5.786\%$

23. Let x = amount invested in 8% CD; then 25000 - x is the amount in $9\frac{1}{2}\%$ MF.

$$.095(25000 - x) = 60 + .08x$$
$$2375 - .095x = 60 + .08x$$
$$x = \frac{2315}{.175} \approx 13{,}229$$

   Therefore, \$13,229 was invested in the CD, and \$11,771 was invested in the mutual fund.

25. If r is her rate, $r = \dfrac{d}{t} = \dfrac{800}{112} = 7.143$ m/sec

27. They will pass each other when the sum of the distances traveled by each is 600 miles (the total). The sum of their velocities is 135 mph. Since $t = \dfrac{d}{r}$,
   $t = \dfrac{600}{135} = 4\frac{4}{9}$ hours.

29. a) $C = 1650 + 35q$, where q is the number of items
   b) $C = 1650 + 35(215) = \$9175$

31. $TV = \dfrac{V_E}{f} = \dfrac{6.8}{17} = 0.4$ liters of air per breath

33. Let x = amount of 30% solution used; then 40 - x is the amount of 22% solution.

$$.30x + .22(40 - x) = .25(40)$$
$$.30x + 8.8 - .22x = 10$$
$$.08x = 1.2$$
$$x = \frac{1.2}{.08} = 15$$

Therefore, 15g of the 30% solution and 25g of the 22% solution were used.

35. Let x = weight of 18% ore. Then 3x = weight of 12% ore.

$$.12(3x) + .18x = 20$$
$$.36x + .18x = 20$$
$$.54x = 20$$
$$x = \frac{20}{.54} = 37.037$$

$x + 3x = 4x = 4(37.037) = 148.15 \approx 148$ ounces total weight.

37. $U_{n-1} = \frac{y_n - g + 3U_n}{3}$, so $U_{1984} = \frac{.025 - .03 + 3(.075)}{3} = \frac{.22}{3} = .07333$ or $7\frac{1}{3}\%$

39. $h = \frac{PE}{mg} = \frac{361}{(8)9.81} \approx 4.60$ m

41. $t = \frac{v - v_0}{a} = \frac{140 - 20}{30} = \frac{120}{30} = 4$ seconds

43. $C = \frac{5F - 160}{9} = \frac{5(80) - 160}{9} \approx 26.7°C$ (actually $26\frac{2}{3}°C$)

45. $\frac{13}{10} \cdot 100 = 130$

47. If x = actual age, $90 = \frac{20}{x} \cdot 100$, so $x = \frac{2000}{90} = 22\frac{2}{9}$ years

49. Let x = the number of at bats the player had beyond the initial 160. The number of hits he got in the first 160 at bats is .375(160) = 60. The number of hits in the at bats after the first 160 would be .285x.

Therefore, $.325 = \frac{60 + .285x}{160 + x} = \frac{\text{total hits for season}}{\text{total at bats for season}}$

Multiply both sides by 160 + x to obtain $52 + .325x = 60 + .285x$ so .04x = 8 or x = 200.

Therefore, he batted an additional 200 times for a total of 360 at bats.

51. The perimeter of a rectangle is $2l + 2w$. If $l = w + 20$, the perimeter is $4w + 40 = 200$, which means $w = 40$. So the dimensions are 60ft X 40 ft.

53. Let $x =$ the amount of \$4.80 per pound coffee used; then $25 - x$ is the amount of \$3.60 per pound coffee used.

$$4.80x + 3.60(25 - x) = 25(4.25)$$
$$4.80x + 90 - 3.60x = 106.25$$
$$1.20x = 16.25$$
$$x = \frac{16.25}{1.20} = 13.54$$

So 13.54 lbs of the \$4.80/lb and 11.46 lbs of the \$3.60/lb coffee were used.

55. Let $x =$ octane rating of Brand X.

$$150(94) + 250x = 400(89)$$
$$14,100 + 250x = 35,600$$

$$x = \frac{21,500}{250} = 86 \text{ octane of Brand X}$$

57. Let $x =$ length of time it takes them to do the job together. Add the portions of the job they can do in one hour and set equal to the amount they can do together in one hour to obtain $\frac{1}{3} + \frac{1}{4} = \frac{1}{x}$. Multiply both sides by $12x$ to obtain $4x + 3x = 12$.
Therefore $x = \frac{12}{7}$, so together they can do the job in $\frac{12}{7}$ hours, or ≈ 1 hr 43min.

59. Use $\frac{1}{R} = \frac{1}{R_1} + \frac{1}{R_2}$. Let x be the unknown resistance. Then

$\frac{1}{20} = \frac{1}{x} + \frac{1}{30}$. Multiply both sides by $60x$. Obtain $3x = 60 + 2x$, or $x = 60$ ohms.

61. If x is the first integer, then $x + 1$ is the next integer and $x + (x + 1) = 115$ so $2x = 114$ or $x = 57$. Therefore, the numbers are 57 and 58.

63. Odd integers are often represented by 2k + 1, k an integer. So the sum of four consecutive odd integers could be written as (2k + 1) + (2k + 3) + (2k + 5) + (2k + 7) = 104. 8k + 16 = 104, so k = 11 and the numbers are 23, 25, 27, and 29. (This could also be done using x + (x + 2) + (x + 4) + (x + 6) = 104 and if x turned out to be anything but an odd integer, the problem could not be solved).

65. Let x = Dot's current age. Then x + 8 is Sheila's current age.

(x + 8) + 4 = 2(x + 4) so x + 12 = 2x + 8, so x = 4, meaning Dot is currently 4 years old.

67. Recall $t = \frac{d}{r}$. The total distance they travel is 120 feet, total rate is 27 ft/sec so $t = \frac{120}{27} = \frac{40}{9}$ sec is the time each runs until they meet. Peter's speed is 15 ft/sec, so he travels $15 \cdot \left(\frac{40}{9}\right) = \frac{200}{3}$ or $66\frac{2}{3}$ feet.

69. Each train travels 75 miles until they collide, since they are traveling at the same rate of speed. Since $t = \frac{d}{r}$, each train travels for $\frac{75}{60} = \frac{5}{4}$ hr. The fly also travels for $\frac{5}{4}$ hr. Multiply this by his speed of 80 mph to obtain 100 miles traveled by the fly.

71. There are a lot that you could make up. Here's one:

1. Choose a number
2. Add 5
3. Square your number
4. Subtract 25
5. Subtract the square of your number
6. Divide by your number
7. The result is 10. Why?

73. a) The bird will lose 45 miles each hour, so it will never fly forward but will take $2\frac{2}{9}$ hour to be blown 100 miles backword.

b) There may be another way to look at a problem that leads to a different and possibly easier w way to solve it.

Problems 2.3

1. $(x - 3)(x - 3) = 0$.   Therefore, x = 3 is a double root.

3. $(z + 5)(z + 5) = 0$.   Therefore, z = -5 is a double root.

5. $(x + 3)(x - 2) = 0$.   The roots are x = -3 and x = 2.

7. $(x + 5)(x + 2) = 0$.   The roots are x = -5 and x = -2.

9. $x^2 - 2x - 24 = 0$

   $(x - 6)(x + 4) = 0$.   The roots are x = 6 and x = -4.

11. $-2(n - 2)(n - 1) = 0$. The roots are n = 2 and n = 1.

13. $(2x + 1)(2x + 3) = 0$. The roots are $x = -\frac{1}{2}$ and $x = -\frac{3}{2}$

15. $(6y - 1)(2y + 3) = 0$.   The roots are $y = \frac{1}{6}$ and $y = -\frac{3}{2}$

17. $y^2 - 3y + \frac{9}{4} - \frac{9}{4}$, or $\left(y - \frac{3}{2}\right)^2 - \frac{9}{4}$

19. $r^2 + 7r + \frac{49}{4} - \frac{49}{4}$, or $\left(r + \frac{7}{2}\right)^2 - \frac{49}{4}$

21. $x^2 + 3.2x + 2.56 - 2.56$, or $(x + 1.6)^2 - 2.56$

23. $u^2 + 15.\,206\,u + 57.805609 - 57.805609$, or $(u + 7.603)^2 - 57.805609$

25. $x^2 + 2x + 1 = 2 + 1$

   $(x + 1)^2 = 3$

   $x + 1 = \pm \sqrt{3}$

   $x = -1 \pm \sqrt{3}$

27. $(x + 2)^2 - 4 - 3 = 0$

    $(x + 2)^2 = 7$

    $x + 2 = \pm\sqrt{7}$

    $x = -2 \pm\sqrt{7}$

29. $u^2 - u + \frac{1}{4} = 1 + \frac{1}{4}$

    $\left(u - \frac{1}{2}\right)^2 = \frac{5}{4}$

        $u = \frac{1}{2} \pm \frac{\sqrt{5}}{2} = \frac{1 \pm \sqrt{5}}{2}$

31. $r^2 - 11r + \frac{121}{4} = -7 + \frac{121}{4}$

    $\left(r - \frac{11}{2}\right)^2 = \frac{93}{4}$

        $r = \frac{11}{2} \pm \frac{\sqrt{93}}{2} = \frac{11 \pm \sqrt{93}}{2}$

33. $s^2 - \frac{1}{2}s + \frac{1}{16} = -\frac{1}{4} + \frac{1}{16}$

    $\left(s - \frac{1}{4}\right)^2 = -\frac{3}{16}$, so there are no real solutions (no square can be negative)

35. $y^2 - .0032y = .0156$

    $(y - .0016)^2 = .0156 + .00000256 = .01560256$

      $y - .0016 = \pm .1249102$

           $y = .0016 \pm .1249102$

           $y = 0.1265102$ and $y = -0.1233102$ are the two roots

37. $D = 3^2 - 4(1)(1) = 9 - 4 = 5 > 0$, so there are two real roots.

39. $D = (-6)^2 - 4(1)(9) = 36 - 36 = 0$, so one real root.

41. $D = (-5)^2 - 4(1)(10) = 25 - 40 = -15 < 0$, so there are no real roots.

43. $5r^2 + 12r + 10 = 0$. $D = 12^2 - 4(5)(10) = 144 - 200 = -56 < 0$, so there are no real roots.

45. $D = (-8.9)^2 - 4(1.6)(7.4) = 79.21 - 47.36 = 31.85 > 0$, so there are two real roots.

47. $x = \dfrac{-5 \pm \sqrt{5^2 - 4(1)(-6)}}{2(1)} = \dfrac{-5 \pm \sqrt{25 + 24}}{2} = \dfrac{-5 \pm \sqrt{49}}{2} = \dfrac{-5 \pm 7}{2} = \dfrac{2}{2}$ or $\dfrac{-12}{2}$

    so $x = 1$, $x = -6$ are the two roots.

49. $z = \dfrac{-4 \pm \sqrt{16 - 4(3)(-10)}}{6} = \dfrac{-4 \pm \sqrt{136}}{6} = \dfrac{-4 \pm 2\sqrt{34}}{6} = \dfrac{-2 \pm \sqrt{34}}{3}$

51. $4v^2 - 6v + 5 = 0$

    Since $D = (-6)^2 - 4(4)(5) = 36 - 80 = -44 < 0$, there are no real solutions.

53. Multiply through by 2 to get $w^2 + 2w + 20 = 0$

    $D = 2^2 - 4(1)(20) = 4 - 80 = -76$, so there are no real solutions

55. Multiply through by 12 to get $3x^2 + 4x - 2 = 0$

    $x = \dfrac{-4 \pm \sqrt{4^2 - 4(3)(-2)}}{6} = \dfrac{-4 \pm \sqrt{16 + 24}}{6} = \dfrac{-4 \pm \sqrt{40}}{6} = \dfrac{-4 \pm 2\sqrt{10}}{6} = \dfrac{-2 \pm \sqrt{10}}{3}$

57. $u = \dfrac{15{,}106 \pm \sqrt{(15{,}106)^2 - 4(37{,}502)(-23{,}208)}}{75{,}004} = \dfrac{15{,}106 \pm 60{,}906}{75{,}004}$

    so 1.013 and -0.6106 are the roots.

59. $x^2 - 7x + 10 = 0$

    $(x - 5)(x - 2) = 0$ so the roots are $x = 5$ and $x = 2$

61. $0 = z^2 + 8z - 11$

    $z = \dfrac{-8 \pm \sqrt{64 - 4(-11)}}{2} = \dfrac{-8 \pm \sqrt{108}}{2} = \dfrac{-8 \pm 6\sqrt{3}}{2} = -4 \pm 3\sqrt{3}$

63. $w^2 + 2w = w^2 - 2w + 1$

      $4w = 1$

      $w = \frac{1}{4}$ (reduced to a linear equation, so only one root).

65. $1 - x^2 = x^2 - 2x + 1$

      $0 = 2x^2 - 2x$

      $0 = 2x(x - 1)$

      roots are $x = 0$ and $x = 1$

67. $y^2 + 12y + 36 = y^2 - 4y + 4$

          $16y = -32$

            $y = -2$ (reduced to a linear equation)

69. sum $= -\frac{b}{a} = -\frac{4}{2} = -2$, product $= \frac{c}{a} = \frac{-6}{2} = -3$

71. sum $= -\frac{49}{7} = -7$, product $= \frac{-1}{7}$

73. sum $= -\frac{-50}{1} = 50$, product $= \frac{137}{1} = 137$

75. sum $= -\frac{8056}{1} = -8056$, product $= \frac{-1137}{1} = -1137$

77. $\frac{24}{c} = -3$ so $c = -8$

79. $-\frac{c}{1} = 6$, so $c = -6$

81. $\frac{c}{1} = -7$, so $c = -7$

83. $\frac{4}{c} = 5$, so $c = \frac{4}{5}$

85. Let r_1 and r_2 be the two roots. Then $r_1 = 2r_2$. Now $\frac{c}{1} = r_1 r_2$.

    $\frac{9}{1} = r_1 + r_2 = 2r_2 + r_2 = 3r_2$, so $r_2 = 3$, $r_1 = 6$, and $c = 6\cdot3 = 18$

87. The product of the roots must be -1, so $\frac{c}{1} = -1$ which means $c = -1$

89. $(x - 1)(x + 3) = x^2 + 2x - 3$

91. $(x + 1)(x + 3) = x^2 + 4x + 3$

93. $(x - 6)^2 = x^2 - 12x + 36$

95. $(x + \frac{1}{2})^2 = x^2 + x + \frac{1}{4}$

97. $(x - \frac{5}{2})(x + \frac{3}{2}) = x^2 - x - \frac{15}{4}$

99. $(x - 1.206)(x + 2.451) = x^2 + 1.245x - 2.955906$

101. $(x + 2137)(x + 4916) = x^2 + 7053x + 10, 505, 492$

103. $\dfrac{1.7 \pm \sqrt{(1.7)^2 - 4(1)(-3.5)}}{2(1)} = \dfrac{1.7 \pm \sqrt{2.89 + 14}}{2} = \dfrac{1.7 \pm \sqrt{16.89}}{2} = \dfrac{1.7 \pm 4.109744518}{2}$
$= 2.9049$ or -1.2049

105. $\dfrac{-3.2 \pm \sqrt{(3.2)^2 - 4(2.1)(-5.7)}}{2(2.1)} = \dfrac{-3.2 \pm \sqrt{58.12}}{4.2} = \dfrac{-3.2 \pm 7.623647421}{4.2} = 1.0532, -2.5771$

107. $\dfrac{-7.8 \times 10^{-3} \pm \sqrt{(7.8 \times 10^{-3})^2 - 4(1)(1.2) \times 10^{-6}}}{2(1)} = \dfrac{-7.8 \times 10^{-3} \pm 7.485986909 \times 10^{-3}}{2}$

$= -1.5701 \times 10^{-4}, -7.6430 \times 10^{-3}$

109. From the quadratic formula obtain roots of the polynomial are 4 and $-\frac{5}{2}$;
since the polynomial has leading coefficient 2, obtain $2(x + \frac{5}{2})(x - 4)$ or $(2x + 5)(x - 4)$

111. The quadratic formula gives roots of $-\frac{4}{5}$ and $-\frac{7}{2}$; therefore, the polynomial can be factored
as $10(x + \frac{4}{5})(x + \frac{7}{2}) = 5(x + \frac{4}{5})2(x + \frac{7}{2}) = (5x + 4)(2x + 7)$

113. The quadratic formula give roots of $\frac{9}{5}$ and $-\frac{5}{3}$ so factoring is $15(x - \frac{9}{5})(x + \frac{5}{3})$
$= (5x - 9)(3x + 5)$

Problems 2.4

Recall that if $a > 0$, $\sqrt{-a} = \sqrt{a}\, i$

1. $5i$

3. $\sqrt{5}\, i$

The conjugate of $a + bi$ is $a - bi$

5. $3 + 7i$ 7. 5 9. $-2 + \frac{1}{3}i$ 11. $-\sqrt{73}\, i$ 13. $\frac{4}{7} - \frac{\sqrt{2}\, i}{7}$ 15. $3 - 8i$

17. $\frac{-4 + 8i}{8} = -\frac{1}{2} + i$

19. $2 + 4 = 6$, $3 + 5 = 8$, so the answer is $6 + 8i$

21. $\frac{1}{2} - \frac{1}{20}i$

23. $6 - 6 = 0$, $\sqrt{3} - (-\sqrt{3}) = 2\sqrt{3}$, so the answer is $2\sqrt{3}i$

25. $1 + i + 2i + 2i^2 = 1 + 3i - 2 = -1 + 3i$

27. $1 - 2i + i^2 = 1 - 2i - 1 = -2i$

29. $16 + 16i + 4i^2 = 16 + 16i - 4 = 12 + 16i$

31. $21 - 35i + 6i - 10i^2 = 21 - 29i + 10 = 31 - 29i$

33. $-\frac{3}{2} + \frac{1}{8}i - 3i + \frac{1}{4}i^2 = -\frac{3}{2} - \frac{23}{8}i - \frac{1}{4} = -\frac{7}{4} - \frac{23}{8}i$

35. $10i + 4i^2 - 15 - 6i = 4i - 4 - 15 = -19 + 4i$

37. $i^{11} = (i^4)^2 i^3 = 1 \cdot i^3 = -i$

39. $(-i)^{15} = -i^{15} = -(i^4)^3 i^3 = -(1)(-i) = i$

41. $(1 - i)^2 (1 - i)^2 = (1 - 2i + i^2)(1 - 2i + i^2) = (1 - 2i - 1)^2 = (-2i)^2 = 4i^2 = -4$

43. $\frac{1}{1 + i} \cdot \frac{1 - i}{1 - i} = \frac{1 - i}{1 - i^2} = \frac{1 - i}{1 + 1} = \frac{1 - i}{2} = \frac{1}{2} - \frac{1}{2}i$

45. $\frac{8 - 3i}{2i} \cdot \frac{2i}{2i} = \frac{16i - 6i^2}{-4} = \frac{16i + 6}{-4} = -\frac{3}{2} - 4i$

47. $\frac{1 + i}{1 - i} \cdot \frac{1 + i}{1 + i} = \frac{1 + 2i + i^2}{1 - i^2} = \frac{1 + 2i - 1}{1 + 1} = \frac{2i}{2} = i$

49. $\frac{3 + 2i}{2 - 3i} \cdot \frac{2 - 3i}{2 - 3i} = \frac{6 - 5i - 6i^2}{4 + 9i^2} = \frac{6 - 5i + 6}{4 - 9} = \frac{-5i}{-5} = i$

51. $\frac{5 + 7i}{2 + 4i} \cdot \frac{2 - 4i}{2 - 4i} = \frac{10 - 6i - 28i^2}{4 + 16} = \frac{38 - 6i}{20} = \frac{19}{10} - \frac{3}{10}i = 1.9 - 0.3i$

53. $\frac{1}{2i} \cdot \frac{2i}{2i} = \frac{2i}{4i^2} = \frac{2i}{-4} = -\frac{1}{2}i$

55. $\frac{1}{1 - i} \cdot \frac{1 + i}{1 + i} = \frac{1 + i}{1 - i^2} = \frac{1 + i}{1 + 1} = \frac{1}{2} + \frac{1}{2}i$

57. $\frac{1}{2 - 5i} \cdot \frac{2 + 5i}{2 + 5i} = \frac{2 + 5i}{4 - 25i^2} = \frac{2 + 5i}{29} = \frac{2}{29} + \frac{5}{29}i$

59. $\frac{1}{4 - 6i} \cdot \frac{4 + 6i}{4 + 6i} = \frac{4 + 6i}{16 + 36} = \frac{1}{13} + \frac{3}{26}i$

61. $\frac{1}{a - bi} \cdot \frac{a + bi}{a + bi} = \frac{a}{a^2 + b^2} + \frac{b}{a^2 + b^2}i$

63. $u + v = 5$ and $-2 = 2v$ are obtained by equating the real parts and the imaginary parts, respectively. The second equation implies $v = -1$, so $u = 6$ from first.

65. $2u = v$ and $-3v = 7u - 2$. Substituting the first equation into the second, one

obtains $-3(2u) = 7u - 2$

so $-6u = 7u - 2$

$-13u = -2$

$u = \frac{2}{13}$ and so $v = \frac{4}{13}$

67. $x^2 = -4$, so $x = \pm\sqrt{-4} = \pm 2i$

69. $z = \frac{-1 \pm \sqrt{1^2 - 4(1)(2)}}{2} = \frac{-1 \pm \sqrt{-7}}{2} = \frac{-1 \pm \sqrt{7}\,i}{2}$

71. $p = \frac{6 \pm \sqrt{36 - 40}}{2} = \frac{6 \pm 2i}{2} = 3 \pm i$ 73. $y = \frac{3 \pm \sqrt{9 - (4)(2)(5)}}{4} = \frac{3 \pm \sqrt{31}\,i}{4}$

75. Multiply through by 12 to obtain $6x^2 + 4x + 3 = 0$

$x = \frac{-4 \pm \sqrt{16 - 72}}{12} = \frac{-4 \pm \sqrt{56}\,i}{12} = -\frac{1}{3} \pm \frac{\sqrt{14}}{6}\,i$

77. $z = \frac{8.06 \pm \sqrt{64.9636 - 843.5904}}{25.44} = \frac{8.06 \pm 27.903885i}{25.44} \approx 0.317 \pm 1.097i$

79. Let $a + bi$ be a complex number. If the number equals its conjugate, then $a + bi = a - bi$, so $a = a$ and $b = -b$. But $b = -b$ only if $b = 0$, so the number must be real.

81. The third equation does not imply the fourth equation: $\sqrt{\frac{a}{b}} = \frac{\sqrt{a}}{\sqrt{b}}$ is true only when a and b are both positive.

Problems 2.5

1. Square both sides to obtain $x - 3 = 16$, which has solution $x = 19$

3. Cube both sides to obtain $z + 5 = 125$, which has solution $z = 120$

5. Raise both sides to the power $\frac{2}{3}$ (always use the reciprocal of the given one) to obtain y - 1 = 9, which has solution y = 10.

7. Let $y = z^2$ to obtain the equation y^2 - 7y + 12 = 0

$$(y - 4)(y - 3) = 0$$

which has solutions y = 4 and y = 3.

Therefore, $z^2 = 4$ or $z^2 = 3$, and so the solutions to the original equation are z = 2, z = -2, z= $\sqrt{3}$, and z = -$\sqrt{3}$.

9. Let $y = v^2$ to obtain the equation y^2 - 81 = 0

$$(y - 9)(y + 9) = 0$$

Therefore, y = 9 or y = -9

$v^2 = 9$ or $v^2 = -9$, so the solutions to the original equation are

v = 3, v = -3, v= 3i, and v = -3i

11. Let $y = x^2$ to obtain the equation y^2 - 25y + 144 = 0

$$(y - 9)(y - 16) = 0$$

Therefore y = 9 or y = 16, which says $x^2 = 9$ or $x^2 = 16$, so the solutions to the original equation are x = 3, x = -3, x = 4, and x = -4

13. $(p - 4)(p^2 + 4p + 16) = 0$ So either p - 4 = 0 or $p^2 + 4p + 16 = 0$

So either p = 4 or p = $\dfrac{-4 \pm \sqrt{16 - 64}}{2} = \dfrac{-4 \pm \sqrt{-48}}{2}$ = -2 $\pm$ 2$\sqrt{3}$ i

15. Factors as $(x^3 - 8)(x^3 - 1) = 0$

$$(x - 2)(x^2 + 2x + 4)(x - 1)(x^2 + x + 1) = 0$$

Either x - 2 = 0, $x^2 + 2x + 4 = 0$, x - 1 = 0, or $x^2 + x + 1 = 0$

Now $x^2 + 2x + 4 = 0$ when x = $\dfrac{-2 \pm \sqrt{4 - 16}}{2}$ = $\dfrac{-2 \pm \sqrt{-12}}{2} = \dfrac{-2 \pm 2\sqrt{3}\,i}{2}$ = -1 $\pm$ $\sqrt{3}$ i

and $x^2 + x + 1 = 0$ when x = $\dfrac{-1 \pm \sqrt{1 - 4}}{2} = \dfrac{-1 \pm \sqrt{3}\,i}{2}$.

So the solutions are 2, -1 $\pm$ $\sqrt{3}$ i , 1, and -$\frac{1}{2}$ $\pm$ $\frac{\sqrt{3}}{2}$ i

17. This factors as $(x^3 + 8)(x^3 + 1) = 0$

$$(x + 2)(x^2 - 2x + 4)(x + 1)(x^2 - x + 1) = 0$$

Noting that the factors are identical to those in problem 15 except for signs,

it follows that the solutions are -1, $1 \pm \sqrt{3}$ i, -2, and $\frac{1}{2} \pm \frac{\sqrt{3}}{2}$ i.

19. This problem factors as $(y^4 - 16)(y^4 - 1) = 0$

$$(y^2 - 4)(y^2 + 4)(y^2 - 1)(y^2 + 1) = 0$$

The solutions are y = 2, -2, 2i, -2i, 1, -1, i, -i

21. Let $y = w^{2/7}$. The equation becomes $y^2 - 3y - 10 = 0$

$$(y - 5)(y + 2) = 0$$

So y = 5 or y = -2, or $w^{2/7} = 5$ or $w^{2/7} = -2$

$w^2 = 5^7 = 78{,}125$ or $w^2 = (-2)^7 = -128$

$w = \pm \sqrt{78{,}125} = \pm 125\sqrt{5}$ or $w = \pm \sqrt{-128} = \pm 8\sqrt{2}$ i

23. Let $y = u^{0.1}$. Then the equation becomes $y^2 + 13y + 42 = 0$

$$(y + 6)(y + 7) = 0$$

So y = -6 or y = -7 . But the tenth root of a number cannot be negative which means -6 = $u^{0.1}$ and $u^{0.1}$ = -7 have no solutions, nor does the original problem.

25. Multiply both sides of the equation by (y - 2)(y + 1) to give

$3y + 3 = 4y - 8 + 5(y^2 - y - 2)$

$0 = 5y^2 - 4y - 21$

$$y = \frac{4 \pm \sqrt{16 - 4(5)(-21)}}{10} = \frac{4 \pm \sqrt{436}}{10} = \frac{4 \pm 2\sqrt{109}}{10} = \frac{2 \pm \sqrt{109}}{5}$$

27. Multiply through by z to give $1 + z^2 = 2z$

$z^2 - 2z + 1 = 0$

$(z - 1)^2 = 0$

$z = 1$

29. Raise both sides to the third power to obtain $x^2 - 3x - 2 = 8$

$$x^2 - 3x - 10 = 0$$
$$(x - 5)(x + 2) = 0$$
$$x = 5 \text{ or } x = -2$$

31. Square both sides to obtain $3z + 4 = z^2$

$$0 = z^2 - 3z - 4$$
$$0 = (z - 4)(z + 1)$$

$z = 4$ is a solution to the original equation but $z = -1$ is not (extraneous)

33. Square both sides to obtain $x + 9 = x^2 - 6x + 9$

$$0 = x^2 - 7x$$
$$0 = x(x - 7)$$

$x = 7$ is a solution to the original equation but $x = 0$ is not (extraneous).

35. Square both sides to obtain $3p + 3 = p^2 + 2p + 1$

$$0 = p^2 - p - 2$$
$$0 = (p - 2)(p + 1)$$

$p = 2$ and $p = -1$ are both solutions to the original equation.

37. Square both sides, obtain $3x + 4 = 4x^2 - 4x + 1$

$$0 = 4x^2 - 7x - 3$$

$x = \dfrac{7 \pm \sqrt{49 - 4(4)(-3)}}{8} = \dfrac{7 \pm \sqrt{97}}{8}$ but only $+$ works, the one with the $-$ does not (extraneous). So, the only solution is $x = \dfrac{7 + \sqrt{97}}{8}$.

39. Square and obtain $2z - 3 = z^2 - 6z + 9$

$$0 = z^2 - 8z + 12$$
$$0 = (z - 6)(z - 2)$$

$z = 6$ is

41. Subtract $\sqrt{s + 2}$ from both sides to obtain $\sqrt{s + 1} = 1 - \sqrt{s + 2}$

Square both sides to obtain $s + 1 = 1 - 2\sqrt{s + 2} + s + 2$

$$- 2 = -2\sqrt{s + 2}$$

$$1 = \sqrt{s + 2}$$

Square again to obtain $\qquad 1 = s + 2$

$$s = -1$$

43. Add $\sqrt{2x - 1}$ to both sides to obtain $\sqrt{2x - 1} = 3 + \sqrt{x + 5}$

Now square both sides to obtain $\qquad 2x - 1 = 9 + \sqrt{6x+5} + 14x +5$

$$x - 15 = 6\sqrt{x + 5}$$

Square again and obtain $\qquad x^2 - 30x + 225 = 36(x + 5)$

$$x^2 - 66x + 45 = 0$$

$$x = \frac{66 \pm \sqrt{66^2 - 4(1)(45)}}{2} = \frac{66 \pm \sqrt{4176}}{2} = \frac{66 \pm 12\sqrt{29}}{2} = 33 \pm 6\sqrt{29}$$

Take $+$ because $-$ is extraneous. Therefore, the solution is $x = 33 + 6\sqrt{29}$

45. Square both sides to obtain $2x + 1 + 2\sqrt{(2x + 1)(3x - 5)} + 3x - 5 = 4x + 7$

$$2\sqrt{6x^2 - 7x - 5} = -x + 11$$

Square again, obtain $24x^2 - 28x - 20 = x^2 - 22x + 121$

$$23x^2 - 6x - 141 = 0$$

$$x = \frac{6 \pm \sqrt{36 - 4(23)(-141)}}{46} = \frac{6 \pm \sqrt{36 + 12972}}{46} = \frac{6 \pm \sqrt{13008}}{46}$$

$$= \frac{6 \pm 4\sqrt{813}}{46} = \frac{3 \pm 2\sqrt{813}}{23} \qquad$$ Must take $+$, since $-$ is extraneous, so $x = \dfrac{3 + 2\sqrt{813}}{23}$.

47. $x = x^2 - 6 \qquad\qquad$ or $\qquad\qquad x = -(x^2 - 6) = x = -x^2 + 6$

$\quad 0 = x^2 - x - 6 \qquad\qquad\qquad\qquad\qquad x^2 + x - 6 = 0$

$\quad 0 = (x - 3)(x + 2) \qquad\qquad\qquad\qquad (x + 3)(x - 2) = 0$

$\quad x = 3$ or $x = -2 \qquad\qquad\qquad\qquad\; x = -3$ or $x = 2$

Of the four above solutions, only $x = \pm 3$ work, because ± 2 make the right side negative.

49. $x^2 + 1 = x^2 - 1$ or $x^2 + 1 = -(x^2 - 1) = -x^2 + 1$

     $1 = -1$ $2x^2 = 0$

     none here $x = 0$

So 0 is the only solution.

51. $x^2 + 7x = x + 7$ or $x^2 + 7x = -(x + 7) = -x - 7$

     $x^2 + 6x - 7 = 0$ $x^2 + 8x + 7 = 0$

     $(x + 7)(x - 1) = 0$ $(x + 7)(x + 1) = 0$

Combining the solutions from the above, we get -7, 1, and -1 as the solutions.

53. Let $u = x^3$. This leads to the equation $u^2 + 17u - 35 = 0$

$$u = \frac{-17 \pm \sqrt{289 - 4(1)(-35)}}{2(1)} = \frac{-17 \pm 20.71231518}{2}, \text{ so } u = -18.856 \text{ or } 1.857.$$

To obtain the real roots to the equation, take the cube roots of the above and obtain

$x = -2.6617$ and $x = 1.2290$

55. Let $u = x^{1/8}$ and obtain $u^2 - 6.72u + 1.84 = 0$

$$u = \frac{6.72 \pm \sqrt{(6.72)^2 - 4(1)(1.84)}}{2(1)} = \frac{6.72 \pm 6.148040338}{2}, \text{ so } u = 6.43020169 \text{ or } 0.285979831$$

Raise each of these numbers to the eight power, obtain 2,936,698.381 and 0.000044739

57. Multiply the equation through by x^4 and rearrange to obtain $4.3x^4 - 15x^2 + 1 = 0$

Let $u = x^2$ and then the quadratic formula to get $u = \dfrac{15 \pm \sqrt{225 - 4(1)(4.3)}}{2(1)} = \dfrac{15 \pm 14.41526968}{2}$

So $u = 3.420380196$ or 0.067991897. Take square roots, $x = \pm 1.8494$ or ± 0.2608

59. Solve $x^3 = -1$ or $x^3 + 1 = 0$

     $(x + 1)(x^2 - x + 1) = 0$. Solutions are $x = -1$ and $x = \dfrac{1 \pm \sqrt{1 - 4(1)(1)}}{2(1)} = \dfrac{1}{2} \pm \dfrac{\sqrt{3}}{2}i$

61. $x^3 = 5$

$x^3 - 5 = 0$ Use the difference of cubes formula where a = x and

$b = \sqrt[3]{5}$ (the cube root of 5)

to factor as $(x - \sqrt[3]{5})(x^2 + \sqrt[3]{5}x + \sqrt[3]{25}) = 0$

Solutions are obtained by setting each factor to zero; one solution is $x = \sqrt[3]{5}$

and the other two come from $x^2 + \sqrt[3]{5}x + \sqrt[3]{25} = 0$. Use the quadratic formula.

$$x = \frac{-\sqrt[3]{5} \pm \sqrt{\sqrt[3]{25} - 4\sqrt[3]{25}}}{2} = \frac{-\sqrt[3]{5} \pm \sqrt{-3\sqrt[3]{25}}}{2} = \frac{-\sqrt[3]{5} \pm \sqrt{3}\sqrt[6]{25}\,i}{2}$$

$$= \frac{-\sqrt[3]{5} \pm \sqrt{3}\sqrt[3]{5}\,i}{2} \text{ (follows since } 5^2 = 25) = \frac{-\sqrt[3]{5}\,(1 \pm \sqrt{3}\,i)}{2}$$

63. $A^3 = -1$ does not imply that $A = -1$, since -1 has 3 cube roots.

Problems 2.6

1. $17,125 = 5q^2 - 200q + 1125$

    $3425 = q^2 - 40q + 225$

      $0 = q^2 - 40q - 3200$

      $0 = (q - 80)(q + 40)$

     $q = 80$ units is the only feasible solution

3. $9694.72 = -0.12q^2 + 100q$

   $.12q^2 - 100q + 9694.72 = 0$

$$q = \frac{100 \pm \sqrt{10,000 - 4(.12)(9694.72)}}{.24} = \frac{100 \pm \sqrt{5346.5344}}{.24} = \frac{100 \pm 73.12}{.24}$$

The two solutions are $\frac{173.12}{.24} = 721\frac{1}{3}$ items and $\frac{26.88}{.24} = 112$ items

5.   a) $R = 14q$

   b) $P = 14q - (250 + 3q + 0.01q^2) = -0.01q^2 + 11q - 250$

   c) $1406 = -.01q^2 + 11q - 250$

     $.01q^2 - 11q + 1656 = 0$ $q = \frac{11 \pm \sqrt{121 - 4(.01)(1656)}}{.02} = \frac{11 \pm \sqrt{121 - 66.24}}{.02} = \frac{11 \pm 7.4}{.02}$ to

   obtain 920 or 180. The only one in the stated domain is 180 dolls.

7. $19,466 = 5000 + 250q - .01q^2$

$.01q^2 - 250q + 14466 = 0$

$$q = \frac{250 \pm \sqrt{62500 - 4(.01)(14466)}}{.02} = \frac{250 \pm 248.84003}{.02}.$$ Taking + gives a

number much larger than 300; taking − gives the solution, approximately 58 sets.

9. a) $P = (400q - .02q^2) - (5000 + 250q - .01q^2)$

$\quad = -0.01q^2 + 150q - 5000$

b) $18,010 = -.01q^2 + 150 q - 5000$

$\quad .01q^2 - 150q + 23,010 = 0$

$$q = \frac{150 \pm \sqrt{22,500 - 4(.01)(23010)}}{.02} = \frac{150 \pm 146.89997}{.02}.$$

Must take − to get the feasible solution of 155 sets.

11. $s = 16t^2$

$\quad 200 = 16t^2$

$\quad t^2 = 12.5$ so $t = \sqrt{12.5} \approx 3.54$ sec. (since t must be positive)

13. $s = 4.9t^2$ so $200 = 4.9t^2$

$t = \sqrt{\dfrac{200}{4.9}} \approx 6.39$ sec

15. As in #13, $t = \sqrt{\dfrac{2500}{4.9}} \approx 22.6$ sec

17. $16t^2 + 50t = 500$

$\quad 16t^2 + 50t - 500 = 0$

$\quad 8t^2 + 25t - 250 = 0$

$$t = \frac{-25 \pm \sqrt{625 - 4(8)(-250)}}{16} = \frac{-25 \pm 92.87}{16}.$$ Taking +, obtain $\sim$4.24 secs.

19. $s = 1.96t^2$ so $t = \sqrt{\dfrac{1000}{1.96}} = 22.59$ secs

- 69 -

21. Let i be the interest rate. Then add the interest for the two investments:

$$10{,}000(1 + i)^2 + 25{,}000(1 + i) = 40{,}125$$

$$10{,}000 + 20{,}000i + 10{,}000i^2 + 25{,}000 + 25{,}000i = 40{,}125$$

$$10{,}000i^2 + 45{,}000i - 5125 = 0.$$

To make the numbers smaller, you can divide both sides by 10,000.

$$i^2 + 4.5i - .5125 = 0$$

Use the quadratic formula to obtain $i = \dfrac{-4.5 \pm \sqrt{20.25 + 2.05}}{2} = \dfrac{-4.5 \pm \sqrt{22.30}}{2}$
$= \dfrac{-4.5 \pm 4.7222876}{2}$. Take $+$ to obtain .111143, or approximately 11.11%.

23. Let i be the interest rate for one-half year.

$$2500(1 + i)^2 = 2691$$

$$1 + 2i + i^2 = 1.0764$$

$$i^2 + 2i - 0.0764 = 0$$

Use the quadratic formula to obtain $i = \dfrac{-2 \pm \sqrt{4 + 4(.0764)}}{2} = \dfrac{-2 \pm 2.0749936}{2}$

Taking plus, obtain $\dfrac{.0749939}{2}$ per half year, or $\sim 7.5\%$ per year.

25. If x = width of courtyard, then 2x + 3 = length of courtyard

$$\text{Area} = 27596 = x(2x + 3) = 2x^2 + 3x$$

$$0 = 2x^2 + 3x - 275$$

$$0 = (x - 11)(2x + 25)$$

Only the solution x = 11 makes sense. Dimensions are 11ft. by 25 ft.

27. Let x = one of the integers; then the other = x + 2

$$x(x + 2) = 783$$

$$x^2 + 2x = 783$$

$$x^2 + 2x - 783 = 0$$

$$(x + 29)(x - 27) = 0$$

x = 27 yields that the two integers are 27 and 29; x = -29 gives -29 and -27.

29. Let x = the number. Then $x - \frac{1}{x} = \frac{39}{40}$.

Multiply through by 40x to obtain $40x^2 - 40 = 39x$

$40x^2 - 39x - 40 = 0$ or $(8x + 5)(5x - 8) = 0$

The number and its reciprocal are $\frac{8}{5}$ and $\frac{5}{8}$ or $-\frac{5}{8}$ and $\frac{8}{5}$

31. Let R = unknown resitance of one resistor; the other has resistance R + 2

    From the equation in Example 11, $\frac{9}{40} = \frac{1}{R} + \frac{1}{R+2}$

    Multiply through by 40R(R + 2) to obtain $9R(R+2) = 40(R + 2) + 40R$

$$9R^2 + 18R = 40R + 80 + 40R$$

$$9R^2 - 62R - 80 = 0$$

$$(R - 8)(9R + 10) = 0$$

    R cannot be negative , so R = 8. The two resistances are 8 and 10 ohms.

33. Let h = height of triangle. Then its base is h + 1

    Area $= \frac{1}{2}$ X base X height $= \frac{1}{2}h(h + 1) = 15$

$$h^2 + h = 30$$

$$h^2 + h - 30 = 0$$

$$(h + 6)(h - 5) = 0$$

    Since h cannot be negative, h = 5, so the height is 5 cm and base is 6 cm.

35. Let t = time it would take Mary alone. Then t + 2 = time for Jeff alone.

    Then $\frac{1}{t} + \frac{1}{t+2} = \frac{1}{8}$ (adding the amount each could do in one hour)

    Multiply through by 8t(t + 2) to obtain $8(t + 2) + 8t = t(t + 2)$

$$16t + 16 = t^2 + 2t$$

$$0 = t^2 - 14t - 16$$

    So $t = \dfrac{14 \pm \sqrt{196 - 4(1)(-16)}}{2} \approx \dfrac{14 \pm 16.125}{2}$. Take + to obtain

    t = 15.062. So Mary's time alone is approximately 15 hours and Jeff's is 17 hrs.

Problems 2.7

1. [-1, 6]   3. (-3, 6]   5. [0, ∞)   7. (-∞, 7)   9. $[\frac{1}{2}, \frac{7}{5}]$ 11. [39, 429)

13. (1, 2)   15. (0, 8)   17. [-2, 0]   19. (0, 5]   21. [-1.32, 4.16)   23. (-∞, 0)

25. (-∞, 2]   27. (-∞, -5)

29. x - 4 < 7.   Add 4 to both sides to obtain

   x < 11   or (-∞, 11)

31. x - 4 > $\frac{7}{2}$. Add 4 to both sides to obtain

   x > $\frac{15}{2}$ or $\left(\frac{15}{2}, \infty\right)$

33. -x + 2 ≤ 3

   -x ≤ 1   Now multiply both sides by -1 to obtain

   x ≥ -1 (recall that mutliplying by a negative number reverses the inequality) or [-1, ∞)

35. 2x - 7 ≤ 2

   2x ≤ 9

   x ≤ $\frac{9}{2}$ or $\left(-\infty, \frac{9}{2}\right]$

37. 2 ≥ 6x + 14

   -12 ≥ 6x

   -2 ≥ x   or (-∞, -2]

39. -2 < 4 - 8x

   -6 < -8x

   $\frac{3}{4}$ > x   or $\left(-\infty, \frac{3}{4}\right)$

41. 1 ≤ x + 2 ≤ 4

   -1 ≤ x ≤ 2   or [-1, 2]

43. -2 < -x + 3 ≤ 7

   -5 < -x ≤ 4

   5 > x ≥ -4 or [-4, 5)

45. $\frac{1}{5}$ ≤ 2x + $\frac{2}{5}$ < $\frac{4}{5}$

   1 ≤ 10x + 2 < 4

   -1 ≤ 10x < 2

   - $\frac{1}{10}$ ≤ x < $\frac{1}{5}$ or [- $\frac{1}{10}$, $\frac{1}{5}$)

47. $-1 \leq 2x + 5 < 7$

   $-6 \leq 2x < 2$

   $-3 \leq x < 1$ or $[-3, 1)$

49. $-2 < -2x - 4 \leq 10$

   $2 < -2x \leq 14$

   $-1 > x \geq -7$ or $[-7, -1)$

51. Multiply through by 3 and obtain

   $-12 < 2x - 4 \leq 21$ so $-8 < 2x \leq 25$

   $-4 < x \leq \frac{25}{2}$ or $(-4, \frac{25}{2}]$

53. $a \leq \frac{b\,x + c}{d} < e$

   $ad \leq bx + c < de$

   $ad - c \leq bx < de - c$

   $\frac{ad - c}{b} \leq x < \frac{de - c}{b}$ or $[\frac{ad - c}{b}, \frac{de - c}{b})$

55. Subtract 21.562 from both sides to obtain $x \geq 15.733$,   or $[15.733, \infty)$

57. Divide both sides by 0.137 to obtain $x > 3.438$, or $(3.438, \infty)$

59. Divide both sides by -2×10^{-5} to obtain $x < -15$ or $(-\infty, -15)$

61. $2 \times 10^{-6} \leq \frac{4 \times 10^{-5}x - 7.2 \times 10^{-7}}{3.1 \times 10^{-3}} \leq 6.5 \times 10^{-5}$

   $6.2 \times 10^{-9} \leq 4 \times 10^{-5}x - 7.2 \times 10^{-7} \leq 2.015 \times 10^{-7}$

   $7.262 \times 10^{-7} \leq 4 \times 10^{-5}x \leq 9.215 \times 10^{-7}$

   $1.816 \times 10^{-2} \leq x \leq 2.304 \times 10^{-2}$

   $[1.816 \times 10^{-2}, 2.304 \times 10^{-2}]$

Problems 2.8

1. $3 - 2 = 1$ 3. $2 - 3 = -1$ 5. $10 - 12 = -2$ 7. $\pi - 2$ 9. $-(\pi - 7) = 7 - \pi$

11. (-1, 1)   13. (-∞, -4] or [4, ∞)

15. All real numbers since absolute value is never negative   (-∞, ∞)

17. No real numbers.

19. -1 < x - 2 < 1

   1 < x < 3   or (1, 3)

21. x - 3 > 2 or x - 3 < -2

   x > 5 or x < 1

   (5, ∞) or (-∞, 1)

23. -5 < 4x + 1 < 5

   -6 < 4x < 4

   $-\frac{3}{2} < x < 1$ or $\left(-\frac{3}{2}, 1\right)$

25. 5 - x ≥ 1 or 5 - x ≤ -1

   4 ≥ x or 6 ≤ x   or (-∞, 4] or [6, ∞)

27. Add 1 to both sides and obtain

   |-2x - 4| > 6

   -2x - 4 > 6 or -2x - 4 < -6

   -2x > 10 or -2x < -2

   x < -5   or x > 1; that is, $\left(-\infty, -5\right)$ or $\left(1, \infty\right)$

29.   First suppose that both expressions inside the | | are positive or zero. Then

   6 - 4x > 0   and   x - 2 > 0

   6 > 4x and   x > 2

   $x < \frac{3}{2}$ and   x > 2   This is not possible.

   (continued on next page)

Now suppose both are negative or zero.

Then $6 - 4x \leq 0$ and $x - 2 \leq 0$.

So $x \geq \frac{3}{2}$ and $x \leq 2$.

The inequality becomes $-6 + 4x \geq -x + 2$, since $|a| = -a$ when $a < 0$.

$$5x \geq 8$$
$$x \geq \frac{8}{5}$$

The sets $x \geq \frac{3}{2}$, $x \leq 2$, and $x \geq \frac{8}{5}$ must be intersected for the solution $\frac{8}{5} \leq x \leq 2$.

The third case is where the first expression is positive, the second is negative, or both are zero. Then get $x \leq \frac{3}{2}$ and $x \leq 2$. So $x \leq \frac{3}{2}$.

The inequality becomes $6 - 4x \geq -x + 2$

$$4 \geq 3x$$
$$x \leq \frac{4}{3} \quad \text{Since } \frac{4}{3} \leq \frac{3}{2}$$

now we have obtained that all $x \leq \frac{4}{3}$ are solutions of this case.

The last case is where the first expression is negative and the second positive, or both are zero. Obtain $x \geq \frac{3}{2}$ and $x \geq 2$. So $x \geq 2$.

The inequality becomes $-6 + 4x \geq x - 2$

$$3x \geq 4$$
$$x \geq \frac{4}{3} \quad \text{Since } 2 \geq \frac{4}{3}$$

solutions obtained here are $x \geq 2$.

Taking the union of the solutions to all 4 cases, obtain $x \leq \frac{4}{3}$ or $x \geq \frac{8}{5}$

31. $\frac{3x + 17}{4} > 9$ or $\frac{3x + 17}{4} < -9$

$3x + 17 > 36$ or $3x + 17 < -36$

$3x > 19 \qquad$ or $\quad 3x < -53$

$x > \frac{19}{3} \qquad$ or $\quad x < -\frac{53}{3} \qquad$ or $\left(-\infty, -\frac{53}{3}\right)$ or $\left(\frac{19}{3}, \infty\right)$

33. $ax + b \geq c$ or $ax + b \leq -c$

$ax \geq -b + c \qquad$ or $\qquad ax \leq -b - c$

$x \leq \frac{-b + c}{a} \qquad$ or $\qquad x \geq \frac{-b - c}{a} \quad$ since $a < 0 \quad$ Solution: $\left(-\infty, \frac{-b + c}{a}\right]$ or $\left[\frac{-b - c}{a}, \infty\right)$

35. As in # 29, we have to work in cases.

First case is where both expressions inside | | are positive or 5 - 2x is zero.

$2x > 0$ and $5 - 2x \geq 0$

$x > 0$ and $-2x \geq -5$

$x > 0$ and $x \leq \frac{2}{5}$

The inequality becomes $2x > 5 - 2x$

$$4x > 5$$
$$x > \frac{5}{4}$$

Taking the intersection of the above solutions results in $\frac{5}{4} < x \leq \frac{5}{2}$.

Next case is where both expressions are negative or 5 - 2x is zero.

Obtain $x < 0$ and $x \geq \frac{5}{2}$. This is not possible, so no solutions here.

Third case is where first expression is positive and second negative or 5 - 2x = 0.

Then $x > 0$ and $x \geq \frac{5}{2}$ so $x \geq \frac{5}{2}$

The inequlaity becomes $2x > -5 + 2x$ or $0 > -5$, which is satisfied by all x.

Therefore, solutions from this case are $x \geq \frac{5}{2}$

Last case is where the first expression is negative and the second positive or zero.

$x < 0$ and $x \leq \frac{5}{2}$ so $x \leq 0$

$-2x > 5 - 2x$

$0 > 5$ is impossible, so no solutions from this case.

Take the union of the solutions of all to all cases to obtain $\frac{5}{4} < x$ as the solution.

Many of the following problems require the use of a test value to determine if an expression is positive or negative for all x in an indicated interval. The test value can be any number in the interval. Recall that a positive number times a positive number yields a positive number; a negative number times a negative number yields a positive number; and a positive number times a negative number results in a negative number. The same results hold for division. When solving non-linear inequalities, you must keep in mind that you cannot multiply both sides of an inequality by a linear expression; you don't know if the expression has a positive or negative value and so you don't know whether to switch the inequality sign or not.

37. $x^2 - 7x + 10 > 0$

$(x - 5)(x - 2) > 0$

Interval	Test Value	Sign of x - 5	Sign of x - 2	Sign of Result
$(-\infty, 2)$	0	-	-	+
$(2, 5)$	3	-	+	-
$(5, \infty)$	6	+	+	+

Want positive, so solutions are $(-\infty, 2)$ or $(5, \infty)$

39. $4s^2 < 1$

$4s^2 - 1 < 0$

$(2s - 1)(2s + 1) < 0$

Interval	Test Value	Sign of 2s - 1	Sign of 2s + 1	Sign of Product
$\left(-\infty, -\frac{1}{2}\right)$	-1	-	-	+
$\left(-\frac{1}{2}, \frac{1}{2}\right)$	0	-	+	-
$\left(\frac{1}{2}, \infty\right)$	1	+	+	+

Want expression to be negative, so solution is $\left(-\frac{1}{2}, \frac{1}{2}\right)$

41. $w^2 - 5w \geq 0$

$w(w - 5) \geq 0$

Interval	Test Value	Sign of w	Sign of w - 5	Sign of Product
$(-\infty, 0)$	-1	-	-	+
$(0, 5)$	1	+	-	-
$(5, \infty)$	6	+	+	+

Want expression to be greater than or equal to zero, so $(-\infty, 0]$ or $[5, \infty)$

43. $x^2 - 4x + 4 > 0$

$(x - 2)^2 > 0$

The square of an expression is always positive except when the expression is zero, so the solutions are $(-\infty, 2)$ or $(2, \infty)$

45. $x^2 - 4x + 4 \leq 0$

$(x - 2)^2 \leq 0$

$x = 2$ is the only solution, since $(x - 2)^2$ cannot be negative.

47. $x^2 - 4x + 9 < 0$

Use the quadratic formula to find where the expression at left equals 0:

$x = \dfrac{4 \pm \sqrt{16 - 4(1)(9)}}{2}$, which has no real roots. Therefore, the expression on the left of the inequality is either always positive or always negative. If $x = 0$ is plugged in, 9 is obtained. Therefore, the expression is always positive and the inequality has no solutions.

49. Look at #47. The solution to this problem is all real numbers.

51. $x^2 - 9x + 14 < 0$

$(x - 7)(x - 2) < 0$

Interval	Test Value	Sign of x - 7	Sign of x - 2	Sign of Product
$(-\infty, 2)$	0	-	-	+
$(2, 7)$	3	-	+	-
$(7, \infty)$	8	+	+	+

Want the expression to be negative, so the solution is $(2, 7)$

53. $-x^2 + x + 6 > 0$

$(2 + x)(3 - x) > 0$

Interval	Test Value	Sign of 2 + x	Sign of 3 - x	Sign of Product
$(-\infty, -2)$	-3	-	+	-
$(-2, 3)$	0	+	+	+
$(3, \infty)$	4	+	-	-

Want the expression to be positive, so solution: $(-2, 3)$

55. The expression at left equals 0 when $x = \dfrac{-2 \pm \sqrt{4 - 4(-7)}}{2} = \dfrac{-2 \pm 4\sqrt{2}}{2} = -1 \pm \sqrt{2}$

Interval	Test Value	Sign of Product
$\left(-\infty, -1 - 2\sqrt{2}\right)$	-5	+
$\left(-1 - 2\sqrt{2}, -1 + 2\sqrt{2}\right)$	0	-
$\left(-1 + 2\sqrt{2}, \infty\right)$	4	+

So, solution is $[-1 - 2\sqrt{2} \; , \; -1 + 2\sqrt{2}]$

57. $(3x + 7)(x - 1) \geq 0$

Interval	Test Value	Sign of 3x + 7	Sign of x - 1	Sign of Product
$\left(-\infty, -\frac{7}{3}\right)$	-3	-	-	+
$\left(-\frac{7}{3}, 1\right)$	0	+	-	-
$\left(1, \infty\right)$	2	+	+	+

Solutions are $[-\infty, -\frac{7}{3}]$ or $[1, \infty)$

59.

Interval	Test Value	Sign of 3x	Sign of 6 - 5x	Sign of Result
$\left(-\infty, 0\right)$	-1	-	+	-
$\left(0, \frac{6}{5}\right)$	1	+	+	+
$\left(\frac{6}{5}, \infty\right)$	2	+	-	-

Want positive, so solution: $(0, \frac{6}{5})$

61. $x^3 - x^2 < 0$

$x^2(x - 1) < 0$

You can do this problem with test values, but if you realize that x^2 cannot be negative, the solution can be obtained by considering when $x - 1 < 0$, and that is $(-\infty, 1)$

63. $x^3 + 2x^2 - 15x \leq 0$

$x(x + 5)(x - 3) \leq 0$

Interval	Test Value	Sign of x	Sign of x + 5	Sign of x - 3	Sign of Product
$(-\infty, -5)$	-6	-	-	-	-
$(-5, 0)$	-1	-	+	-	+
$(0, 3)$	1	+	+	-	-
$(3, \infty)$	4	+	+	+	+

$(-\infty, -5]$ or $[0, 3]$

65. $2 - \dfrac{1}{5 - x} < 0$ Put over a common denominator and obtain $2 \cdot \dfrac{5 - x}{5 - x} - \dfrac{1}{5 - x} < 0$

$\dfrac{10 - 2x - 1}{5 - x} < 0$ or $\dfrac{9 - 2x}{5 - x} < 0$

Interval	Test Value	Sign of 9 - 2x	Sign of 5 - x	Sign of Quotient
$\left(-\infty, \frac{9}{2}\right)$	0	+	+	+
$\left(\frac{9}{2}, 5\right)$	$\frac{19}{4}$	-	+	-
$\left(5, \infty\right)$	6	-	-	+

Want negative, so solutions are $\left(\frac{9}{2}, 5\right)$

67. $\dfrac{1}{x - 2} - \dfrac{2}{x + 3} > 0$

$\dfrac{1}{x - 2} \cdot \dfrac{x + 3}{x + 3} - \dfrac{2}{x + 3} \cdot \dfrac{x - 2}{x - 2} > 0$

$\dfrac{x + 3 - 2x + 4}{(x - 2)(x + 3)} > 0$ or $\dfrac{- x + 7}{(x - 2)(x + 3)} > 0$ (continued next page)

Interval	Test Value	Sign of x - 2	Sign of x + 3	Sign of -x + 7	Result
$(-\infty, -3)$	-4	-	-	+	+
$(-3, 2)$	0	-	+	+	-
$(2, 7)$	5	+	+	+	+
$(7, \infty)$	8	+	+	-	-

Want positive, so the solutions are $(-\infty, -3)$ or $(2, 7)$

69. $\dfrac{x + 1}{x - 3} + 4 > 0$

$\dfrac{x + 1}{x - 3} + 4 \cdot \dfrac{x - 3}{x - 3} > 0$

$\dfrac{x + 1 + 4x - 12}{x - 3} > 0$

$\dfrac{5x - 11}{x - 3} > 0$

Interval	Test Value	Sign of 5x - 11	Sign of x - 3	Sign of Quotient
$\left(-\infty, \frac{11}{5}\right)$	0	-	-	+
$\left(\frac{11}{5}, 3\right)$	$\frac{13}{5}$	+	-	-
$\left(3, \infty\right)$	4	+	+	+

Want positive, so the solutions are $\left(-\infty, \frac{11}{5}\right)$ or $(3, \infty)$

71. $\dfrac{2x - 4}{x + 1} \geq 0$

Interval	Test Value	Sign of x + 1	Sign of 2x - 4	Sign of Quotient
$(-\infty, -1)$	-2	-	-	+
$(-1, 2)$	0	+	-	-
$(2, \infty)$	3	+	+	+

Want positive (and = 0), so the solutions are $(-\infty, -1)$ or $[2, \infty)$

73. $\dfrac{x - 3}{(x - 4)(x + 4)} < 0$

Interval	Test Value	Sign of x - 3	Sign of x - 4	Sign of x + 4	Result
$(-\infty, -4)$	-5	-	-	-	-
$(-4, 3)$	0	-	-	+	+
$(3, 4)$	$\frac{7}{2}$	+	-	+	-
$(4, \infty)$	5	+	+	+	+

Want negative, so solutions are $(-\infty, -4)$ or $(3, 4)$

75. Case I. $x \geq 0, y \geq 0$

Then $xy \geq 0$ and so $|x| = x$, $|y| = y$, $|xy| = xy$.

So $|xy| = xy = |x||y|$

Case II. $x \geq 0, y < 0$.

Then $xy \leq 0$ and so $|xy| = -xy$, $|x| = x$, $|y| = -y$.

So $|xy| = -xy = x(-y) = |x||y|$

Case III. $x < 0, y < 0$

Then $xy > 0$ and so $|xy| = xy$, $|x| = -x$, $|y| = -y$.

$|xy| = xy = (-x)(-y) = |x||y|$

Case IV. $x < 0, y \geq 0$

Then $xy < 0$ and so $|xy| = -xy$, $|x| = -x$, $|y| = y$

$|xy| = -xy = (-x)(y) = |x||y|$

77. We are given that $x > 0$ and that $y < 0$. Therefore, $|x| = x$, $|y| = -y$.

Two cases are needed.

Case I. $|x| \geq |y|$. In this case, $x + y$ is positive.

$|x + y| = x + y < x + -y = |x| + |y|$

Case II. $|x| \leq |y|$. In this case, $x + y$ is negative.

$|x + y| = -(x + y) = -x + -y < x + -y = |x| + |y|$

79. All that is still needed is to show that if $x < 0$ and $y > 0$, then

$|x + y| < |x| + |y|$.

The proof of this is identical to that of #77 with the roles of x and y interchanged.

81. For each problem, we will use cases.

   i) $x \geq 2$. Then $2 - x$ is negative and $2 + x$ is positive, so $|2 - x| = x - 2$
   and $|2 + x| = 2 + x$

   ii) $-2 < x < 2$. Then both $2 - x$ and $2 + x$ are positive and so
   $|2 - x| = 2 - x$ and $|2 + x| = 2 + x$

   iii) $x \leq -2$. Then $2 - x$ is positive and $2 + x$ is negative.
   $|2 - x| = 2 - x$ and $|2 + x| = -2 - x$

a) i) $-2 + x + 2 + x \leq 10$ ii) $2 - x + 2 + x \leq 10$ iii) $2 - x - 2 - x \leq 10$

   $2x \leq 10$ $4 \leq 10$ $-2x \leq 10$

   $x \leq 5$ $x \geq -5$

   $[2, 5]$ $(-2, 2)$ $[-5, -2]$

   Take union of above to obtain $[-5, 5]$

b) i) $-2 + x + 2 + x > 6$ ii) $2 - x + 2 + x > 6$ iii) $2 - x - 2 - x > 6$

   $2x > 6$ $4 > 6$ $-2x > 6$

   $x > 3$ $x < -3$

   $(3, \infty)$ none $(-\infty, -3)$

   Combine and obtain $(-\infty, -3)$ or $(3, \infty)$

c) i) $2x \leq 4$ ii) $4 \leq 4$ iii) $-2x \leq 4$

   $x \leq 2$ $x \geq -2$

   $\{2\}$ $(-2, 2)$ $\{-2\}$

   Combine to obtain $[-2, 2]$

d) i) $2x \leq 3.99$ ii) $4 \leq 3.99$ iii) $-2x \leq 3.99$

   $x \leq 1.995$ $x \geq -1.995$

   none   none   none

   So there are no solutions.

83. a) The midpoint of the interval (0, 5) is $\frac{5}{2}$. Any point to the right of $\frac{5}{2}$ is closer to 5 then to 0. So x $> \frac{5}{2}$.

    b) 0 is the midpoint of (-2, 2), so y < 0.

85. Suppose s is smaller than t. Then s - t is negative and so $|s - t| = -(s - t)$
    $= t - s$. Then $\dfrac{s + t - |s - t|}{2} = \dfrac{s + t - (t - s)}{2} = \dfrac{2s}{2} = s$.

    Now suppose t is smaller thans s. Then s - t is positive and $|s - t| = s - t$
    So $\dfrac{s + t - |s - t|}{2} = \dfrac{s + t - (s - t)}{2} = \dfrac{s + t - s + t}{2} = \dfrac{2t}{2} = t$.   QED

87. Subtract 3.72s from both sides to obtain 0 $> .34s$

    This is true only when s is negative; $(-\infty, 0)$.

89. No. The range for w, if accurate to one decimal place, is $1.25 \leq w \leq 1.35$
    Square each term to obtain $1.5625 \leq w^2 \leq 1.8225$. To one decimal place,
    w^2 could equal 1.6, 1.7, or 1.8

91. We have to solve the inequality $-16t^2 + 120t \geq 125$

    $-16t^2 + 120t - 125 \geq 0$

    $-(4t - 5)(4t - 25) \geq 0$

    or $(4t - 5)(4t - 25) \leq 0$

    The product is negative or 0 on the interval $[\frac{5}{4}, \frac{25}{4}]$

93. $-4.5 \leq 1.6x - 2.7 \leq 4.5$

    $-1.8 \leq 1.6x \leq 7.2$

    $-1.125 \leq x \leq 4.5$ $[-1.125, 4.5]$

95. $326.5x + 242 > 3.8$ or $326.5x + 242 < -3.8$

    $326.5x > -238.2$ or $326.5x < -245.8$

    $x > -.729555895$ or $x < -.752833078$ $(-\infty, -0.752833078)$ or $(-0.729555895, \infty)$

97. $-0.001 < x - 3 < 0.001$

    $2.999 < x < 3.001$ $(2.999, 3.001)$

99. $-0.02 < 2x + 7 < 0.02$

   $-7.02 < 2x < -6.98$

   $-3.51 < x < -3.49$ $(-3.51, -3.49)$

101. $-0.01 < x^2 - 1 < 0.01$

   $0.99 < x^2 < 1.01$ $(\sqrt{0.99}, \sqrt{1.01}) \cup (-\sqrt{1.01}, -\sqrt{0.99})$

103. $|x + 3| < \frac{1}{3}$

   $3|x + 3| < 1$

   $|3(x + 3)| < 1$

   $|3x + 9| < 1$

   $|3x + 10 - 1| < 1$

105. $|x - 2| < 0.005$

   $2|x - 2| < 0.01$

   $|2(x - 2)| < 0.01$

   $|2x + 1 - 5| < 0.01$

107. $* |x - 2| < 0.2$

   $|x - 2||x + 2| < 0.2|x + 2|$

   $|x^2 - 4| < 0.2(4.2) = 0.84 < 1$ ($*$ implies that x is betwwen 1.8 and 2.2)

109. c; [-1, 3]   111. i; $(-\infty, -2] \cup [3, \infty)$ 113. a; [1, 9]   115. f; [0, 1]   117. d; $[-\infty, -1] \cup [3, \infty)$

Review Exercises for Chapter Two

1. $x = -5$ 3. $8x = -9$ so $x = -\frac{9}{8}$

5. Multiply both sides by $7(y - 3)$ to obtain $7y + 7 = 2y - 6$

$$5y = -13$$
$$y = -\frac{13}{5}$$

7. Multiply through by x to obtain $1 - 4 = 7x$

$$-3 = 7x$$

$$x = -\frac{3}{7}$$

9. Multiply through by $(x - 3)(x + 4)$ to obtain $x^2 + 6x + 8 = x^2 - 2x - 3$

$$8x = -11$$

$$x = -\frac{11}{8}$$

11. $a - d = bc$ or $\frac{a - d}{b} = c$

13. Multiply both sides by $\frac{r^2}{F}$ to obtain $r^2 = \frac{Gm_1m_2}{F}$. $r = \sqrt{\frac{Gm_1m_2}{F}}$

15. Let $x = $ original price of car. Then $x + 0.075x = 5160$

$$1.075x = 5160$$

$$x = \frac{5160}{1.075} = 4800$$

The original price of the car was $4800.

17. a) $C = 2300 + 60q$ if q is the number of units produced

    b) $C = 2300 + 60(135) = 2300 + 8100 = \$10,400$

19. Let x be test score needed on fifth test. Then $\dfrac{61 + 78 + 83 + 82 + x}{5} = 80$

$$\frac{304 + x}{5} = 80$$

$304 + x = 400$ or $x = 96$ So the student needs a 96 on the fifth test.

21. $(x - 3)(x + 3) = 0$

    $x = 3$ and $x = -3$ are the solutions

23. $(x + 7)(x - 2) = 0$

    $x = -7$ and $x = 2$ are the solutions

25. $(y - 6)(y - 1) = 0$

    $y = 6$ and $y = 1$ are the solutions

27. $u = \dfrac{-5 \pm \sqrt{25 - 4(1)(2)}}{2} = \dfrac{-5 \pm \sqrt{17}}{2}$

29. $v^2 + 10v - 7 = 0$

$$v = \dfrac{-10 \pm \sqrt{100 + 28}}{2} = \dfrac{-10 \pm 8\sqrt{2}}{2} = -5 \pm 4\sqrt{2}$$

31. $w = \dfrac{-6 \pm \sqrt{36 + 60}}{10} = \dfrac{-6 \pm 4\sqrt{6}}{10} = \dfrac{-3 \pm 2\sqrt{6}}{5}$

33. $z = \dfrac{-1 \pm \sqrt{1 - 4}}{2} = \dfrac{-1 \pm \sqrt{3}\, i}{2}$

35. $\dfrac{-2 \pm \sqrt{4 - 12}}{2} = \dfrac{-2 \pm 2\sqrt{2}\, i}{2} = -1 \pm \sqrt{2}\, i$

37. $\text{sum} = -\dfrac{b}{a} = -\dfrac{10}{3} \qquad \text{product} = \dfrac{c}{a} = \dfrac{-4}{3}$

39. $D = 16 - 4(1)(-11) = 16 + 44 = 60 > 0$ so 2 real roots.

41. $D = .16 - 4(1)(.04) = .16 - .16 = 0$ so one double root.

43. $D = 16 - 4(1)(\sqrt{2}) = 16 - 4\sqrt{2} \approx 10.34 > 0$ so 2 real roots.

45. $-5 + 3i$ 　　　　　　　　　　　　　47. $7 - 8i$

49. $12 + 30i - 6i - 15i^2 = 12 + 24i + 15 = 27 + 24i$

51. $i^7 = i^4 i^3 = 1 \cdot (-i) = -i$

53. $\dfrac{2 + i}{2 - i} \cdot \dfrac{2 + i}{2 + i} = \dfrac{4 + 4i + i^2}{4 - i^2} = \dfrac{4 + 4i - 1}{4 + 1} = \dfrac{3}{5} + \dfrac{4}{5}i$

55. Square both sides to obtain $x + 2 = 49$ which has solution $x = 47$

57. Factor as $(z^2 - 4)^2 = 0$ or $(z - 2)^2 (z + 2)^2 = 0$
So the solutions are $z = 2$ and $z = -2$

59. Factor as $(u^2 - 4)(u^2 + 2) = 0$

   The solutions are $\pm 2, \pm \sqrt{2}\,i$

   (you could also do this by substituting $w = u^2$ and then solving the quadratic equation, finally substituting at the end)

61. Multiply through by $(y + 1)(y - 3)$ to obtain $4(y - 3) = 7(y + 1) - 2(y + 1)(y - 3)$

$$4y - 12 = 7y + 7 - 2y^2 + 4y + 6$$
$$2y^2 - 7y - 25 = 0$$

$$y = \frac{7 \pm \sqrt{49 + 200}}{4} = \frac{7 \pm \sqrt{249}}{4}$$

63. Write as $\sqrt{2v + 1} = 1 + \sqrt{v}$ and then square both sides to obtain

   $2v + 1 = 1 + 2\sqrt{v} + v$.   Rewrite as $v = 2\sqrt{v}$ and square both sides to obtain

   $v^2 = 4v$ or $v^2 - 4v = 0$ or $v(v - 4) = 0$.   Solutions are 0 and 4.

65. $7824 = 2000 + 27q - .005q^2$

   $.005q^2 - 27q + 5824 = 0$

$$q = \frac{27 \pm \sqrt{729 - 4(.005)(5824)}}{.01} = \frac{27 \pm \sqrt{615.52}}{.01} = \frac{27 \pm 24.74914}{.01}$$

   Take $-$ above to obtain 225 ($+$ gives an answer that is too big)

67. a) $P = (42 q - .015q^2) - (2000 + 27q - .005q^2) = -0.01q^2 + 15q - 2000$

   b) $-0.01q^2 + 15q - 2000 = 1776$

   $-0.01q^2 + 15q - 3776 = 0$

$$q = \frac{-15 \pm \sqrt{225 - 4(.01)(3776)}}{.01}.$$ Take $+$ to obtain 320 items.

69. Using $s = \frac{1}{2}at^2 + v_0 t$, obtain the equation $15{,}000 = 16t^2 + 80t$

   To make the equation simpler, subtract 15,000 from both sides, divide

   both sides by 8 to obtain $0 = 2t^2 + 10t - 1875$

   Use the quadratic formula, obtain $t = \dfrac{-10 \pm \sqrt{100 + 4(2)(1875)}}{4}$

   Take $+$ to obtain ~ 28.2 seconds.

71. $[3, 7]$

73. $(0, 5]$

75. $[4, \infty)$

77. $x < 7$ $(-\infty, 7)$

79. $2x \leq 5$
    $x \leq \frac{5}{2}$ $(-\infty, \frac{5}{2}]$

81. $-2 \leq x < 5$ $[-2, 5)$

83. $2 > x - 5 > -4$
    $7 > x > 1$ $(1, 7)$

85. $x - 4 > 5$ or $x - 4 < -5$
    $x > 9$ or $x < -1$ $(9, \infty)$ or $(-\infty, -1)$

87. $-3 < 2x - 5 < 3$
    $2 < 2x < 8$
    $1 < x < 4$ $(1, 4)$

89. The absolute value of an expression can never be negative, so no solutions.

91. $6 - 2x > 1$ or $6 - 2x < -1$
         $-2x > -5$ $-2x < -7$
           $x < \frac{5}{2}$ $x > \frac{7}{2}$

    Solution $(-\infty, \frac{5}{2})$ or $(\frac{7}{2}, \infty)$

93. $\frac{5x - 8}{3} > 6$ or $\frac{5x - 8}{3} < -6$

     $5x - 8 > 18$ or $5x - 8 < -18$

       $5x > 26$ or $5x < -10$

         $x > \frac{26}{5}$ or $x < -2$ $(-\infty, -2)$ or $\left(\frac{26}{5}, \infty\right)$

95. $x^2 - 7x + 6 > 0$

   $(x - 6)(x - 1) > 0$

Interval	Test Value	Sign of x - 6	Sign of x - 1	Sign of Product
$(-\infty, 1)$	0	-	-	+
$(1, 6)$	2	-	+	-
$(6, \infty)$	7	+	+	+

Want positive, so the solutions are $(-\infty, 1)$ or $(6, \infty)$

97. $x^2 - 7x + 6 < 0$

   $(x - 6)(x - 1) < 0$

Interval	Test Value	Sign of x - 6	Sign of x - 1	Sign of Product
$(-\infty, 1)$	0	-	-	+
$(1, 6)$	2	-	+	-
$(6, \infty)$	7	+	+	+

Solution is $(1, 6)$

99. $x^2 + 3x - 10 \geq 0$

   $(x + 5)(x - 2) \geq 0$

Interval	Test Value	Sign of x - 2	Sign of x + 5	Sign of Product
$(-\infty, -5)$	-6	-	-	+
$(-5, 2)$	0	-	+	-
$(2, \infty)$	3	+	+	+

Want greater than or equal to 0; the solutions are $(-\infty, -5]$ or $[2, \infty)$

101. First write as $x^2 - x - 12 \leq 0$

Factors as $(x - 4)(x + 3)$.   Intervals are $(-\infty, -3)$, $(-3, 4)$ and $(4, \infty)$. Using test points, can find solution to be $[-3, 4]$.

103. $\dfrac{1}{x + 2} - \dfrac{2}{x - 1} > 0$

$\dfrac{1}{x + 2} \cdot \dfrac{x - 1}{x - 1} - \dfrac{2}{x - 1} \cdot \dfrac{x + 2}{x + 2} > 0$

$\dfrac{x - 1 - 2x - 4}{(x + 2)(x - 1)} > 0$

$\dfrac{-x - 5}{(x + 2)(x - 1)} > 0$

Interval	Test Value	Sign of $x + 2$	Sign of $x - 1$	Sign of $-x - 5$	Result
$(-\infty, -5)$	-6	-	-	+	+
$(-5, -2)$	-3	-	-	-	-
$(-2, 1)$	0	+	-	-	+
$(1, \infty)$	2	+	+	-	-

Want positive; the solutions are $(-\infty, -5)$ or $(-2, 1)$

CHAPTER THREE
Functions and Graphs

Chapter Objectives

The first thing with which you will become familiar is the Cartesian coordinate system, which is also called the x-y coordinate system. This enables you to represent pictorially many of the important concepts that you will be studying. You should learn how to plot points, find the distance between two points, and work with the equation of a circle.

Straight lines are very important and easy to work with, so you'll next learn how to find the slope of a straight line and different forms of the equation of a straight line. You will also look at the relationships between parallel and perpendicular lines, and applications of straight lines.

Functions come next. An understanding of functions is essential to any further study in mathematics. You should learn to become comfortable with evaluating functions, finding domains and ranges of functions, graphing functions, creating functions from combinations of other functions, and determining the inverse of a function (if there is one).

At the end of the chapter you will learn the concept of proportionality, both direct and inverse.

Chapter Summary

- **Distance Formula:** The distance, d, between the points (x_1, y_1) and (x_2, y_2) is given by

$$d = \sqrt{(x_1 - x_2)^2 + (y_1 - y_2)^2}$$

- **Midpoint Formula:** The midpoint of the line joining (x_1, y_1) and (x_2, y_2) is

$$\left(\frac{x_1 + x_2}{2}, \frac{y_1 + y_2}{2} \right)$$

- **Graph:** The graph of an equation in two variables is the set of all points in the xy-plane whose coordinates satisfy that equation.

- **Circles:** The unit circle is centered at $(0, 0)$, and has radius 1. Its equation is $x^2 + y^2 = 1$. The equation $(x - h)^2 + (y - k)^2 = r^2$ is the equation of a circle centered at (h, k) with radius r.

- **Linear Equation in Two Variables.** This is an equation of the form $ax + dy = c$, where a and d are not both equal to zero.

- **Slope:** the slope, m, of a line is given by $m = \frac{y_2 - y_1}{x_2 - x_1}$, where (x_1, y_1) and (x_2, y_2) are any two points on the line and $x_2 \neq x_1$.

  If the line is vertical $(x_1 = x_2)$ then the slope is undefined. If the slope, m, of a line is positive, then the graph of the line will rise as x increases. If m is negative, then the graph of the line will fall as x increases.

  Horizontal lines have a slope of zero.

  Two lines are parallel if and only if their slopes are equal.

  Two nonvertical lines are perpendicular if and only if their slopes are negative reciprocals of one another $\left(m_1 = -\frac{1}{m_2}\right)$.

- **Equations of a Line:** Let (x_1, y_1) be a point on a line with slope m and y-intercept b.

  point-slope equation: $y - y_1 = m(x - x_1)$

  slope-intercept equation: $y = mx + b$

  standard equation: $Ax + By = C$

- A **function**, f, is a rule that assigns to each memeber of one set (called the domain of the function) a unique member of another set (called the range of the function). We often write $y = f(x)$. Here y is called the **image** of x. The variable x is called the independent variable and y is the dependent variable.

- **Vertical Lines Test:** An equation in two variables defines a function if every vertical line in the plane intersects the graph of the equation in at most one point.

- The function $f(x)$ is **even** if $f(-x) = f(x)$ or **odd** if $f(-x) = -f(x)$. The graph of a function is the set of points $\{(x, f(x)): x \ \varepsilon \ \text{domain of } f\}$.

- If $f(x)$ is even, then its graph is symmetric about the y-axis.

- If $f(x)$ is odd, then its graph is symmetric about the origin.

- **Shifting Graphs**

  The graph of $f(x) + c$ is the graph of $f(x)$ shifted up c units if $c > 0$ or down $|c|$ units if $c < 0$.

  The graph of $f(x - c)$ is the graph of $f(x)$ shifted c units to the right if $c > 0$ or $|c|$ units to the left if $c < 0$.

  The graph of $-f(x)$ is the graph of $f(x)$ reflected about the x-axis.

  The graph of $f(-x)$ is the graph of $f(x)$ reflected about the y-axis.

- **Function Operations**

  Sum: $(f + g)(x) = f(x) + g(x)$

  Difference: $(f - g)(x) = f(x) - g(x)$

  Product: $(fg)(x) = f(x)g(x)$

  Quotient: $\frac{f}{g}(x) = \frac{f(x)}{g(x)}$

  Composition: $(f \circ g)(x) = f(g(x))$

- **Inverse Functions:** f and g are inverse functions if

  (i) for every $x \; \varepsilon$ domain of g, $g(x) \; \varepsilon$ domain of f and $f(g(x)) = x$.

  (ii) for every $x \; \varepsilon$ domain of f, $f(x) \; \varepsilon$ domain of g and $g(f(x)) = x$.

- **One-to-One Function:** $f(x)$ is one-to-one on $[a, b]$ if $f(x_1) = f(x_2)$ implies that $x_1 = x_2$ whenver x_1 and x_2 are in $[a, b]$.

  A one to one function has an inverse which is denoted by f^{-1}.

- **Horizontal lines test:** f is one-to-one on its domain if every horizontal line in the plane intersects the graph of f in at most one point.

- f is increasing on $[a, b]$ if $f(x_2) > f(x_1)$ when $x_2 > x_1$.
- f is decreasing on $[a, b]$ if $f(x_2) < f(x_1)$ when $x_2 > x_1$.
- If f is increasing or decreasing, then f is one-to-one.

- **Reflection Property:** The graphs of f and f^{-1} are reflections of one another about the line $y = x$.

- **Direct Variation:** y varies directly as x or y is directly proportional to x if $y = kx$ for some constant k (called the constant of proportionality).

- **Inverse Variation:** y varies inversely or is inversely proportional to x if $y = \frac{k}{x}$ for some constant k

- **Joint Variation:** z varies jointly as x and y if $z = kxy$ for some constant k.

SOLUTIONS TO CHAPTER THREE PROBLEMS

Problems 3.1

1.

IV

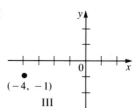

IV

(3, −2)

3.

on *x*-axis

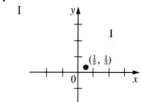

(2, 0)

5.

III

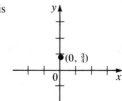

(−4, −1)

III

7.

I

I

$(\frac{1}{2}, \frac{1}{3})$

9.

on *y*-axis

$(0, \frac{3}{4})$

Problems 11-19 use the distance formula.

11. $d = \sqrt{(4 - 1)^2 + (7 - 3)^2} = \sqrt{3^2 + 4^2} = \sqrt{9 + 16} = \sqrt{25} = 5$

13. $\sqrt{(-9 - 0)^2 + (4 - (-7))^2} = \sqrt{81 + 121} = \sqrt{202}$

15. $\sqrt{4 + 25} = \sqrt{29}$ 17. $\sqrt{c^2 + d^2}$

19. $\sqrt{(2.94)^2 + (7.87)^2} = \sqrt{8.6436 + 61.9369} = \sqrt{70.5805} \approx 8.40$

Problems 21 and 23 use the Pythagorean Theorem.

21. The distance between $(0, 3)$ and $(1, 4)$ is $\sqrt{2}$.

The distance between $(1, 4)$ and $(5, 0)$ is $\sqrt{32}$.

The distance between $(5, 0)$ and $(0, 3)$ is $\sqrt{34}$.

$\sqrt{2}$ and $\sqrt{34}$ are the smallest, and $(\sqrt{2})^2 + (\sqrt{32})^2 = 2 + 32 = 34 = (\sqrt{34})^2$

Therefore, the points are the verticies of a right triangle.

23. The distance between $(-2, 8)$ and $(1, 3)$ is $\sqrt{34}$.

The distance betweem $(1, 3)$ and $(2, 7)$ is $\sqrt{17}$.

The distance between $(2, 7)$ and $(-2, 8)$ is $\sqrt{17}$.

$(\sqrt{17})^2 + (\sqrt{17})^2 = (\sqrt{34})^2$ so the given points are vertices of a right triangle.

25. The two legs of the triangle are each of length 4. The third side is of length $\sqrt{32} = 4\sqrt{2}$ and since $4^2 + 4^2 = 32$, it is a right triangle (that is obvious anyway).

27. Yes. The three sides are of length $\sqrt{10}$, $\sqrt{40} = 2\sqrt{10}$, $\sqrt{50} = 5\sqrt{2}$, and $(\sqrt{10})^2 + (\sqrt{40})^2 = (\sqrt{50})^2$

29. No. The sides of the triangle are of length 4, $\sqrt{16 + \frac{1}{4}} = \frac{\sqrt{65}}{2}$, $\sqrt{16 + \frac{49}{4}} = \frac{\sqrt{113}}{2}$, and they do not satisfy the Pythagorean Theorem.

31. See picture and note that the two legs of the triangle are horizontal and vertical line segments respectively, which is why a = 1 works.

Algebraically; the distance between (1, -1) and (2, 3) is $\sqrt{17}$.

The distance between (2, 3) and (a, 3) is $\sqrt{(a-2)^2}$

The distance between (a, 3) and (1, -1) is $\sqrt{(a-1)^2 + 16}$

By the Pythagorean Theorem, $(a-2)^2 + (a-1)^2 + 16 = 17$

$a^2 - 4a + 4 + a^2 - 2a + 1 + 16 = 17$

$2a^2 - 6a + 4 = 0$

$2(a^2 - 3a + 2) = 2(a-2)(a-1) = 0$ so a = 2 or a = 1.

But a = 2 gives the point (2, 3) again. Therefore, a = 1

Note that instead of the right angle being at the point (a, 3)

it could be at the point (1, -1) (see picture again).

Now we also have the lengths as above but the Pythagorean Theorem gives

$(a-2)^2 = (a-1)^2 + 16 + 17$

$a^2 - 4a + 4 = a^2 - 2a + 1 + 16 + 17$

$a = -15$

So the problem has two solutions: a = 1 and a = -15

Problems 33-37 use the midpoint formula

33. $\left(\frac{4+8}{2}, \frac{-7+5}{2}\right) = \left(6, -1\right)$ 35. $\left(\frac{3}{2}, 7\right)$

37. $\left(\frac{3.2+4.6}{2}, \frac{-1.2+2.8}{2}\right) = \left(\frac{7.6}{2}, \frac{1.6}{2}\right) = (3.9, 0.8)$

39. Let the consecutive vertices be (0, 0), (a, 0) , (a, b), (0, b). The length

of each of the two diagonals is $\sqrt{a^2 + b^2}$.

41. To go from the point (6, 1) to the point (2, -1) involves a decrease of 4 in the x coordinate and a decrease of 2 in the y coordinate. To go from (4, 5) the fourth point must involve the same changes since that will form a side opposite the previous one, and must be parallel and of the same length. So, if the x coordinate decreases by 4 and the y coordinate by 2, the other point is (0, 3).

Recall the standard form of an equation of a circle: $(x - h)^2 + (y - k)^2 = r^2$

43. $(x - 0)^2 + (y - 2)^2 = 1$

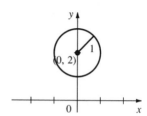

45. $(x - 1)^2 + (y - 1)^2 = 2$

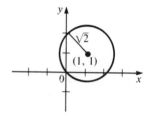

47. $(x + 1)^2 + (y - 4)^2 = 25$

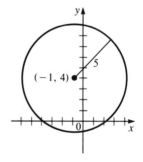

49. $(x - \pi)^2 + (y - 2\pi)^2 = \pi$

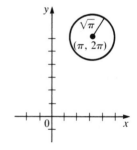

51. $(x - 3)^2 + (y + 2)^2 = 16$

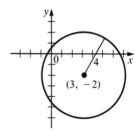

Problems 53 and 55 require completing the square.

53. $x^2 + y^2 - 6y + 3 = 0$

$x^2 + y^2 - 6y + 9 = -3 + 9$ or $x^2 + (y - 3)^2 =$

Center $(0, 3)$ radius $\sqrt{6}$

55. $x^2 + x + y^2 - \frac{1}{2}y = \frac{61}{16}$ so $x^2 + x + \frac{1}{4} + y^2 - \frac{1}{2}y + \frac{1}{16} = \frac{61}{16} + \frac{1}{4} + \frac{1}{16}$

$\left(x + \frac{1}{2}\right)^2 + \left(y - \frac{1}{4}\right)^2 = \frac{66}{16}$

Center is $\left(-\frac{1}{2}, \frac{1}{4}\right)$ Radius is $\sqrt{\frac{66}{16}} = \frac{\sqrt{66}}{4}$

57. The general equation of a circle is $(x - h)^2 + (y - k)^2 = r^2$

Plug in coordinates of each point to obtain

$(0 - h)^2 + (-2 - k)^2 = r^2$ $(6 - h)^2 + (-12 - k)^2 = r^2$

$(-2 - h)^2 + (-4 - k)^2 = r^2$

Since each equation $= r^2$, we can set any pair of them equal:

$(6 - h)^2 + (-12 - k)^2 = (0 - h)^2 + (-2 - k)^2$

$36 - 12h + h^2 + 144 + 24k + k^2 = h^2 + 4 + 4k + k^2$

$176 = 12h - 20k$

Another pair: $(-2 - h)^2 + (-4 - k)^2 = (0 - h)^2 + (-2 - k)^2$

$4 + 4h + h^2 + 16 + 8k + k^2 = h^2 + 4 + 4k + k^2$

$16 = -4h - 4k$

Solve the system $176 = 12h - 20k$ simultaneously.

$16 = -4h - 4k$

Multiply the second equation by 3 to obtain $48 = -12h - 12k$.

Add to the first equation and obtain $224 = -32k$ or $k = -\frac{224}{32} = -7$

Substitute into one of the original equations, say the second:

$16 = -4h - 4(-7)$ so $16 = 28 - 4h$ and $-12 = -4h$ or $h = \frac{-12}{-4} = 3$

The circle's equation is $(x - 3)^2 + (y + 7)^2 = r^2$. To find r^2, plug in any point

given, say $(0, -2)$: $(0 - 3)^2 + (-2 + 7)^2 = r^2$

$9 + 25 = r^2$ or $r^2 = 34$

Therefore, the equation is $(x - 3)^2 + (y + 7)^2 = 34$

59. a) $\frac{1}{2}|2\cdot4 + 0\cdot(-6) + 3\cdot1 - 2\cdot(-6) - 0\cdot1 - 3\cdot4| = \frac{1}{2}|8 + 0 + 3 + 12 - 0 - 12| = \frac{11}{2}$

b) $\frac{1}{2}|-20 - 3 + 14 - 12 + 2 + 35| = \frac{16}{2} = 8$

Problems 3.2

1. Slope $= \frac{4 - 6}{2 - 1} = -2$

3. $m = \frac{-4 - (-2)}{-3 - (-1)} = \frac{-2}{-2} = 1$

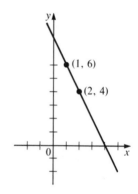

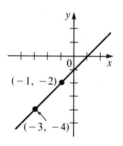

5. $m = \frac{-2 - 3}{8 - (-4)} = -\frac{5}{12}$

7. $m = \frac{-3 - (-3)}{2 - 5} = 0$

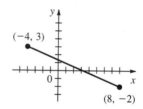

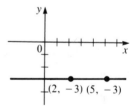

9. $m = \frac{0 - a}{a - 0} = -1$

11. $m = \frac{d - b}{c - a} = \frac{b - d}{a - c}$

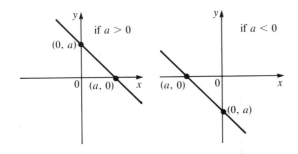

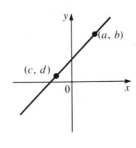

13. $(0, 6)$ and $(5, 0)$ are two points on the line. Slope is therefore $-\frac{6}{5}$.

15. A horizontal line has slope zero.

17. Two points on the line are $(1, 10)$ and $(3, 1)$. The slope is $-\frac{9}{2}$.

19. Two points on the line are $(1, 1)$ and $(7, 2)$. The slope is $\frac{1}{6}$.

21. $m_1 = \frac{5}{5} = 1$, $m_2 = \frac{-2}{2} = -1$

product is -1; perpendicular

23. $m_1 = \frac{-6}{2} = -3$ $m_2 = \frac{3}{-1} = -3$ slopes are

equal, so lines are parallel.

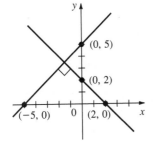

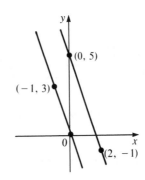

- 101 -

25. $m_1 = \frac{6}{1} = 6$ $m_2 = \frac{1}{-6}$

slopes are negative reciprocals, so

lines are perpendicular

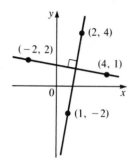

27. $m_1 = \frac{6}{0}$ is undefined, so line is vertical

$m_2 = \frac{0}{3} = 0$, so line is horizontal

lines are perpendicular

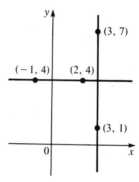

29. $\frac{6.25 - 4}{2.5 - 2} = \frac{2.25}{0.5} = 4.5$

31. $\frac{(2 + h)^2 - 4}{2 + h - 2} = \frac{4 + 4h + h^2 - 4}{h} = \frac{4h + h^2}{h} = \frac{h(4 + h)}{h} = 4 + h$

33. $\frac{0.5 - 0.4}{2 - 2.5} = \frac{0.1}{-0.5} = -0.2$

35. $\frac{\frac{1}{2 + h} - \frac{1}{2}}{2 + h - 2} = \frac{\frac{1}{2 + h} - \frac{1}{2}}{h} \cdot \frac{(2 + h)2}{(2 + h)2} = \frac{2 - (2 + h)}{h(2 + h)2} = \frac{-h}{h(2 + h)2} = \frac{-1}{2(2 + h)}$

37. $\frac{\sqrt{4.1} - 2}{4.1 - 4} \approx 0.248$

39. $m = \frac{\sqrt{a + h} - \sqrt{a}}{a + h - a} \cdot \frac{\sqrt{a + h} + \sqrt{a}}{\sqrt{a + h} + \sqrt{a}} = \frac{a + h - a}{h\left(\sqrt{a + h} + \sqrt{a}\right)} = \frac{1}{\sqrt{a + h} + \sqrt{a}}$

41. $m = \frac{-4 - (-1)}{2 - 3} = 3$ so point slope is either $y + 4 = 3(x - 2)$ or $y + 1 = 3(x - 3)$.

Multiply out either of these; the slope-intercept form is $y = 3x - 10$

Put x term on left to obtain a standard form of $-3x + y = -10$.

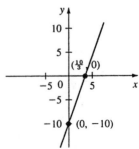

43. Point-slope is $y - 3 = 0(x - 8)$.

    Slope-intercept is $y = 0x + 3$ or $y = 3$.

    Standard form is $y = 3$.

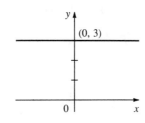

45. Slope is $\dfrac{-1/2 - 0}{3 - 1/3} = \dfrac{-1/2}{8/3} = -\dfrac{1}{2} \cdot \dfrac{3}{8} = -\dfrac{3}{16}$.

    The point-slope form is $y - 0 = -\dfrac{3}{16}(x - \dfrac{1}{3})$ or $y + \dfrac{1}{2} = -\dfrac{3}{16}(x - 3)$.

    The slope-intercept form is $y = -\dfrac{3}{16}x + \dfrac{1}{16}$.

    A standard form is $3x + 16y = 1$.

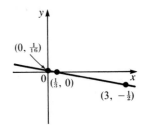

47. Use the slope formula to find the slope is 1.

    Point-slope forms could be $y + 1 = 1(x - 5)$ or $y - 2 = 1(x - 8)$

    Slope-intercept form is $y = x - 6$.   (Could also leave out 1's above).

    A standard form is $x - y = 6$.

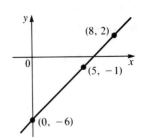

49. Point-slope form is $y - 1 = \dfrac{3}{7}(x + 5)$.

    Slope-intercept form is $y = \dfrac{3}{7}x + \dfrac{22}{7}$.

    A standard form is $3x - 7y = -22$.

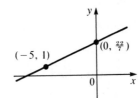

51. Point-slope form is y - b = c(x - a)

    Slope-intercept form is y = cx + (b - ac).

    A general form is cx - y = ac - b.

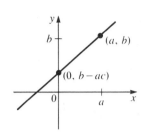

53. Point slope form is y - .0058 = 12.611(x - .0146).

    Slope-intercept form is y = 12.611x - .1783206

    A general form is 12.611x - y = .1783206

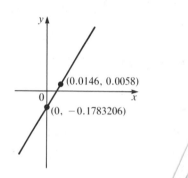

55. Parallel lines have the same slope.

    The slope of 5x - 7y = 3 is found by solving for y; $y = \frac{5}{7}x - \frac{3}{7}$.

    The given line has slope $\frac{5}{7}$, so the desired line also has slope $\frac{5}{7}$.

    Point-slope form of the desired line is $y - 5 = \frac{5}{7}(x - 2)$.

    Slope-intercept form of the desired line is $y = \frac{5}{7}x + \frac{25}{7}$.

    $$y = \frac{5}{7}x + \frac{25}{7}$$

57. Solving the given line for y, one obtains $y = \frac{4}{3}x - \frac{14}{3}$.

    The line given has slope $\frac{4}{3}$. The desired line is perpendicular, so has slope

    negative reciprocal of the one given, which would be $-\frac{3}{4}$. $$y = -\frac{3}{4}x + \frac{13}{4}$$

    Point-slope form of the equation of the desired line is $y - 4 = -\frac{3}{4}(x + 1)$.

    Slope-intercept form is $y = -\frac{3}{4}x + \frac{13}{4}$.

59. Add the two equations to obtain 3x = 12. Therefore, x = 4 and so y = 1. (4, 1) is point.

61. Multiply the first equation by $\frac{3}{2}$ and see that you have the same equation. Solutions are points of the form $(x, -2x + \frac{3}{2})$.

63. Multiply the first equation by $-\frac{3}{2}$, add to the second equation to obtain $-\frac{17}{2}y = \frac{5}{2}$. Therfore, $y = -\frac{5}{17}$ and $x = \frac{50}{17}$.

65. Multiply the first equation by $\frac{0.8196}{1.609}$ to obtain $1.17515053x - 0.8196y = 2.80569099$

Add to the second equation to obtain $1.34935053x = 166.0456910$

or $x = \frac{166.0456910}{1.34935053} = 123.0560092$. Now substitute this into any of the

other equations, say the first: $2.307(123.0560092) - 1.609y = 5.508$

$283.8902128 - 1.609y = 5.508$

$-1.609y = -278.3822128$ or $y = \frac{-278.3822128}{-1.609} = 173.0156701$

The point of interesection is $(123.0560092, 173.0156701)$ rounded to $(123.056, 173.016)$.

67. The distance between (x_1, y_1) and Q is $\sqrt{\left(\frac{2x_1 + x_2}{3} - x_1\right)^2 + \left(\frac{2y_1 + y_2}{3} - y_1\right)^2}$

$= \sqrt{\left(\frac{x_2 - x_1}{3}\right)^2 + \left(\frac{y_2 - y_1}{3}\right)^2} = \frac{1}{3}\sqrt{(x_2 - x_1)^2 + (y_2 - y_1)^2}$

The distance between R and Q is $\sqrt{\left(\frac{x_1 + 2x_2}{3} - \frac{2x_1 + x_2}{3}\right)^2 + \left(\frac{y_1 + 2y_2}{3} - \frac{2y_1 + y_2}{2}\right)^2}$

$= \sqrt{\left(\frac{x_2 - x_1}{3}\right)^2 + \left(\frac{y_2 - y_1}{3}\right)^2} = \frac{1}{3}\sqrt{(x_2 - x_1)^2 + (y_2 - y_1)^2}$

The distance between R and (x_2, y_2) is $\sqrt{\left(x_2 - \frac{x_1 + 2x_2}{3}\right)^2 + \left(y_2 - \frac{y_1 + 2y_2}{3}\right)^2}$

$= \sqrt{\left(\frac{x_2 - x_1}{3}\right)^2 + \left(\frac{y_2 - y_1}{3}\right)^2} = \frac{1}{3}\sqrt{(x_2 - x_1)^2 + (y_2 - y_1)^2}$.

Therefore, all 3 distances are equal and are equal to $\frac{1}{3}$ the distance between (x_1, y_1) and (x_2, y_2).

Problems 3.3

1. $f(0) = \frac{1}{1+0} = 1$ $f(1) = \frac{1}{1+1} = \frac{1}{2}$ $f(-2) = -1$ $f(-5) = -\frac{1}{4}$ $f(x^2) = \frac{1}{1+x^2}$ $f(\sqrt{x}) = \frac{1}{1+\sqrt{x}}$

3. $f(0) = 2(0)^2 - 1 = -1$ $f(2) = 2(2)^2 - 1 = 7$ $f(-3) = 17$ $f(\frac{1}{2}) = -\frac{1}{2}$

   $f(\sqrt{w}) = 2(\sqrt{w})^2 - 1 = 2w - 1$ $f(w^5) = 2(w^5)^2 - 1 = 2w^{10} - 1$

5. $f(0) = 0^4 = 0$ $f(2) = 2^4 = 16$ $f(-2) = 16$ $f(\sqrt{5}) = 25$ $f(s^{1/5}) = (s^{1/5})^4 = s^{4/5}$

   $f(s-1) = (s-1)^4 = (s^2 - 2s + 1)^2 = s^4 - 4s^3 + 6s^2 - 4s + 1$

7. $g(0) = \sqrt{0+1} = 1$ $g(-1) = \sqrt{0} = 0$ $g(3) = \sqrt{4} = 2$ $g(7) = \sqrt{8} = 2\sqrt{2}$

   $g(n^3 - 1) = \sqrt{n^3 - 1 + 1} = \sqrt{n^3} = n\sqrt{n} = n^{3/2}$ $g(\frac{1}{w}) = \sqrt{\frac{1}{w} + 1} = \sqrt{\frac{1+w}{w}} = \frac{\sqrt{w(w+1)}}{w}$

9. $h(0) = 0^2 - 0 + 1 = 0 + 0 + 1 = 1$ $h(2) = 2^2 - 2 + 1 = 3$

   $h(10) = 10^2 - 10 + 1 = 91$ $h(-5) = 31$ $h(n^2) = (n^2)^2 - n^2 + 1 = n^4 - n^2 + 1$

   $h\left(\frac{1}{n^3}\right) = \frac{1}{n^6} - \frac{1}{n^3} + 1$

11. Yes

13. No; a cannot be mapped to both u and w.

15. Yes

17. No; w is not mapped to any point.

19. Yes

21. Yes; $y = \frac{6 - 2x}{3}$

23. Yes; $y = \frac{5 - x^2}{2}$

25. No, since y is squared. $y = \pm\sqrt{4 - 4x^2} / 2$

27. Yes; squaring both sides gives $x + y = 1$ or $y = 1 - x$; however, the domains are not the same so this isn't an equivalent function.

29. Yes; $y^3 = x$ or $y = \sqrt[3]{x}$

31. Yes; it already is.

33. $y^n = x$ has 2 solutions if n is even, but only one if n is odd. So if n is even, there are 2 x's for each y; that is not permitted for a function.

35. Yes   37. Yes   39. Yes   41. No   43. No; there are two point with x coordinate 0

45. Domain is all real numbers, since the function is a polynomial.
    Range is all real numbers, which is the case for all linear functions,
    except constants. This is because linear functions have no largest
    nor smallest values (the y coordinates take on all real numbers as values).

47. Domain is all real numbers except zero, since denominator cannot be 0.
    Range is all positve real numbers, since the quotient cannot be zero
    or negative, but the denominator (and hence the whole fraction) can
    be made as small or as large as possible. Dividing by large numbers gives
    small quotients; dividing by small numbers gives large quotients.

49. Domain is all real numbers except -1, since denominator cannot be 0.
    Range is all real numbers except zero (a fraction equals zero only
    when its numerator equals zero).

51. To find the domain, realize what is under a square root must be non-negative.
    Therefore, $x^3 - 1 \geq 0$, or $x^3 \geq 1$. That means the domain is all numbers
    greater than or equal to 1.
    To find the range, realize that if x = 1, f(x) = 0. As x gets larger, so does
    f(x) and without upper bound. Therefore, the range is all non-negative real numbers.

53. The domain is all real numbers except zero, since denominator cannot be 0.
    The range is all positive real numbers.

55. Since each piece is a polynomial, and all real numbers are either ≥ 0 or
    < 0, then the domain is all real numbers.
    To find the range, realize that when $x = 0$, $y = 0$, that when $x > 0$
    so is y (since $y = 2x$) and that y gets arbitrarily large when x does.
    Also realize that when $x < 0$, then $-x > 0$, so the range is all real numbers
    greater than or equal to 0.

57. The domain is all real numbers, since every number is either > 0 or ≤ 0.
    The range is all real numbers greater than or equal to zero, since $x^2 \geq 0$ for all x.

59. The domain is all real numbers, and the range is all integers.

61. $g(0.18) = 1.419936$ $g(3.95) = 54.372225$ $g(-11.62) = -1676.34304$

63. $f(5.8) = -0.980796045$ $f(-23.4) = 0.227833604$

65. $f(x + \Delta) = (x + \Delta)^3 = x^3 + 3x^2\Delta x + 3x(\Delta x)^2 + (\Delta x)^3$

$$\frac{f(x + \Delta x) - f(x)}{\Delta x} = \frac{3x^2\Delta x + 3x(\Delta x)^2 + (\Delta x)^3}{\Delta x} = \frac{\Delta x(3x^2 + 3x\Delta x \ (\Delta x)^2}{\Delta x} = 3x^2 + 3x\Delta x + (\Delta x)^2$$

67. Case 1. $x > 0$. Then $|x| = x$, so $\frac{|x|}{x} = \frac{x}{x} = 1$

    Case 2. $x < 0$. Then $|x| = -x$, so $\frac{|x|}{x} = \frac{-x}{x} = -1$

    Domain = Reals, $x \neq 0$. Range is $\{-1, 1\}$.

69. The perimeter of a rectangle of base b and width w is $2b + 2w$.
    If the perimeter is 50 and the width is W, then $50 = 2b + 2W$, or $b = 25 - W$.
    The area of a rectangle is bh. In our case, get $A = f(w) = (25 - W)W = 25w - w^2$.
    The domain is $(0, 25)$ and the range is $(0, \frac{625}{4})$

71. Since the runner is traveling at 30 ft/sec and each base is 90 feet apart, the runner will get to first base after 3 seconds, to second base after 6 seconds, to third base after 9 seconds, and home after 12 seconds.

First consider the interval $0 \leq t < 3$. The runner has travelled 30t feet from home toward first. He is therefore 90 - 30t feet from first. His distance from second is the hypoteneuse of a right triangle with legs 90 feet and 90 - 30t feet. By the Pythagorean Theorem, the hypotneuse of a right triangle has length equal to the square root of the sum of the squares of the legs; in this case, obtain

$$\sqrt{(90)^2 + (90 - 30t)^2} \quad \text{which simplifies to} \quad 30\sqrt{t^2 - 6t + 18} \ .$$

Now consider the interval $3 \leq t \leq 6$. At $t = 3$ the runner is at first base, 90 feet from second. At $t = 6$ he is 0 feet from second. Since he is running in a straight line toward second, his distance from second after t seconds is $90 - 30(t - 3) = 180 - 30t$.

Now consider the interval $6 < t \leq 9$. At $t = 6$, he is 0 feet from second. At $t = 9$, he is 90 feet from second. He is running at 30 feet per second directly away from second base, so his distance from second is $30(t - 6)$ which simplifies to $30t - 180$.

In the interval $9 < t \leq 12$, he will be $30(t - 9)$ feet from third base, and the distance from second to third is 90 feet, so his distance to second is the hypoteneuse of a right triangle with given legs. By the Pythagorean Theorem, this distance is

$$\sqrt{(90)^2 + [30(t - 9)]^2} \quad \text{which simplifies to} \quad 30\sqrt{t^2 - 18t + 90}.$$

73. This problem is just a matter of reading the chart correctly.

$A(1) = 1183 \quad A(8) = 1187 \quad A(30) = 1203 \quad A(60) = 1248 \quad A(88) = 1204$

To help in determining if a function is even or odd, keep the following in mind:

a) All polynomial functions with no odd exponents are even.

b) All polynomial functions with no even exponents and no constant term are odd.

c) All other polynomial functions are neither even nor odd.

d) The absolute value function is even.

e) The sum or difference of even functions is even; sum or difference of odds is odd.

75. $f(-x) = (-x)^2 - 1 = x^2 - 1 = f(x)$. Therefore, it is an even function (see a) also).

77. $f(-x) = (-x)^4 + (-x)^2 = x^4 + x^2 = f(x)$. Therefore, even (see a)).

79. Neither. $f(-x) = (-x)^4 + (-x)^2 + 2(-x) = x^4 + x^2 - 2x$. You can see that $f(-x)$ $\neq f(x)$ and $f(x) \neq -f(x) = -x^3 - x^2 - 2$ ($f(x)$ has both even and odd exponents).

81. $f(-x) = \dfrac{1}{(-x)^2} = \dfrac{1}{x^2} = f(x)$ so even.

83. $f(-x) = (-x)^4 - (-x)^2 + 2 = x^4 - x^2 + 2 = f(x)$ so even. (a)

85. $f(-x) = \dfrac{(-x)^2 - 2}{(-x)^2 + 5} = \dfrac{x^2 - 2}{x^2 + 5} = f(x)$ so even.

87. $f(-x) = (-x)^{3/5} = -x^{3/5} = -f(x)$ (works since $(-x)^3 = -x^3$). Therefore, odd.

89. $f(-x) = \dfrac{(-x)^3}{(-x)^2 + 12} = \dfrac{-x^3}{x^2 + 12} = -\dfrac{x^3}{x^2 + 12} = -f(x)$, so odd.

91. Neither. $f(-x) = (-x)^3 + (-x) + 1$. $f(-x) \neq f(x)$ and $f(x) \neq -f(x) = -x^3 - x - 1$.

Problems 3.4

1. Slope is 0.12. 12 cents is gained for

   each dollar each year (so is the rate of interest).

$m = 0.12$ represents the rate of interest

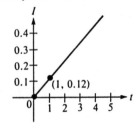

3. $I = 1000(0.13)t$. Slope is 130.

I increases by 130 for each increase in t of 1 unit ($130 annual simple interst on $1000 at 13%.)

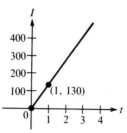

$m = 130$ represents annual simple interest on a $1000 investment

5. Slope is 70.

Each increase of t by 1 unit means an increase in s of 70 (velocity).

$m = 70$ represents velocity

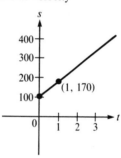

7. Slope is -0.025.

For each increase in q of 1, p decreases

by 0.025 (price decreases for each increase of 1 unit sold)

$m = -0.025$ represents the amount the price is reduced for each unit sold

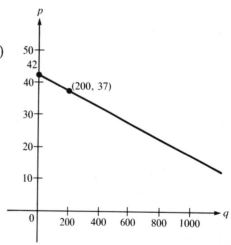

9. Slope is 0.015.

For each increase in q by 1,

s increases by 0.015

(price increase per unit sales increase)

$m = 0.015$ represents how the price increases as the number of units (q) produced increases

- 111 -

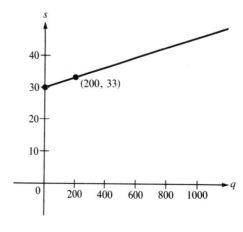

11. Set 42 - 0.025q = 30 + 0.015q. Obtain 12 = 0.040q, or q = $\frac{12}{0.040}$ = 300
     The equilibrium quantity is 300. Substitute into either equation. If
     the first equation is used, obtain 42 - 0.025(300) = 42 - 7.50 = \$34.50
     for the equilibrium price (second equation yields same result).

13. Set 500 - .15q = 450 + .35q, or 50 = .50q, meaning q = 100 is the
     equilibrium quantity. Substitute into either equation, obtain p = \$485 for
     the equilibrium price.

15. For the endurance, use the points (0, 0) and (8, -42). The slope is - $\frac{42}{8}$ = -5.25.
     This means that the concentration of lactic acid decreases 5.25 mg/100ml
     per week. For the sprint, use (0, 0) amd (8, -20) as the points. The slope is
     $\frac{-20 - 0}{8}$ = -2.5, which means the concentration of acid decreases 2.5 mg/100ml per week.

17. Use the points (1, 8) and (2, 4), obtain slope is $\frac{-4}{1}$ = -4. The equilibrium price
     for wheat is \$4 per bushel.

19. a) Let h be the number of kWh used. Then h - 30 is the number of kWh used
     over 30. The cost equation is therefore C = 6 + .07(h - 30) = 6 + .07h - 2.1,
     or C = .07h + 3.90 as long as h $\geq$ 30.

     b)

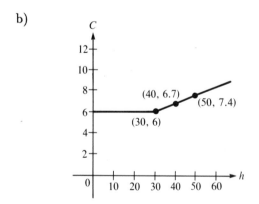

     c) C = .07(75) + 3.90 = 5.25 + 3.90 = \$9.15

21. a) Let I = amount of income, T = tax owed.

    Then T = 7434 + .44(I - 28,800) = 7434 + 0.44I - 12,672 = 0.44I - 5238

    provided $28,800 \leq I \leq 34,100$.

b)

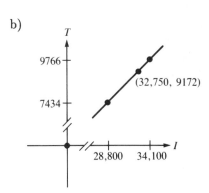

c) T = .44(32,750) - 5238 = 14,410 - 5,238 = \$9,172.

23. Reading the coefficient of the variable from the proper equation, obtain .34.

25. As the octane increased by 5 (from 85 to 90), the cost increased by $6\frac{1}{2}$ cents.

    Since the relationship is linear, another increase in octane of 5 (from 90 to

    95) means the cost of the gas should increase by another $6\frac{1}{2}$ cents to a total of \$1.56

Problems 3.5

1.

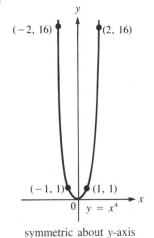

symmetric about y-axis

3.

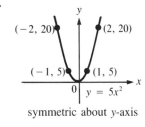

symmetric about y-axis

5.

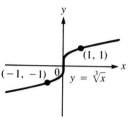

symmetric about origin

7.

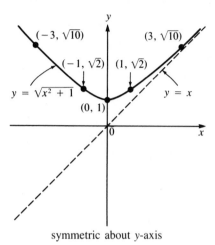

symmetric about y-axis

9. This is the graph of $y = x^2$ shifted down 5 units

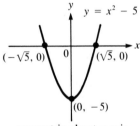

symmetric about y-axis

11. This is the graph of $y = x^2$ shifted right 4 units

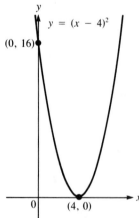

symmetric about line $x = 4$

13. The graph of $y = x^2$ shifted right 1 then up 3

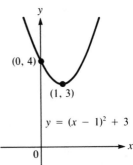

$(0, 4)$

$(1, 3)$

$y = (x - 1)^2 + 3$

0

symmetric about line $x = 1$

15.

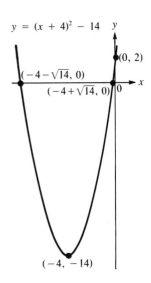

$y = (x + 4)^2 - 14$

$(0, 2)$

$(-4-\sqrt{14}, 0)$

$(-4+\sqrt{14}, 0)$ 0

$(-4, -14)$

17.

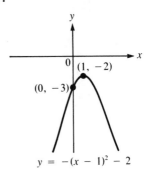

0 $(1, -2)$

$(0, -3)$

$y = -(x - 1)^2 - 2$

19. Shifted up 1

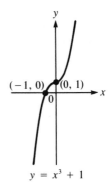

$(-1, 0)$ $(0, 1)$

0

$y = x^3 + 1$

21. shift left 3

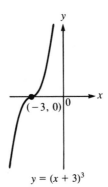

$(-3, 0)$ 0

$y = (x + 3)^3$

23. reflect along x axis then shift left 2

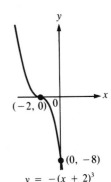

$(-2, 0)$ 0

$(0, -8)$

$y = -(x + 2)^3$

- 115 -

25. a) shift left 3

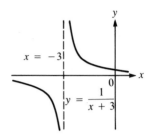

b) shift right 4

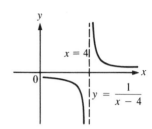

c) shift up 3

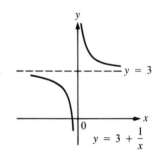

d) reflect along x axis

   shift up 2

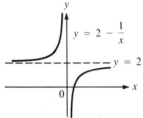

e)

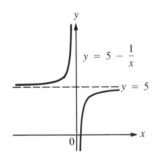

27. up 2

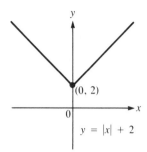

$y = |x| + 2$

29. left 5

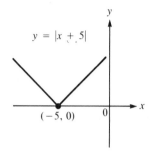

$y = |x + 5|$

31. reflect and shift up 5, left 4

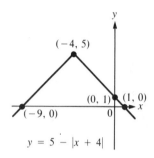

$y = 5 - |x + 4|$

33.

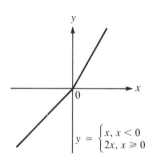

$y = \begin{cases} 4, & x > 3 \\ -2, & x \leq 3 \end{cases}$

35.

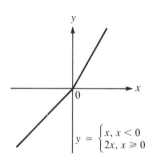

$y = \begin{cases} x, & x < 0 \\ 2x, & x \geq 0 \end{cases}$

37.

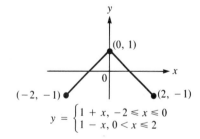

$y = \begin{cases} 1 + x, & -2 \leq x \leq 0 \\ 1 - x, & 0 < x \leq 2 \end{cases}$

39.

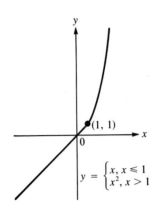

$$y = \begin{cases} x, x \leqslant 1 \\ x^2, x > 1 \end{cases}$$

41.

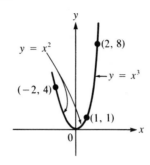

43.

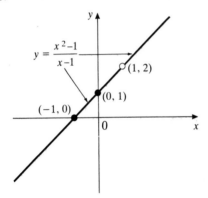

45.

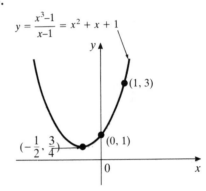

47. a) right 2

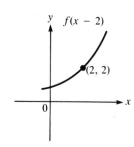
$f(x - 2)$
(2, 2)

b) left 3

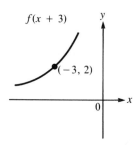
$f(x + 3)$
(−3, 2)

c) reflect along x axis

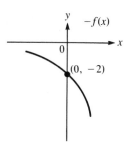
$-f(x)$
(0, −2)

d) reflect along y axis

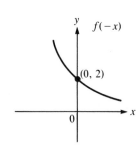
$f(-x)$
(0, 2)

e) write as $y = f(-(x - 2)) + 3$

reflect across y-axis to obtain $y = f(-x)$,

then shift right 2, getting $y = f(-(x - 2))$,

then shift up 3

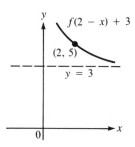
$f(2 - x) + 3$
(2, 5)
$y = 3$

49. Same shifts as #47

(a)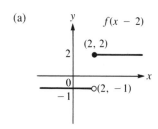

(b)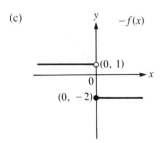

(c)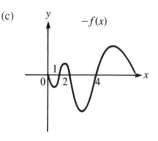

(d)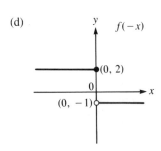

(e)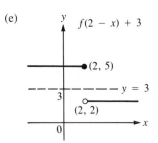

51. Same shifts as #47

(a)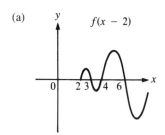

(b)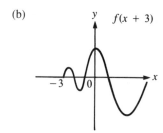

(c)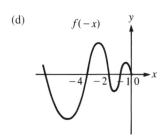

(d)

(e)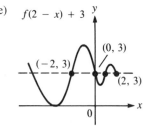

53. Again, same shifts

(a)
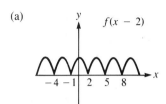
$f(x - 2)$

(b)
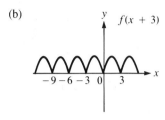
$f(x + 3)$

(c)
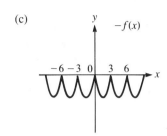
$-f(x)$

(d)
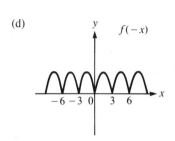
$f(-x)$

(e)
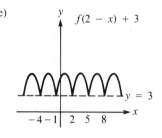
$f(2 - x) + 3$

55.
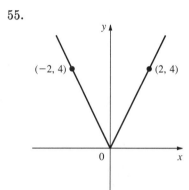

57.
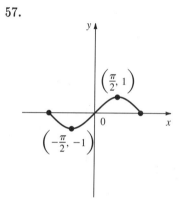

59.
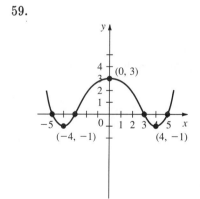

61.
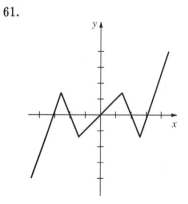

Problems 3.6

In 1-8, just add, subtract, multiply, and divide the expressions given.

1. $(f + g)(x) = 2x + 3 + -3x = -x + 3$; Domain is all real numbers, since the domain of f and the domain of g are both all real numbers.

    $(f - g)(x) = 2x + 3 - (-3x) = 5x + 3$; Domain is all real numbers

    $(fg)(x) = (2x + 3)(-3x) = -6x^2 - 9x$; Domain is all real numbers

    $(\frac{f}{g})(x) = -\frac{2x + 3}{3x}$; Domain is all real numbers except zero (no division by 0).

3. $(f + g)(x) = 4 + 10 = 14$

    $(f - g)(x) = 4 - 10 = -6$

    $(fg)(x) = (4)(10) = 40$

    $(\frac{f}{g})(x) = \frac{4}{10} = 0.4$

    The domain of each is all real numbers

5. $(f + g)(x) = x + \frac{1}{x} = \frac{x^2 + 1}{x}$; Domain is all real numbers except zero.

    $(f - g)(x) = x - \frac{1}{x} = \frac{x^2 - 1}{x}$; Domain is all real numbers except zero.

    $(fg)(x) = x\left(\frac{1}{x}\right) = 1$; The domain is still all real numbers except zero.

    $(\frac{f}{g})(x) = \frac{x}{\left(\frac{1}{x}\right)} = x \cdot x = x^2$; Domain all real numbers except zero.

7. $(f + g)(x) = \sqrt{x + 1} + \sqrt{1 - x}$; The domain of f is all real numbers greater than or equal to -1, and the domain of g is all real numbers less than or equal to 1. The domain of f + g is the intersection of the two, hence [-1, 1].

    $(f - g)(x) = \sqrt{x + 1} - \sqrt{1 - x}$; Domain [-1, 1]

    $(fg)(x) = \sqrt{x + 1} \sqrt{1 - x} = \sqrt{1 - x^2}$; Domain is [-1, 1]

    $(\frac{f}{g})(x) = \frac{\sqrt{x + 1}}{\sqrt{1 - x}} = \frac{\sqrt{1 - x^2}}{1 - x}$ Cannot include 1 in domain, so Domain is [-1, 1)

9. $(f + g)(x) = 1 + x^5 + 1 - |x| = x^5 - |x| + 2$;   Domain is all real numbers

   $(f - g)(x) = 1 + x^5 - (1 - |x|) = x^5 + |x|$;   Domain is all real numbers

   $(fg)(x) = 1 - |x| + x^5 - x^5|x|$;   Domain is all real numbers

   $(\frac{f}{g})(x) = \frac{1 + x^5}{1 - |x|}$;   Domain is all real numbers except ± 1.

11. $(f + g)(x) = \sqrt[5]{x + 2} + \sqrt[4]{x - 3}$;   Domain of f is all real numbers, domain of g is

    all real numbers greater than or equal to 3, so domain f + g is $[3, \infty)$

    $(f - g)(x) = \sqrt[5]{x + 2} - \sqrt[4]{x - 3}$;   Domain is $[3, \infty)$

    $(fg)(x) = \sqrt[5]{x + 2} \sqrt[4]{x - 3}$;   Domain is $[3, \infty)$

    $(\frac{f}{g})(x) = \frac{\sqrt[5]{x + 2}}{\sqrt[4]{x - 3}}$;   Domain is $(3, \infty)$

13. $(f \circ g)(x) = f(g(x)) = f(x - 2) = 3(x - 2) = 3x - 6$;   Domain is all real numbers.

    $(g \circ f)(x) = g(f(x)) = g(3x) = 3x - 2$;   Domain is all real numbers.

15. $(f \circ g)(x) = f(g(x)) = f(8) = 5$;   Domain is all real numbers.

    $(g \circ f)(x) = g(f(x)) = g(5) = 8$;   Domain is all real numbers.

17. $(f \circ g)(x) = f(g(x)) = f(\frac{1}{2x}) = \frac{1}{2x}$;   Domain is all real numbers except zero.

    $(g \circ f)(x) = g(f(x)) = g(x) = \frac{1}{2x}$; Domain is all real numbers except zero.

19. $(f \circ g)(x) = f(x + 3) = 2(x + 3) - 4 = 2x + 2$; Domain is all real numbers.

    $(g \circ f)(x) = g(2x - 4) = 2x - 4 + 3 = 2x - 1$; Domain all real nos.

21. $(f \circ g)(x) = f\left(\dfrac{x+1}{x}\right) = \dfrac{\frac{x+1}{x}}{2 - \frac{x+1}{x}}$. To simplify, multiply top and bottom by x to obtain

$\dfrac{x+1}{2x - (x+1)} = \dfrac{x+1}{x-1}$. This expression is undefined only when x = 1, and the domain of g is all

real numbers except 0, so the domain of the composition is all real numbers except 0 and 1.

$(g \circ f)(x) = \dfrac{\frac{x}{2-x} + 1}{\frac{x}{2-x}}$. To simplify, multiply top and bottom by 2 - x to obtain

$\dfrac{x + (2-x)}{x} = \dfrac{2}{x}$. This is not defined for zero, the domain of f is all real

numbers except 2, so composition domain is all real numbers except 0 and 2.

23. $(f \circ g)(x) = \sqrt{1 - \sqrt{x-1}}$; x must be greater than or equal to 1 if what is

under the smaller radical is to be non-negative; but x must be less than

or equal to 2 in order for $\sqrt{x-1}$ to be less than or equal to 1, allowing what is

under the larger radical to be non-negative. The domain is [1, 2].

$(g \circ f)(x) = \sqrt{\sqrt{1-x} - 1}$; Domain is $(-\infty, 0]$

25. $(f \circ g)(x) = 2(\frac{1}{2}x - 2) + 4 = x - 4 + 4 = x$

$(g \circ f)(x) = \frac{1}{2}(2x + 4) - 2 = x + 2 - 2 = x$

27. Need two expressions that when squared give $x^2 - 10x + 25$.

They are $g_1(x) = x - 5$ and $g_2(x) = 5 - x$

29. Domain is $[0, \infty)$ since what is under $\sqrt{\ }$ must not be negative.

One assignment; $h(x) = \sqrt{x}$ $\quad g(x) = 1 + x$ $\quad f(x) = x^{5/7}$.

31. Want $(f \circ g)(x) = a(cx + d) + b = c(ax + b) + d = (g \circ f)(x)$

$acx + ad + b = acx + bc + d$, or $ad + b = bc + d$

33. a) $C(x) = 8x + 8000$, so $C(p) = 8(400(50 - p)) + 8000 = 3200(50 - p) + 8000$

$$= 160,000 - 3200p + 8000 = 168,000 - 3200p$$

b) $R(p) = px = p(400(50 - p)) = 400p(50 - p) = 20,000p - 400p^2$

c) $P(p) = R(p) - C(p) = (20,000p - 400p^2) - (168,000 - 3200p)$

$$= -400p^2 + 23,200p - 168,000$$

d) $P(p) = -400(p^2 - 58p) - 168,000$. Half of 58 is 29, so

$P(p) = -400(p - 29)^2 + 168,400$. Therefore, the price giving the largest

profit is $29, and the maximum profit is $168,400 (since has + above and a = -400).

35. We must show that $h(-x) = -h(x)$. Now $h(-x) = \frac{1}{2}[f(-x) - f(x)]$

$-h(x) = -\frac{1}{2}[f(x) - f(-x)] = -\frac{1}{2}f(x) + \frac{1}{2}f(-x)$ by the distributive property

$= \frac{1}{2}f(-x) - \frac{1}{2}f(x)$ by the commutative property

$= \frac{1}{2}[f(-x) - f(x)]$ by the distributive property

$= h(-x)$.   QED

Problems 3.7

1. The function passes the horizontal line test, as do all linear functions.

   To find the inverse; set $f(x) = y = 2x - 1$

   Interchange x and y $x = 2y - 1$

   Solve for y $x + 1 = 2y$

   $$y = \frac{x + 1}{2} = f^{-1}(x).$$

3. Passes horizontal line test, since linear.

   $y = \frac{2}{3}x - \frac{1}{4}$

   $x = \frac{2}{3}y - \frac{1}{4}$ or $x + \frac{1}{4} = \frac{2}{3}y$ so $y = \frac{3}{2}x + \frac{3}{8} = f^{-1}(x)$

5. Passes horizontal line test, since linear.

$$y = \frac{3 - 2x}{11}$$

$$x = \frac{3 - 2y}{11}$$

$11x = 3 - 2y$ or $11x - 3 = -2y$

$$y = \frac{11x - 3}{-2} = \frac{3}{2} - \frac{11}{2}x = f^{-1}(x).$$

7. Passes horizontal line test.

$y = \frac{3}{x}$ so $x = \frac{3}{y}$

$xy = 3$

$y = \frac{3}{x} = f^{-1}(x)$

9. Passes HLT.

$x = \frac{1}{y + 1}$.   Multiply both sides by $y + 1$ and obtain

$xy + x = 1$

$xy = -x + 1$

$y = \frac{-x + 1}{x} = -1 + \frac{1}{x} = f^{-1}(x)$

11. Passes HLT.

$x = \frac{3}{4 - y}$

$4x - xy = 3$

$4x - 3 = xy$

$y = \frac{4x - 3}{x} = 4 - \frac{3}{x} = f^{-1}(x)$

13. Passes HLT

$x = \frac{4}{2 + y^3}$

$2x + xy^3 = 4$

$xy^3 = 4 - 2x$

$y^3 = \frac{4 - 2x}{x}$

$y = \sqrt[3]{\frac{4 - 2x}{x}}$

15. Passes HLT $x = \sqrt{y + 2}$

$x^2 = y + 2$ by squaring both sides.

$y = x^2 - 2 = f^{-1}(x)$. From first line, this applies only when $x \geq 0$.

17. Passes HLT.

$x = \sqrt{1 - 2y}$

$x^2 = 1 - 2y$

$2y = 1 - x^2$

$y = \frac{1 - x^2}{2} = f^{-1}(x)$. As in #15, x must not be negative since the square root of an expression cannot be negative.

19. Passes HLT.

$x = \dfrac{1}{\sqrt[3]{y - 7}}$

$\sqrt[3]{y - 7} = \frac{1}{x}$

$y - 7 = \frac{1}{x^3}$ or $y = \frac{1}{x^3} + 7 = f^{-1}(x)$

21. Passes HLT.

$x = \dfrac{y}{y + 1}$

$xy + x = y$

$xy - y = -x$

$y(x - 1) = -x$

$y = \frac{-x}{x - 1} = \frac{x}{1 - x} = f^{-1}(x)$

23. Passes HLT

$x = (y + 3)^3$

$\sqrt[3]{x} = y + 3$

$y = \sqrt[3]{x} - 3 = f^{-1}(x)$

25. f is one-to-one on $[0, \infty)$

$y = 1 + x^2$

$x = 1 + y^2$ or $y^2 = x - 1$

$y = \sqrt{x-1} = f^{-1}(x)$ (Take the positive square root because the range of $f^{-1}(x)$ is the domain of $f(x)$; which is $[0, \infty)$).

f is also one-to-one on on $(-\infty, 0]$. Do the same calculation, except

take $y = -\sqrt{x-1}$ in the last line because range must be $(-\infty, 0]$.

27. f is one-to-one on $[4, \infty)$

$y = (4 - x)^2$

$x = (4 - y)^2$

$-\sqrt{x} = 4 - y$ (Take negative square root) so $y = 4 + \sqrt{x}$

f is also one-to-one on $(-\infty, 4]$

Do the same calculation except take positive square root; that is, $\sqrt{x} = 4 - y$, or $y = 4 - \sqrt{x}$

29. f is one-to-one on $[0, \infty)$

$y = x^4$ so $x = y^4$

$y = \sqrt[4]{x}$ (take positive root).

f is also one-to-one on $(-\infty, 0]$; here take the negative root and obtain $y = -\sqrt[4]{x}$

31. f is one-to-one on $[0, \infty)$

$x = y^4 - 5$ so $y^4 = x + 5$

$y = \sqrt[4]{x + 5}$ (take positive root)

f is also one-to-one on $(-\infty, 0]$; here take the negative root to obtain $y = -\sqrt[4]{x + 5}$

33. Complete the square to rewrite $f(x) = \left(x - \frac{7}{2}\right)^2 + 6 - \frac{49}{4} = \left(x - \frac{7}{2}\right)^2 - \frac{25}{4}$

So f is one-to-one on $[\frac{7}{2}, \infty)$

$x = \left(y - \frac{7}{2}\right)^2 - \frac{25}{4}$ so $x + \frac{25}{4} = \left(y - \frac{7}{2}\right)^2$

$\sqrt{x + \frac{25}{4}} = |y - \frac{7}{2}|$. If $y > \frac{7}{2}$, then $|y - \frac{7}{2}| = y - \frac{7}{2}$ and $y = \sqrt{x + \frac{25}{4}} + \frac{7}{2}$ is the inverse.

f is also one-to-one on $(-\infty, \frac{7}{2}]$. Then $y < \frac{7}{2}$ and $|y - \frac{7}{2}| = \frac{7}{2} - y$

$y = -\sqrt{x + \frac{25}{4}} + \frac{7}{2}$ is the inverse there.

35. $y = (x^2 + 2)$; f is one-to-one on $[0, \infty)$

   $x = (y^2 + 2)^2$ which means $\sqrt{x} = y^2 + 2$ and so $\sqrt{x} - 2 = y^2$

   $|y| = \sqrt{\sqrt{x} - 2}.$

   The inverse on $[0, \infty)$ is $\sqrt{\sqrt{x} - 2} = y$

   The function f is also one-to-one on $(-\infty, 0]$, and there the inverse is $y = -\sqrt{\sqrt{x} - 2}$.

37. f is one-to-one on $[0, \infty)$. There $y = x$

   Interchange and get $x = y$ which means $y = x$ is the inverse.

   f is also one-to-one on $(-\infty, 0]$. There $y = -x$.

   Interchange and get $x = -y$ or $y = -x$, which is the inverse on this interval.

39. As x increases, so does x^3 and so f is increasing throughout its domain and hence one-to-one.

41. As x increases, x^3 increases, so $\dfrac{1}{x^3 + x + 1}$ decreases. The function

   is always decreasing

   and hence one-to-one.

43. Yes   45. No   47. No   49. Yes   51. Yes

53.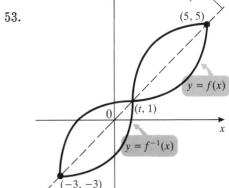

55.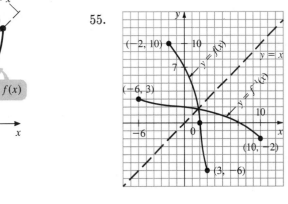

57. Let g and h both be inverses of the function f.

   If $g \neq h$, then there is some h such that $g(x) \neq h(x)$. Since f is one-to-one,

   $f(g(x)) \neq f(h(x))$. But $f(g(x)) = x$ and $f(h(x)) = x$ since both g and h are

   inverses of f. Since $x = x$, the assumption that there are 2 distinct inverses

   of f is false; therefore, the inverse is unique.

59. a) is a direct consequence of the Intermediate Value Theorem

    b) Since f is increasing, f is one-to-one and has an inverse

    c) Since f^{-1} is one-to-one, it is either increasing or decreasing.

    f(a) < f(b) since f is increasing

    $a = f^{-1}(f(a)) < f^{-1}(f(b)) = b$ so f^{-1} is increasing

Problems 3.8

1. $y = 10x$ 3. $z = \frac{3.3}{w}$ 5. $x = kz^3$

7. $w = k\sqrt{z}$ 9. $y = \frac{k}{v^{1/4}}$ 11. $R = kx^3\sqrt{z}$

13. $w = \frac{ku^2}{v^3}$

15. Know that $C = kq^2$

    $12.80 = k(80)^2$

    $12.80 = 6400k$

    $k = .002$

    a) So $C = 0.002q^2$

    b) $C(200) = 0.002(200)^2 = (.002)(40000) = \$80$

17. Since $S = kr^2$, $16\pi = k(2)^2$

    $16\pi = 4k$ or $k = 4\pi$

    a) So $S = 4\pi r^2$

    b) $S = 4\pi(3)^2 = 36\pi$ square units.

19. $z = kw^3$ is the equation. If w triples, now you have 3w.

    $z = k(3w)^3 = 27kw^3$, which is 27 times what it was before.

21. $v = \frac{k}{u^2}$ is the equation. If u is halved, now you have $\frac{1}{2}u$.

    $v = \frac{k}{\left(\frac{1}{2}u\right)^2} = \frac{k}{u^2/4} = \frac{4k}{u^2}$. So v quadruples.

23. From problem 22 obtain that $y = (4.07 \times 10^{-10})d^{3/2}$. Plug in $d = 36,000,000$ and obtain $y \approx 87.91$ days.

25. In problem 24 you obtain $P = \frac{1500}{V}$. Plug in $V = 20$ and get P increases to 75 lbs/in^2

27. $I = \frac{k}{a^2}$

$100 = \frac{k}{5^2} = \frac{k}{25}$. Therefore, $k = 2500$ and the equation is $I = \frac{2500}{a^2}$, where

a is the distance to the source.

a) $I = \frac{2500}{2^2} = \frac{2500}{4} = 625$ ft-candles.   b) $I = \frac{2500}{10^2} = 25$ ft-candles.

29. In problem 28 the relationship $F = \frac{kv^2}{r} = \frac{120v^2}{r}$ is obtained. If $F = 20,000$

and $r = 75$, then $20,000 = \frac{120v^2}{75}$ and $v^2 = 12500$. Therefore, $v \approx 111.8$ ft/sec

31. The formula for the volume of a cone is obtained in Problem 30 and found to be $V = \frac{1}{3}\pi r^2 h$. If the radius doubles (now giving $2r$) and the height is cut in third (giving $\frac{h}{3}$) we get $V = \frac{1}{3}\pi(2r)^2\frac{h}{3} = \frac{1}{3}\pi r^2 h\frac{4}{3}$, which means the volume has been multiplied by $\frac{4}{3}$.

33. $R = \frac{kl}{r^4}$ from problem 32. If both R and l double, get $2R = \frac{k2l}{r^4}$; if both sides are divided by 2, the original equation is obtained. So, r doesn't change.

Chapter 3 Review Problems

1. $d = \sqrt{(3-2)^2 + (2-(-1))^2} = \sqrt{1^2 + 3^2} = \sqrt{1+9} = \sqrt{10}$

3. $d = \sqrt{2^2 + (-2)^2} = \sqrt{4+4} = \sqrt{8} = 2\sqrt{2}$

5. $\left(\frac{1 + -5}{2}, \frac{2+7}{2}\right) = \left(-2, \frac{9}{2}\right)$

7. $x^2 + 4x + y^2 - 8y = -16$

$(x + 2)^2 + (y - 4)^2 = -16 + 4 + 16$

$(x + 2)^2 + (y - 4)^2 = 4$

Circle with center (-2, 4) and radius $\sqrt{4} = 2$

9. a) $y = -x + 2$; if $x = -3$, then $y = 3 = 2 = 5$

b) if $x = 0$ then $y = 2$ and (0, 2) is the y-intercept.

if $y = 0$ then $x = 2$ and (2, 0) is the x-intercept.

c) Just place the x- and y-intercepts on the graph and

draw the straight line through them.

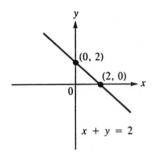

11. a) If $x = -2$ then $-6 + 5y = 15$ which means $5y = 21$ or $y = \frac{21}{5}$

b) If $x = 0$ then $5y = 15$ or $y = 3$ which means (0, 3) is the y-intercept.

If $y = 0$ then $3x = 15$ or $x = 5$ which means (5, 0) is the x-intercept.

c)

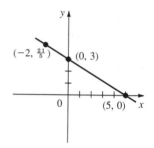

13. $\frac{3 - 5}{-1 - 2} = \frac{-2}{-3} = \frac{2}{3}$.

A point-slope form of the line could be $y - 5 = \frac{2}{3}(x - 2)$ or $y - 3 = \frac{2}{3}(x + 1)$

Solve for y to obtain $y = \frac{2}{3}x + \frac{11}{3}$ as the slope-intercept form.   Multiply through by -3, add the

x term to left to obtain $2x - 3y = -11$ as one possibility for the standard form.

15. $m = \frac{-4 - (-2)}{2 - 4} = 1$

Either $y + 2 = 1(x - 4)$ or $y + 4 = 1(x - 2)$ are point-slope forms

$y = x - 6$ is the slope intercept form

$x - y = 6$ is one possibility for the standard form.

17. Since both points have the same x-coordinate, the line is vertical and has undefined slope. The only equations of the line are $x = 1$ or the standard one ($x - 1 = 0$ being an example).

19. Solving the equation for y we obtain $y = -\frac{2}{5}x + \frac{4}{5}$. Therefore, the given line has slope $-\frac{2}{5}$ and since the desired line is parallel, it has slope $-\frac{2}{5}$ also. The point-slope form of the desired line is $y + 3 = -\frac{2}{5}(x - 2)$. Solve

for y to obtain the slope-intercept form $y = \frac{2}{5}x - \frac{11}{5}$.

21. The slope is -.04. If the quantity increases by 1 unit, the price decreases by \$.04.

 $p = 80 - 0.04q$

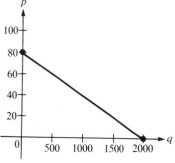

23. Set $80 - .04q = 50 + .02q$

$$30 = .06q$$

$q = \frac{30}{.06} = 500$ units. Substitute into either of the above, say the first and obtain $80 - .04(500) = 80 - 20 = 60$; that is, \$60.

25. Yes, it is a nonvertical straight line. Both the domain and range are all real numbers.

27. Multiply both sides by x to obtain $y = x$. That is definitely a function.

   $y = x$ has domain all real numbers, but the original expression is not defined when $x = 0$, so domain is really all real numbers except zero; 0 divided by x can never equal 1 no matter what x is, so y cannot be zero and so the range is really all real numbers except zero.

29. It is a function. The domain is $[-3, \infty)$ since what is under the radical cannot be negative. The range is $[0, \infty)$ because $\sqrt{0} = 0$ and there is no largest value the expression can take on.

31. As before, this is a function since it is written in the form $y =$ expression involving x. The domain is all real numbers, since a quotient of polynomials is undefined only when the denominator is zero, and $x^2 + 1$ is never 0. To find the range, we can try solving for y in terms of x. Multiply both sides of the equation by $x^2 + 1$ to obtain $x^2y + y = x$ or $yx^2 - x + y = 0$. Apply the quadratic formula where x is the variable and $a = y$, $b = -1$ and $c = y$ to obtain

$$x = \frac{1 \pm \sqrt{1 - 4y^2}}{2}.$$ Now $1 - 4y^2$ must be greater

than or equal to zero. This is the case only when $y \; \varepsilon \; [-\frac{1}{2}, \frac{1}{2}]$ so that is the range.

33. It is a function. It must be that $x^2 - 6$ must be greater than or equal to zero, so the domain of the function is $(-\infty, -\sqrt{6}$ or $[\sqrt{6}, \infty)$. The range is $(0, \infty)$ as in problem number 29.

35. $\dfrac{f(x + \Delta x) - f(x)}{\Delta x} = \dfrac{\frac{1}{x + \Delta x} - \frac{1}{x}}{\Delta x} = \dfrac{\frac{1}{x + \Delta x} - \frac{1}{x}}{\Delta x} \cdot \dfrac{(x + \Delta x)x}{(x + \Delta x)x} = \dfrac{x - (x + \Delta x)}{\Delta x(x + \Delta x)x}$

$$= \dfrac{-\Delta x}{\Delta x(x + \Delta x)x} = -\dfrac{1}{x(x + \Delta x)} \quad \text{QED}$$

37. $f(-x) = \dfrac{1}{(-x)^4} = \dfrac{1}{x^4} = f(x)$ so even.

39. $f(-x) = \dfrac{(-x)^3}{(-x)^5 + (-x)} = \dfrac{-x^3}{-x^5 - x} = \dfrac{-x^3}{-(x^5 + x)} = \dfrac{x^3}{x^5 + x} = f(x)$ so even.

Problems 41-47 are shifts and/or reflections of $y = x^2$ or $y = x^3$

41.

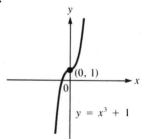

$(0, 1)$

$y = x^3 + 1$

43.

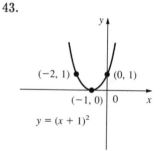

$(-2, 1)$ $(0, 1)$

$(-1, 0)$ 0

$y = (x + 1)^2$

45.

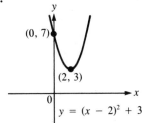

$(0, 7)$
$(2, 3)$
$y = (x - 2)^2 + 3$

47.

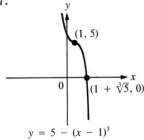

$(1, 5)$
$(1 + \sqrt[3]{5}, 0)$
$y = 5 - (x - 1)^3$

49.

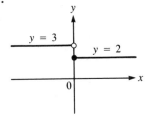

$y = 3$
$y = 2$

51.

(a)

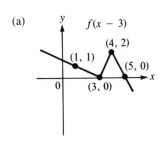

$f(x - 3)$
$(4, 2)$
$(1, 1)$
$(5, 0)$
$(3, 0)$

(b)

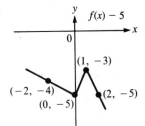

$f(x) - 5$
$(1, -3)$
$(-2, -4)$
$(2, -5)$
$(0, -5)$

(c)

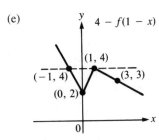

$f(-x)$
$(-1, 2)$
$(2, 1)$
$(-2, 0)$

(d)

$-f(x)$
$(-2, -1)$
$(2, 0)$
$(1, -2)$

(e)

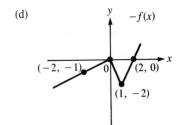

$4 - f(1 - x)$
$(1, 4)$
$(-1, 4)$
$(3, 3)$
$(0, 2)$

53. $(f + g)(x) = \sqrt{x + 1} + x^3$; Domain $[-1, \infty)$

$(f - g)(x) = \sqrt{x + 1} - x^3$; Domain $[-1, \infty)$

$(fg)(x) = x^3\sqrt{x + 1}$; Domain $[-1, \infty)$

$(\frac{g}{f})(x) = \frac{x^3}{\sqrt{x + 1}}$; Domain $(-1, \infty)$

$(f \circ g)(x) = f(g(x)) = f(x^3) = \sqrt{x^3 + 1}$ domain $[-1, \infty)$

$(g \circ f)(x) = g(f(x)) = g(\sqrt{x + 1}) = (\sqrt{x + 1})^3 = (x + 1)^{3/2}$ domain $[-1, \infty)$

55. Any non-vertical straight line function is one-to-one one.

$y = 4x - 1$ Now interchange

$x = 4y - 1$ Now solve for y

$x + 1 = 4y$

$y = \frac{x + 1}{4} = f^{-1}(x)$.

57. f passes the horizontal line test, so it is one-to-one.

$y = \frac{2}{x}$ Interchange, obtain $x = \frac{2}{y}$. Multiply both sides by y to obtain $xy = 2$; then divide

both sides by x to obtain $y = \frac{2}{x}$, which is the inverse.

59. f is increasing throughout the domain, so one-to-one.

$y = \sqrt{x - 2}$ Interchange, obtain $x = \sqrt{y - 2}$

Square both sides, obtain $x^2 = y - 2$ or $y = x^2 + 2$, which is the inverse

and is defined only for $x \geq 0$ (which is the range of f(x)).

61. f is one-to-one on $[0, \infty)$

$y = 3 + 2x^2$

$x = 3 + 2y^2$

$2y^2 = x - 3$

$y^2 = \frac{x - 3}{2}$ or $y = \sqrt{\frac{x - 3}{2}}$ since you want $y \; \varepsilon \; [0, \infty)$

f is also one-to-one on $(-\infty, 0]$ and in that case take $-\sqrt{}$ above to

obtain the inverse of f as

$$y = -\sqrt{\frac{x - 3}{2}}$$

63.

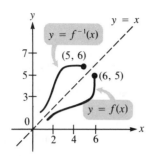

65. $v = \frac{k}{u}$

67. $P = \frac{ks}{t^3}$

69. $F = kx$ is the equation. Plug in $F = 10$ and $x = 1$ to obtain

$10 = k$. Therefore

$F = 10x$ is the equation for this situation. Now plug in $\frac{3}{12} = \frac{1}{4}$ (since 3 inches

is $\frac{1}{4}$ ft.) to obtain

$F = 10 \cdot \frac{1}{4} = 2.5$ lbs.

CHAPTER FOUR
Polynomials and Zeros of Polynomials

Chapter Objectives

In chapter four you will be introduced to the world of polynomials, the so called "nicely behaved" mathematical functions. You will learn how to sketch their graphs beginning with polynomials of degree two. You will see how important the x-intercepts (zeros) of a function are in the creation of its graph; the search for x-intercepts leads us to explore some of the theory dealing with solving equations of degree greater than two. You will see that it is possible to determine rational zeros when they exist with pencil and paper but electronic assistance is almost a necessity when using the bisection method or Newton's method to appoximate zeros that are real but not rational.

Chapter Summary

- A quadratic function is a function of the form $p(x) = ax^2 + bx + c$. The graph of a quadratic function is a parabola. It opens upward if $a > 0$ and downward if $a < 0$. The vertex of the parabola is $\left(\frac{-b}{2a}, p\left(-\frac{b}{2a}\right) \right)$.

- The polynomial $ax^2 + bx + c$ is irreducible if the quadratic equation $ax^2+bx + c=0$ has no real roots (this occurs if $b^2-4ac < 0$). In this case, its graph does not intersect the x-axis.

- The function $p(x) = ax^n$ is called a **power function**. If n is even, the graph resembles the graph of ax^2. If n is odd, the graph resembles the graph of ax^3.

- r is a **zero** or **root** of p(x) if $p(r) = 0$. In this case $(x - r)$ is a factor of p and $(r,0)$ is an x-intercept of the graph of $y = p(x)$.

- If $p(a) > 0$ and $p(b) < 0$ (or vice versa), then p(x) has a zero between a and b.

- $f(x) \to \infty$ as $x \to \infty$ means that f(x) increases without bound as x increases without bound.

- **Division algorithm**: if the degree of the polynomial $d(x) <$ degree of the polynomial $p(x)$ and $d(x) \neq 0$, then there exist unique polynomials q(x) and r(x) such that $p(x) = d(x)q(x) + r(x)$. q(x) is called the quotient and r(x) is called the remainder. Here degree $r(x) <$ degree $d(x)$.

- **Quotient theorem:**

$$\frac{p(x)}{d(x)} = q(x) + \frac{r(x)}{d(x)}$$

where $r(x) = 0$ or the degree of $r(x)$ < degree of $d(x)$

- **Remainder Theorem:** If $p(x)$ is divided by $(x - c)$, then the remainder $r = p(c)$.

- **Factor Theorm:** $(x - c)$ is a factor of the polynomial $p(x)$ if and only if $p(c) = 0$.

- **Synthetic Division** is a process for quickly dividing a polynomial by $x - c$.

- **Horner's Method** is a technique for evaluating a polynomial quickly on a calculator by factoring out x repeatedly.

- **Complex zeros** of polynomials with real coefficients occur in complex conjugate pairs.

- **Rational Zeros Theorem**
    Let $p(x) = a_n x^n + a_{n-1} x^{n-1} + \cdots + a_1 x + a_0$, where $a_n \neq 0$ and the coefficients a_0, a_1, ... ,a_n are all integers. Suppose $\frac{m}{k}$ is a rational zero of $p(x)$ and $\frac{m}{k}$ is reduced to lowest terms. Then
    i. m is a factor of a_0
    ii. k is a factor of a_n.

- If $(x - c)^k$ is a factor of the polynomial $p(x)$ but $(x - c)^{k+1}$ is not, then c is a zero of $p(x)$ of multiplicity k.

- **Fundamental Theorem of Algebra:** Every polynomial $p(x)$ of degree n has, counting multiplicities and complex zeros, exactly n zeros.

- **Descartes' Rule of Signs:** Let $p(x)$ be a polynomial with a non-zero constant term.
 (i) The number of positive real zeros, P, is either equal to V, the number of variations of sign in $p(x)$, or is less than V, and V - P is an even integer.
 (ii) The number of negative real zeros, N, is either equal to the number of variations of sign ,V_n, in $p(-x)$ or is less than V_n, and V_n - N is an even integer.

- **Upper and Lower Bound Theorem**   Let $p(x) = a_n x^n + a_{n-1} x^{n-1} + \cdots + a_1 x + a_0$ have $a_n > 0$ and all coefficients real. Divide $p(x)$ by $x - c$ using synthetic division.

    (i) c is an upper bound for the zeros of $p(x)$ if $c > 0$ and all the numbers in the bottom row of the synthetic division are nonnegative.

    (ii) c is a lower bound for the zeros of $p(x)$ if $c < 0$ and the numbers in the bottom row alternate in sign (where 0 can be considered positive or negative as required).

- **The Bisection Method for Finding a Zero of a Function $p(x)$**

    Step 1: Find two numbers a and b such that $p(a) < 0$ and $p(b) > 0$ or $p(a) > 0$ and $p(b) < 0$. Then there is a zero in (a, b) by the intermediate value theorem. Let's assume here that $p(a) < 0$ and $p(b) > 0$.

    Step 2: Compute p at the midpoint $\frac{a + b}{2}$; this is your first approximation to the zero.

    Step 3: a. If $p(\frac{a + b}{2}) > 0$, then there is a zero in $(a, \frac{a + b}{2})$.

    b. If $p(\frac{a + b}{2}) < 0$, then there is a zero in $(\frac{a + b}{2}, b)$.

    Step 4: Continue to cut the interval containing a zero in half until the length of the interval is less than the accuracy you seek.

- **Newton's Method for Finding Real Zeros of a Polynomial $P(x)$**

    Step 1 Find an interval [a,b] that contains a zero of $p(x)$. To do so, find a and b such that either $p(a) > 0$ and $p(b) < 0$ or $p(a) < 0$ and $p(b) > 0$.

    Step 2 Choose a number x_0 in [a,b].

    Step 3 Set $x_1 = x_0 - \dfrac{p(x_0)}{d(x_0)}$

    Step 4 Set $x_2 = x_1 - \dfrac{p(x_1)}{d(x_1)}$

    Step 5 Continue as in steps 3 and 4. If x_k is chosen, then $x_{k+1} = x_k - \dfrac{p(x_k)}{d(x_k)}$.

    The numbers $x_0, x_1, x_2, \cdots$ are called **iterates**.

    Step 6 Stop when two successive iterates agree with the number of decimal places carried in the computation.

Solutions to Chapter Four Problems

Problems 4.1

1. $f(x) = 3x^2$ sketch:

table: x: -2 -1 0 1 2

    $f(x)$: 12 3   0 3 12

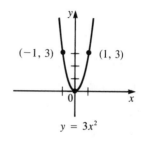

3. $f(x) = -3x^2$ sketch:

table: x: -2 -1 0 1 2

    $f(x)$: -12 -3   0 -3 -12

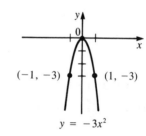

5. $f(x) = 50x^2$ sketch:

table: x: -1 0 1

    $f(x)$: 50 0 50

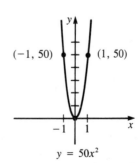

7. $y = x^2 + 2x$: half of 2 is 1; 1^2 is 1 so add and subtract 1

$= x^2 + 2x + 1 - 1$: regroup

$= (x^2 + 2x + 1) - 1$: now factor

$= (x + 1)^2 - 1$: visualize as $[x - (-1)]^2 - 1$ so the vertex is (-1, -1) ie, (h,k)

The x-intercepts are found by solving $x^2 + 2x = 0$

$$x(x + 2) = 0$$

so x = 0 or x = -2

One other "random" point might be helpful so let's let x = 2.

Then $y = 2^2 + 2 \cdot 2 = 8$ so (2,8) is on the graph.

Sketch: $y = (x + 1)^2 - 1$; vertex $(-1, -1)$

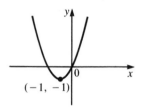

9. $y = x^2 + 5x + 4$: half of 5 is $\frac{5}{2}$ and $\frac{5}{2}$ squared is $\frac{25}{4}$, what we add and subtract

$= x^2 + 5x + \frac{25}{4} + 4 - \frac{25}{4}$

$= (x + \frac{5}{2})^2 - \frac{9}{4}$ so vertex is $(-\frac{5}{2}, -\frac{9}{4})$

y-intercept: let x = 0 ; get y = 4

x-intercepts: set $x^2 + 5x + 4 = 0$; $(x + 4)(x + 1) = 0$ so x = -4 or x = -1

sketch:

$$y = \left(x + \frac{5}{2}\right)^2 - \frac{9}{4}; \text{ vertex } \left(-\frac{5}{2}, -\frac{9}{4}\right)$$

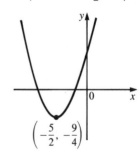

11. $y = 2x^2 + 4x - 5$: factor 2 out of the first two terms only

    $= 2(x^2 + 2x) - 5$: half of 2 is 1; 1^2 is 1 so add & subtract 1 (inside)

    $= 2(x^2 + 2x + 1 - 1) - 5$: regroup

    $= 2(x^2 + 2x + 1) - 2 - 5$: factor

    $= 2(x + 1)^2 - 7$ so the vertex is (-1, -7)

y-intercept is -5 (by letting $x = 0$)

x-intercepts: set $2x^2 + 4x - 5 = 0$: won't factor, will it?

so let's not bother with them since they weren't explicitly asked for.

Sketch:

$y = 2(x + 1)^2 - 7$; vertex $(-1, -7)$

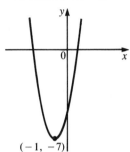

$(-1, -7)$

13. $y = -3x^2 + 9x + 12$: factor -3 (it opens down!) from 1st two terms

    $= -3(x^2 - 3x) + 12$: half of -3 is (-3/2) and $(-3/2)^2$ is 9/4

    $= -3(x^2 - 3x + 9/4 - 9/4) + 12$

    $= -3(x - 3/2)^2 + 27/4 + 12$: $\frac{27}{4} + \frac{12}{1} = \frac{27+48}{4} = \frac{75}{4}$

    $= -3(x - \frac{3}{2})^2 + \frac{75}{4}$

                      so the vertex is $(\frac{3}{2}, \frac{75}{4})$ or (1.5, 18.75)

The y intercept is +12 and the vertex and the y-intercept will suffice.

Sketch:

$y = -3\left(x - \frac{3}{2}\right)^2 + \frac{75}{4}$; vertex $\left(\frac{3}{2}, \frac{75}{4}\right)$

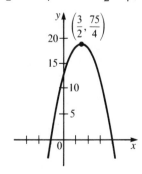

15. $y = -7x^2 + 3x - 2$

   $= -7(x^2 - \frac{3}{7}x) - 2$: half of $\frac{-3}{7}$ is $\frac{-3}{14}$ which when squared is $\frac{9}{196}$

   $= -7(x^2 - \frac{3}{7}x + \frac{9}{196} - \frac{9}{196}) - 2$

   $= -7(x - \frac{3}{14})^2 + \frac{7 \cdot 9}{196} - 2$: $\frac{9}{28} - \frac{2}{1} = \frac{9 - 56}{28} = \frac{-47}{28}$

   $= -7(x - \frac{3}{14})^2 - \frac{47}{28}$ so the vertex is $(\frac{3}{14}, -\frac{47}{28})$ and the y intercept is -2

   Sketch:
   $$y = -7\left(x - \frac{3}{14}\right)^2 - \frac{47}{28}; \text{ vertex } \left(\frac{3}{14}, -\frac{47}{28}\right)$$

   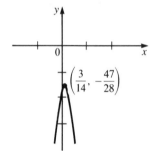

17. $y = \frac{x^2}{10} + \frac{x}{5} - 1$: factor $\frac{1}{10}$ from 1st two terms

   $= \frac{1}{10}(x^2 + 2x) - 1 = \frac{1}{10}(x^2 + 2x + 1 - 1) - 1 = \frac{1}{10}(x + 1)^2 - \frac{1}{10} - 1$

   $= \frac{1}{10}(x + 1)^2 - \frac{11}{10}$ so vertex is $(-1, -\frac{11}{10})$ and the y-intercept is -1

   Sketch:
   $$y = \frac{1}{10}(x + 1)^2 - \frac{11}{10}; \text{ vertex } \left(-1, -\frac{11}{10}\right)$$

   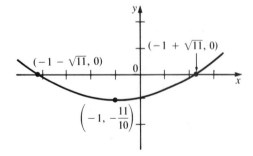

19. $y = x^2 - 4$: visualize as $1 \cdot x^2 - 0 \cdot x - 4$ (opens up)

vertex: $h = -\dfrac{b}{2a} = -\dfrac{0}{2 \cdot 1} = 0$

$k = 0^2 - 4 = -4$ so the vertex is $(0, -4)$ (also the y-intercept)

x-intercepts: set $x^2 - 4 = 0$ and get $(x - 2)(x + 2) = 0$ so $x = 2$ and $x = -2$ are the x-intercepts

Since the parabola opens up, it has a minimum value of -4

Sketch:

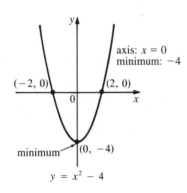

axis: $x = 0$
minimum: -4

$(-2, 0)$ $(2, 0)$

minimum $(0, -4)$

$y = x^2 - 4$

21. $f(x) = x^2 - 5x + 4$ (opens up)

vertex: $x = -\dfrac{-5}{2(1)} = \dfrac{5}{2}$; $y = f\left(\dfrac{5}{2}\right) = \left(\dfrac{5}{2}\right)^2 - 5\left(\dfrac{5}{2}\right) + 4 = \dfrac{25}{4} - \dfrac{25}{2} + \dfrac{4}{1} = \dfrac{25-50+16}{4} = -\dfrac{9}{4}$

so the vertex is $\left(\dfrac{5}{2}, -\dfrac{9}{4}\right)$

y-intercept: $f(0) = 4$ so $(0,4)$ is the y-intercept

x-intercepts: $x^2 - 5x + 4 = 0$: $(x - 4)(x - 1) = 0$ so $x = 4$ and $x = 1$

Since the parabola opens up, it has a minimum value of $-\dfrac{9}{4}$

Sketch:

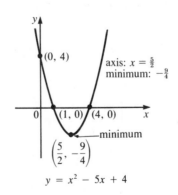

$(0, 4)$

axis: $x = \frac{5}{2}$
minimum: $-\frac{9}{4}$

$(1, 0)$ $(4, 0)$

minimum

$\left(\dfrac{5}{2}, -\dfrac{9}{4}\right)$

$y = x^2 - 5x + 4$

23. $f(x) = 1 - 4x^2 = -4x^2 + 1$ (opens down)

vertex: $x = \dfrac{-0}{2(-4)} = 0$; $y = f(0) = 1$ so vertex is the point $(0,1)$ (also the y-int)

x-intercepts: set $1 - 4x^2 = 0$; $(1 - 2x)(1 + 2x) = 0$; $1 - 2x = 0$ or $1 + 2x = 0$

$$1 = 2x \text{ or } 2x = -1$$

$$1/2 = x \text{ or } x = -1/2$$

Since the parabola opens down, the vertex y value, 1, is a maximum value.

Sketch:

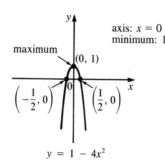

27. $g(u) = 3u^2 + 6u + 3$ (opens up)

vertex: $u = \dfrac{-6}{2(3)} = -1$; $g(-1) = 3 - 6 + 3 = 0$ so $(-1,0)$ is the vertex and also doubles as the u-intercept.

vertical axis intercept: $g(0) = 3$

25. $s(t) = 2t^2 + 4t - 6$ (opens up)

vertex: $t = \dfrac{-4}{2(2)} = -1$; $s(-1) = 2 - 4 - 6 = -8$ so the vertex is $(-1,-8)$

s-intercept: $s(0) = -6$

t-intercepts: set $2t^2 + 4t - 6 = 0$

$2(t + 3)(t - 1) = 0$ so $t = -3$ and $t = 1$ are the t-intercepts

Since the parabola opens up, -8 is the minimum value.

Sketch:

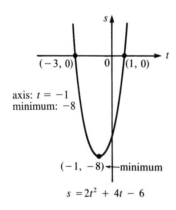

- 146 -

27. (continued)

Since the parabola opens up, 0, the y-value of the vertex, is the minimum.

Sketch:

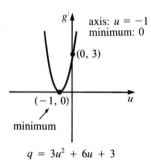

axis: $u = -1$
minimum: 0

$(0, 3)$

$(-1, 0)$

minimum

$q = 3u^2 + 6u + 3$

29. $f(x) = \frac{1}{4}x^2 + x - 1$ (opens up)

vertex: $x = \dfrac{-1}{2\left(\frac{1}{4}\right)} = \dfrac{-1}{1/2} = -2$; $y = f(-2) = \frac{1}{4}(4) - 2 - 1 = -2$ so the vertex is $(-2,-2)$.

y-intercept: $f(0) = -1$

x-intercepts: $\frac{1}{4}x^2 + x - 1 = 0$: multiply by 4

$\qquad x^2 + 4x - 4 = 0$: won't factor so use quadratic formula

$$x = \frac{-4 \pm \sqrt{16 - 4 \cdot 1 \cdot (-4)}}{2 \cdot 1}$$

$$x = \frac{-4 \pm \sqrt{32}}{2} = \frac{-4 \pm 4\sqrt{2}}{2} = -2 \pm 2\sqrt{2} \text{ and since } \sqrt{2} \text{ is approximately } 1.4, \text{ the x-}$$

intercepts are approximately -2 ± 2.8,   ie., 0.8 or -4.8

Since the parabola opens up, -2 is the minimum value.

Sketch:

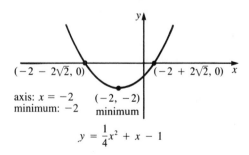

$(-2 - 2\sqrt{2}, 0)$ $(-2 + 2\sqrt{2}, 0)$ x

axis: $x = -2$ $(-2, -2)$
minimum: -2 minimum

$y = \frac{1}{4}x^2 + x - 1$

31. since 2 and 6 are x-intercepts we can say that $y = a(x - 2)(x - 6)$

The y- intercept is 4 so the point (0,4) must satisfy the above equation

which becomes $4 = a(0 - 2)(0 - 6)$ or $4 = 12a$ so $a = 1/3$ and the equation is

$$y = \tfrac{1}{3}(x - 2)(x - 6) \text{ or } y = \tfrac{1}{3}(x^2 - 8x + 12) \text{ or } y = \tfrac{1}{3}x^2 - \tfrac{8}{3}x + 4$$

completing the square on the middle form above:

$$y = \tfrac{1}{3}(x^2 - 8x + 16 - 16 + 12) = \tfrac{1}{3}(x^2 - 8x + 16) - \tfrac{4}{3} = \tfrac{1}{3}(x - 4)^2 - \tfrac{4}{3}$$

33. $y = a(x - 3)(x + 3)$ and (0,3) must satisfy so $3 = a(-3)(3)$ or $3 = -9a$ so $a = -\tfrac{1}{3}$

and equation is $y = -\tfrac{1}{3}(x^2 - 9)$ or $y = -\tfrac{1}{3}x^2 + 3$ or $y = -\tfrac{1}{3}(x - 0)^2 + 3$

35. $y = a(x - 0)(x - 3)$ and (-1, -8) must satisfy so $-8 = a(-1 - 0)(-1 -3) = 4a$ so $a = -2$

Therefore, $y = -2(x)(x - 3)$ or $y = -2x^2 + 6x$ is the equation.

Completing the square: $y = -2(x^2 - 3x + \tfrac{9}{4} - \tfrac{9}{4}) = -2(x - \tfrac{3}{2})^2 + \tfrac{9}{2}$

37. $C(q) = 5q^2 - 200q + 3125, \quad q \leq 100$

a) $q = \dfrac{-b}{2a} = \dfrac{-(-200)}{2(5)} = 20$ (units)

b) minimum cost $= C(20) = 5(20)^2 - 200(20) + 3125 = 2000 - 4000 + 3125 = \$1125$

c) $C(0) = 3125$ so 3125 is the "vertical" intercept

(20,1125) is the vertex

(100, 33125) is the endpoint

Sketch:

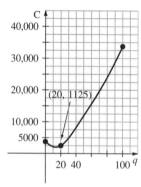

39. $C(q) = 5000 - 30q + 0.2q^2$, $q \leq 300$

    a) $q = \dfrac{-(-30)}{2(0.2)} = \dfrac{30}{0.4} = 75$ (units)

    b) $C = 5000 - 30(75) + 0.2(75)^2 = 5000 - 2250 + 1125 = \$3875$

    c) C-intercept is 5000

    (75 , 3875) is the vertex

    $C(300) = 5000 - 30(300) + 0.2(300)^2 = 14{,}000$ so (300 , 14000) is endpoint

    sketch:

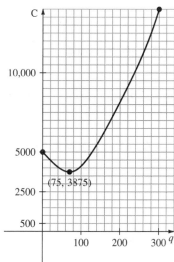

41. Use $s = -16t^2 + v_0 t$ with $v_0 = 40$ ft/sec to get $s = -16t^2 + 40t$ (feet) as the height function.

    a) if object is hitting the ground, its height must be 0 so substitute 0 for s:

      $0 = -16t^2 + 40t$

      $0 = -8t(2t - 5)$ so $-8t = 0$ or $(2t - 5) = 0$ and the latter gives the answer

        $t = 5/2$ or 2.5 sec ($t = 0$ is the moment the object was thrown upward)

    b) same as finding y value of vertex of any parabola:

      $t = \dfrac{-b}{2a} = \dfrac{-40}{2(-16)} = \dfrac{40}{32} = \dfrac{5}{4}$ sec so $\quad s_{max} = -16\left(\dfrac{5}{4}\right)^2 + 40\left(\dfrac{5}{4}\right) = -25 + 50 = 25$ feet

43. $s = -16t^2 + 5280t$ (feet)

    a) $0 = -16t^2 + 5280t$

      $= -16t(t - 330)$ so time to hit ground must be 330 seconds

43. (continued)

b) $t = \frac{-5280}{2(-16)} = 165$

$s_{max} = -16(165)^2 + 5280(165) = 435,600$ feet

45. metric: $s = -4.9t^2 + 200t$ (meters)

a) $0 = -4.9t^2 + 200t$

$= -t(4.9t - 200)$ so $4.9t = 200$ or $t \approx 40.8$ seconds

b) time for maximum height $= \frac{-200}{2(-4.9)} \approx 20.408$ seconds

maximum height $\approx -4.9(20.408)^2 + 200(20.408) \approx 2040.8$ meters

47. $h(t) = -1.96t^2 + 100t$ (meters)

time for maximum height $= \frac{-100}{2(-1.96)} \approx 25.51$ seconds

maximum height $\approx -1.96(25.51)^2 + 100(25.51) \approx 1276$ meters

49. $f(x) = 2.37x^2 - 1.08x - 31.24$

$x = \frac{-(-1.08)}{2(2.37)} \approx 0.2278$

$y \approx 2.37(0.2278)^2 - 1.08(0.2278) - 31.24 = -31.36304$

so the vertex is $(0.2278 , -31.36304)$ and the minimum value is about -31.36

51. $f(x) = 10^{-4}x^2 - 10^{-5}x + 2 \times 10^{-3}$

$x = \frac{-(-10^{-5})}{2(10^{-4})} = \frac{1(10^{-5})}{2(10^{-4})} = 0.5 \times 10^{-1} = 5 \times 10^{-2} = 0.05$

$$y = 10^{-4}(5 \times 10^{-2})^2 - 10^{-5}(5 \times 10^{-2}) + 2 \times 10^{-3}$$

$$= 25 \times 10^{-8} - 5 \times 10^{-7} + 2 \times 10^{-3}$$

$$= -2.5 \times 10^{-7} + 2 \times 10^{-3} \quad \text{which is still practically } 2 \times 10^{-3} \text{ , i.e., } \approx 0.002$$

(more precisely: $-0.00000025 + 0.002$ which equals 0.00199975)

so the vertex is approximately $(0.05, 0.002)$ and 0.002 is the minimum value.

Problems 4.2

1. $p(x) = \frac{1}{3}x^3$ (same basic shape as x^3 only "flatter")

    table: x: -2 -1 0 1 2   sketch:

    $p(x)$: $-\frac{8}{3}$ $-\frac{1}{3}$ 0 $\frac{1}{3}$ $\frac{8}{3}$

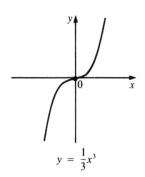

$$y = \frac{1}{3}x^3$$

3. $p(x) = \frac{-x^5}{10}$

                                sketch:

    table: x: -2 -1 0 1 2

           $p(x)$: 3.2 0.1 0 -0.1 -3.2

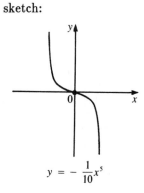

$$y = -\frac{1}{10}x^5$$

- 151 -

5. $p(x) = x^3 + 2$ (same as graph of x^3 only raised up 2 blocks)

table: x: -2 -1 0 1 2   sketch:

p(x): -6   1   2   3   10

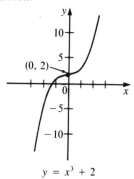

$$y = x^3 + 2$$

7. $p(x) = (x - 3)^3$ (graph of x^3 only shifted 3 blocks to the right!)

table: x: 1   2   3   4   5   [the x-values are "centered around"

p(x): -8 -1   0   1   8   the x-value 3 which makes (x - 3) = 0]

sketch:

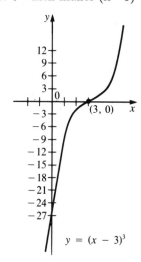

$$y = (x - 3)^3$$

9. $p(x) = x^4 + 1$ [same as graph of x^4 (which is close to graph of x^2) only raised

one block]

table: x: -2 -1 0 1 2   sketch:

p(x): 17   2   0   2   17

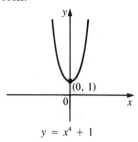

$$y = x^4 + 1$$

11. $p(x) = (x - 1)^4$ (same as graph of x^4 only shifted one block to the right)

   table: x: -1 0 1 2 3   (centered about 1)

   p(x): 16 1 0 1 16   sketch:

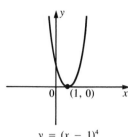

$$y = (x - 1)^4$$

13. $p(x) = -(x - 2)^3$ (graph of x^3 shifted 2 blocks to the right AND reflected about the horizontal axis)

   table: x: 0 1 2 3 4   sketch:

   p(x): 8 1 0 -1 -8

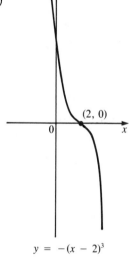

$$y = -(x - 2)^3$$

15. $p(x) = 2(x - 1)^4 - 5$ graph of x^4 shifted 1 block to the right [by the $(x - 1)$]

   5 blocks down [by the - 5]

   made steeper by the factor 2

   table: x: -1 0 1 2 3   sketch:

   p(x): 27 -3 -5 -3 27

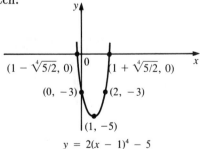

$$y = 2(x - 1)^4 - 5$$

17. $p(x) = 3^4(x + 1)^4$ graph of x^4 shifted 1 to the left and made ridiculously steep by the 81 factor

table: x: -3   -2 -1   0   1   sketch:

p(x): 1296 81   0 81 1296

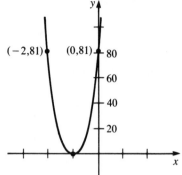

$y = (3x + 3)^4 = 81(x + 1)^4$

19. $p(x) = (x - 1)(x - 2)(x - 3)$

x-intercepts: set $p(x) = 0$ and solve to get x = 1, x = 2, x = 3

y-intercept: find $p(0) = (-1)(-2)(-3) = -6$

signs of p(x) on intervals created by x-intercepts:

interval	signs of factors	sign of p(x)	where graph is
$(-\infty,1)$	(-)(-)(-)	negative	below x-axis
(1,2)	(+)(-)(-)	positive	above x-axis
(2,3)	(+)(+)(-)	negative	below
$(3,\infty)$	(+)(+)(+)	positive	above

note: to determine the signs of the factors, simply pick a test value from somewhere in each interval. For example, in the interval $(-\infty,1)$ we could pick x = -5 and then check out the signs of (-5 - 3), (-5 - 2) and (-5 - 1) to see where the (-)(-)(-) in the table above comes from.

role of dominating term: if p(x) were multiplied out, the dominating (highest degreed) term would be x^3 so....

as x → ∞, x^3→ ∞ (graph goes up on right); as x → -∞, x^3→ -∞ (down on left!)

sketch:

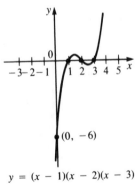

$y = (x - 1)(x - 2)(x - 3)$

21. $p(x) = (x - 4)(x + 5)(x - 6)$

x-intercepts: set $p(x) = 0$ to easily obtain $x = 4$, $x = -5$, and $x = 6$

y-intercept: $p(0) = (-4)(5)(-6) = 120$

signs:

interval	signs of factors	sign of p(x)	position of graph
$(-\infty,-5)$	$(-)(-)(-)$	negative	below x-axis
$(-5,4)$	$(-)(+)(-)$	positive	above
$(4,6)$	$(+)(+)(-)$	negative	below
$(6,\infty)$	$(+)(+)(+)$	positive	above

dominating term is x^3 and as $x \to \infty$, $x^3 \to \infty$; also as $x \to -\infty$, $x^3 \to -\infty$ so this graph goes up on the right and down on the left side.

sketch:

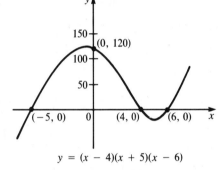

$y = (x - 4)(x + 5)(x - 6)$

23. $p(x) = -(x + 2)(x + 6)(x - 2)(x + 5)$

x-intercepts: -2, -6, 2, and -5

y-intercept: $p(0) = -(2)(6)(-2)(5) = 120$

signs:

interval	signs of factors	sign of p(x)	position of graph
$(-\infty,-6)$	$-(-)(-)(-)(-)$	negative	below x-axis
$(-6,-5)$	$-(-)(+)(-)(-)$	positive	above
$(-5,-2)$	$-(-)(+)(-)(+$	negative	below
$(-2,2)$	$-(+)(+)(-)(+$	positive	above
$(2,\infty)$	$-(+)(+)(+)(+)$	negative	below

dominating term is $-x^4$ which decreases without bound as x either increases

or decreases without bound; ie, as $x \to \infty$ and as $x \to -\infty$, $-x^4 \to -\infty$

sketch:

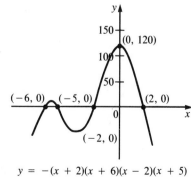

$y = -(x + 2)(x + 6)(x - 2)(x + 5)$

25. $p(x) = (x + 1)(x - 1)(x + 2)(x - 2)$

x-intercepts: -1, 1, -2 and 2

y-intercept: $(1)(-1)(2)(-2) = 4$

signs:

interval	signs of factors	sign of p(x)	position of graph
$(-\infty,-2)$	(-)(-)(-)(-)	positive	above x-axis
$(-2,-1)$	(-)(-)(+)(-)	negative	below
$(-1,1)$	(+)(-)(+)(-)	positive	above
$(1,2)$	(+)(+)(+)(-)	negative	below
$(2,\infty)$	(+)(+)(+)(+)	positive	above

dominating term is x^4 which approaches $+\infty$ as x approaches either $+\infty$ or $-\infty$

sketch:

(a) $-2, -1, 1, 2$ (b) positive $(-\infty, -2)$, $(-1, 1)$ and $(2, \infty)$; negative $(-2, -1)$ and $(1, 2)$

(c)

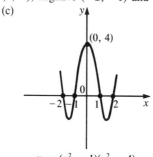

$y = (x^2 - 1)(x^2 - 4)$

27. $p(x) = x^3 - 9x^2 = x^2(x - 9)$

x intercepts: set $x^2(x - 9) = 0$; get $x^2 = 0$ which implies $x = 0$ or $x - 9 = 0$, $x = 9$

y-intercept: $p(0) = 0$ (graph goes through the origin)

signs:

interval	signs of factors	sign of p(x)	position of graph
$(-\infty,0)$	(+)(-)	negative	below x-axis
$(0,9)$	(+)(-)	negative	below again
$(9,\infty)$	(+)(+)	positive	above

dominating term is x^3 and as $x \to \infty$, $x^3 \to \infty$; also as $x \to -\infty$, $x^3 \to -\infty$

sketch: (a) 0, 9 (b) positive $(9, \infty)$; negative $(-\infty, 0)$ and $(0, 9)$

(c)

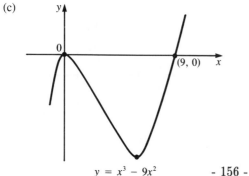

$y = x^3 - 9x^2$

- 156 -

29. $p(x) = x^3 + 6x^2 + 8x = x(x^2 + 6x + 8) = x(x + 4)(x + 2)$

x-intercepts: $x = 0$, $x = -4$, $x = -2$ by setting factors to zero

y-intercept: $p(0) = 0$

signs:

interval	signs of factors	sign of p(x)	position of graph
$(-\infty,-4)$	$(-)(-)(-)$	negative	below x-axis
$(-4,-2)$	$(-)(+)(-)$	positive	above
$(-2,0)$	$(-)(+)(+)$	negative	below
$(0,\infty)$	$(+)(+)(+)$	positive	above

dominating term is x^3; as $x \rightarrow \infty$, $x^3 \rightarrow \infty$ and as $x \rightarrow -\infty$, $x^3 \rightarrow -\infty$

sketch: (a) $-4, -2, 0$ (b) positive $(-4, -2)$ and $(0, \infty)$; negative $(-\infty, -4)$ and $(-2, 0)$

(c)

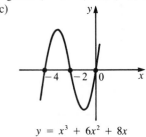

$y = x^3 + 6x^2 + 8x$

31. $p(x) = x^2(x^2 - 3x + 2) = x^2(x - 2)(x - 1)$

x-intercepts: $x = 0$, $x = 2$, $x = 1$

y-intercept: $p(0) = 0$

signs:

interval	signs of factors	sign of p(x)	position of graph
$(-\infty,0)$	$(+)(-)(-)$	positive	above x-axis
$(0,1)$	$(+)(-)(-)$	positive	above again
$(1,2)$	$(+)(-)(+)$	negative	below
$(2,\infty)$	$(+)(+)(+)$	positive	above

dominating term is x^4 which approaches $+\infty$ as x approaches either $\pm\infty$

sketch: (a) $0, 1, 2$ (b) positive $(-\infty, 0)$, $(0, 1)$ and $(2, \infty)$; negative $(1, 2)$

(c)

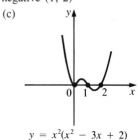

$y = x^2(x^2 - 3x + 2)$

33. $p(x) = x^3 + x = x(x^2 + 1)$ sketch:

x- and y-intercept is the origin (0,0)

since 0 is a root of p(x)=0 and p(0) = 0

table: x: -2 -1 1 2

   p(x): -10 -2 2 10

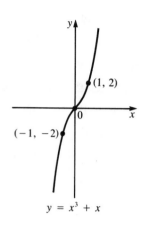

$y = x^3 + x$

35. $p(x) = x^3 + x^2 + 5x = x(x^2 + x + 5)$ sketch:

x and y-intercept is (0,0)

table: x: -2 -1 1   2

   p(x): -14 -5 7   22

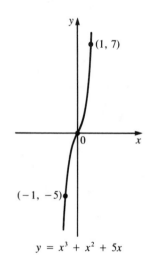

$y = x^3 + x^2 + 5x$

37. $p(x) = x^4 + 3x^2 + 2 = (x^2 + 2)(x^2 + 1)$

x-intercepts: NONE since none of the factors have real roots

y-intercept is 2 sketch:

table: x: -2 -1 1 2

   p(x): 30 6 6 30

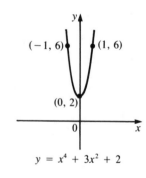

$y = x^4 + 3x^2 + 2$

39. $p(x) = x^3 + 1 = (x + 1)(x^2 - x + 1)$

x-intercept: set p(x) = 0 ; this will happen if x + 1 = 0 so x = -1 is it

y-intercept: p(0) = 1

39. (continued)   sketch:

    table: x: -2   1   2

    p(x): -7   2   9

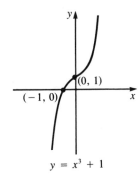

$y = x^3 + 1$

41. $p(x) = 4x^3 - 5x^2 + 4x - 7$

   $p(1) = 4 - 5 + 4 - 7 = -4$ while $p(2) = 32 - 20 + 8 - 7 = 13$ and since $p(1)$ is

   negative and $p(2)$ is positive, there must be some number r between 1 and 2

   for which $p(r)$ will be 0.

      some trials: $p(1.4) = -0.223$ and $p(1.45) = +0.482$ so it's between 1.4 and 1.45

         $p(1.42) = 0.05$ (close enough!) so the estimated zero is 1.42

43. $p(x) = 3x^5 + x^4 - 9x^2 + 3x - 4$

   $p(1) = 3 + 1 - 9 + 3 - 4 = -6$ while $p(2) = 96 + 16 - 36 + 6 - 4 = 78$

   These results have opposite signs so there must be a root between 1 & 2

      some trials: $p(1.3) = -1.31$

         $p(1.4) = 2.5$ so the root is between 1.3 and 1.4

         $p(1.35) = 0.42$ so the root is between 1.3 and 1.35

         $p(1.34) = 0.04$ (close enough!)

         so the estimated zero is 1.34

45. If - 4 and 7 are the only zeros of a polynomial then that polynomial must consist solely of the

   factors $(x + 4)$ and $(x - 7)$. Therefore $p(x) = (x + 4)(x - 7) = x^2 - 3x - 28$.

47. {0, 1, 2} being the only zeros implies (x), $(x - 1)$, and $(x - 2)$ are the only factors and then we can

   say that $p(x) = x(x - 1)(x - 2) = x(x^2 - 3x + 2) = x^3 - 3x^2 + 2x$.

49. $\{-1, 1, 1 + \sqrt{3}, 1 - \sqrt{3}\}$ as zeros $\Rightarrow$ $p(x) = (x + 1)(x - 1)\left(x - (1 + \sqrt{3})\right)\left(x - (1 - \sqrt{3})\right)$

$$= (x^2 - 1)\left(\left((x - 1) - \sqrt{3}\right)\left((x - 1) + \sqrt{3}\right)\right)$$

$$= (x^2 - 1)\left((x - 1)^2 - 3\right)$$

$$= (x^2 - 1)(x^2 - 2x - 2)$$

$$= x^4 - 2x^3 - 2x^2 - x^2 + 2x + 2 = x^4 - 2x^3 - 3x^2 + 2x + 2$$

51. $\{\pm\sqrt{2}, \pm\sqrt{3}, \pm\sqrt{5}\}$ as zeros $\Rightarrow (x - \sqrt{2})(x + \sqrt{2})(x - \sqrt{3})(x + \sqrt{3})(x - \sqrt{5})(x + \sqrt{5}) = p(x)$

so $p(x) = (x^2 - 2)(x^2 - 3)(x^2 - 5) = (x^2 - 2)(x^4 - 8x^2 + 15)$

$$= x^6 - 8x^4 + 15x^2 - 2x^4 + 16x^2 - 30$$

$$= x^6 - 10x^4 + 31x^2 - 30$$

53. Degree is n so there must be n factors; k is the only zero so each factor is of the form $(x - k)$; therefore we can say that $p(x) = (x - k)(x - k) \cdots (x - k)$ \qquad n times

$$= (x - k)^n$$

55. Know: p(x) is a third degree polynomial

   $\{-3, 0, \text{ and } 2\}$ are only zeros and $p(1) = 1$

   Leading coefficient is not necessarily 1; let's call it a.

   So, $p(x) = a(x + 3)(x)(x - 2) = ax(x^2 + x - 6) = ax^3 + ax^2 - 6ax$

   Since $p(1) = 1$ we can substitute as follows:

   $$p(1) = a(1)^3 + a(1)^2 - 6a(1) = -4a \text{ and this must equal 1 by the given fact so } a = -\frac{1}{4}$$

   and $p(x) = -\frac{1}{4}x(x + 3)(x - 2)$

57. $q(p) = 1000 - 5p - 0.1p^3$ where $p \geq 0$

   some trials: (done most efficiently with a programmable calculator)

   $$p(20) = 100$$

   $$p(21) = -31.1 \text{ so it's between 20 and 21}$$

   $$p(20.77) = 0.14 \text{ (best we can do to two decimal places)}$$

59. $P_3(x) = \frac{5x^3 - 3x}{2} = \frac{x(5x^2 - 3)}{2}$ will equal zero if $x = 0$ or $5x^2 - 3 = 0$. The latter condition is

   true if $x = \pm\frac{\sqrt{3}}{\sqrt{5}} = \pm\frac{\sqrt{15}}{5} \approx \pm 0.775$. So the x-intercepts are 0 and $\pm\frac{\sqrt{15}}{5}$. Y-int. $= 0$.

59. (continued)   finding out where $P_3(x) = \dfrac{x(5x^2 - 3)}{2}$ is positive and negative:

   interval   signs of factors   sign of expression   position of graph

interval	signs of factors	sign of expression	position of graph	
$(-1, -\frac{\sqrt{15}}{5})$	$(-)(+)$	negative	below x-axis	(test value here was - 0.8)
$(-\frac{\sqrt{15}}{5}, 0)$	$(-)(-)$	positive	above x-axis	(- 0.5)
$(0, \frac{\sqrt{15}}{5})$	$(+)(-)$	negative	below	(0.5)
$(\frac{\sqrt{15}}{5}, 1)$	$(+)(+)$	positive	above	(0.8)

61. We are given the relationship $(n+1)P_{n+1}(x) + nP_{n-1}(x) = (2n+1)xP_n(x)$ which relates 3 consecutive Legendre polynomials. Since we know $P_3(x)$ and $P_4(x)$ and wish to find $P_2(x)$, we should let $n = 3$ in the above relationship:

   we'd get: $4P_4(x) + 3P_2(x) = 7xP_3(x)$ and we can solve this for $P_2(x)$ to get

$$P_2(x) = \frac{7xP_3(x) - 4P_4(x)}{3} = \frac{1}{3}\left[7x\left(\frac{5x^3 - 3x}{2}\right) - 4\left(\frac{35x^4 - 30x^2 + 3}{8}\right) \right]$$

$$= \frac{1}{3}\left[\frac{35x^4 - 21x^2}{2} - \frac{35x^4 - 30x^2 + 3}{2} \right]$$

$$= \frac{1}{3}\left[\frac{9x^2 - 3}{2} \right] = \frac{3x^2 - 1}{2}$$

   sketch:

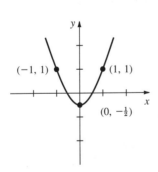

63.   a.)   Let s = one of the sides of one of the square ends.

   Then s + 2 = the length of the box.

   Since volume = length X width X height for a box, substituting in our symbols leads to

   volume = $(s + 2)(s)(s) = s^3 + 2s^2$

b.) The domain for s for which the graph would make sense is s > 0; Plotting a few points

(e.g., for s = 1/2, 1, 2,...) leads to the sketch:

(a) $V(s) = s^2(s + 2) = s^3 + 2s^2,\ s \geq 0$

(b)

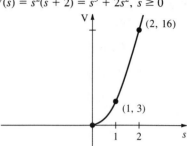

Problems 4.3

1.

$$
\begin{array}{r}
x - 2 \\
x - 1 \overline{\smash{\big)}\ x^2 - 3x + 2} \\
\underline{x^2 - x} \\
-2x + 2 \\
\underline{-2x + 2}
\end{array}
$$

(since we're subtracting, this row becomes "$-x^2 + x$")

(visualize this row as "$+2x - 2$")

so: $q(x) = x - 2$ and $r(x) = 0$

3.

$$
\begin{array}{r}
3x + 4 \\
2x - 5 \overline{\smash{\big)}\ 6x^2 - 7x - 20} \\
\underline{6x^2 - 15x} \\
+8x - 20 \\
\underline{+8x - 20}
\end{array}
$$

so $q(x) = 3x + 4$

$r(x) = 0$

5.

$$
\begin{array}{r}
x + 3 \\
2x + 1 \overline{\smash{\big)}\ 2x^2 + 7x - 4} \\
\underline{2x^2 + x} \\
6x - 4 \\
\underline{6x + 3} \\
-7
\end{array}
$$

so $q(x) = x + 3$

$r(x) = -7$

7.
$$\begin{array}{r} 4x + 8 \\ 3x - 5 \overline{\smash{\big)}\ 12x^2 + 4x + 40} \\ \underline{12x^2 - 20x} \\ 24x + 40 \\ \underline{24x - 40} \\ 80 \end{array}$$

so $q(x) = 4x + 8$

$r(x) = 80$

9.
$$\begin{array}{r} 2x \\ 4x + 2 \overline{\smash{\big)}\ 8x^2 + 4x - 1} \\ \underline{8x^2 + 4x} \\ -1 \end{array}$$

so $q(x) = 2x$

$r(x) = -1$

11.
$$\begin{array}{r} x^2 + x + 4 \\ x + 3 \overline{\smash{\big)}\ x^3 + 4x^2 + 7x + 12} \\ \underline{x^3 + 3x^2} \\ 1x^2 + 7x \\ \underline{x^2 + 3x} \\ 4x + 12 \\ \underline{4x + 12} \end{array}$$

so $q(x) = x^2 + x + 4$ and $r(x) = 0$

13.
$$\begin{array}{r} 2x^2 + 9x + 21 \\ x - 2 \overline{\smash{\big)}\ 2x^3 + 5x^2 + 3x + 35} \\ \underline{2x^3 - 4x^2} \\ 9x^2 + 3x \\ \underline{9x^2 - 18x} \\ 21x + 35 \\ \underline{21x - 42} \\ 77 \end{array}$$

so $q(x) = 2x^2 + 9x + 21$ and $r(x) = 77$

15.
$$
\begin{array}{r}
x^2 - (1/2)x + 3/4 \\
2x + 1 \overline{) 2x^3 + 0x^2 + 1x + 3} \\
\underline{2x^3 + 1x^2} \\
-1x^2 + 1x \\
\underline{-1x^2 - (1/2)x} \\
(3/2)x + 3 \\
\underline{(3/2)x + 3/4} \\
2\tfrac{1}{4} \text{ or } \tfrac{9}{4}
\end{array}
$$
so $q(x) = x^2 - \tfrac{1}{2}x + \tfrac{3}{4}$ and $r(x) = \tfrac{9}{4}$

17.
$$
\begin{array}{r}
x \\
x^3 + 1 \overline{) x^4 + 0x^3 + 1x^2 + 0x + 0} \\
\underline{x^4 \qquad\qquad + 1x} \\
1x^2 \qquad -x
\end{array}
$$
so $q(x) = x$ and $r(x) = x^2 - x$

(done since degree is lower than divisor)

19.
$$
\begin{array}{r}
4x^2 + 2x - 17 \\
x^2 - x + 5 \overline{) 4x^4 - 2x^3 + x^2 - x + 4} \\
\underline{4x^4 - 4x^3 + 20x^2} \\
2x^3 - 19x^2 - x \\
\underline{2x^3 - 2x^2 + 10x} \\
-17x^2 - 11x + 4 \\
\underline{-17x^2 + 17x - 85} \\
-28x + 89
\end{array}
$$
so $q(x) = 4x^2 + 2x - 17$

$r(x) = -28x + 89$

21. represent $x^2 - 5x + 6$ as "1 -5 6" and represent $x - 3$ as "3"; dividing synthetically we have:

$$
\begin{array}{r}
3 \,\overline{)\, 1 \quad -5 \quad\; 6} \\
\underline{ 3 \;\; -6} \\
1 \quad -2 \quad \boxed{0}
\end{array}
$$

piecing the answer together:

The "1 -2" are the coefficients of $q(x)$ which must have a degree 1 less than the degree of the original dividend, in this case 2. So $q(x)$ must be a polynomial of degree $(2 - 1)$ or 1 and have coefficients 1 and -2 so $q(x)$ must be $1x - 2$. The number in the box is the remainder and in this case it is 0.

23. $(5x^2 - 4x + 3) \div (x + 5)$:

$$-5 \overline{)\;5 \quad -4 \quad 3}$$
$$\underline{\quad\; -25 \quad 145}$$
$$5 \quad -29 \quad \boxed{148}$$

so $q(x) = 5x - 29$ (one less in degree than $5x^2 - 4x + 3$) and $r(x) = 148$

25. $(x^4 + 16) \div (x + 2)$:

$$-2 \overline{)\;1 \quad 0 \quad 0 \quad 0 \quad 16}$$
$$\underline{\quad\; -2 \quad 4 \quad -8 \quad 16}$$
$$1 \quad -2 \quad 4 \quad -8 \quad \boxed{32} \quad \text{so } q(x) = 1x^3 - 2x^2 + 4x - 8$$
$$\text{and } r(x) = 32$$

27. $(2x^3 - 5x^2 + 3x - 4) \div (x - 3)$:

$$3 \overline{)\;2 \quad -5 \quad 3 \quad -4}$$
$$\underline{\quad\quad\; 6 \quad 3 \quad 18}$$
$$2 \quad 1 \quad 6 \quad \boxed{14} \quad \text{so } q(x) = 2x^2 + 1x + 6$$
$$\text{and } r(x) = 14$$

29. $(2x^5 - 1) \div (x - 1)$:

$$1 \overline{)\;2 \quad 0 \quad 0 \quad 0 \quad 0 \quad -1}$$
$$\underline{\quad\;\; 2 \quad 2 \quad 2 \quad 2 \quad 2}$$
$$2 \quad 2 \quad 2 \quad 2 \quad 2 \quad \boxed{1} \quad q(x) = 2x^4 + 2x^3 + 2x^2 + 2x + 2$$
$$r(x) = 1$$

31. $(-3x^4 - 2x^3 + 3) \div (x + \tfrac{1}{2})$:

$$-\tfrac{1}{2} \overline{)\;-3 \quad -2 \quad 0 \quad 0 \quad 3}$$
$$\underline{\qquad\quad \tfrac{3}{2} \quad \tfrac{1}{4} \quad -\tfrac{1}{8} \quad \tfrac{1}{16}}$$
$$-3 \quad -\tfrac{1}{2} \quad \tfrac{1}{4} \quad -\tfrac{1}{8} \quad \boxed{\tfrac{49}{16}}$$

so $q(x) = -3x^3 - \tfrac{1}{2}x^2 + \tfrac{1}{4}x - \tfrac{1}{8}$ and $r(x) = \tfrac{49}{16}$

33. $(x^5 + x^4 + x^3 + x^2 + x + 1) \div (x - \tfrac{1}{2})$:

$$\tfrac{1}{2} \overline{)\;1 \quad 1 \quad 1 \quad 1 \quad 1 \quad 1}$$
$$\underline{\qquad\; 1/2 \quad 3/4 \quad 7/8 \quad 15/16 \quad 31/32}$$
$$1 \quad 3/2 \quad 7/4 \quad 15/8 \quad 31/16 \quad \boxed{63/32}$$

so $q(x) = x^4 + \tfrac{3}{2}x^3 + \tfrac{7}{4}x^2 + \tfrac{15}{8}x + \tfrac{31}{16}$ and $r(x) = \tfrac{63}{32}$

35. $(x^{10} - 1) \div (x - 1)$:

$$1 \overline{)\ 1\ \ 0\ \ 0\ \ 0\ \ 0\ \ 0\ \ 0\ \ 0\ \ 0\ \ 0\ \ -1}$$

$$\underline{\ \ \ \ \ \ \ 1\ \ 1\ \ 1\ \ 1\ \ 1\ \ 1\ \ 1\ \ 1\ \ 1\ \ 1}$$

$$1\ \ 1\ \ 1\ \ 1\ \ 1\ \ 1\ \ 1\ \ 1\ \ 1\ \ 1\ \ \boxed{0}$$

so $q(x) = x^9 + x^8 + x^7 + x^6 + x^5 + x^4 + x^3 + x^2 + x + 1$ and $r(x) = 0$

37. $(x^4 + x^2 - 3) \div (x + \frac{3}{4})$:

$$-\tfrac{3}{4} \overline{)\ 1\ \ \ \ \ \ 0\ \ \ \ \ \ \ 1\ \ \ \ \ \ \ \ 0\ \ \ \ \ \ \ \ \ \ -3}$$

$$\underline{\ \ \ \ \ \ \ \ -3/4\ \ 9/16\ \ -75/64\ \ \ 225/256}$$

$$1\ \ -3/4\ \ 25/16\ \ -75/64\ \ \boxed{-543/256}$$

so $q(x) = x^3 - \frac{3}{4}x^2 + \frac{25}{16}x - \frac{75}{64}$ and $r(x) = -\frac{543}{256}$

39. $(-3x^4 + 3x^3 - 2x^2 - 2x + 5) \div (x + 2)$:

$$-2 \overline{)\ -3\ \ \ 3\ \ \ -2\ \ \ -2\ \ \ 5}$$

$$\underline{\ \ \ \ \ \ \ \ \ \ 6\ \ \ -18\ \ \ 40\ \ -76}$$

$$-3\ \ \ 9\ \ \ -20\ \ \ 38\ \ \boxed{-71}$$

so $q(x) = -3x^3 + 9x^2 - 20x + 38$ and $r(x) = -71$

41. $(x^{200} + 1) \div (x + 1)$:

$$-1 \overline{)\ 1\ \ \ 0\ \ \ 0\ \ \ 0\ \ \ 0\ \ \ 0\ \cdots\ 0\ \ \ 1}$$

$$\underline{\ \ \ \ \ \ \ -1\ \ \ 1\ \ -1\ \ \ 1\ \ -1\cdots -1\ \ \ 1}$$

$$1\ \ -1\ \ \ 1\ \ -1\ \ \ 1\ \ \ -1\cdots -1\ \ \boxed{2}$$

so $q(x) = x^{199} - x^{198} + x^{197} - \cdots -x^2 + x - 1$ and $r(x) = 2$

43. $p(x) = x^3 + 2x^2 - x + 3$ and we want $p(-2)$ which is the remainder when $p(x)$ is divided by $x + 2$ which means we do synthetic division with -2 and note the remainder:

$$-2 \overline{)\ 1\ \ \ 2\ \ \ -1\ \ \ 3}$$

$$\underline{\ \ \ \ \ \ \ \ -2\ \ \ 0\ \ \ 2}$$

$$1\ \ \ 0\ \ \ -1\ \ \boxed{5}\quad \text{so } p(-2) = 5$$

45. $p(x) = 5x^3 - 2x^2 + 4x + 9$, $p(-3)$:

$$-3 \overline{)\ 5\ \ \ -2\ \ \ 4\ \ \ \ \ \ 9}$$

$$\underline{\ \ \ \ \ \ \ \ \ -15\ \ \ 51\ \ \ -165}$$

$$5\ \ -17\ \ \ 55\ \ \boxed{-156}\quad \text{so } p(-3) = -156$$

47. $p(x) = x^5 + 2x^3 - 5x + 1$ and we want $p(2)$:

$$2 \overline{)\begin{array}{cccccc} 1 & 0 & 2 & 0 & -5 & 1 \end{array}}$$
$$\begin{array}{cccccc} & 2 & 4 & 12 & 24 & 38 \end{array}$$
$$\begin{array}{cccccc} 1 & 2 & 6 & 12 & 19 & \boxed{39} = p(2) \end{array}$$

49. $p(x) = 3x^4 + 2x^3 + 1$ and we want $p(-\frac{1}{2})$:

$$-\frac{1}{2} \overline{)\begin{array}{ccccc} 3 & 2 & 0 & 0 & 1 \end{array}}$$
$$\begin{array}{ccccc} & -\frac{3}{2} & -\frac{1}{4} & \frac{1}{8} & -\frac{1}{16} \end{array}$$
$$\begin{array}{ccccc} 3 & \frac{1}{2} & -\frac{1}{4} & \frac{1}{8} & \boxed{\frac{15}{16}} = p\left(-\frac{1}{2}\right) \end{array}$$

51. Since 2,3, and -1 are zeros, $(x - 2)$, $(x - 3)$, and $(x + 1)$ must be factors.   We could say that

$p(x) = a(x - 2)(x - 3)(x + 1)$ and then try to find a. That's what $p(0) = 12$ will help us do:

$p(0) = a(-2)(-3)(1) = 6a = 12$ so $a = 2$ which means that $p(x) = 2(x - 2)(x - 3)(x + 1)$.

53. $p(x) = a(x - 2)(x - i)(x + i) = a(x - 2)(x^2 + 1)$

Since $p(0) = \frac{1}{4}$ we can say that $p(0) = a(-2)(1) = -2a$ which must $= \frac{1}{4}$ so $a = -\frac{1}{8}$ which means

that $p(x) = -\frac{1}{8}(x - 2)(x^2 + 1)$.

55. $p(x) = a(x - 3)(x + 2)(x)$; since $p(1) = 6$ we know that $a(-2)(3)(1) = -6a = 6$

so $a = -1$ and $p(x) = -1(x - 3)(x + 2)(x)$ or $-x^3 + x^2 + 6x$.

57. $p(x) = a(x - 1)(x + 2)(x - 3)(x + 4)$; $p(0) = a(-1)(2)(-3)(4) = 24a$ which must $= 2$ so $a = \frac{1}{12}$

and $p(x) = \frac{1}{12}(x - 1)(x + 2)(x - 3)(x + 4)$.

59. let b stand for the unknown zero (there must be 3 zeros all together)

$p(x) = a(x - 1)(x - 2)(x - b)$

$p(0) = a(-1)(-2)(-b) = -2ab = 2$

$p(3) = a(2)(1)(3 - b) = 6a - 2ab = 14$

substituting the fact that $-2ab = 2$ (from the $p(0)$ equation) into the $p(3)$

equation gives us $6a + 2 = 14$ or $6a = 12$ or $a = 2$.

if $a = 2$, $-2ab = 2$ becomes $-2(2)b = 2$ so $b = -\frac{1}{2}$

Therefore, $p(x) = 2(x - 1)(x - 2)(x + \frac{1}{2})$

61. $0.45 \overline{)\ 1 \quad +3 \quad\quad -4}$

$\quad\quad\quad\quad \underline{0.45 \quad 1.5525}$

$\quad\quad\quad 1 \quad 3.45 \quad \boxed{-2.4475} = p(0.45)$

63. $4.58 \overline{)\ 3 \quad -2 \quad\quad 1 \quad\quad\quad -1}$

$\quad\quad\quad\quad \underline{13.74 \quad 53.7692 \quad\quad 250.84294}$

$\quad\quad\quad 3 \quad 11.74 \quad 54.7692 \quad\quad \boxed{249.84294} = p(4.58)$

65. $-0.177 \overline{)\ -0.01 \quad 0.126 \quad\quad 0 \quad\quad\quad\quad -5.84}$

$\quad\quad\quad\quad\quad \underline{0.00177 \quad -0.0226153 \quad 0.0040029}$

$\quad\quad\quad -0.01 \quad 0.12777 \quad\quad -0.0226153 \boxed{-5.8359971} = p(-0.177)$

67. $p(x) = 4x^3 - 7x^2 + 5x + 3;\ x = 2.8$

First step is to progressively factor out x's:

$p(x) = 4x^3 - 7x^2 + 5x + 3 = (4x^2 - 7x + 5)x + 3 = [(4x - 7)x + 5]x + 3$

Now use whatever process necessary to store 2.8 in your calculator's memory; quite possibly it's to enter 2.8 and then hit $\boxed{\text{sto}}$. Once this is accomplished it's easy to evaluate p(2.8):

Use the final factored form, start in the innermost group and try to remember that each time you encounter x, you can just recall it from your calculators memory (here I used $\boxed{\text{mr}}$ for "memory recall"; here are the keystrokes for many calculators and some intermediate results:

$\boxed{4}\ \boxed{\text{X}}\ \boxed{\text{mr}}\ \boxed{-}\ \boxed{7}\ \boxed{=}$ (should see 4.2)

$\boxed{\text{X}}\ \boxed{\text{mr}}\ \boxed{+}\ \boxed{5}\ \boxed{=}$ (should see 16.76)

$\boxed{\text{X}}\ \boxed{\text{mr}}\ \boxed{+}\ \boxed{3}\ \boxed{=}$ final result is 49.928 = p(2.8)

69. Use exactly the same keystrokes as in problem 67 only store 2.74×10^{12} first; here are the strokes and some intermediate results:

  1.096   13   (ie., 1.096×10^{13})

  3.003   25

  8.2283   37   which is the desired result.

Note: to store the value in scientific notation, I had to enter $2.74\ \boxed{\text{X}}\ 12\ \boxed{10^{\text{X}}}\ \boxed{=}\ \boxed{\text{sto}}$

71. $p(x) = -2x^4 - x^3 + 5x^2 + 12x - 8 = (-2x^3 - x^2 + 5x + 12)x - 8 = [(-2x^2 - x + 5)x + 12]x - 8$
$$= \{[(-2x - 1)x + 5]x + 12\}x - 8$$

We wish to find p(2) so store 2 (2 $\boxed{\text{sto}}$); the keystrokes and intermediate results:

$\boxed{2}$ $\boxed{\pm}$ $\boxed{\text{X}}$ $\boxed{\text{mr}}$ $\boxed{-}$ $\boxed{1}$ $\boxed{=}$ (should see - 5)

           $\boxed{\text{X}}$ $\boxed{\text{mr}}$ $\boxed{+}$ $\boxed{5}$ $\boxed{=}$ (- 5)

           $\boxed{\text{X}}$ $\boxed{\text{mr}}$ $\boxed{+}$ $\boxed{12}$ $\boxed{=}$ (2)

           $\boxed{\text{X}}$ $\boxed{\text{mr}}$ $\boxed{-}$ $\boxed{8}$ $\boxed{=}$ results in - 4 which must equal p(2).

73. Use the same keystrokes as in problem 71 only store 0.0182 initially. Your results along the way:

     - 1.0364, 4.9811375, 12.090657, and finally p(0.0182) is seen to be - 7.7799501.

75. $p(x) = 8x^5 - 6x^3 + x^2 - 9 = (8x^3 - 6x + 1)x^2 - 9 = [(8x^2 - 6)x + 1]x^2 - 9$

     (a few modifications are necessary here) store 4.3125×10^{-7} first; keystrokes and results:

           $\boxed{8}$ $\boxed{\text{X}}$ $\boxed{\text{mr}}$ $\boxed{x^2}$ $\boxed{-}$ $\boxed{6}$ $\boxed{=}$ (may see - 6; the other stuff is so small!)

           $\boxed{\text{X}}$ $\boxed{\text{mr}}$ $\boxed{+}$ $\boxed{1}$ $\boxed{=}$ (how about 0.9999975 ?)

           $\boxed{\text{X}}$ $\boxed{\text{mr}}$ $\boxed{x^2}$ $\boxed{-}$ $\boxed{9}$ $\boxed{=}$ after which you'll probably just see - 9 which is $\approx p(4.3125 \times 10^{-7})$

77. If x = 1 is a zero of $x^3 + x^2 - 6x + 4$ (x - 1) is a factor of it.

Dividing by (x - 1) synthetically will produce the other factor which we

can search for zeros:

```
   1 ) 1   1   -6    4
              1   2   -4
       1   2   -4   0
```

So this other factor is $x^2 + 2x - 4$ which does not factor so the equation

$x^2 + 2x - 4 = 0$ is solved by the quadratic formula:

$$x = \frac{-(2) \pm \sqrt{(2)^2 - 4(1)(-4)}}{2(1)} = \frac{-2 \pm \sqrt{20}}{2} = -1 \pm \sqrt{5} \text{ and these are the other two zeros.}$$

79.
```
   1/2 ) -8   12   -14    5
               -4    4   -5
        -8    8   -10   0
```
so now we must solve $-8x^2 + 8x - 10 = 0$

         divide both sides by -2 to get $4x^2 - 4x + 5 = 0$

79. (continued) $x = \dfrac{4 \pm \sqrt{16 - 4(4)(5)}}{2(4)} = \dfrac{4 \pm \sqrt{-64}}{8} = \dfrac{4 \pm 8i}{8} = \dfrac{1}{2} \pm i$

81.

$-1 \overline{)\ -2\quad 3\quad -1\quad -6}$

$\ \ 2\quad -5\quad 6$

$\ -2\quad 5\quad -6\quad \boxed{0}$ so look at $-2x^2 + 5x - 6 = 0$ or $2x^2 - 5x + 6 = 0$

$$x = \frac{5 \pm \sqrt{25 - 4(2)(6)}}{2(2)} = \frac{5 \pm \sqrt{-23}}{4} = \frac{5}{4} \pm \frac{\sqrt{23}}{4}i$$

83.

$4 \overline{)\ 1\quad -6\quad 16\quad -32}$

$\ \ 4\quad -8\quad 32$

$\ 1\quad -2\quad 8\quad \boxed{0}$ now, $x^2 - 2x + 8 = 0$ has as its zeros:

$$x = \frac{2 \pm \sqrt{4 - 4(1)(8)}}{2} = \frac{2 \pm \sqrt{-28}}{2} = 1 \pm \sqrt{7}\, i$$

85.

$0.37 \overline{)\ 1\quad -1.82\quad -5.14\quad 2.100305}$

$\ \ 0.37\quad -0.5365\quad -2.100305$

$\ 1\quad -1.45\quad -5.6765\quad \boxed{0}$ now solve $x^2 - 1.45x - 5.6765 = 0$

$$x = \frac{1.45 \pm \sqrt{2.1025 - 4(1)(-5.6765)}}{2}$$

$$= \frac{1.45 \pm \sqrt{24.8085}}{2} = 3.215407 \text{ or } -1.765407$$

87. First divide by $x - 1$:

$1 \overline{)\ 1\quad 1\quad -13\quad -1\quad 12}$

$\ 1\quad 2\quad -11\quad -12$

$\ 1\quad 2\quad -11\quad -12\quad \boxed{0}$

Now divide the result by $x + 1$:

$-1 \overline{)\ 1\quad 2\quad -11\quad -12}$

$\ 2\quad -10\quad 12$

$\ 1\quad 1\quad -12\quad \boxed{0}$

Now solve $x^2 + x - 12 = 0$ by factoring it into $(x - 3)(x + 4) = 0$

So $x = 3$ and $x = -4$ are the last remaining zeros.

89.

$$3 \overline{) \; 2 \quad 3 \quad -23 \quad -27 \quad 45}$$

$$\underline{\quad \quad 6 \quad 27 \quad 12 \quad -45}$$

$$2 \quad 9 \quad 4 \quad -15 \quad \boxed{0}$$

$$-3 \overline{) \; 2 \quad 9 \quad 4 \quad -15}$$

$$\underline{\quad \quad -6 \quad -9 \quad 15}$$

$$2 \quad 3 \quad -5 \quad \boxed{0}$$

solving $2x^2 + 3x - 5 = 0$ yields $(2x + 5)(x - 1) = 0$ so $x = -\frac{5}{2}$ and $x = 1$.

91. Carry on with synthetic division and then "insist" that the remainder be 0.

$$2 \overline{) \; 1 \quad 3 \quad b \quad \quad -8}$$

$$\underline{\quad \quad 2 \quad 10 \quad \quad 20 + 2b}$$

$$1 \quad 5 \quad (10+b) \quad \boxed{(12+2b)}$$

now set $12 + 2b = 0$ which means that $2b = -12$ so $b = -6$

93.

$$1 \overline{) \; 2 \quad 0 \quad 0 \quad -b \quad \quad 4}$$

$$\underline{\quad \quad 2 \quad 2 \quad 2 \quad (2 - b)}$$

$$2 \quad 2 \quad 2 \quad (2 - b) \quad \boxed{(6 - b)}$$

since we want the remainder to be 2

set $6 - b = 2$; $4 = b$

95. Try dividing by $x - r$ and then seeing if the remainder can be 0.

$$r \overline{) \; 1 \quad 0 \quad 1 \quad \quad 0 \quad \quad 1}$$

$$\underline{\quad \quad r \quad r^2 \quad (r^3 + r) \quad (r^4 + r^2)}$$

$$1 \quad r \quad (r^2+1) \quad (r^3 + r) \quad \boxed{(r^4 + r^2 + 1)}$$

The question now is: can $r^4 + r^2 + 1 = 0$?   Answer: It can't because x^2 and x^4 are non-negative and 1 is positive so their sum could never be 0. This means that $x - r$ cannot be a factor of $x^4 + x^2 + 1$.

97. Let $p(x) = x^n - a^n$

Since $p(a) = a^n - a^n = 0$, $x - a$ must be a factor of $p(x)$.

99. Let $p(x) = x^n + a^n$ where n is odd.

$p(-a) = (-a)^n + a^n = -a^n + a^n = 0$ so $x + a$ must be a factor of $p(x)$.

101. a) using synthetic substitution:

$$300 \overline{) \; -0.001 \quad 0.5 \quad 80 \quad \quad 10,000}$$

$$\underline{\quad \quad \quad \quad -0.3 \quad 60 \quad \quad 42,000}$$

$$-0.001 \quad 0.2 \quad 140 \quad 52,000 \quad \text{so } C(300) = 52,000 \text{ dollars}$$

101. (continued)

   b) We want $- 0.001q^3 + 0.5q^2 + 80q + 1000$ to $= 52,000$.

     In other words we must find a solution of $- 0.001q^3 + 0.5q^2 + 80q - 42,000 = 0$ other than 300.

     Scan values of the function $f(q) = - 0.001q^3 + 0.5q^2 + 80q - 42000$ and look for a change of sign.

     Note: a programmable calculator comes in handy for this!

     The closest whole number value for q came out to be 487 tons.

Problems 4.4

1. $p(x) = (x + 5)(x - 4)^2$

    $x = -5$ is a zero of multiplicity 1   (the exponent on $x + 5$)

    $x = 4$ is a zero of multiplicity 2   (the exponent on $x - 4$)

3. $p(x) = (x + 1)(x + 2)(x - 7)$ setting each factor to zero gives

    $x = -1$, $x = -2$, and $x = 7$ all zeros of multiplicity 1

5. $p(x) = (x^2 + 4)^5 = [(x + 2i)(x - 2i)]^5 = (x + 2i)^5(x - 2i)^5$ setting these factors to 0

    gives $x = - 2i$ and $x = 2i$ as zeros of multiplicity 5

7. $p(x) = (x^2 - 4)^3(x^2 - 3x - 18)^5 = [(x + 2)(x - 2]^3[(x + 3)(x - 6)]^5$
                 $= (x + 2)^3(x - 2)^3(x + 3)^5(x - 6)^5$

    so $x = - 2$ and $x = 2$ are zeros of multiplicity 3 and $x = - 3$ and $x = 6$ are zeros of multiplicity 5

9. $p(x) = x^2 + 2x + 3$ has no sign changes from term to term. Therefore it has NO positive zeros.

    $p(-x) = (-x)^2 + 2(-x) + 3 = x^2 - 2x + 3$ has 2 sign changes (1st & 2nd, 2nd & 3rd terms) so it

     either has two or no negative zeros. (just keep subtracting 2 from the number of sign changes

     until the result is ≤ 1)

     SHORT CUT: When determining $p(-x)$ even degreed terms remain unchanged in sign and odd

                  degreed terms will have the opposite sign from their counterpart in $p(x)$ and of

                  course constant terms will stay the same.

11. $p(x) = 7x^3 + 1$ has no sign changes so it has no positive zeros.

$p(-x) = 7(-x)^3 + 1 = -7x^3 + 1$ which has one sign change so it will have one negative zero.

13. $p(x) = 2x^3 - 3x^2 + 4x - 1$ has three changes of sign so $p(x)$ might have either three positive zeros or one positive zero.

$p(-x) = 2(-x)^3 - 3(-x)^2 + 4(-x) - 1 = -2x^3 - 3x^2 - 4x - 1$ has no change of sign so no negative zeros.

15. $p(x) = -3x^3 - 2x^2 - x - 8$ has no changes in sign so $p(x)$ has no positive zeros.

$p(-x) = 3x^3 - 2x^2 + x - 8$ has 3 changes in sign so $p(x)$ will have 3 or 1 negative zeros.

17. $p(x) = 7x^5 + x^3 + 4x + 3$ has no changes in sign so $p(x)$ has no positive zeros.

$p(-x) = -7x^5 - x^3 - 4x + 3$ has one change in sign so $p(x)$ will have one negative zero.

19. $p(x) = 2x^5 + 4x^4 - 3x^3 - 2x^2 - 5x + 1$ has 2 changes so $p(x)$ will either have 2 or 0 positive zeros.

$p(-x) = -2x^5 + 4x^4 + 3x^3 - 2x^2 + 5x + 1$ has 3 changes of sign so $p(x)$ has either 3 or 1 negative zeros.

21. $p(x) = x^4 + ax^2 + b$ has no changes in sign (a and b are positive) so $p(x)$ has no positive zeros.

$p(-x) = x^4 + ax^2 + b$ also has no changes in sign so $p(x)$ has no negative zeros.

$p(0) = b$ so 0 is not a zero. Therefore $p(x)$ has no real zeros.

23. $p(x) = x^3 - x^2 - 4x + 4$ factors of 4 (the constant): $\pm 4, \pm 2, \pm 1$

factors of 1 (the leading coefficient): ± 1

so $p(x)$ has as possible rational zeros: $\pm \frac{4}{1}, \frac{2}{1}$, or $\frac{1}{1}$ ie., $\pm 4, \pm 2$, or ± 1

dividing by $x - 1$ (ie, testing to see if $+1$ is a root):

$$
\begin{array}{r}
1 \,)\,\overline{1 \;\; -1 \;\; -4 \;\;\; 4} \\
\underline{1 \;\;\;\; 0 \;\; -4} \\
1 \;\;\; 0 \;\; -4 \;\; \boxed{0}
\end{array}
$$

Since division by $x - 1$ has produced a remainder of 0, $(x - 1)$ is a factor of $p(x)$. Another factor can be "decoded" from the leftmost three numbers 1 0 - 4 as $x^2 + 0 \cdot x - 4$ which can be factored as $(x - 2)(x + 2)$ So $p(x) = (x - 1)(x - 2)(x + 2)$ and has three real zeros: 1, 2, and - 2. (Note: the zeros 2 and - 2 are in the list and would also have been found by more synthetic divisions had the factor $x^2 - 4$ not been "decoded".

25. $p(x) = x^3 - 2x^2 - x + 2$ has $\frac{\pm 1, \pm 2}{1}$ or $+1, -1, +2, -2$ as its rational root candidates.

trying $+1$: $1\,\overline{)\,1\ \text{-}2\ \text{-}1\ \ 2}$

$\underline{\phantom{\text{-}2}\ 1\ \text{-}1\ \text{-}2}$

$1\ \text{-}1\ \text{-}2\ \boxed{0}$ great!

so $p(x) = (x - 1)(x^2 - x - 2) = (x - 1)(x - 2)(x + 1)$

and its real zeros are 1, 2 and -1.

27. $p(x) = 2x^3 + 11x^2 + 9x + 2$ has $\frac{\pm 1, \pm 2}{1, 2}$ or $+1, -1, +\frac{1}{2}, -\frac{1}{2}, +2, -2$ as its rational root possibilities.

note: since $p(x)$ has no changes in sign, no positive roots are possible!

trying -1: $-1\,\overline{)\,2\ \ 11\ \ 9\ \ \ 2}$

$\underline{\text{-}2\ \text{-}9\ \ 0}$

$2\ \ \ 9\ \ \ 0\ \ \boxed{2}$ (no go!)

trying $-\frac{1}{2}$: $-\frac{1}{2}\,\overline{)\,2\ \ 11\ \ 9\ \ \ 2}$

$\phantom{-\frac{1}{2}\,)\,2\ \ }\underline{\text{-}1\ \text{-}5\ \text{-}2}$

$\phantom{-\frac{1}{2}\,)\,}2\ \ 10\ \ 4\ \ \boxed{0}$ (ok!) so $p(x) = (x + \frac{1}{2})(2x^2 + 10x + 4)$

$$= (x + \tfrac{1}{2}) \cdot 2 \cdot (x^2 + 5x + 2)$$

$$= (2x + 1)(x^2 + 5x + 2)$$

finding the zeros of $x^2 + 5x + 2$ requires the quadratic formula which gives

$$x = \frac{-5 \pm \sqrt{5^2 - 4 \cdot 1 \cdot 2}}{2} = \frac{-5 \pm \sqrt{17}}{2}$$

so all in all, the real zeros of $p(x)$ are $-\frac{1}{2}, \frac{-5 \pm \sqrt{17}}{2}$

29. $p(x) = 6x^3 + 13x^2 + 32x + 5$ has $\frac{\pm 1, \pm 5}{1, 2, 3, 6}$ or $\pm 1, \pm \frac{1}{2}, \pm \frac{1}{3}, \pm \frac{1}{6}, \pm 5, \pm \frac{5}{2}, \pm \frac{5}{3}, \pm \frac{5}{6}$ as its rational root possibilities.

29. (continued

According to Desartes, testing positive numbers would be a waste as p(x) has no sign changes.

trying $-\frac{1}{6}$ (after many other tries) $-\frac{1}{6}$) $\overline{6 \quad 13 \quad 32 \quad 5}$

$$\underline{\quad -1 \quad -2 \quad -5 \quad}$$

$$6 \quad 12 \quad 30 \quad \boxed{0}$$

So at this point p(x) can be factored as $(x + 1/6)(6x^2 + 12x + 30)$. Factoring further gives $(x + 1/6) \cdot 6 \cdot (x^2 + 2x + 5)$ or $(6x + 1)(x^2 + 2x + 5)$. Using the quadratic formula on $x^2 + 2x + 5 = 0$ doesn't give any more real roots because the discriminant, $b^2 - 4ac$, is -16. Therefore, $-\frac{1}{6}$ is the only real zero.

31. $p(x) = x^3 - x^2 - 4x + 4 = (x - 1)(x - 2)(x + 2)$

x-intercepts: 1, 2, and - 2 (same as a real zeros)

y-intercept: p(0) = 4   sketch:

table: x: -3   1.5   3

p(x): -20 - 0.875   10

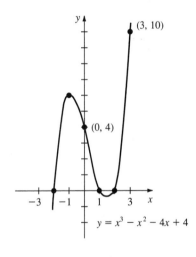

33. $p(x) = 6x^3 + 13x^2 + 32x + 5 = (6x + 1)(x^2 + 2x + 5)$

x-intercept: $-\frac{1}{6}$ y-intercept: p(0) = 5

table: x: -2   -1   1

p(x): -55   -20   56

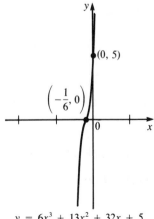

35. If - 2 and i are zeros, then - i must also be a zero (conjugates!); so p(x) in factored form is

$(x + 2)(x - i)(x + i) = (x + 2)(x^2 + 1) = x^3 + 2x^2 + x + 2$ which satisfies the requirement that 1 be the leading coefficient.

37. If the <u>only</u> zeros are 0 and ± i, that means that 0 must be a zero of multiplicity 2 since there must be 4 zeros altogether (if say i repeated, then - i would have to repeat again as well). So p(x) in factored form would look like $(x - 0)(x - 0)(x - i)(x + i) = x^2(x^2 + 1) = x^4 + x^2$ which has the required leading coefficient of one.

39. Since - 1 + 2i is a zero so is - 1 - 2i. The leading coefficient is not necessarily 1 so p(x) in factored form would appear as $a(x - 1)(x + 1)[x - (-1 + 2i)][x - (-1 - 2i)]$.

So $p(x) = a(x^2 - 1)(x + 1 - 2i)(x + 1 + 2i) = a(x^2 - 1)[(x + 1)^2 - 4i^2] = a(x^2 - 1)(x^2 + 2x + 5)$.

The problem stated that $p(0) = - 20$; therefore

$p(0) = a(-1)(+5) = - 5a$ and this expression must equal - 20 so a is forced to be 4 and...

$p(x) = 4(x^2 - 1)(x^2 + 2x + 5) = 4x^4 + 8x^3 + 16x^2 - 8x - 20$

41. $p(x) = x^4 + 3x^3 - x - 3$ has rational root possibilities ±3 and ±1

trying +1:   1) 1 3 0 -1 -3   (the 0 is in there for the missing x^2 term)

                  1 4 4 3

        1 4 4 3 [0]   yea! 1 is a zero.

now we can use the quotient 1 4 4 3 for the continuing search which will include only negative tries due to the absence of sign changes (1 is an upperbound for the real zeros)

trying -3:   -3) 1 4 4 3

                -3 -3 -3

       1 1 1 [0]   so -3 is another real zero.

the "other" factor, $x^2 + x + 1$, (from the 1 1 1) has no real zeros (negative discriminant) so the rational zeros are 1 and -3.

43. $p(x) = x^4 + x^3 - 3x^2 - 4x - 4$ has ±4, ±2, and ±1 as its rational root possibilities.

+1 doesn't give a 0; the bottom numbers were 1 2 -1 -5 (not all positive) so try higher;

43. (continued)

trying 2: $2\,)\overline{1\ \ 1\ \ -3\ \ -4\ \ -4}$

$\underline{\ \ \ 2\ \ \ 6\ \ \ 6\ \ \ 4}$

$1\ \ \ 3\ \ \ 3\ \ \ 2\ \ \boxed{0}$ so we now know that 2 is a zero and an upper bound for the zeros.

- 1 didn't work; trying -2: $-2\,)\overline{1\ \ 3\ \ 3\ \ 2}$

$\underline{-2\ -2\ -2}$

$1\ \ \ 1\ \ \ 1\ \ \boxed{0}$ voila!

trying -4 produced numbers alternating in sign, the signal for a lower bound for the zeros. Any other zeros would have to be solutions to the "decoded" quadratic $1x^2 + 1x + 1 = 0$ and there are no real zeros to this as the discriminant is - 4. So the rational zeros are 2 and - 2.

45. $p(x) = x^4 - 10x^2 + 9$ (don't forget about normal factoring!)

$= (x^2 - 9)(x^2 - 1)$

$= (x + 3)(x - 3)(x + 1)(x - 1)$ so the rational zeros are -3, 3, -1, and 1

3 is an upper bound and -3 a lower bound for the real zeros.

47. $p(x) = 9x^4 + 8x^2 - 1 = (9x^2 - 1)(x^2 + 1) = (3x + 1)(3x - 1)(x^2 + 1)$

so its rational zeros are $-\frac{1}{3}$ and $\frac{1}{3}$. ($x^2 + 1 = 0$ has imaginary roots)

$\frac{1}{3}$ is an upper bound and $-\frac{1}{3}$ is a lower bound for the real zeros.

49. $p(x) = x^4 + 2x^3 + 6x^2 + 5x + 6$ has ± 6, ± 3, ± 2, and ± 1 as its rational zero possibilities.

Since all the coefficients are positive, all the positive possibilities above can be ruled out. When -1 and -2 are tried the bottom numbers are 1 1 5 0 6 and 1 0 6 -7 20 (not ending with a 0 (signal for a zero) or alternating (signal for a lower bound). But...

trying -3: $-1)\overline{1\ \ \ 2\ \ \ 6\ \ \ 5\ \ \ 6}$

$\underline{-3\ \ \ 3\ \ -27\ \ 66}$

$1\ \ -1\ \ \ 9\ \ -22\ \ \boxed{72}$ which alternates in sign so -3 is a lower bound for the real roots and there are simply no rational zeros for this function.

51. $p(x) = x^5 - 3x^3 + 2x^2 = x^2(x^3 - 3x + 2)$ so $x = 0$ is definitely one rational zero.

Now, concentrating on $x^3 - 3x + 2$:

Its rational root possibilities are $\pm 1, \pm 2$; trying 1:

$$1 \overline{)\, 1 \quad 0 \quad -3 \quad 2}$$
$$ \quad 1 \quad 1 \quad -2$$
$$\overline{\, 1 \quad 1 \quad -2 \quad \boxed{0}}$$

and now a quadratic, $x^2 + x - 2$ is visible from the bottom row and it factors as $(x + 2)(x - 1)$ and so the five zeros of $p(x)$ are 0(multiplicity 2), 1(multiplicity 2), and -2. 2 is certainly an upper bound and -2 a lower bound for the real zeros.

53. $p(x) = 6x^6 + x^5 - x^4 + 18x^2 + 3x - 3$ has $\dfrac{\pm 1, \pm 3}{\pm 1, \pm 2, \pm 3, \pm 6}$ as possibilities.

trying $\frac{1}{3}$:

$$\tfrac{1}{3} \overline{)\, 6 \quad 1 \quad -1 \quad 0 \quad 18 \quad 3 \quad -3}$$
$$\phantom{\tfrac{1}{3})} \quad 2 \quad 1 \quad 0 \quad 0 \quad 6 \quad 3$$
$$\overline{\phantom{\tfrac{1}{3})}\, 6 \quad 3 \quad 0 \quad 0 \quad 18 \quad 9 \quad \boxed{0}}$$

The numbers 6 3 0 0 18 9 show no sign changes ($\frac{1}{3}$ is an upper bound on the positive zeros)

Trying $\frac{1}{6}$ did not produce a zero, only wild fractions; there are no nore positive roots.

trying $-\frac{1}{2}$:

$$-\tfrac{1}{2} \overline{)\, 6 \quad 3 \quad 0 \quad 0 \quad 18 \quad 9}$$
$$\phantom{-\tfrac{1}{2})} \quad -3 \quad 0 \quad 0 \quad 0 \quad -9$$
$$\overline{\phantom{-\tfrac{1}{2})}\, 6 \quad 0 \quad 0 \quad 0 \quad 18 \quad \boxed{0}}$$

residual polynomial is $6x^4 + 18$ (no real solutions here!)

55. $p(x) = x^4 - 10x^3 + 27x^2 + 84x - 73$ supposedly has no rational zero. Why?

By the rational root theorem, any rational zero would have to have the composition $\dfrac{\text{factor of } 73}{\text{factor of } 1}$ which means the only possibilities are ± 1 and ± 73. Try synthetic division with these four numbers and you will find that none produces 0 (1 produces 29, -1 produces -119, 73 and -73 produce numerically huge results which can't possibly be zero. Since there are only four possibilities and none of them pan out, $p(x)$ must not have any rational zeros.

57. $x^{10} - 11x + 378,929$ cannot have a rational zero for the following reasons:

(1). If there were any rational zeros, they would be of the form $\dfrac{\text{factor of } 378,929}{\text{factor of } 1}$ which means

they are integers and there are at least four possibilities: ± 1 and $\pm 378,929$.

(2). Since $p(x)$ has 2 sign changes, there are either 2 positive zeros or NO positive zeros. Since $p(-x) = x^{10} + 11x + 378,929$ has NO sign changes, there are NO negative zeros at all. This narrows our list of possible rational roots to at least $+1$ or $+378,929$.

(3). By direct substitution, $p(1) = 1 - 11 + 378,929 = 378,919$ which certainly isn't 0 so the rational zero isn't 1.

(4). There are two things we'd prefer not to have to do: find $p(378,929)$ either directly or synthetically or try to track down other factors of $378,929$. What we'll hope to do is use the upper bound theorem to tell us there's nothing to look for; ie, let's try synthetic division with 2:

```
2 ) 1  0  0   0   0   0   0    0     -11   378,929
       2  4   8   16  32  64   128   256   512    1002
    ────────────────────────────────────────────────────
    1  2  4   8   16  32  64   128   256   501   ┌───────┐
                                                  379,931
                                                 └───────┘
```

Now all the bottom numbers are positive which means there CAN'T be a positive zero past 2. This means we've ruled out the negatives, we've ruled out $+1$, and no number > 2 can be a zero, rational or not. No rational candidates are left so $p(x)$ has no rational zeros.

59. Main idea in argument: Odd integers have only odd factors; e.g., the factors of 27 are $\{1,3,9,27\}$.

The argument: Any rational zeros of $p(x)$ must be of the form $\dfrac{\text{factor of } a_0}{\text{factor of } a_n}$. Since the problem states that a_0 and a_n are odd, it follows that their factors are all odd integers. If $\frac{m}{k}$ is a rational zero of $p(x)$, then it must have resulted from one of the odd factors of a_0 over one of the odd factors of a_n. Perhaps dividing these two odd integers by some odd common factor resulted in the final reduced form $\frac{m}{k}$ but even then, m and k are the remaining factors of a_0 and a_n respectively and must be odd as well because odd integers have <u>only</u> odd factors.

61. By problem 60 we know that $k^n p\left(\frac{m}{k}\right) = 0$ (anything times 0 equals 0)

What we'll do now is create the "long form" for $k^n p\left(\frac{m}{k}\right)$:

Substitute $\frac{m}{k}$ for x in p(x) to get:

$$p\left(\frac{m}{k}\right) = a_n\left(\frac{m}{k}\right)^n + a_{n-1}\left(\frac{m}{k}\right)^{n-1} + \ldots + a_1\left(\frac{m}{k}\right) + a_0 \quad \text{(apply exponent property)}$$

$$= a_n \cdot \frac{m^n}{k^n} + a_{n-1} \cdot \frac{m^{n-1}}{k^{n-1}} + \ldots + a_1 \cdot \frac{m}{k} + a_0 \quad \text{(now multiply both sides by } k^n\text{)}$$

$$k^n p\left(\frac{m}{k}\right) = k^n a_n \frac{m^n}{k^n} + k^n a_{n-1}\frac{m^{n-1}}{k^{n-1}} + \ldots + k^n a_1 \frac{m}{k} + k^n a_0 \quad \text{(now cancel powers of k out)}$$

$$= a_n m^n + k\, a_{n-1} m^{n-1} + \ldots + k^{n-1} a_1 m + k^n a_0$$

Now this "long form" is still equal to $k^n p\left(\frac{m}{k}\right)$ which we know must equal 0; therefore our " long form" is equal to 0 as well.

63. From problem 59 we know that m and k are odd integers; so it's true that any positive integral power of either of these odd integers is also odd (ie., m^r is odd and k^s is odd); it is also true that a product of odd integers is odd so the product of these two powers of odd numbers must be odd as well, ie, $m^r k^s$ is odd.

65. Problem 64 says that $p(1) = a_n(1 - m^n) + a_{n-1}(1 - m^{n-1}k) + \ldots + a_1(1 - mk^{n-1}) + a_0(1 - k^n)$

In each set of parentheses there is an expression of the form $1 - m^r k^s$ where the m and k are odd. So what happens is that if these odd products are subtracted from 1, an even number results each time and these even numbers are added together to create the value of p(1). Since the sum of any number of even numbers is even, we can conlclude that p(1) is even.

67. $\sqrt{5}$ is a root of $x^2 - 5 = 0$

The rational root possibilities are ± 5 and ± 1 and $\sqrt{5}$ is not in this list. Therefore, $\sqrt{5}$ must be an irrational number. (If it were rational and a root it would be in the list.)

69. $p(x) = x^3 + ax^2 + bx + c$

$p(y - \frac{a}{3}) = (y - \frac{a}{3})^3 + a(y - \frac{a}{3})^2 + b(y - \frac{a}{3}) + c$

$$= y^3 - 3y^2(\frac{a}{3}) + 3y(\frac{a}{3})^2 - \frac{a^3}{27} + a(y^2 - \frac{2ya}{3} + \frac{a^2}{9}) + by - \frac{ba}{3} + c$$

$$= y^3 - ay^2 + \frac{a^2y}{3} - \frac{a^3}{27} + ay^2 - \frac{2a^2y}{3} + \frac{a^3}{9} + by - \frac{ba}{3} + c$$

$$= y^3 + \left(by + \frac{a^2y}{3} - \frac{2a^2y}{3} \right) + \left(c - \frac{a^3}{27} + \frac{a^3}{9} - \frac{ab}{3} \right)$$

$$= y^3 + \left(b - \frac{a^2}{3} \right)y + \left(c + \frac{2a^3}{27} - \frac{ab}{3} \right)$$

71. $x^3 + 3x^2 + 6x - 4 = 0$

If you did problem 70 correctly you got the equation $y^3 + 3y = 8$

so now m = 3 and n = 8 in Cardano's formula gives the wild result:

$$y = \sqrt[3]{\sqrt{16 + 1} + 4} - \sqrt[3]{\sqrt{16 + 1} - 4} = \sqrt[3]{\sqrt{17} + 4} - \sqrt[3]{\sqrt{17} - 4}$$

and since y - 1 took the place of x we have x = y - 1 so subtract 1 from the answer above and you have one of the roots for x for the original equation:

$$x = \sqrt[3]{\sqrt{17} + 4} - \sqrt[3]{\sqrt{17} - 4} - 1$$

Problems 4.5

1. $p(x) = x^3 + x + 1$ must have a zero between - 1 and 0 because the graph obviously crosses the
   x axis somewhere in between; to further support this, $p(-1) = -1$ and $p(0) = +1$ (opposite signs!)
   If we used the middle point, - 0.5, as our guess for the zero, the most this estimate could be off
   from the true answer is 0.5 (half the width of the interval known to contain the zero).
   But we must continue the process of halving the interval known to contain the zero:
   since $p(-0.5) = 0.375$ (positive!) while $p(-1)$ is still - 1, the zero's between - 1 and - 0.5;
   using the midpoint, - 0.75, as a guess would make the maximum error at this stage ± 0.25;
   since $p(-0.75) = -0.1719$ and $p(-0.5)$ is still 0.375, there's a zero between - 0.75 and - 0.5;
   using the midpoint, - 0.625, would make the maximum error ± 0.125 (half the width)
   (Note: to calculate midpoints, ADD the endpoints and divide by 2; to calculate maxi-
   maximum error in using the midpoint of an interval, SUBTRACT the absolute values of
   the endpoints (smaller from larger) and divide by 2);
   since $p(-0.625) = 0.13086$ and $p(-0.75)$ is still negative, there's a zero between
   - 0.75 and - 0.625;
   using the midpoint, - 0.6875, makes the maximum error .0625 (we're getting there!)
   since $p(-0.6875) = -0.01245$ (negative!), and $p(-0.625)$ is the last positive value around, the
   zero must be between - 0.6875 and - 0.625;
   using the midpoint, - 0.65625, makes the maximum error .03125;
   since $p(-0.65625) = 0.06113$ and $p(-0.6875)$ is still the last negative value around, the zero
   must be between - 0.6875 and - 0.65625;
   using the midpoint, - 0.671875, makes the maximum error ± 0.01563 (the difference of the
   last two endponts divided by 2)
   since $p(-0.671875) = +0.02483$ and $p(-0.6875)$ was negative, the zero's between them;
   using the midpoint, - 0.6796875 makes the maximum error $\pm .00782$ (one more time!)
   since $p(-0.6796875) = +0.0063135$ and $p(-0.6875)$ was negative, the zero is between them;
   using the midpoint, - 0.6835938, makes the maximum error 0.0039 which is less than .005
   and since the error is less than 0.005 we can round to the nearest hundreth and feel secure
   about stating that the zero is approximately - 0.68.

3. $p(x) = x^3 + 3x^2 + 6x + 2$ and there's a zero somewhere between -1 and 0.
   $p(-1) = -2$ and $p(0) = 2$ so the first midpoint will be - 0.5

3. (continued)

   p(- 0.5) = - 0.375

   p(0) = 2

                 so next midpoint will be - 0.25   (max error = 0.25)

   p(- 0.25) = 0.67188

   p(- 0.5) = - 0.375

                 so next midpoint will be - 0.375 (max error = 0.125, half the width)

   p(- 0.375) = 0.11914

   p(- 0.5) = - 0.375

                 so the next midpoint must be - 0.4375   (max error = 0.0625)

   p(- 0.4375) = - 0.13452

   p(- 0.375) = 0.11914

                 so next midpoint is - 0.40625   (max error = 0.03125)

   p(- 0.40625) = - 0.00943

   p(- 0.375) = 0.11914

                 so next midpoint is - 0.39063   (max error = 0.01562)

   p(- 0.39063) = 0.05439

   p(- 0.40625) = - 0.00943

                 so the next midpoint is - 0.39844   (max error = 0.00781)

   p(- 0.39844) = 0.02238

   p(- 0.40625) = - 0.00943

                 so the next midpoint is - 0.40234   (max error = 0.0039)

Since the error in using - 0.40234 is less than the required 0.005 we are done. The approximate
zero of p(x) is - 0.40234 which can safely be rounded to - 0.40.

5. $p(x) = - 5x^3 - 3x + 2$ and there is a zero between 0 and 1.

   p(0) = 2 while p(1) = - 6 so the first midpoint approximation will be 0.5

a) p(0.5) = - 0.125   b)   p(0.25) = 1.17188

   p(0) = 2 so the next midpoint is 0.25   p(0.5) = - 0.125 so next midpoint is 0.375

c) p(0.375) = 0.61133   d)   p(0.4375) = 0.26880

   p(0.5) = - 0.125 so next midpt is 0.4375   p(0.5) = - 0.125 so next midpt is 0.46875

5. (continued)

e) p(0.46875) = 0.00788

p(0.5) = - 0.125 so next midpt is 0.48438   (error in using this would be (0.5 - 0.46875)/2 ≈ .016)

f) p(0.48438) = - 0.00214

p(0.46875) = 0.00788 so next midpt is 0.47656

g) p(0.47656) = 0.02916

p(0.48438) = - 0.00214 so the next (and last) midpoint is 0.48047 (error < 0.00391   which is half the width of this last interval being halved; if the error is less than 0.00391 it is certainly less than the 0.005 reguired in the problem).   Rounded zero: 0.48.

7. $p(x) = x^3 - 6x^2 - 15x + 4$ apparently has three zeros: one between - 3 and - 2, one between 0 and 1, and one between 7 and 8. Detail will be shown on the first; only successive intervals, midpoints, and functional values on the latter two;

Between - 3 and - 2:

p(- 3) = - 32 and p(- 2) = 2 and the first midpoint approximation will be - 2.5

i) p(- 2.5) = - 11.625 and p(- 2) = 2 so next midpoint will be - 2.25

ii) p(- 2.25) = - 4.01563 and p(-2) = 2 so next midpoint is - 2.125

iii) p(- 2.125) = - 0.81445 and p(-2) = 2 so next midpoint is - 2.0625

iv) p(- 2.0625) = 0.64038 and p(- 2.125) = - 0.81445 so next midpoint is - 2.09375

v) p(- 2.09375) = - 0.07504 and p(- 2.0625) = 0.64038 so next midpoint is - 2.07813

(note: error if this last midpoint were to be used is (2.09375 - 2.0625)/2 ≈ 0.01562)

vi) p(- 2.07813) = 0.28554 and p(- 2.09375) = - 0.07504 so next midpt is - 2.08594

vii) p(- 2.08594) = 0.10600 and p(- 2.09375) = - 0.07504 so the next midpt is - 2.08984 and, in using this, the error can't be bigger than half the interval width which is 0.00391 again.

Approximate zero between - 3 and -2: - 2.09

Between 0 and 1:

p(0) = 4 and p(1) = -16 and the first midpoint approximation is 0.5

The successive intervals formed by watching for change of sign were: (0, 0.5), (0, 0.25), (0.125, 0.25), (0.1875, 0.25), (0.21875, 0.25), (0.23438, 0.25), (0.24219, 0.25)

The successive midpoints which led to that choice of intervals: 0.25, 0.125, 0.1875, 0.21875, 0.23438, 0.24219

7. (continued)   The functional values: $p(0.5) = - 4.875$, $p(0.25) = - 0.10934$, $p(0.125) = 2.03320$,

$p(0.1875) = 0.98315$, $p(0.21875) = 0.44211$, $p(0.23438) = 0.16757$, $p(0.24219) = 0.02942$

Half the width of the last interval is 0.00391 so the last midpoint will be 0.24609 and it will

be our choice for the zero of $p(x)$ between 0 and 1.

Approximate zero between 0 and 1: 0.25

Between 7 and 8 where $p(7) = - 52$ and $p(8) = 12$:

Successive intervals: (7.5, 8), (7.75, 8), (7.75, 7.875), (7.8125, 7.875), (7.84375, 7.875),

(7.84375, 7.85938), (7.84375, 7.85156)

Successive midpoints: 7.75, 7.875, 7.8125, 7.84375, 7.85938, 7.85156, and the last: 7.84766

The functional values: $p(7.5) = - 24.125$, $p(7.75) = - 7.14063$, $p(7.875) = 2.15430$, $p(7.8125)$

$= - 2.56128$, $p(7.85938) = 0.96291$, $p(7.85156) = 0.36973$.

Approximate zero between 7 and 8: 7.85

9. $p(x) = x^3 - x^2 - 1$ with a zero between 1 and 2; $p(1) = - 1$ and $p(2) = 3$; first midpoint is 1.5

 i) $p(1.5) = 0.125$

 $p(1) = -1$ so next midpoint: 1.25

 ii) $p(1.25) = - 0.60938$

 $p(1.5) = 0.125$ next midpoint: 1.375

 iii) $p(1.375) = - 0.29102$

 $p(1.5) = 0.125$ so next midpoint: 1.4375

 iv) $p(1.4375) = - 0.09595$

 $p(1.5) = 0.125$ so next midpoint 1.46875

 v) $p(1.46875) = 0.01120$ while $p(1.4375) = - 0.09595$ so next midpont: 1.45313

 vi) $p(1.45313) = - 04318$ while $p(1.46875) = 0.01120$ so next midpoint: 1.46094

 vii) $p(1.46094) = - 0.01619$ while $p(1.46875) = 0.01120$ so next midpoint is 1.46484

 viii) $p(1.46484) = - 0.002567$ while $p(1.46875) = 0.01120$ so next midpoint is 1.46680

 ix) $p(1.46680) = 0.00432$ while $p(1.46484) = - 0.002567$ so next midpoint is 1.46582

 x) $p(1.46582) = 0.0008738$ while $p(1.46484) = - 0.002567$ so next midpoint is 1.46533

At this moment, the error is less than $\frac{1.46582 - 1.46484}{2} = 0.00049$ which is < 0.0005 so

an answer for the zero that shows the correct thousandths place is 1.465

11. $p(x) = 3x^5 + 2x^4 + x^2 + 3$ with a zero between - 2 and -1; $p(-2) = -57$ while $p(-1) = +3$;

First midpoint is - 1.5

i) $p(-1.5) = -7.406$ while $p(-1) = 3$ so the next midpoint is - 1.25

ii) $p(-1.25) = 0.290$ while $p(-1.5) = -7.406$ so next midpoint is - 1.375

iii) $p(-1.375) = -2.705$ while $p(-1.25) = 0.290$ so next midpoint is - 1.3125

iv) $p(-1.3125) = -1.027$ while $p(-1.25) = 0.290$ so next midpoint is - 1.28125

v) $p(-1.28125) = -0.327$ while $p(-1.25) = 0.290$ so next midpoint is - 1.26563

vi) $p(-1.26563) = -0.00869$ while $p(-1.25) = 0.290$ so next midpoint is - 1.25781

vii) $p(-1.25781) = 0.143$ while $p(-1.26563) = -0.00869$ so next midpoint is - 1.26172

viii) $p(-1.26172) = 0.068$ while $p(-1.26563) = -0.00869$ so next midpoint is - 1.26367

ix) $p(-1.26367) = 0.0298$ while $p(-1.26563) = -0.00869$ so next midpoint is - 1.26465

x) $p(-1.26465) = 0.0106$ while $p(-1.26563) = -0.00869$ so next midpoint is - 1.26514

xi) $p(-1.26514) = 0.00097$ while $p(-1.26563) = -0.00869$ so next and last midpoint is - 1.265137

which when rounded to the nearest thousandth yields - 1.265 as our approximation for this zero.

13. $p(x) = 6x^4 - 2x^3 + 3x^2 - 4x - 2$ and $p(0)$ is - 2 while $p(1)$ is 1 so there's a zero between 0 and 1.

i) $p(0.5) = -3.125$ and $p(1) = 1$ so the next midpoint is 0.75

ii) $p(0.75) = -2.2578$ and $p(1) = 1$ so the next midpoint is 0.875

iii) $p(0.875) = -1.0259$ and $p(1)$ is still 1 so the next midpoint is 0.9375

iv) $p(0.9375) = -0.1264$ and $p(1)$ is still positive so the next midpoint is 0.96875

v) $p(0.96875) = 0.4066$ and $p(0.9375)$ was negative at - 0.1264 so next midpoint is 0.95313

vi) $p(0.95313) = 0.13286$ and $p(0.9375) = -0.1264$ so the next midpoint is 0.94531

vii) $p(0.94531) = 0.00136$ and $p(0.9375)$ is still negative so the next midpoint is 0.94141

viii) $p(0.94141) = -0.0629$ and $p(0.94531)$ is positive so the next midpoint is 0.94366

ix) $p(0.94366) = -0.0309$ and $p(0.94531)$ 0.00136 so the next midpoint is 0.94434

x) $p(0.94434) = -0.0147$ and $p(0.94531)$ is positive so the next and last midpoint is 0.94482

rounding safely to the nearest thousandth gives us the zero 0.945

15. $p(x) = x^3 - 8x^2 + 2x - 14$ and we must find all the zeros with errors less than 0.01.

By using synthetic substitution to find the values of p at different integers, a sign change

occurs between 7 and 8 as $p(7) = -49$ and $p(8) = +2$.

First midpoint is 7.5

i.) $p(7.5) = -27.125$ while $p(8) = +2$ so the next midpoint is 7.75

15. (continued)

ii.) p(7.75) = - 13.5156 while p(8) is still positive so the next midpoint is 7.875

iii.) p(7.875) = - 6.00195 while p(8) remains positive so the next midpoint is 7.9375

iv.) p(7.9375) = - 2.06274 and p(8) hasn't changed so the next midpoint is 7.96875

v.) p(7.96875) = - 0.046906 and p(8) is positive 2 so the next midpoint is 7.98438

vi.) p(7.98438) = 0.97297 (finally a positive one!) so the next midpoint is halfway between 7.96875 and 7.98438 at 7.97656.

vii.) p(7.97656) = 0.46176 and p(7.96875) = - 0.046906; these last two x-values are less than 0.01 apart so their midpoint can be used as our zero to accuracy 0.01. It is 7.97266.

<u>Problems 4.6</u>

These problems were designed to be done with computer assistance. We used the BASIC program listed below (which has lines 10 and 20 containing the function from problem 1 and its derivative function and line 80 stipulating the desired accuracy; obviously you must change lines 10 and 20 as you do each homework problem).

```
10  DEF FNP(X) = X^2 - 3*X - 10
20  DEF FND(X) = 2*X - 3
30  I = 0
40  INPUT "1ST NUMBER"; A
50  LET R = FNP(A)/FND(A)
60  LPRINT I; A; TAB(14); FNP(A); TAB(32); FND(A); TAB(50); R; TAB(68); A - R
70  I = I + 1
80  IF ABS(R) > 0.000001 AND I < 10 THEN LET A = A - R: GO TO 50
90  END
```

1. $p(x) = x^2 - 3x - 10$ and so $d(x) = 2x - 3$

Here is the computer's output:

n	x_n	$p(x_n)$	$d(x_n)$	$r = p(x_n)/d(x_n)$	$x_{n-1} = x_n - r$
0	4	-6	5	- 1.2	5.2
1	5.2	1.439998	7.4	0.1945943	5.005406
2	5.005406	0.037867	7.010811	0.0054012	5.000005
3	5.000005	0.000029	7.000009	0.0000041	5.000000
4	5.000000	0.000000	7.000000	0.000000	5.000000

and so the fourth and fifth iterates will agree (to six decimal places) and we'll take 5 as our zero.

Factoring $p(x)$ into $(x - 5)(x + 2)$ would have yielded the same answer!

3. $p(x) = 3x^2 - 5x - 2$ and so $d(x) = 6x - 5$

(change declared functions in lines 10 and 20 to $3*x^2 - 5*x - 2$ and $6*x - 5$ respectively)

output:

n	x_n	$p(x_n)$	$d(x_n)$	$r = p(x_n)/d(x_n)$	$x_{n-1} = x_n - r$
0	-1	6	-11	- 0.5454546	-0.4545455
1	-0.4545455	0.8925619	-7.727273	- 0.115508	-0.3390374
2	-0.3390374	0.0400262	-7.034225	-0.0056902	-0.3333472
3	-0.3333472	0.0000970	-7.000083	- 0.000014	-0.3333334
4	-0.3333334	0.0000000	-7.000000	-0.000000	-0.3333334

and so our zero appears to be - 0.333333 and factoring $p(x)$ into $(3x + 1)(x - 2)$ gives $-\frac{1}{3}$ as a zero.

5. $p(x) = x^2 + x - 1$ and so $d(x) = 2x + 1$ (be sure to change lines 10 and 20 accordingly)

output:

n	x_n	$p(x_n)$	$d(x_n)$	$r = p(x_n)/d(x_n)$	$x_{n-1} = x_n - r$
0	0	-1	1	- 1	1
1	1	1	3	0.3333334	0.6666666
2	0.6666666	0.1111111	2.333333	0.0476190	0.6190476
3	0.6190476	0.0022675	2.238095	0.0010131	0.6180345
4	0.6180345	0.0000001	2.236069	0.0000005	0.618034
5	0.618034	0.000000	2.23068	0.000000	0.618034

and so our zero will be taken as ≈ 0.618034; the quadratic formula would have yielded $\frac{-1 + \sqrt{5}}{2}$ as one of the zeros and a six decimal place approximation of this is 0.618034 as well!

7. $p(x) = x^2 - 7x + 5 = 0$ and so $d(x) = 2x - 7$; we need to look for some changes in sign in $p(x)$ so that we can determine the intervals, if any, which contain the zeros. Trial and error determined that:

 p(0) was positive but p(1) was negative so [0, 1] is one of our intervals

 p(6) was negative but p(7) was positive so [6,7] is the other.

output:

n	x_n	$p(x_n)$	$d(x_n)$	$r = p(x_n)/d(x_n)$	$x_{n-1} = x_n - r$
0	0	5	-7	- 0.7142858	0.7142858
1	0.7142858	0.5102043	-5.571429	- 0.0915751	0.8058609
2	0.8058609	0.0083857	-5.388278	- 0.0015563	0.8074171
3	0.8074171	0.0000019	-5.385166	- 0.0000004	0.8074175
4	0.8074175	0.0000005	-5.385165	- 0.00000009	0.8074176

and so the zero between 0 and 1 could be approximated by 0.807418 to six places;

now for the zero between 6 and 7

0	6	- 1	5	- 0.2	6.2
1	6.2	0.0400009	5.4	0.0074076	6.192592
2	6.192592	0.0000534	5.385185	0.0000099	6.192582
3	6.192582	- 0.0000038	5.385165	- 0.0000007	6.192583
4	6.192583	0.000000	5.385165	0.000000	6.192583

and so the other zero is approximately 6.192583

9. $p(x) = x^3 - x^2 - 1$ and so $d(x) = 3x^2 - 2x$ (x^3 - x^2 - 1 and 3*x^2 - 2*x in BASIC)

n	x_n	$p(x_n)$	$d(x_n)$	$r = p(x_n)/d(x_n)$	$x_{n+1} = x_n - r$
0	1	-1	1	-1	2
1	2	3	8	0.375	1.625
2	1.625	0.6503906	4.671875	0.139214	1.485786
3	1.485786	0.0724015	3.651108	0.198300	1.465956
4	1.465956	0.0035183	3.515169	0.000385	1.465571
5	1.465571	0.0000007	3.512556	0.0000002	1.465571

and so the 5th and 6th iterates agree to six decimal places so the zero is approximately 1.465571

11. $p(x) = 3x^5 + 2x^4 + x^2 + 3$ and therefore $d(x) = 15x^4 + 8x^3 + 2x$

n	x_n	$p(x_n)$	$d(x_n)$	$r = p(x_n)/d(x_n)$	$x_{n+1} = x_n - r$
0	- 2	- 57	172	- 0.3313954	- 1.668605
1	- 1.668605	- 17.51683	75.77641	- 0.2311647	- 1.43744
2	- 1.43744	- 4.805803	37.40425	- 0.1284828	- 1.308957
3	- 1.308957	- 0.9432161	23.47476	- 0.0401800	- 1.268777
4	- 1.268777	- 0.0712056	19.99429	- 0.0035613	- 1.265216
5	- 1.265216	- 0.0005190	19.70402	- 0.0000263	- 1.26519
6	- 1.26519	0	19.70189	0	- 1.26519

which means the 6th and 7th iterates agree to six places so the zero is approximately - 1.265190

13. $p(x) = 6x^4 - 2x^3 + 3x^2 - 4x - 2$ and so $d(x) = 24x^3 - 6x^2 + 6x - 4$

Note: enter 1 for the first number as its closer to the root.

the output:

n	x_n	$p(x_n)$	$d(x_n)$	$r = p(x_n)/d(x_n)$	$x_{n+1} = x_n - r$
0	1	1	20	0.5	0.95
1	0.95	0.0797875	16.862	0.0047318	0.9452681
2	0.9452681	0.0006638	16.58147	0.0000003	0.9452281
3	0.9452281	0.0000005	16.57911	0.00000003	0.9452281

and so our approximate zero of $p(x)$ is 0.945228

15. $p(x) = x^3 + 14x^2 + 60x + 78$; trial and error evaluation of $p(x)$ with negative integers (they're the only ones that have a chance because all the coefficients are positive!) detected changes of sign between - 3 and - 2, between - 5 and - 4 and between - 8 and - 7. $d(x) = 3x^2 + 28x + 60$

15. (continued)   computer output:

n	x_n	$p(x_n)$	$d(x_n)$	$r = p(x_n)/d(x_n)$	$x_{n+1} = x_n - r$
0	- 3	- 3	3	- 1	- 2
1	- 2	6	16	0.375	- 2.375
2	- 2.375	1.072266	10.42188	0.1028861	- 2.477886
3	- 2.477886	0.716858	9.038948	0.0079308	- 2.485817
4	- 2.485817	0.000412	8.934986	0.0000461	- 2.485863
5	- 2.485863	0.000008	8.934383	0.0000009	- 2.485864
6	- 2.485864	- 0.000053	8.934368	- 0.0000017	- 2.485862
7	- 2.485862	0.000008	8.934395	0.00000085	- 2.485863
8	- 2.485863	0	8.934379	0	- 2.485863

so one of the zeros is approximately - 2.485863

now for the one between - 5 and - 4:

n	x_n	$p(x_n)$	$d(x_n)$	r	x_{n+1}
0	- 5	3	- 5	- 0.6	- 4.4
1	- 4.4	- 0.1439972	- 5.120003	0.0281244	- 4.428125
2	- 4.428125	0.0005951	- 5.162628	- 0.0001153	- 4.428009
3	- 4.428009	0	- 5.162461	0	- 4.428009

so the next zero is approximately - 4.428009

now for the zero between - 8 and - 7:

n	x_n	$p(x_n)$	$d(x_n)$	r	x_{n+1}
0	- 8	- 18	28	- 0.6428571	- 7.357143
1	- 7.357143	- 3.867005	16.38266	- 0.2360425	- 7.121101
2	- 7.121101	- 0.436554	12.73941	- 0.0342680	- 7.086833
3	- 7.086833	- 0.008637	12.23828	- 0.0000569	- 7.086127
4	- 7.086127	- 0.000031	12.22803	- 0.0000025	- 7.086129
5	- 7.086129	0	12.22806	0	- 7.086129

and so the last zero is approximately - 7.086129

17. The program went into an infinite loop; the iterates did not converge because there are no real zeroes for $p(x) = x^2 + 5x + 7$. Try to use the quadratic formula on it!

19. Setting the volume function from problem 63 in section 4.2 equal to 60 gives us the equation:
$s^3 + 2s^2 = 60$; that is, $s^3 + 2s^2 - 60 = 0$ and we can have our little BASIC program help us locate the zero(s) of the function $p(s) = s^3 + 2s^2 - 60$. Since $p(3) = - 45$ and $p(4) = + 36$, there is a zero between 3 and 4 and we'll let the program start at 3.

19. (continued)

Lines 10 and 20 were changed to reflect this function and its derivative:

10 DEF FNP(X) = X^3 + 2*X^2 - 60

20 DEF FND(X) = 3*X^2 + 4*X

The iterates essentially agreed at 3.349142 which rounded to the nearest hundredth is 3.35 ft. So the square ends of the box would measure about 3.35 ft. on a side and the length of the box is approximately 5.35 ft. long.

21. Setting C(q) to 200,000 yields the equation $-0.002q^3 + 1.4q^2 + 200q + 25,000 = 200,000$

or equivalently $-0.002q^3 + 1.4q^2 + 200q - 175,000 = 0$. Once again we'll let the BASIC program track down the zeros of the related function $p(q) = -0.002q^3 + 1.4q^2 + 200q - 175,000$ the derivative of which is $-0.006q^2 + 2.8q + 200$ (lines 10 and 20 must be changed to these) Letting the program start at 100 produced the approximate solution 397.24 and letting the program start at 600 produced the approximate answer 644.51. These were the only solutions detected between 0 and 750.

Chapter 4 Review

1. $f(x) = \frac{1}{4}x^2$ table: x: -2   -1   0   1   2   sketch:

          f(x):   1 $\frac{1}{4}$ 0 $\frac{1}{4}$ 1

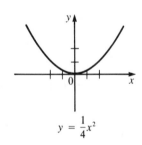

$$y = \frac{1}{4}x^2$$

3. $y = x^2 + 4x$: half of 4 is 2 and $2^2 = 4$ so add and subtract 4

$y = x^2 + 4x + 4 - 4$

$y = (x + 2)^2 - 4$ so the vertex is (-2, -4)

5. $y = 2x^2 + 6x - 8$: factor 2 out of first two terms

$y = 2(x^2 + 3x) - 8$: half of 3 is $\frac{3}{2}$ and $\left(\frac{3}{2}\right)^2 = \frac{9}{4}$, which is what we add and subtract

$y = 2(x^2 + 3x + \frac{9}{4} - \frac{9}{4}) - 8$

$y = 2(x^2 + 3x + \frac{9}{4}) - \frac{9}{2} - 8$

$y = 2(x + \frac{3}{2})^2 - \frac{25}{2}$ so the vertex is $\left(-\frac{3}{2}, -\frac{25}{2}\right)$

7. $f(x) = 4 - x^2$ (opens down because a $= -1$)

vertex: $x = \frac{-b}{2a} = \frac{-0}{2(-1)} = 0$; $y = f(0) = 4$ (also the y-intercept)

x-intercepts: set $4 - x^2 = 0$ and solve: $(2 - x)(2 + x) = 0$ so x = 2 or -2

The maximum (since it opens down) value of f(x) is 4, the y-value of the vertex.

sketch:

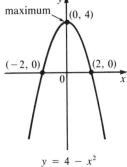

$y = 4 - x^2$

9. Use $s = -16t^2 + v_0 t$ with $v_0 = 80$ ft/sec to get $s = -16t^2 + 80t$ as the height function.

a) The ball will hit the ground when s $= 0$ so set the function equal to zero and solve for t:

$-16t^2 + 80t = 0$; $-16t(t - 5) = 0$; so either t$= 0$ sec or t $= 5$ sec are times

when the ball is on the ground so it must hit the ground after 5 seconds.

b) The ball will be at its highest point at the vertex. At that point $t = \frac{-b}{2a} = \frac{-80}{2(-16)} = \frac{5}{2}$ or 2.5

and we will have $s(2.5) = -16(2.5)^2 + 80(2.5)$ ft $= 100$ ft.

11. $p(x) = \frac{1}{2}x^3$ sketch:

table: x: -2   -1   0   1   2

p(x): -4 $-\frac{1}{2}$ 0 $\frac{1}{2}$ 4

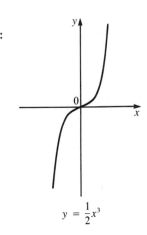

$y = \frac{1}{2}x^3$

13. $p(x) = (x + 2)^5$ (same basic shape as x^3 but shifted 2 blocks to the LEFT)

table: (x values centered around -2)   sketch:

x: -4 -3 -2 -1 0

p(x): -32 -1 0 1 32

$y = (x + 2)^5$

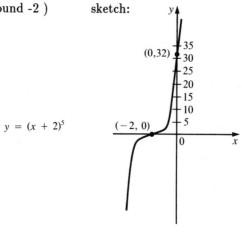

15. $p(x) = (x^2 - 1)(x - 2) = (x - 1)(x + 1)(x - 2)$

x-int: x = 1, x = -1 and x = 2 from the factors

y-int: p(0) = (-1)(1)(-2) = 2

Interval	test value	signs of factors	sign of p(x)	position of graph
(-∞,-1)	-3	(-)(-)(-)	negative	below x-axis
(-1,1)	0	(-)(+)(-)	positive	above
(1,2)	1.5	(+)(+)(-)	negative	below
(2,∞)	3	(+)(+)(+)	positive	above

sketch:

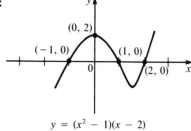

$y = (x^2 - 1)(x - 2)$

17. $p(x) = (x - 1)(x^2 - 3x - 4) = (x - 1)(x - 4)(x + 1)$

x-int: 1, 4, and -1

y-int: p(0) = (-1)(-4)(1) = 4

Interval	test value	signs of factors	sign of p(x)	position of graph
(-∞,-1)	-2	(-)(-)(-)	negative	below x-axis
(-1,1)	0	(-)(-)(+)	positive	above
(1,4)	3	(+)(-)(+)	negative	below
(4,∞)	5	(+)(+)(+)	positive	above

17. (continued)

sketch:

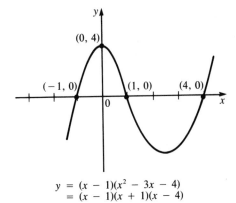

$$y = (x - 1)(x^2 - 3x - 4)$$
$$= (x - 1)(x + 1)(x - 4)$$

19. $p(x) = x^4 + 4x^2 + 4$ (has y-axis symmetry as $p(-x) = p(x)$)

table: x: 0   1   2

   p(x): 4   9   36

as $x \to \infty$, p(x) does the same so graph goes up on the right side

sketch:

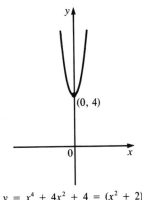

$$y = x^4 + 4x^2 + 4 = (x^2 + 2)^2$$

21.

$$
\begin{array}{r}
x - 3 \ \text{(quotient)} \\
x^2 + 2x \ ) \overline{x^3 - x^2 + x + 1} \\
\underline{x^3 + 2x^2} \\
-3x^2 + x \\
\underline{-3x^2 - 6x} \\
+ 7x + 1 \ \text{(remainder)}
\end{array}
$$

23.

$$\begin{array}{r} x^2 + 4x + 15 \quad \text{(quotient)} \end{array}$$

$$x^3 - 4x^2 + 5 \overline{\smash{\big)}\ x^5 + 0x^4 - x^3 + 0x^2 + 2x + 3}$$

$$\underline{x^5 - 4x^4 \qquad\quad + 5x^2}$$

$$+ 4x^4 - x^3 - 5x^2 + 2x$$

$$\underline{+ 4x^4 - 16x^3 \qquad + 20x}$$

$$15x^3 - 5x^2 - 18x + 3$$

$$\underline{15x^3 - 60x^2 \qquad + 75}$$

$$55x^2 - 18x - 72 \quad \text{(remainder)}$$

25. $-1 \overline{\smash{\big)}\ 3 \quad 3 \quad -1 \quad 4 \quad -9}$

$\underline{\qquad -3 \quad 0 \quad 1 \quad -5}$

$3 \quad 0 \quad -1 \quad 5 \quad \boxed{-14}$ so quotient $= 3x^3 - x + 5$ and the remainder is -14

27. $-2 \overline{\smash{\big)}\ 4 \quad 0 \quad -1 \quad 0 \quad 2 \quad -1}$

$\underline{\qquad -8 \quad 16 \quad -30 \quad 60 \quad -124}$

$4 \quad -8 \quad 15 \quad -30 \quad 62 \quad \boxed{-125}$ quotient is $4x^4 - 8x^3 + 15x^2 - 30x + 62$

remainder is -125

29. $5 \overline{\smash{\big)}\ 4 \quad -3 \quad 0 \quad -1 \quad 3}$

$\underline{\qquad 20 \quad 85 \quad 425 \quad 2120}$

$4 \quad 17 \quad 85 \quad 424 \quad \boxed{2123}$ so p(5) = 2123

31. $p(x) = 4x^3 - 7x^2 - 3x - 8$ has 1 sign change so it will have 1 positive zero

$p(-x) = 4(-x)^3 - 7(x)^2 - 3(-x) - 8 = -4x^3 - 7x^2 + 3x - 8$ has 2 sign changes so

p(x) will either have 2 negative zeros or no negative zeros.

33. $p(x) = 4x^4 + 12x^3 - 11x^2 - 5x + 3$ has 2 sign changes so it will have either 2 or no positive roots;

$p(-x) = 4x^4 - 12x^3 - 11x^2 + 5x + 3$ has 2 sign changes so p(x) will have either 2 or no negative

roots.

35. $p(x) = x^3 - 7x^2 + 14x - 8$ has rational root possibilities $\pm 8, \pm 4, \pm 2, \pm 1$

35. (continued)

      trying 1: (always try 1 first!)

$$1\ )\ \overline{1\quad -7\quad 14\quad -8}$$

$$\underline{\quad\quad 1\quad -6\quad 8\quad}$$

        1   -6   8 $\boxed{0}$ factorization of p(x) at this point: $(x-1)(x^2-6x+8)$

$x^2 - 6x + 8$ factors further into $(x - 4)(x - 2)$ so the zeros are 1, 4, and 2.

37. $p(x) = 2x^3 + x^2 + 2x + 1$ has possibilites $-1, -\frac{1}{2}$ because of no sign changes;

    -1 didn't work; trying -1/2:

$$-1/2\ )\ \overline{2\quad 1\quad 2\quad 1}$$

$$\underline{\quad\quad -1\quad 0\quad -1\quad}$$

        2   0   2 $\boxed{0}$ factorization so far: $(x + \frac{1}{2})(2x^2 + 2)$

solving $2x^2 + 2 = 0$: $2x^2 = -2$; $x^2 = -1$; $x = \pm i$ so the zeros are $-\frac{1}{2}$, i, and -i.

39. $p(x) = x^4 - x^3 - 10x^2 + 4x - 24$ has $\pm 1, \pm 2, \pm 3, \pm 4, \pm 6, \pm 8, \pm 12$ and ± 24 as its rational root

    possibilities. Using the synthetic substitution approach and starting at 1 (it didn't work; nor

    did 2 or 3) and the bottom row of numbers all turned positive when 4 was substituted.

    Therefore, 6,8,12, and 24 have no chance. -1 and -2 and -3 didn't work either and -4 made the

    bottom numbers alternate in sign flip-flop so no need to look at -6,-8,-12 or -24. Therefore,

    this p(x) has no rational zeros.

41. $p(x) = x^5 - 4x^4 + 4x^3 - 16x^2 - 5x + 20$ poss: $\pm 1,2,4,5,10,20$

    sooner or later you'll try 1: $1\ )\ \overline{1\quad -4\quad 4\quad -16\quad -5\quad 20}$

$$\underline{\quad\quad\quad 1\quad -3\quad 1\quad -15\quad -20\quad}$$

                1   -3   1   -15   -20 $\boxed{0}$

Now we can concentrate on the polynomial $x^4 - 3x^3 + x^2 - 15x - 20$ (which still has same possi-

bilities as original). Trying to synthetically substitute 4 into it yielded the value 0. At this

point, 1 and 4 are zeros and the polynomial under consideration is $x^3 + x^2 + 5x + 5$ (from the

bottom row of your last synthetic substitution of 4). The only possibilities for this new third

degree polynomial could be - 1 or -5; the - 1 works and that's it. The rational zeros are ± 1 and

4; the other two zeros must be irrational.

43. $p(x) = x^3 + 2x - 2$; $p(0) = -2$ and $p(1) = 1$ so the first midpoint is 0.5.

$p(0.5) = -0.875$ while $p(1)$ is 1 so the next midpoint is 0.75

$p(0.75) = -0.078125$ while $p(1)$ is still +1 so the next midpoint is 0.875

$p(0.875) = 0.4199219$ and $p(0.75)$ was negative so the next midpoint is 0.8125

$p(0.8125) = 0.161377$ and $p(0.75)$ was the last with a negative value so next midpoint is 0.78125

$p(0.78125) = 0.039337$ and $p(0.75)$ is still negative so next midpoint is 0.765625

$p(0.765625) = -0.019955$ and $p(0.78125) = +0.0039337$ so next midpoint is 0.7734375

$p(0.7734375) = 0.0095496$ and $p(0.765625)$ was negative so next midpoint is 0.7695313

$p(0.7695313) = -0.0052376$ and $p(0.7734375)$ was positive so the next midpoint is 0.7714844

and this will be our approximate zero because the last two endpoints, 0.7695313 and 0.7734375 are less than 0.005 apart so our answer must be within 0.005 of the true one.

45. $p(x) = x^4 + 3x^3 + x - 10$ and $p(1) = -5$ while $p(2) = 48$ so there is a zero between 1 and 2; the first midpoint is 1.5.

$p(1.5) = 9.21875$ while $p(1)$ was -5 so the next midpoint is 1.25

$p(1.25) = 0.1611328$ while $p(1)$ was -5 so the next midpoint is 1.125

$p(1.125) = -2.801483$ while $p(1.25)$ was positive so the next midpoint is 1.1875

$p(1.1875) = -1.427426$ and $p(1.25)$ is still positive so the next midpoint is 1.121875

$p(1.121875) = -0.6615429$ while $p(1.25)$ remains positive so the next midpoint is 1.234375

$p(1.234375) = -0.2575092$ and $p(1.25) = 0.1611328$ so the next midpoint us 1.242188

$p(1.242188) = -0.050028$ and $p(1.25)$ is positive so the next and last midpoint is 1.246094

47. $p(x) = x^3 + 2x - 2$ with a zero between 0 and 1; the derivative function would be $3x^2 + 2$ and lines 10 and 20 in the program would have to be changed to:

        10 DEF FNP(X) = X^3 + 2*X - 2
        20 DEF FND(X) = 3*X^2 + 2

and input 0 at the prompt. The program produced iterates agreeing at approximately 0.770917

49. $p(x) = x^5 + 3x^3 + x - 10$ with a zero between 1 and 2

    the derivative function would be $5x^4 + 9x^2 + 1$ Change lines 10 and 20 to:

        10 DEF FNP(X) = X^5 + 3*X^3 + X - 10
        20 DEF FND(X) = 5*X^4 + 9*X^2 + 1

and input 1 at the prompt. The iterates should agree at approximately 1.244051

CHAPTER FIVE
Rational Functions and Conic Sections

Chapter Objectives

In chapter five you will learn about rational (that is, fractional-in-appearance) functions. The emphasis will be on how to go about sketching their graphs relatively accurately. Many of their graphs exhibit what is called "asymptotic behavior" which means that the graph will get closer and closer to certain dotted lines in certain situations. These lines are called asymptotes. The remainder of the chapter is devoted to studying three curves called the conic sections; the term "conic" is used because these curves can be seen on the intersection of a plane and a two-napped cone (see page 291 in text). The three curves are called parabolas, ellipses, and hyperbolas.

Chapter Summary

- A **rational function** is a function of the form $f(x) = \dfrac{p(x)}{q(x)}$ where p(x) and q(x) are polynomials.

- The vertical line $x = c$ is a **vertical asymptote** of the graph of $y = f(x)$ if either $f(x) \to \infty$ or $f(x) \to -\infty$ as $x \to c^+$ or $x \to c^-$. It is a vertical asymptote of $f(x) = \dfrac{p(x)}{q(x)}$ if $q(c) = 0$ but $p(c) \neq 0$.

- The horizontal line $y = c$ is a **horizontal asymptote** of the graph of $y = f(x)$ if $f(x) \to c$ as $x \to \infty$ or $x \to -\infty$.

- Horizontal Asymptotes Theorem:

$$\text{If } f(x) = \frac{p(x)}{q(x)} = \frac{a_n x^n + a_{n-1} x^{n-1} + \cdots + a_1 x + a_0}{b_m x^m + b_{m-1} x^{m-1} + \cdots + b_1 x + b_0} \text{ , } a_n \text{ and } b_m \neq 0, \text{ then}$$

i. if $n < m$, $y = 0$ is a horizontal asymptote for f.

ii. if $n = m$, then $y = \dfrac{a_n}{b_m}$ is a horizontal asymptote for f.

iii. if $n > m$, then f has no horizontal asymptote.

- An **ellipse** is the set of points (x,y) such that the sum of the distances from (x,y) to two given points is fixed. Each of the two points is called a focus of the ellipse.

- The standard equation of an ellipse is $\frac{x^2}{a^2} + \frac{y^2}{b^2} = 1$

- If $a > b$, the line segment from $(-a,0)$ to $(a,0)$ is the **major axis** and the line from $(0,-b)$ to $(0,b)$ is the **minor axis**, and the points $(-a,0)$ and $(a,0)$ are the **vertices** of the ellipse. If $b > a$, the major and minor axes are reversed. If $a = b$, the ellipse is a **circle**. The intersection of the axes is the **center** of the ellipse.

- The **eccentricity**, e, of an ellipse is given by $e = \frac{c}{a}$ if $a \geq b$ and $e = \frac{c}{b}$ if $b \geq a$, where 2c is the distance between the foci.

- A **translated ellipse** has the standard equation $\frac{(x - x_0)^2}{a^2} + \frac{(y - y_0)^2}{b^2} = 1$

- A **parabola** is the set of points (x,y) equidistant from a fixed point called the **focus** and a fixed line that does not contain the focus called the **directrix**.

- The **standard equations of a parabola** are $x^2 = 4cy$ (which opens upward if $c > 0$ and downward if $c < 0$) and $y^2 = 4cx$ (which opens to the right if $c > 0$ and to the left if $c < 0$).

- The line about which a parabola is symmetric is the **axis** of the parabola.

- The point at which the axis and the parabola intersect is the **vertex** of the parabola.

- A **translated parabola** takes the standard form $(x - x_0)^2 = 4c(y - y_0)$ or $(y - y_0)^2 = 4c(x - x_0)$

- A **hyperbola** is a set of points (x, y) with the property that the positive difference between the distances from (x, y) and each of the two distinct points, called **foci**, is a constant.

- The **principal axis** of a hyperbola is the line containing the foci. The points of intersection of the principal axis and the hyperbola are the **vertices**, and the line segment joining the vertices is the **transverse axis**. The midpoint of the line segment joining the foci is the **center** of the hyperbola.

- The **standard equations of a hyperbola** are

$$\frac{x^2}{a^2} - \frac{y^2}{b^2} = 1 \quad \text{and} \quad \frac{y^2}{a^2} - \frac{x^2}{b^2} = 1$$

- A **translated hyperbola** takes the standard form

$$\frac{(x - x_0)^2}{a^2} - \frac{(y - y_0)^2}{b^2} = 1 \quad \text{or} \quad \frac{(y - y_0)^2}{a^2} - \frac{(x - x_0)^2}{b^2} = 1$$

SOLUTIONS TO CHAPTER FIVE PROBLEMS

Problems 5.1

1. $f(x) = \dfrac{1}{x^2 - 1}$

vertical asymptotes: (occur at x-values which make only the denominator 0)

to find, solve the equation $x^2 - 1 = 0$

$$(x + 1)(x - 1) = 0$$

$x = -1$ and $x = 1$ are the equations

horizontal asymptote: (check out what happens to the value of f(x) as $x \to \infty$)

method I (common sense): as x grows without bound, $x^2 - 1$ does likewise

and the fraction $\dfrac{1}{x^2 - 1}$ gets SMALLER, closer to 0

so we'd say that the horizonatal line $y = 0$, the x-axis, is the

horizontal asymptote.

method II (comparing degrees of numerator and denominator): here n,

the degree of the numerator, is 0 and m, the degree of the de-

nominator, is 2. We have $n < m$ so $y = 0$ is the asymptote. (see

chapter summary, the horizontal asymptote theorem)

3. $f(x) = \dfrac{x - 1}{x^2 + 5x + 6} = \dfrac{x - 1}{(x + 3)(x + 2)}$

verticals: the x-values -3 and -2 make only the denominator 0 so $x = -3$ and

$x = -2$ are the vertical asymptotes (VA's).

horizontal: here $n = 1$ and $m = 2$ so $n < m$ and $y = 0$ must be the horizontal

asymptote (HA).

5. $f(x) = \dfrac{x + 2}{x^2 + 7x + 6} = \dfrac{x + 2}{(x + 6)(x + 1)}$

vertical: $x = -6$ and $x = -1$ make only the denominator 0 so they are the locations

of the vertical asymptotes.

horizontal: $y = 0$ since $n < m$

7. $f(x) = \dfrac{x^2 - 9}{(x - 9)^2} = \dfrac{(x + 3)(x - 3)}{(x - 9)^2}$

    vertical: x = 9 since 9 makes only the denominator 0

    horizontal: y = 1 (since n = 2 and m = 2 we take the quotient of the leading

    coefficients which in this problem would be $\frac{1}{1} = 1$)

9. $f(x) = \dfrac{3x + 5}{x^2 - 2x} = \dfrac{3x + 5}{x(x - 2)}$

    vertical: x = 0 and x = 2

    horizontal: y = 0 since n < m (1 is less than 2)

11. $f(x) = \dfrac{3}{x - 2}$

    vertical: x = 2 as 2 is the only number that makes only the denominator 0

    horizontal: y = 0 since n < m (0 is less than 1)

13. $f(x) = \dfrac{x^2 + 2x + 1}{3x + 4}$

    vertical: set 3x + 4 = 0 and solve to get x = - $\frac{4}{3}$

    horizontal: none since n > m (2 is greater than 1)

    Note: this function WOULD have an oblique asymptote!

15. $f(x) = \dfrac{x^3 + 1}{1000x^2 + 50x + 100} = \dfrac{(x + 1)(x^2 - x + 1)}{50(20x^2 + x + 2)}$

    vertical: none since $20x^2 + x + 2 = 0$ doesn't have any real solutions (its discriminant is -159!)

    horizontal: none since n > m (oblique again)

17. $f(x) = \dfrac{(x + 1)(x - 3)(x + 5)}{(x - 1)(x + 2)(x - 4)}$

    vertical: x = 1, x = - 2, and x = 4 all make only the denominator 0

    horizontal: y = 1 because m = n = 3 and the quotient of the leading coefficients, 1 and 1, is 1

19. $f(x) = \dfrac{(x^2 - 1)^6}{1 - x^7}$ here n = 12 and m = 7

reducing is a little chore here because both numerator and denominator are 0 at x = 1:

$$f(x) = \frac{(x + 1)^6 (x - 1)^6}{(1 - x)(1 + x + x^2 + x^3 + x^4 + x^5 + x^6)} = \frac{-(x + 1)^6 (x - 1)^5}{1 + x + x^2 + x^3 + x^4 + x^5 + x^6}$$

vertical: none that can be found

horizontal: none since n > m

21. $f(x) = \dfrac{x^3 - 2}{x^2 + 2}$ which equals $\dfrac{x - \dfrac{2}{x^2}}{1 + \dfrac{2}{x^2}}$ if numerator and denominator are divided by x^2 with $x \neq 0$

   (x^2 is the highest power of x in the denominator of f(x))

and if x gets really big in absolute value ($|x| \to \infty$) the $\dfrac{2}{x^2}$ fractions each approach 0 in value

and $f(x) \approx \dfrac{x}{1} = x$ and so y = x is an oblique asymptote.

23. $f(x) = \dfrac{2x^3 - 3x + 2}{1 - 5x^2} = \dfrac{2x - \dfrac{3}{x} + \dfrac{2}{x^2}}{\dfrac{1}{x^2} - 5}$ if numerator and denominator are divided by x^2;

If we think about x values that are very large in absolute value, we can ignore all those fractions

with x's in the denominator (their values approach 0) and see that $f(x) \approx \dfrac{2x}{-5}$

and so $y = -\dfrac{2}{5}x$ is an oblique asymptote for this f(x).

25. $f(x) = \dfrac{x^5 - x^3 + 2x}{2x^4 - 3x^2 + 3} = \dfrac{x - \dfrac{1}{x} + \dfrac{2}{x^3}}{2 - \dfrac{3}{x^2} + \dfrac{3}{x^4}}$ if numerator and denominator are divided by x^4

Ignoring all those fractions whose values dwindle as $|x| \to \infty$ shows that $f(x) \approx \dfrac{x}{2}$ and there-

fore $y = \dfrac{1}{2} x$ is the equation of the oblique asymptote here.

27. $f(x) = \dfrac{1}{x + 2}$ would have a vertical asymptote at x = -2, a horizontal asymptote at y = 0

and a y-intercept at $(0, \dfrac{1}{2})$; this information coincides only with graph e.

29. $f(x) = \dfrac{2}{1 - x^2}$ would have vertical asymptotes at $x = 1$ and at $x = -1$; it would have $y = 0$ as

the horizontal asymptote and would cross the y-axis at $(0, 2)$. Also, since $f(x) = f(-x)$, the graph must have y-axis symmetry. All this information points to graph a.

31. $f(x) = \dfrac{x + 1}{x^2 + x - 2} = \dfrac{x + 1}{(x + 2)(x - 1)}$ has vertical asymptotes at $x = - 2$ and at $x = 1$. It has a

horizontal asymptote at $y = 0$ because the larger degree is in the denominator. The y-intercept is $(0, - \frac{1}{2})$ so graph h must be the one.

33. $f(x) = \dfrac{x}{x^2 - 1}$ has vertical asymptotes at $x = 1$ and at $x = -1$ and $y = 0$ serves as the hori-

zontal asymptote; the y- intercept here however is $(0, 0)$ and this indicates graph i is the one.

35. $f(x) = \dfrac{x^2 - 1}{(x - 1)^2} = \dfrac{(x + 1)(x - 1)}{(x - 1)(x - 1)} = \dfrac{x + 1}{x - 1}$ if $x \neq 1$

The graph of f(x) would have a vertical asymptote at $x = 1$, a horizontal asymptote at $y = 1$
(divide numerator and denominator by x and then let x get big), and would have a y-intercept
at $(0, - 1)$. It must be graph f.

37. $f(x) = \dfrac{1}{x - 3}$

VA: $x = 3$ HA: $y = 0$ sketch:

x-int: none since there is no x-value
which makes only the numerator 0

y-int: $f(0) = - \frac{1}{3}$

table: x: 1   2   4   5

f(x): $-\frac{1}{2}$ -1   1 $\frac{1}{2}$

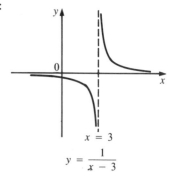

$x = 3$

$y = \dfrac{1}{x - 3}$

39. $f(x) = \dfrac{1}{x^2}$

sketch:

VA: x = 0 makes only the denominator 0

HA: y = 0 because n < m

x-int: none

y-int: none because f(0) is undefined

table: x: -2 -1   1   2

f(x): $\frac{1}{4}$ 1   1 $\frac{1}{4}$

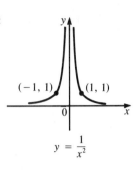

$y = \dfrac{1}{x^2}$

41. $f(x) = \dfrac{1}{x^2} - 2$ (this is the graph of $\dfrac{1}{x^2}$, #39 above, lowered two blocks!)

sketch:

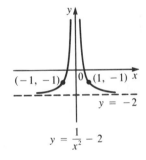

$y = \dfrac{1}{x^2} - 2$

43. $f(x) = \dfrac{1}{x^3}$

vertical: x = 0

horizontal: y = 0 (n < m)

x-int: none since $1 \neq 0$

y-int: none since f(0) is undefined

table: x: -2   -1   1   2

f(x): $-\frac{1}{8}$ -1   1 $\frac{1}{8}$

sketch:

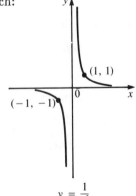

$y = \dfrac{1}{x^3}$

45. $f(x) = \frac{1}{x^3} + 3$

this would be the graph of $\frac{1}{x^3}$, #43 above, raised 3 blocks

sketch:

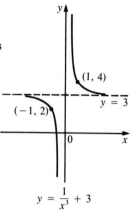

$$y = \frac{1}{x^3} + 3$$

47. $f(x) = \frac{1}{x^4}$ (symmetric to y-axis because $f(-x) = \frac{1}{(-x)^4} = \frac{1}{x^4} = f(x)$)

vertical: $x = 0$

horizontal: $y = 0$

neither x- nor y-intercepts exist

table: x: -2 -1   1   2

        f(x): $\frac{1}{16}$ 1   1 $\frac{1}{16}$

sketch:

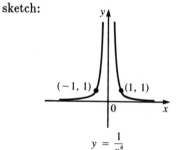

$$y = \frac{1}{x^4}$$

49. $f(x) = \frac{1}{x - x^2} = \frac{1}{x(1 - x)}$

vertical: $x = 0$ and $x = 1$ make only the denominator 0

horizontal: $y = 0$ because degree of numerator is less than the degree of the
        denominator

x-intercepts: none because no value of x makes only the numerator 0

y-intercept: none because $f(0)$ is undefined

table: (choose x-values on all sides of all vertical asymptotes)

      x: -2   -1 1/2   2   3

   f(x): -1/6   -1/2 4 -1/2 -1/6

sketch:

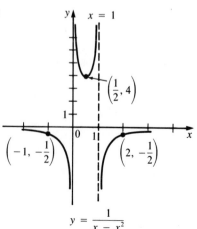

$$y = \frac{1}{x - x^2}$$

- 206 -

51. $f(x) = \dfrac{-2}{x^2 - 3x + 2} = \dfrac{-2}{(x - 2)(x - 1)}$ sketch:

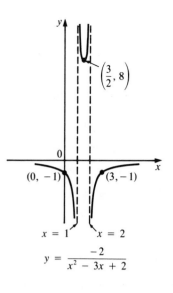

$\left(\dfrac{3}{2}, 8\right)$

$(0, -1)$ $(3, -1)$

$x = 1$ $x = 2$

$y = \dfrac{-2}{x^2 - 3x + 2}$

vertical: $x = 2$ and $x = 1$ are the VA's

horizontal: $y = 0$ (n < m again)

NO x-intercepts because the numerator can't be 0

y-intercept: $f(0) = \dfrac{-2}{2} = -1$

table:

x:	-1	1.5	3	4
f(x):	$-\frac{1}{3}$	8	-1	$-\frac{1}{3}$

53. $f(x) = \dfrac{2}{(x + 5)(x - 3)(x + 7)}$ sketch:

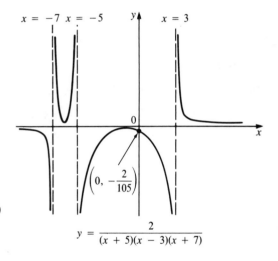

$x = -7$ $x = -5$ $x = 3$

$\left(0, -\dfrac{2}{105}\right)$

$y = \dfrac{2}{(x + 5)(x - 3)(x + 7)}$

vertical: $x = -5$, $x = 3$, and $x = -7$ qualify

horizontal: $y = 0$ is approached by $f(x)$

NO x-intercepts

y-intercept: $f(0) = \dfrac{2}{(5)(-3)(7)} = -\dfrac{2}{105}$

table:

x:	-8	-6	-2	1	4
f(x):	-2/33	2/9	-2/75	-1/48	2/99

55. $f(x) = \dfrac{4}{x^2 + 4x + 5} = \dfrac{4}{(x + 2)^2 + 1}$

vertical: none since $(x + 2)^2 + 1 \neq 0$ ever   sketch:

horizontal: $y = 0$ because n < m (0 < 2)

x-intercepts: none because $4 \neq 0$

y-intercept: $f(0) = \dfrac{4}{5}$

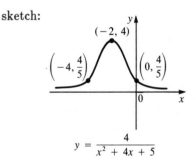

$(-2, 4)$

$\left(-4, \dfrac{4}{5}\right)$ $\left(0, \dfrac{4}{5}\right)$

$y = \dfrac{4}{x^2 + 4x + 5}$

Note: $f(x)$ will be greatest when its denominator

is its smallest and this occurs when $x = -2$ which

is seen from the completed square version.

table:

x:	-4	-3	-2	-1
f(x):	4/5	2	4	2

57. $f(x) = \dfrac{x - 1}{(x - 2)^2}$

sketch:

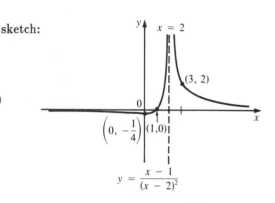

vertical: $x = 2$ makes only the denominator 0

horizontal: $y = 0$ because top has lower degree

x-intercept: $x = 1$ makes only the numerator 0

y-intercept: $f(0) = \dfrac{-1}{(-2)^2} = -\dfrac{1}{4}$

table:

x:	-3	-2	-1	3	4
f(x):	-4/25	-3/16	-2/9	2	3/4

59. $f(x) = \dfrac{x^2 - 3}{x^2 - 3x + 2} = \dfrac{x^2 - 3}{(x - 2)(x - 1)}$

vertical: the lines $x = 2$ and $x = 1$

horizontal: the line $y = 1$ since degrees are the same and $\frac{1}{1} = 1$

x-intercepts: setting $x^2 - 3 = 0$ gives the values $\pm\sqrt{3}$ or approximately ± 1.73

y-intercept: $f(0) = -\dfrac{3}{2}$ sketch:

table:

x:	-1	1.5	3	4
f(x):	-1/3	3	3	13/6

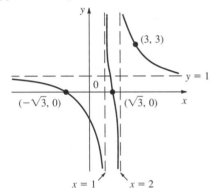

61. $f(x) = \dfrac{(x + 3)^2}{(x + 3)^2 - 4} = \dfrac{(x + 3)^2}{x^2 + 6x + 5} = \dfrac{(x + 3)^2}{(x + 5)(x + 1)}$

vertical: $x = -5$ and $x = -1$ are vertical asymptotes

horizontal: $y = \frac{1}{1} = 1$ (same degree in top and bottom)

x-intercept: $(x + 3)^2 = 0$ if $(x + 3) = 0$ if $x = -3$

y-intercept: $f(0) = 9/5$ sketch:

table:

x:	-7	-6	-4	-2	1
f(x):	4/3	9/5	-1/3	-1/3	4/3

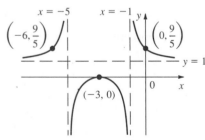

63. $f(x) = \dfrac{x^2 - 1}{(x + 2)^2} = \dfrac{(x + 1)(x - 1)}{(x + 2)^2}$

vertical: the line x = -2 is the only one

horizontal: degrees same so y = 1 (quotient of leading coefficients)

x-intercepts: the x-values -1 and 1 make only the numerator 0

y-intercept: f(0) = -1/4   sketch:

table: x: -4   -3   2

      f(x): 15/4   8   3/16

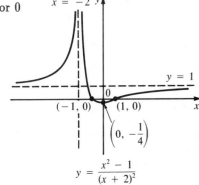

65. $f(x) = \dfrac{x - 3}{x^2 + 2x - 15} = \dfrac{x - 3}{(x + 5)(x - 3)} = \dfrac{1}{x + 5}$ provided $x \neq 3$

Note: Whenever you divide out a common factor from a rational function, you'll

     have to delete a point from the graph that results from the reduced ver-

     sion. Here, we'll have to delete the point which seems to correspond to

     the x-value 3.

     vertical: only x = -5   sketch:

     horizontal: y = 0

     x- intercepts: none

     y-intercept: f(0) = 1/5

     table: x: -7   -6   -4   -3

          f(x): -1/2   -1   1   1/2

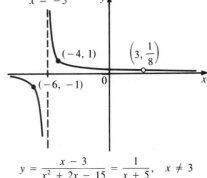

67. $f(x) = \dfrac{x^2 - 1}{x^2} = \dfrac{(x - 1)(x + 1)}{x^2}$ sketch:

vertical: x = 0 makes only the denominator 0

horizontal: y = 1 (same degree)

x-intercepts: x = 1 and x = -1 make the numerator 0

NO y-intercept because f(0) is undefined

table: x: -2   2

     f(x): 3/4   3/4 (y-axis symmetry)

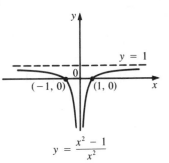

69. $f(x) = \dfrac{x^2 - 2x + 4}{x - 1}$ vertical: $x = 1$ makes only the denominator zero

oblique:

$$
\begin{array}{r}
x - 1 \\
x - 1 \overline{) x^2 - 2x + 4} \\
\underline{x^2 - x} \\
-x + 4 \\
\underline{-x + 1} \\
3
\end{array}
$$

so $y = x - 1$ is the oblique asymptote

x-intercepts: none since $x^2 - 2x + 4 = 0$ has a negative discriminant and therefore can't have any real zeros.

y-intercept: $f(0) = -4$

sketch:

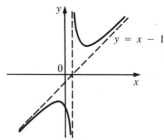

table:

x:	-2	-1	2	3
f(x):	-4	-7/2	4	7/2

71. $F = \dfrac{6}{r^2}$ and we can assume that $r > 0$ since it represents the distance between two particles. There would be a vertical asymptote at $r = 0$ (As the distance between two particles lessens, the gravitational force of attraction betweeen them increases.   There would be a horizontal asymptote in the line $F = 0$ as $n = 0$ and $m = 2$.   sketch:

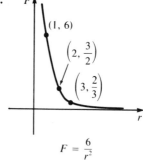

table:

r:	1	2	5	10
F:	6	3/2	6/25	0.06

$F = \dfrac{6}{r^2}$

73. $R = \dfrac{0.75}{r^4} = \dfrac{3}{4r^4}$ vertical asymptote at $r = 0$; horizontal asymptote at $R = 0$

sketch:

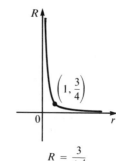

table:

r:	0.5	1	2	10
R:	12	3/4	3/64	3/40000

$R = \dfrac{3}{4r^4}$

Problems 5.3

1. $\frac{x^2}{16} + \frac{y^2}{25} = 1$ Comparing the given equation to $\frac{x^2}{a^2} + \frac{y^2}{b^2} = 1$ lets us know that a = 4 and b = 5
The center is the origin (0,0), and we have b > a (5 > 4) so the major axis must be on the y-axis.
The vertices would be at $(0,\pm b) = (0,\pm 5)$. The foci also lie on the y-axis at points with coordin-
ates $(0,\pm c)$ where (since b > a) $c = \sqrt{b^2 - a^2} = \sqrt{25 - 16} = 3$. Therefore, the foci are $(0,\pm 3)$.
The major axis is the segement connecting (0,-5) and (0,5); the minor axis is the segment connecting
(-a,0) to (a,0), in this case (-4,0) to (4,0). The eccentricity (since b > a) is given by $\frac{c}{b} = \frac{3}{5} = 0.6$

Sketch: $e = \frac{3}{5} = 0.6$

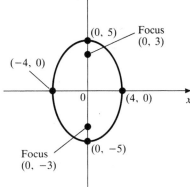

$1 = \frac{x^2}{16} + \frac{y^2}{25} = (\frac{x}{4})^2 + (\frac{y}{5})^2$

3. $x^2 + \frac{y^2}{9} = 1$ has a = 1 and b = 3 so once again b > a making it an "up-down" or "vertical"
ellipse (ie, the major (longer) axis is vertical). The center is (0,0) and $c = \sqrt{b^2 - a^2}$ or
$\sqrt{3^2 - 1^2} = \sqrt{8} = 2\sqrt{2}$ which makes the foci $(0, \pm 2\sqrt{2})$. The vertices are (0,-3) and (0,3) with
the major axis being the segment which connects them. The minor axis is the segment
connecting (-1,0) and (1,0). The eccentricity is given by $e = \frac{c}{b} = \frac{2\sqrt{2}}{3} \approx 0.94$

Sketch:

$e = \frac{2\sqrt{2}}{3} \approx 0.94$

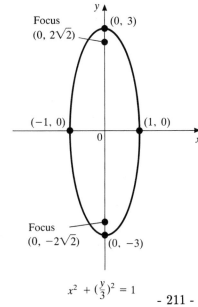

$x^2 + (\frac{y}{3})^2 = 1$

5. $x^2 + 4y^2 = 16$ is equivalent to $\frac{x^2}{16} + \frac{y^2}{4} = 1$ which has a = 4 and b= 2 making the major axis horizontal. The center is (0,0) and $c = \sqrt{a^2 - b^2} = \sqrt{12} = 2\sqrt{3}$ making the foci (on the x-axis) $(\pm 2\sqrt{3}, 0)$. The vertices are at (4, 0) and (-4, 0) and the major axis is the segment connecting them. The minor axis runs vertically connecting (0, -2) and (0, 2). The eccentricity in this case (a>b) is given by $e = \frac{c}{a} = \frac{2\sqrt{3}}{4} = \frac{\sqrt{3}}{2} \approx 0.87$ $e = \frac{\sqrt{3}}{2} \approx 0.87$

Sketch:

$(\frac{x}{4})^2 + (\frac{y}{2})^2 = 1$

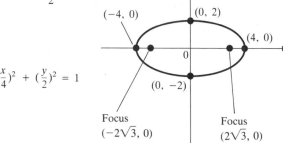

7. $\frac{(x-1)^2}{16} + \frac{(y+3)^2}{25} = 1$ has its center at (1,-3) with a = 4 and b = 5 (an "up-down" ellipse) The vertices must be located 5 blocks above and below the center, ie at (1, -3±5) , ie at (1, 2) and at (1, -8). The major axis connects these points. The ends of the minor axis must be located 4 blocks to the left and right of the the center, ie at (1±4, -3), ie (5, -3) and (-3, -3). Finding c: $c = \sqrt{b^2 - a^2} = \sqrt{9} = 3$ so the foci are located 3 blocks above and below the center which puts them at (1,-3±3) or at (1, 0) and (1, -6). Since b > a the eccentricity is $\frac{c}{b} = \frac{3}{5} = 0.6$

Sketch: $e = \frac{3}{5} = 0.6$

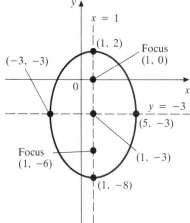

$1 = \frac{(x-1)^2}{16} + \frac{(y+3)^2}{25}$

$= \left(\frac{x-1}{4}\right)^2 + \left(\frac{y+3}{5}\right)^2$

9. $2x^2 + 2y^2 = 2$ is the same as $x^2 + y^2 = 1$.   Sketch:

Since a = b = 1, it's a circle with a radius

of 1 centered at the origin (0,0). In other

words, it's the UNIT CIRCLE.

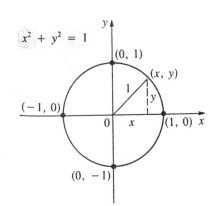

11. $x^2 + 4y^2 = 9$ is the same as $\dfrac{x^2}{9} + \dfrac{4y^2}{9} = 1$ which is the same as $\dfrac{x^2}{9} + \dfrac{y^2}{9/4} = 1$ so now we can

observe that $a = 3$ and $b = 3/2$. Since $a > b$ it is a "left-right" ellipse and the center is the

origin $(0, 0)$. $c = \sqrt{a^2 - b^2} = \sqrt{9 - 9/4} = \sqrt{27/4} = \dfrac{\sqrt{27}}{2}$ which means the foci are on the x-axis

at the points $(\pm\sqrt{27}/2, 0)$ which are appoximated by $(\pm 2.6, 0)$. The major axis runs horizontally

from $(-3,0)$ to $(3,0)$; the minor axis runs vertically from $(0,-3/2)$ to $(0,3/2)$.

$e = \dfrac{c}{a} = \dfrac{\sqrt{27}/2}{3} = \dfrac{\sqrt{27}}{6} \approx 0.87$

Sketch:

$e = \dfrac{\sqrt{3}}{2} \approx 0.87$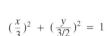

$(\dfrac{x}{3})^2 + (\dfrac{y}{3/2})^2 = 1$

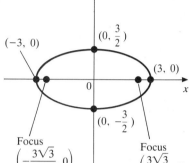

13. $4x^2 + 8x + y^2 + 6y = 3$

View as $4(x^2 + 2x +) + (y^2 + 6y +) = 3$ and then complete the square as follows:

$4(x^2 + 2x + 1) + (y^2 + 6y + 9) = 3 + 4 + 9$ [add (half of coefficient of linear term)^2]

$4(x + 1)^2 + (y + 3)^2 = 16$ or $\dfrac{(x+1)^2}{4} + \dfrac{(y+3)^2}{16} = 1$ [factor and divide by 16]

So the center is at $(-1, -3)$, $a = 2$ and $b = 4$ making the ellpse "vertical". The vertices will be lo-

cated 4 units above and below the center which puts them at $(-1,-3\pm4)$, ie at $(-1,1)$ and at $(-1,-7)$.

The major axis is the segment which connects these points; the minor axis connects points which

are 2 units left or right of the center, i.e. at $(-1\pm2,-3)$, i.e. at $((1,-3)$ and at $(-3,-3)$. Regarding

the foci, $c = \sqrt{16 - 4} = \sqrt{12} = 2\sqrt{3}$ which places the foci at points $2\sqrt{3}$ units above and below

$(-1,-3)$ and these points are $(-1, -3\pm2\sqrt{3})$. The eccentricity, e, is given by $\dfrac{c}{b} = \dfrac{2\sqrt{3}}{4} = \dfrac{\sqrt{3}}{2} \approx 0.87$

Sketch:

$e = \dfrac{\sqrt{3}}{2} \approx 0.87$

$4(x + 1)^2 + (y + 3)^2 = 16$

$(\dfrac{x + 1}{2})^2 + (\dfrac{y + 3}{4})^2 = 1$

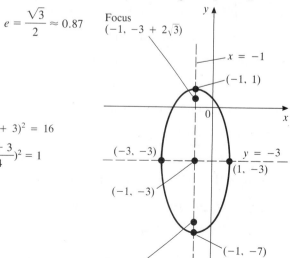

15. $4x^2 + 8x + y^2 - 6y = 3 \Rightarrow 4(x^2 + 2x + \quad) + (y^2 - 6y + \quad) = 3$ and then, completing squares gives $4(x^2 + 2x + 1) + (y^2 - 6y + 9) = 3 + 4 + 9 \Rightarrow \dfrac{(x+1)^2}{4} + \dfrac{(y-3)^2}{16} = 1$ from which it can be seen that the center is at (-1, 3) and a = 2 and b = 4 ; since b>a, it's a "vertical" ellipse. The major axis connects the vertices which are at (-1, 3±4), ie at (-1,7) and at (-1,-1); the minor axis connects the points (-1±2, 3) or (1,3) and (-3,3). The value of c is $\sqrt{b^2 - a^2} = \sqrt{12} = 2\sqrt{3}$ which puts the foci (above and below the center) at (-1, 3±2$\sqrt{3}$). $e = \dfrac{c}{b} = \dfrac{2\sqrt{3}}{4} \approx 0.87$

Sketch:

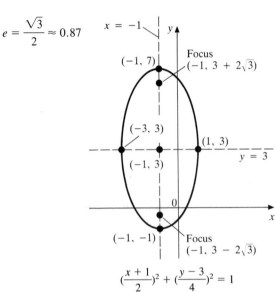

$e = \dfrac{\sqrt{3}}{2} \approx 0.87$

$(\dfrac{x+1}{2})^2 + (\dfrac{y-3}{4})^2 = 1$

17. $3x^2 + 12x + 8y^2 - 4y = 20$

$3(x^2 + 4x + "4") + 8(y^2 - 1/2\, y + "1/16") = 20 + "12" + "1/2"$

$3(x+2)^2 + 8(y - 1/4)^2 = 65/2 \Rightarrow \dfrac{6(x+2)^2}{65} + \dfrac{16(y-1/4)^2}{65} = 1$ or $\dfrac{(x+2)^2}{65/6} + \dfrac{(y-1/4)^2}{65/16} = 1$

so a = $\sqrt{65/6} \approx 3.3$ and b = $\sqrt{65/16} \approx 2$ so a>b and its a "horizontal" ellipse with foci and vertices on the horizontal major axis. The center of the ellipse is at the point (-2, 1/4) and the vertices are located $\sqrt{65/6}$ units to the left and right of the center at (-2± $\sqrt{65/6}$, 1/4). Decimal approximations for the coordinates of these points are (1.3, 0.25) and (-5.3, 0.25). The major axis connects these points. The minor axis connects the points (-2, 1/4 ±$\sqrt{65/16}$) which can be approximated by (-2, 2.25) and (-2, -1.75). Computing c:

$c = \sqrt{a^2 - b^2} = \sqrt{65/6 - 65/16} = $ (lcd is 48) $\sqrt{325/48} \approx 2.6$

This puts the foci this distance to the left and right of the center, that is, at the points

(-2 ± $\sqrt{325/48}$, 1/4) which are approximately the points (0.6, 0.25) and (-4.6, 0.25)

The eccentricity is e = c/a = $\sqrt{325/48} \div \sqrt{65/6} = \sqrt{5/8} = \sqrt{10}/4$ (rationalized) ≈ 0.79

17. (continued)

Sketch:

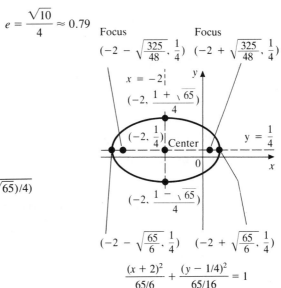

$$e = \frac{\sqrt{10}}{4} \approx 0.79$$

center: $(-2, \frac{1}{4})$
foci: $(-2 \pm \sqrt{\frac{325}{48}}, \frac{1}{4})$
vertices: $(-2 \pm \sqrt{\frac{65}{6}}, \frac{1}{4})$
major axis: $(-2 - \sqrt{\frac{65}{6}}, \frac{1}{4})(-2 + \sqrt{\frac{65}{6}}, \frac{1}{4})$
minor axis: $(-2, (1 - \sqrt{65})/4)(-2, (1 + \sqrt{65})/4)$

Focus $(-2 - \sqrt{\frac{325}{48}}, \frac{1}{4})$ Focus $(-2 + \sqrt{\frac{325}{48}}, \frac{1}{4})$

$x = -2$

$(-2, \frac{1 + \sqrt{65}}{4})$

$(-2, \frac{1}{4})$ Center

$y = \frac{1}{4}$

$(-2, \frac{1 - \sqrt{65}}{4})$

$(-2 - \sqrt{\frac{65}{6}}, \frac{1}{4})$ $(-2 + \sqrt{\frac{65}{6}}, \frac{1}{4})$

$$\frac{(x + 2)^2}{65/6} + \frac{(y - 1/4)^2}{65/16} = 1$$

19. The center must be right in the middle of the foci (or the vertices) which puts the it at $(0,0)$.

The foci, $(0,4)$ and $(0,-4)$ are on the y-axis so the major axis must run vertically which tells us that $b > a$ and $c = \sqrt{b^2 - a^2}$. Meanwhile, from the given points we can deduce that $c = 4$ (from the foci) and that $b = 5$ (from the vertices). Therefore $4 = \sqrt{25 - a^2}$ which implies that $16 = 25 - a^2$; $a^2 = 9$; $a = 3$. The equation for ellipses, $\frac{x^2}{a^2} + \frac{y^2}{b^2} = 1$, becomes $\frac{x^2}{9} + \frac{y^2}{25} = 1$.

21. In problem 19 it was found that $a = 3$ and $b = 5$. If the ellipse is translated, these numbers will not change. Allowing for the new center, $(-1,4)$ yields the equation:

$$\frac{(x+1)^2}{9} + \frac{(y-4)^2}{25} = 1$$

23. Let's complete the square:

$$x^2 + 2x + 2y^2 + 12y = c \Rightarrow (x^2 + 2x + 1) + 2(y^2 + 6y + 9) = c + 1 + 18$$
$$\Rightarrow (x + 1)^2 + 2(y + 3)^2 = c + 19$$

Now, if the number $c + 19$ is positive, dividing both sides by it will produce the equation of an ellipse. So for part a we have the fact that if $c + 19 > 0$ the graph will be an ellipse and this is equivalent to the condition that $c > -19$.

If $c + 19 = 0$ exactly then we can't divide both sides by it and only the point $(-1,-3)$ will satisfy the equation. This is part b.

If $c < -19$ then $c + 19$ will be negative and it will be impossible for the sum of two non-negative terms, $(x+1)^2$ and $2(y+3)^2$ to be negative. Therefore, the solution set and the graph are empty.

25. Assume the ellipse that describes this orbit has a horizontal major axis (a > b).

   Then a = 1/2 of 36.2 = 18.1   and   b = 1/2 of 9.1 = 4.45   The eccentricity sought is given

   in this case by $\frac{c}{a}$ where c = $\sqrt{a^2- b^2}$ = $\sqrt{18.1^2-4.55^2}$ ≈ 17.51. So e ≈ $\frac{17.51}{18.1}$ = 0.9679

27. $T^2= a^3$ and T = 8.4 imply that $8.4^2= a^3$ so that a = $\sqrt[3]{70.56}$ ≈ 4.13 AU which makes the

   major axis have length 2a or 8.3 AU.

29. From the diagram (if we imagine the origin to be in the middle of the tunnel on the roadway)

   a = 56 and b = 20 making the equation of the ellipse $\frac{x^2}{56^2} + \frac{y^2}{20^2} = 1$ or $\frac{x^2}{3136} + \frac{y^2}{400} = 1$

   To determine critical heights for trucks we are interested in y-values associated with particular

   critical x-values. Certainly the extreme ends of lanes 1 and 6 qualify as that is where the tunnel

   is lowest (assuming vehicles stay in one of the six lanes). In the assumed cooridinate system, the

   right end of lane 1 corresponds to an x value of 42. Plugging this into the equation for the ellipse

   and solving for y should give us the most critical height:

   $\frac{42^2}{3136} + \frac{y^2}{400} = 1 \Rightarrow \frac{y^2}{400} = 1 - \frac{1764}{3136} \Rightarrow \frac{y^2}{400} = \frac{1372}{3136} \Rightarrow y^2 = 400 \cdot \frac{1372}{3136} \Rightarrow y \approx 13.2$ ft

   This is the clearance in lanes 1 and 6. To find the clearance in lanes 2 and 5 use x = 28 and to

   find the clearance in lanes 3 and 4 use x = 14.

31. $4x^2+ 25y^2 = 100$ is equivalent to $\frac{x^2}{25} + \frac{y^2}{4} = 1$ so a = 5 and b= 2 and the center is (0,0).

   The major axis is horizontal of length 10 and the minor axis is vertical of length 4. Ellipse

   (b) matches this information. $c^2= a^2- b^2$ gives us $c^2= 21$ so c = $\sqrt{21}$ and the unmarked foci

   are (-$\sqrt{21}$, 0) and ($\sqrt{21}$, 0).

33. $9(x + 2)^2+ 25(y - 1)^2= 225$ is equivalent to $\frac{(x + 2)^2}{25} + \frac{(y - 1)^2}{9} = 1$ which tells us that the

   center is (-2, 1), the major axis is horizontal of length 2·5 = 10 and that the minor axis is vertical

   of length 6. This matches the ellipse (h). The points at the ends of the minor axis would be

   3 units above and below the center at (-2, 1±3) or (-2,4) and (-2,-2).

35. $13x^2+ 4y^2= 52$ is equivalent to $\frac{x^2}{4} + \frac{y^2}{13} = 1$ which is an ellipse with its center at the origin

   and a vertical major axis of length $2\sqrt{13}$ ≈ 7.2 and a horizontal minor axis of length 4. This

   information seems to match ellipse (c). We can ice it by computing the value of $c^2= b^2- a^2$

   which becomes 13 - 4 = 9 so c = 3 and this matches the marked foci on graph (c). The un-

   marked vertices must be (0, -$\sqrt{13}$) and (0, $\sqrt{13}$).

37. $9(x - 3)^2 + 4(y - 1)^2 = 36$ is equivalent to $\dfrac{(x - 3)^2}{4} + \dfrac{(y - 1)^2}{9} = 1$ so the center is $(3,1)$, the major axis is vertical of length 6 and the minor axis is horizontal of length 4. Ellipse (e) is found to match this (the center is the biggest clue). To find the unmarked foci we must calculate c:

$c^2 = 9 - 4 = 5 \Rightarrow c = \sqrt{5}$ and so the foci are on the major axis (vertical) $\sqrt{5}$ units above and below the center which is at $(3,1)$. Therefore, the foci are $(3, 1 \pm \sqrt{5})$.

Problems 5.4

1. $x^2 = 16y$ lines up beautifully with $x^2 = 4cy$ telling us that $4c = 16$ which implies that $c = 4$. The fact that x is squared lets us know that the axis of the parabola is vertical and is in fact the y-axis. Since $c = 4$ we can immediately say that the focus is $(0, c) = (0, 4)$ and the directrix is the line $y = -c$ which is then $y = -4$. The vertex is $(0, 0)$ in this case.

Sketch:

Focus
(0, 4)

0

x

Directrix   y = -4

$x^2 = 16y$

3. $x^2 = -16y$ when compared with $x^2 = 4cy$ informs us that $c = -4$ and so for this parabola:

   a) the axis is vertical (ie, the y-axis since the vertex is the origin)

   b) it opens down due to the "-" sign

   c) focus is at $(0, -4)$ and directrix is $y = 4$

   d) a pair of "fast" points: let $y = -1$; then $x = \pm 4$

Sketch:

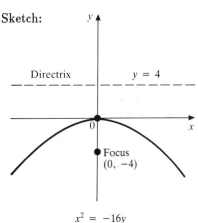

Directrix   y = 4

0

x

Focus
(0, -4)

$x^2 = -16y$

- 217 -

5. $2x^2 = 3y$ is equivalent to $x^2 = 3/2\, y$

    Comparing this to $x^2 = 4cy$ implies that $4c = 3/2$ which implies that $c = 3/8$ and so:

    a) the axis is vertical and passes through the origin which is the vertex

    b) the parabola opens up since c is positive

    c) the focus is at $(0, c)$ or $(0, 3/8)$ and the directrix is the line $y = -c$ which in this case is
    $y = -3/8$

    d) if you really want a pair of points to plot, let $y = 1$ and maybe use a calculator to approximate $\pm\sqrt{1.5}$ as ± 1.22 for the x-values.

    sketch:

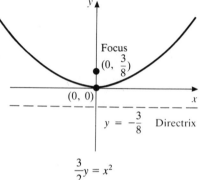

7. $4x^2 = -9y \;\Rightarrow\; x^2 = -9/4\, y$ so if we set $4c = -9/4$ we find that $c = -9/16$; therefore:

    a) the parabola opens downward with its vertex at $(0, 0)$ and the y-axis as its axis

    b) the focus is located at $(0, -9/16)$ and the directrix is the line $y = 9/16$

    c) coordinates of two points could be found by letting $y = -1$ and seeing that $x = \pm\, 3/2$

    Sketch:

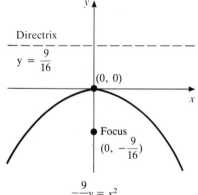

9. $(x - 1)^2 = -16(y + 3)$ has its vertex at the point $(1, -3)$ and since it's x that is squared the axis of the parabola must be vertical: it will be the vertical line $x = 1$. The negative sign alerts us to the fact that the parabola will open down. Comparing the problem to $x^2 = 4cy$ reveals that $4c = -16$ so that $c = -4$ and this number is related to the displacements of the focus and directrix from the vertex $(1, -3)$: the focus must be 4 units below the vertex, ie, at $(1, -7)$ and

9. (continued)

the directrix must be horizontal and 4 units above the vertex, ie, the line y = 1. This graph may also be thought of as the graph of $x^2 = -16y$ shifted 1 unit to the right and 3 units down corresponding to the vertex (1, -3).

Sketch:

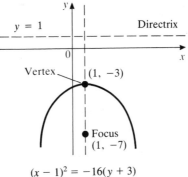

$$(x-1)^2 = -16(y+3)$$

11. $x^2 + 4y = 9$ is the same as $x^2 = -4y + 9$ which is the same as $x^2 = -4(y - 9/4)$ and so the vertex is the point (0, 9/4), the axis is vertical being the y axis (x = 0 is its equation), and the parabola opens downward (from the "-" sign). Since 4c = -4 in this case, we have c = -1 and so the focus is the point (0, 9/4 - 1) which simplifies to (0, 5/4); also, the directrix is 1 unit above the vertex and is given by the equation y = 13/4.

Sketch:

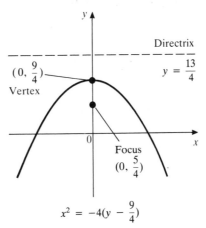

$$x^2 = -4(y - \frac{9}{4})$$

13. $x^2 + 2x + y + 1 = 0$

$x^2 + 2x + 1 = -y$

$(x + 1)^2 = -y$ is basically in the form $x^2 = 4cy$ but with a vertex translated from the origin to the point (- 1, 0). The axis is vertical and passes through the vertex; therefore, its equation is x = - 1. The parabola opens down and 4c must equal - 1 which implies that c = - 1/4.

13. (continued)

The focus is 1/4 of a unit below the vertex at (-1, - 1/4) and the directrix is 1/4 unit above the vertex, the horizontal line y = 1/4.

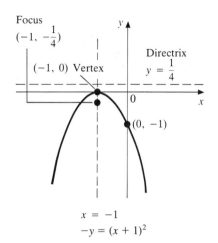

$x = -1$
$-y = (x + 1)^2$

15. $x^2 + 4x + y = 0 \Rightarrow x^2 + 4x = -y \Rightarrow x^2 + 4x + 4 = -y + 4$

$\Rightarrow (x + 2)^2 = -1(y - 4)$ which is in a form we can analyze: the vertex is translated to (- 2, 4), the axis is vertical with with equation x = -2, and the parabola opens down. Setting 4c = - 1 yields c = - 1/4 and so the focus is 1/4 unit below the vertex at (- 2, 15/4); the directrix is 1/4 unit above the vertex being the horizontal line y = 4 + 1/4 , ie, y = 17/4.

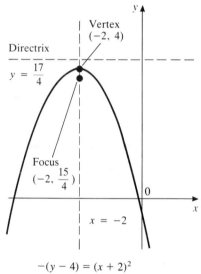

$-(y - 4) = (x + 2)^2$

17. $x^2 + 4x - y = 0$ can have its square completed to become $(x + 2)^2 = 1(y + 4)$. The vertex is at (- 2, - 4) and the axis is vertical having the equation x = - 2 and the parabola opens up. 4c = 1 which implies that c = 1/4. Therefore the focus is 1/4 block above the vertex at (- 2, - 15/4) and the directrix is 1/4 block below the vertex having the equation y = - 17/4.

Sketch:

- 220 -

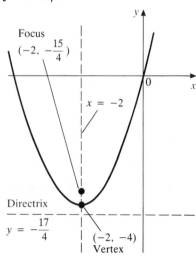

$y + 4 = (x + 2)^2$

19. If the focus is $(0, 4)$ and the directrix is the line $y = -4$ a quick sketch of the information should reveal that the parabola opens up and would have to have its vertex at the origin (midway between focus and directrix, right?) The equation would then have to be of the form $x^2 = 4cy$ and in this case c would have to equal 4 (the distance between focus and vertex) so we now know that the exact equation would have to be $x^2 = 4(4)y$ or simply $x^2 = 16y$.

21. The solution to Problem 20 would have to be $y^2 = -12x$ due to the facts that the parabola must open to the left and c can be shown to be -3. Therefore, if we shift the vertex from $(0, 0)$ to $(-2, 5)$, this will be reflected in the new equation which would have to be $(y - 5)^2 = -12(x + 2)$.

23. In problem 20 it was discovered that the value of c was -3 (putting the focus 3 units to the left of the vertex and the directrix (vertical in this case) 3 units to the right of the vertex. If the vertex is now specified as the point $(3, -1)$, then the new focus must be $(0, -1)$ and the new directrix must be the line $x = 6$

25. Solution for the case $x^2 = 4cy$: from the sketch at the right it seems reasonably clear that the right endpoint of the latus rectum is at the same height as the focus; this height or y-coordinate is c. If we substitute c for y in the equation of the parabola we can determine the x-coordinate of this point and doubling this will give the length of the latus rectum: $x^2 = 4c(c)$ is the same as $x^2 = 4c^2$ which implies that $x = \sqrt{4c^2} = |2c|$ and doubling this produces $2|2c|$ or $|4c|$.

27. In other words, find the equation of a parabola its vertex at $(0, 0)$ and a focus at $(0, 2)$. This implies that the value of c is 2 and since the equation must be of the form $x^2 = 4cy$ we arrive at the equation $x^2 = 8y$.

29. We have, after substituting 1000 for h_0, $h(t) = 1000 - 16t^2$ and, since t refers to elapsed time, $t \geq 0$ would have to hold; setting $h(t) = 0$ and solving for t will give us an idea of the greatest value t might assume: $0 = 1000 - 16t^2 \Rightarrow 16t^2 = 1000 \Rightarrow t^2 = 62.5 \Rightarrow t \approx 7.9$ seconds. This is the time when the ball hits the ground (an exact answer would be $5\sqrt{10}/2$) which is asked for in part b. We graph only the part of the parabola corresponding to the domain $0 \leq t \leq 7.9$

29. (continued)

and perhaps it might be nice to find an additional point by letting t equal, say, 5; we could then compute $h(5) = 1000 - 16(5)^2 = 1000 - 400 = 600$ and the point (5, 600) must lie on the graph.

Sketch:

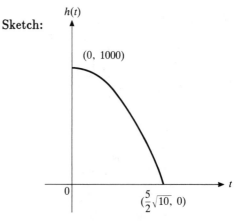

The following matching exercises can be accomplished by putting the equations into one of the standard forms, noting the vertex and direction of opening, and **THEN** making the match.

31. $x^2 + 10x + 3y + 13 = 0$ needs its square completed in x and some rearranging:

$x^2 + 10x \qquad = -3y - 13$

$x^2 + 10x + 25 = -3y - 13 + 25$

$(x + 5)^2 = -3y + 12$ or $(x + 5)^2 = -3(y - 4)$ which permits these conclusions:

opens down (x is squared in conjunction with the "-" sign)

vertex is at (- 5, 4)

This is enough to ice the fact that it must be the parabola shown in figure (f)

33. $4x - y^2 = 0$ is the same as $y^2 = 4x$ and so the vertex is at the origin and the parabola opens to the right since y is squared and c is positive. Only graph (b) matches this information.

35. $2x + y^2 = 0$ is the same as $y^2 = -2x$ from which we can conclude that the vertex is the origin and the parabola opens to the left (y is squared and the negative sign). It's graph (c).

37. $y^2 - 2x - 4y + 6 = 0$ needs its square completed in y and some rearranging:

$y^2 - 4y + 4 = 2x - 6 + 4$

$(y - 2)^2 = 2(x - 1)$ which would be the equation of a parabola which has its vertex at (1, 2) and opens to the right. Only graph (e) has these properties.

Problems 5.5

The easiest way to sketch hyperbolas is definitely the "box" method in which one determines the center of the hyperbola and then uses the numbers a and b to determine distances to move vertically and horizontally from the center and at which distances to draw dotted horizontal and vertical line segments which then create a "box" or rectangle surrounding the center. Drawing dotted lines which contain the diagonals of this rectangle gives you the asymptotes which the graph must approach as the x-(or perhaps y-) values get really large. Determining on which axis the vertices lie is all that remains to be done once the asymptotes are in place: if the term containing x^2 is the positive one, then the vertices are on the x-axis and the branches open left and right; otherwise the vertices are on the y-axis and the branches of the hyperbola will open up and down.

1. $\frac{x^2}{16} - \frac{y^2}{25} = 1$ is in standard form which makes it easy to see that i) the center is at the origin
    ii) the vertices lie on the x-axis and iii) a = 4 and b = 5. The vertices are then the points (4, 0) and (-4, 0) while the points (0, 5) and (0, -5) will help us draw the "box" that will aid in the graphing. The transverse axis is the segment connecting (±4, 0); the conjugate axis is the segment connecting (0, ±5). For hyperbolas we have $c^2 = a^2 + b^2$; here we would have $c^2 = 16 + 25$ which implies that $c = \sqrt{41}$. The foci then are located $\sqrt{41}$ units to the left and right of the center (0, 0) making their coordinates $(-\sqrt{41}, 0)$ and $(\sqrt{41}, 0)$. The asymptotes are given by $y = \pm \frac{a}{b} x$ which become $y = \pm \frac{5}{4}x$.

    Sketch:

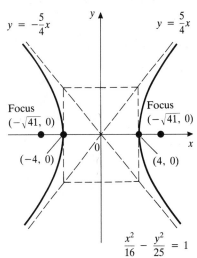

3. $\frac{y^2}{25} - \frac{x^2}{16} = 1$ would have (0,0) as its center and the vertices would be on the y-axis. Here, a = 5 and
    b = 4 so the actual vertices are (0,- 5) and (0, 5) with the transverse axis being the segment which connects them. The conjugate axis connects the points (4,0) and (-4,0) which do

3. (continued)

not lie on the graph itself. $c^2 = a^2 + b^2 = 25 + 16 = 41$ and so $c = \sqrt{41}$ and this puts the foci on the y-axis $\sqrt{41}$ units above and below the center at the points $(0, \pm\sqrt{41})$. The asymptotes, $y = \pm\frac{b}{a}x$, are given by $y = \pm\frac{5}{4}x$ or by $x = \pm\frac{4}{5}y$ if you want to be ornery.

Sketch:

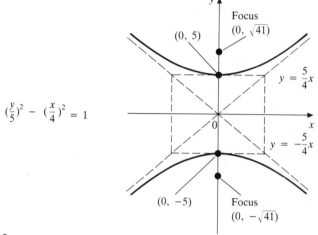

$(\frac{y}{5})^2 - (\frac{x}{4})^2 = 1$

5. $y^2 - x^2 = 1$ can be viewed as $\frac{y^2}{1} - \frac{x^2}{1} = 1$ which would be a hyperbola with its center at the origin, its vertices on the y-axis, and value of c found from $\sqrt{1 + 1} = \sqrt{2}$. The foci then are the points $(0, -\sqrt{2})$ and $(0, \sqrt{2})$. The transverse axis is the segment connecting the vertices which are on the y-axis at the points $(0, 1)$ and $(0, -1)$. The conjugate axis connects the points $(1,0)$ and $(-1,0)$. Since $a = b = 1$ in this problem, the asymptotes are given by $y = \pm\frac{a}{b}x$ which turns into $y = \pm x$.

Sketch:

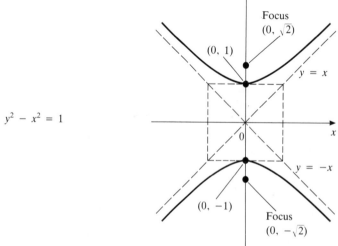

$y^2 - x^2 = 1$

7. $x^2 - 4y^2 = 9$ needs to be put into standard form by dividing by 9 and viewing multiplication as division by the reciprocal:

$$\frac{x^2}{9} - \frac{4y^2}{9} = 1 \text{ or equivalently } \frac{x^2}{9} - \frac{y^2}{9/4} = 1$$

7. (continued)

Now we can see that the center is the origin, the vertices will lie on the x-axis, the values for a and b are 3 and 3/2 respectively and so the actual vertices are (3, 0) and (-3, 0) with the transverse axis connecting them. The conjugate axis connects the points (0, -3/2) and (0, 3/2). We can find c by evaluating the square root of $a^2 + b^2$, ie $\sqrt{9 + 9/4} = \sqrt{45/4} = 3\sqrt{5}/2$ (≈ 3.3) and now we can locate the foci on the x-axis at the points ($-3\sqrt{5}/2$, 0) and ($3\sqrt{5}/2$, 0). The asymptotes are given by $y = \pm\frac{b}{a}x = \pm\frac{3/2}{3}x = \pm\frac{1}{2}x$.

Sketch:

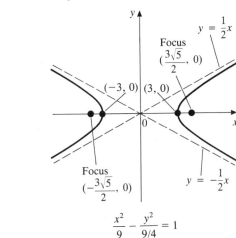

$$\frac{x^2}{9} - \frac{y^2}{9/4} = 1$$

9. $y^2 - 4x^2 = 9$ is very similar to problem 7 above. When put into standard form the equation becomes:

$$\frac{y^2}{9} - \frac{x^2}{9/4} = 1$$

so the vertices are (0, 3) and (0, -3) on the y-axis. The transverse axis connects them. The conjugate axis meanwhile connects (3/2, 0) and (- 3/2, 0). The center is once again the origin and the value of c is $3\sqrt{5}/2$ which puts the foci on the y-axis at (0, $\pm3\sqrt{5}/2$). The value of a is 3 and the value of b is 3/2 which makes the asymptotes in this case $y = \pm$ (a/b) x or $y = \pm$ [3/(3/2)] x which simplify to $y = \pm$ 2x.

Sketch:

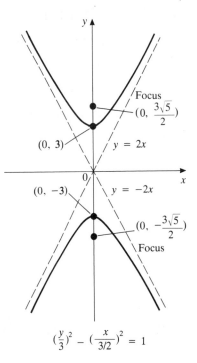

$$\left(\frac{y}{3}\right)^2 - \left(\frac{x}{3/2}\right)^2 = 1$$

11. $2x^2 - 3y^2 = 4 \Rightarrow \dfrac{x^2}{2} - \dfrac{y^2}{4/3} = 1$ so the a and b numbers will be a little messy: $a = \sqrt{2}$ and $b = \sqrt{4/3}$.

But the center is still the origin and the vertices are on the x-axis at $(\pm\sqrt{2}, 0)$ with the transverse axis connecting them. The conjugate axis connects the points $(0, \pm\sqrt{4/3})$. c^2 is $2 + 4/3 = 10/3$ which means $c = \sqrt{10/3}$ and this puts the foci (also on the x-axis) at $(\pm\sqrt{10/3}, 0)$. A decent approximation for $\sqrt{10/3}$ is 1.8. The asymptotes are $y = \pm (b/a) x = \pm(\sqrt{4/3}/\sqrt{2}) = \pm \sqrt{2/3}\, x$.
Sketch:

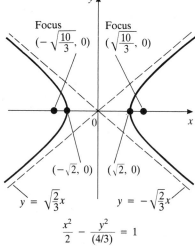

13. $2y^2 - 3x^2 = 4 \Rightarrow \dfrac{y^2}{2} - \dfrac{x^2}{4/3} = 1$ has a and b values as given in problem 11 above. The vertices are on the y-axis at $(0, \sqrt{2})$ and $(0, -\sqrt{2})$ with the transverse axis connecting them. The conjugate axis connects $(\sqrt{4/3}, 0)$ and $(-\sqrt{4/3}, 0)$. The value of c is $\sqrt{10/3}$ as computed above putting the foci at $(0, \pm\sqrt{10/3})$. The asymptotes are now given by $y = \pm a/b\, x$ and they turn out to be $y = \pm (a/b) x = \pm(\sqrt{2}/\sqrt{4/3}) x = \pm \sqrt{3/2}\, x$.

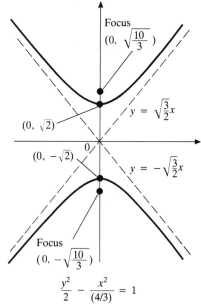

15. $(x-1)^2 - 4(y+2)^2 = 4 \Rightarrow \dfrac{(x-1)^2}{4} - \dfrac{(y+2)^2}{1} = 1$ and so we're dealing with a hyperbola which is not centered at the origin but at $(1, -2)$. We have $a = 2$ and $b = 1$. The vertices will be on the horizontal line (since the x^2 term is positive) $y = -2$; they will be located 2 (square root of 4) units to the left and right of the center $(1, -2)$. This puts the vertices at $((3, -2)$ and $(-1, -2)$

15. (continued)

and the transverse axis is the segment connecting them. The endpoints of the conjugate axis are located 1 unit above and below the new center at the points $(1, -3)$ and $(1, -1)$. The new center also affects the equations of the asymptotes:

$y = \pm (b/a)x$ gives way to $(y + 2) = \pm 1/2 (x - 1)$

which can be put into slope-intercept form and written as $y = (1/2) x - 5/2$ and $y = (-1/2) x - 3/2$

Sketch:

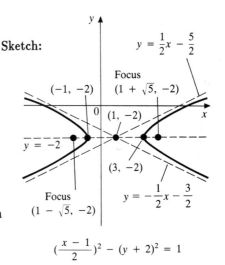

$(\frac{x-1}{2})^2 - (y + 2)^2 = 1$

17. $4x^2 + 8x - y^2 - 6y = 21 \Rightarrow 4(x^2 + 2x + \quad) - (y^2 + 6y + \quad) = 21$

$\Rightarrow 4(x^2 + 2x + 1) - (y^2 + 6y + 9) = 21 + 4 - 9$

$\Rightarrow 4(x + 1)^2 - (y + 3)^2 = 16$

$\Rightarrow \dfrac{(x + 1)^2}{4} - \dfrac{(y + 3)^2}{16} = 1$

and now the following conclusions can be drawn:

center at $(-1, -3)$, the transverse axis is horizontal and connects points 2 units to the left and right of the center, namely $(-3, -3)$ and $(1, -3)$ (the vertices), the conjugate axis connects the points which are 4 units above and below the center, at $(-1, 1)$ and $(-1, -7)$, the values of a^2 and b^2 being 4 and 16 lead to a c^2 value of 20 which makes $c = 2\sqrt{5}$ and this puts the foci on the horizontal line $y = -3$ at distances of $2\sqrt{5}$ units to the left and right of the center. So the foci are at $(-1 -2\sqrt{5}, -3)$ and $(-1 + 2\sqrt{5}, -3)$. The asymptotes will be given by translated versions of $y = \pm (b/a) x$:

$y + 3 = \pm 4/2(x + 1)$ or $y = 2x - 1$ and $y = -2x - 5$

Sketch:

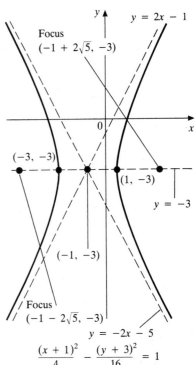

$\dfrac{(x + 1)^2}{4} - \dfrac{(y + 3)^2}{16} = 1$

19. $2x^2 - 16x - 3y^2 + 12y = 45 \Rightarrow 2(x^2 - 8x + 16) - 3(y^2 - 4y + 4) = 45 + 32 - 12$

$$\Rightarrow 2(x - 4)^2 - 3(y - 2)^2 = 65$$
$$\Rightarrow \frac{(x - 4)^2}{65/2} - \frac{(y - 2)^2}{65/3} = 1 \qquad \text{and so the center is at } (4, 2).$$

The values of a and b are $\sqrt{65/2}$ and $\sqrt{65/3}$ respectively (≈ 5.7 and 4.7 in the same order). The vertices must lie on the horizontal line $y = 2$ at points $\sqrt{65/2}$ units to the left and right of the center. This puts them at $(4 \pm \sqrt{65/2}, 2)$. The transverse axis must connect these points while the conjugate axis connects the points $(4, 2 \pm \sqrt{65/3})$. The value of c will be the square root of $65/2 + 65/3$ or $\sqrt{325/6}$ and this means that the foci are this distance to the left and right of the center at $(4 \pm \sqrt{325/6}, 2)$. The asymptotes are given by a translated version of $y = \pm(b/a) x$ which looks quite messy: $y - 2 = \pm \sqrt{65/3} / \sqrt{65/2} (x - 4)$ which when simplified becomes $y - 2 = \pm \sqrt{2/3} (x - 4)$.

Sketch:

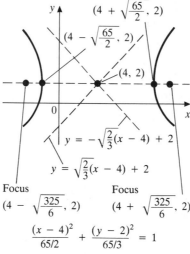

21. The center must be at the midpoint of either the two foci or the two vertices which in this case is obviously $(0, 0)$. Since the foci and vertices are on the horizontal x-axis, the hyperbola must be of the form $\frac{x^2}{a^2} - \frac{y^2}{b^2} = 1$ and we must determine a and b. The foci being $(-4, 0)$ and $(4, 0)$ implies that $c = 4$ and the vertices being $(-3, 0)$ and $(3, 0)$ implies that $a = 3$. Since $c^2 = a^2 + b^2$ we get $16 = 9 + b^2$ which implies that $b^2 = 7$ and since a is 3 we now know the equation:

It's $\frac{x^2}{9} - \frac{y^2}{7} = 1.$

23. The foci and vertices given both lie on the y-axis and their midpoint is the origin. The hyperbola therefore must be of the form $\frac{y^2}{a^2} - \frac{x^2}{b^2} = 1$. The foci being 3 units from the center tells us that $c = 3$; the vertices being 2 units from the center lets us know that $a = 2$ and $c^2 = a^2 + b^2$ allows us to deduce that $9 = 4 + b^2$ so that b^2 is 5. The final equation is therefore $\frac{y^2}{4} - \frac{x^2}{5} = 1.$

- 228 -

25. The midpoint of the foci or the vertices is (2, 1) so this is the center. The foci and vertices lie

on the horizontal line y = 1 so the equation must conform to

$$\frac{(x-2)^2}{a^2} - \frac{(y-1)^2}{b^2} = 1$$

The distance between the foci is 6 which makes c = 3 and the distance between the vertices is 4

which makes a = 2. Solving $c^2 = a^2 + b^2$ with these values for b^2 yields 5.

Answer: $\dfrac{(x-2)^2}{4} - \dfrac{(y-1)^2}{5} = 1$

27. The midpoint of the segment joining the vertices, which lie on the x-axis, is (0, 0) so the equation

must be of the form

$$\frac{x^2}{a^2} - \frac{y^2}{b^2} = 1$$

The fact that a = 2 is clear from the coordinates of the vertices. The asymptotes being y = ± x

tell us that b/a must be 1 (the slope of y = x, the asymptote with positive slope). Therefore,

we can say that b/2 = 1 and this implies that b = 2 as well. The equation is therefore given by

$$\frac{x^2}{4} - \frac{y^2}{4} = 1$$

29. The midpoint of the segment joining the vertices which lie on the horizontal line y = 1 is (3, 1).

The equation must then conform to

$$\frac{(x-3)^2}{a^2} - \frac{(y-1)^2}{b^2} = 1$$

The distance between the vertices is 4 (this would be 2a) so that a = 2 has to be true. The slope

of the asymptote with positive slope is 2 and this must correlate to b/a and since a is 2 we are led

to the fact that b = 4. The equation then must be

$$\frac{(x-3)^2}{4} - \frac{(y-1)^2}{16} = 1$$

31. The answer to problem 21 was $\frac{x^2}{9} - \frac{y^2}{7} = 1$ which had its center at (0, 0). Translating the center to

(4, - 3) would affect the numerators and the result would simply be

$$\frac{(x-4)^2}{9} - \frac{(y+3)^2}{7} = 1$$

33. $e = \dfrac{\text{distance between foci}}{\text{distance between vertices}} = \dfrac{2c}{2a} = \dfrac{c}{a} = \dfrac{\sqrt{a^2 + b^2}}{a}$ which is definitely > 1

Explanation: since $b^2 > 0$ we have $a^2 + b^2 > a^2$ which in turn assures us that $\sqrt{a^2 + b^2} > a$

which makes the numerator of the fraction greater than the denominator and since all the

quantities here are positive ones, that makes the fraction > 1

35. In problem 1 we found a = 4 and c = $\sqrt{41}$ so e = $\dfrac{\sqrt{41}}{4}$.

37. In problem 5 we found a = 1 and c = $\sqrt{2}$ so e = $\sqrt{2}$

39. In problem 9 we found a = 3 and c = $3\sqrt{5}/2$ and so e = c/a = $\sqrt{5}/2$

41. In problem 13, a was $\sqrt{2}$ and c was $\sqrt{10/3}$ so now the eccentricity e would be $\sqrt{10/3}$ / $\sqrt{2}$ = $\sqrt{5/3}$ or if one multiplied by $\sqrt{3/3}$ the rationalized version of e would be $\sqrt{15}/3$.

43. In problem 19 we had a = $\sqrt{65/2}$ and c = $\sqrt{325/6}$. Therefore e = c/a = $\sqrt{325/6} \cdot \sqrt{2/65}$ = $\sqrt{5/3}$

45. Because we are told the hyperbola has its center at (0, 0) and axis parallel to the x-axis (actually, its principal axis would have to BE the x-axis then) we know the equation must look like:

$$\frac{x^2}{a^2} - \frac{y^2}{b^2} = 1$$

and since we know two points which must satisfy this equation, namely (1,2) and (5,12) the plan will be to substitute these values in for x and y above creating two equations in a and b and then solve these equations simultaneously for a^2 and b^2:

inserting (1,2) yields $\frac{1}{a^2} - \frac{4}{b^2} = 1$

inserting (5,12) yields $\frac{25}{a^2} - \frac{144}{b^2} = 1$

To solve this system by the elimination method, multiply both sides of the top equation by - 25 and add the equations together to get

$$\frac{-44}{b^2} = -24 \quad (\text{ now solve for } b^2)$$

$$\frac{-44}{-24} = b^2 \quad \text{so reduce and simplify to see that } b^2 = 11/6$$

To find a^2 we can go back to the system and multiply both sides of the top equation by - 36 and then add the two equations together. $\frac{-11}{a^2} = -35$ should result making $a^2 = 11/35$.

The equation of the hyperbola can now be pieced together as $\frac{x^2}{11/35} - \frac{y^2}{11/6} = 1$

47. The first order of business is to complete the square and put the equation into standard form:

$$x^2 + 4x - 3y^2 + 6y = c \quad \Rightarrow \quad (x^2 + 4x + 4) - 3(y^2 - 2y + 1) = c + 4 - 3$$
$$\Rightarrow \quad (x + 2)^2 - 3(y - 1)^2 = c + 1$$

The next step would be to divide both sides by c + 1 (need a "1" on the right, right?). We would not be able to do this if c were equal to -1 (can't divide by zero, right?) so if c were equal to -1, the equation would become $(x + 2)^2 - 3(y - 1)^2 = 0$ and the left side factors into the difference of two squares; we could create these two equations from those factors:

$$(x + 2) + \sqrt{3}(y - 1) = 0 \qquad \text{and}$$
$$(x + 2) - \sqrt{3}(y - 1) = 0$$

which are linear equations. This answers parts b and c. If c ≠ -1, then nothing would have stopped us from dividing both sides by c + 1 earlier and we could go on to create the standard equation for a hyperbola:

$$\frac{(x + 2)^2}{c + 1} - \frac{(y - 1)^2}{c + 1} = 1 \quad \text{which explains part a of the question.}$$

49. Since the speed of sound in water is 1533 m/sec, in two seconds sound travels 3066 m, making the depth charge 3066 m closer to submarine A than to submarine B. This means that if the subs are con- sidered the foci of a hyperbola containing the point where the depth charge exploded, the difference of the segments to the foci from this point must be 3066 m. This is always 2a when a hyperbola in standard form is being considered. In the drawing at the right, points

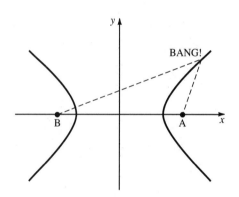

A and B represent the two submarines and they are placed on the x-axis equidistant from the origin. Since the problem states that the subs are 4 km (4000 m) apart, we can give coordin- ates (- 2000, 0) and (2000, 0) to A and B respectively. Since the subs are the foci, this im- plies that c = 2000. By the discusssion above concerning the 3066 m and its being 2a, we see that a = 1533. We can now find b^2 using the relationship $c^2 = a^2 + b^2$ and then form the equa- tion $\frac{x^2}{a^2} - \frac{y^2}{b^2} = 1$ as being the equation of the hyperbola that contains the point at which the

depth charge was dropped: $2000^2 = 1533^2 + b^2 \Rightarrow b^2 = 1,649,911$

So the equation is: $\dfrac{x^2}{2,350,089} - \dfrac{y^2}{1,649,911} = 1$

Chapter 5 Review

1. $f(x) = \dfrac{x}{x^2 - 4} = \dfrac{x}{(x-2)(x+2)}$

    vertical asymptotes: the x-values $x = 2$ and $x = -2$ make only the denominator

    0 so the vertical lines $x = 2$ and $x = -2$ are vertical asymptotes.

    horizontal asymptote: here $n = 1$ and $m = 2$; $n < m$ so $y = 0$ is it.

3. $f(x) = \dfrac{3x^2 - 4}{x^2 - 5x - 4} = \dfrac{3x^2 - 4}{x^2 - 5x - 4}$

    vertical: use quadratic formula to get $x = \dfrac{5 \pm \sqrt{41}}{2}$

    horizontal: here $n = 2$ and $m = 2$ so the height of the horizontal asymptote is the quotient of the

    leading coefficients 3 and 1 or $y = 3/1$ or $y = 3$.

5. $f(x) = \dfrac{2}{x + 3}$ VA: $x = -3$ HA: $y = 0$ since n (which is 0) is less than m (1)

    There is no x-intercept because the numerator can't equal 0. The y-intercept is f(0) which is 2/3.

                table: x: -5   -4   -2   -1

                        f(x):   -1   -2   2   1

    sketch:

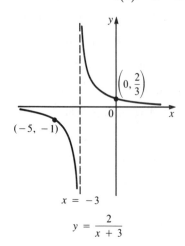

$x = -3$

$y = \dfrac{2}{x + 3}$

7. $f(x) = \dfrac{1}{(x - 2)^3} + 2$ VA: $x = 2$ HA: $y = 2$ (as $x \to \infty$, $f(x) \to 2$ because 1st term approaches 0

     x-int: set $\dfrac{1}{(x - 2)^3} + 2 = 0$: multiply through by the LCD

$$1 + 2(x - 2)^3 = 0 \; : \; \text{isolate } (x - 2)^3$$
$$2(x - 2)^3 = -1$$
$$(x - 2)^3 = -0.5 \quad : \quad \text{take cube roots of both sides}$$
$$x - 2 = \sqrt[3]{-0.5} \; \text{ so } x = \sqrt[3]{-0.5} + 2 \approx 1.21$$

   y-int: $f(0) = 1\frac{7}{8}$ sketch:

table: x: -1   1   3   4

      f(x): $1\frac{26}{27}$ 1   3 $2\frac{1}{8}$

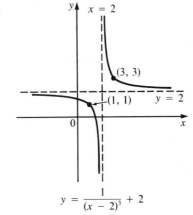

$$y = \dfrac{1}{(x - 2)^3} + 2$$

9. $f(x) = \frac{x - 3}{x}$ VA: $x = 0$ makes only the denominator 0

HA: $n = m = 1$ so the quotient of the coefficients (1 and 1) is the height of the

horizontal asymptote; its equation is $y = 1$

x-int: $x = 3$ makes only the numerator 0   table: x: -2   -1   1   2

NO y-intercept because $f(0)$ is undefined   f(x): 5/2   4   -2   -1/2

sketch:

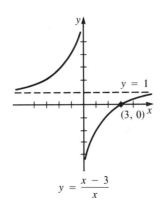

$$y = \dfrac{x - 3}{x}$$

11. $f(x) = \dfrac{x-2}{x^2-2x-8} = \dfrac{x-2}{(x-4)(x+2)}$ VA: $x = 4$ and $x = -2$

HA: $y = 0$ because $n < m$

x-int: $x = 2$ makes only the numerator 0

y-int: $f(0) = 1/4$

table:

x:	-4	-3	-1	1	3	5	6
f(x):	-3/8	-5/7	3/5	1/9	-1/5	3/7	1/4

sketch:

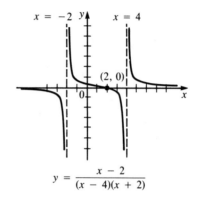

$$y = \dfrac{x-2}{(x-4)(x+2)}$$

13. $\dfrac{x^2}{9} + \dfrac{y^2}{16} = 1$ is an equation of an ellipse with its center at $(0, 0)$ and has a and b values of 3 and 4 respectively. The b value of 4 (from the 16 under y^2) tells us that the major axis is on the y-axis and will connect points 4 units above and below the center, namely $(0,4)$ and $(0, -4)$. The minor axis will connect the points $(3,0)$ and $(-3, 0)$. For ellipses the value of c is found from

whichever of $\sqrt{b^2 - a^2}$ or $\sqrt{a^2 - b^2}$ makes sense. In this case it's $\sqrt{b^2 - a^2} = \sqrt{16-9} = \sqrt{7}$ for c. The foci then are located at $(0, \sqrt{7})$ and $(0, -\sqrt{7})$ (on same axis as the vertices!). For an ellipse the eccentricity is given by the fraction c over the larger of a and b. Here it would be $\sqrt{7}/4$.

Sketch:

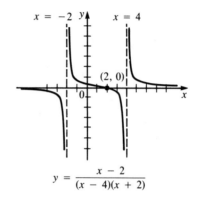

$$\left(\tfrac{x}{3}\right)^2 + \left(\tfrac{y}{4}\right)^2 = 1$$

15. $\frac{x^2}{9} - \frac{y}{16} = 0$ has only an x^2 term so it must be the equation of a parabola with the y-axis as its axis. We'll shoot for the form $x^2 = 4cy$:

add $\frac{y}{16}$ to both sides to get: $\frac{x^2}{9} = \frac{y}{16}$ or $x^2 = \frac{9}{16}y$ which can be viewed as $x^2 = 4(\frac{9}{64})y$.

the following conclusions can be drawn:

since x is squared and coefficient of y is positive the parabola opens up; the vertex is the origin because (0,0) satisfies the equation; the value of c is 9/64 putting the focus at (0, 9/64) and the directrix on the line $y = -9/64$.

Sketch:

Parabola
focus: $(0, \frac{9}{64})$
directrix: $y = -\frac{9}{64}$
axis: $x = 0$
vertex: (0, 0)

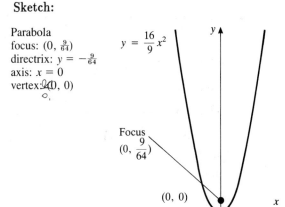

17. $\frac{y^2}{9} - \frac{x^2}{16} = 1$ is a hyperbola (the "-" sign is a tip-off) with a = 3 and b = 4. The vertices would lie on the y-axis (the y^2 term is the positive one) at (0, 3) and (0, - 3) and the transverse axis is the segment connecting them. The conjugate axis connects (4,0) and (-4,0). The box drawn through these four points with sides parallel to the x and y axes will help in the sketch. The center is at (0,0) and the foci depend on the value of c: for hyperbolas $c = \sqrt{a^2 + b^2}$ and here that value would come out to $\sqrt{25}$ or 5. The foci are on the y-axis at (0, 5) and (0,- 5). The eccentricity is given by c/a in this case and equals 5/3. The asymptotes are given by $y = \pm a/b\ x$ or here by $y = \pm(3/4)\ x$. The asymptotes contain the diagonals of the box referred to earlier. Sketch:

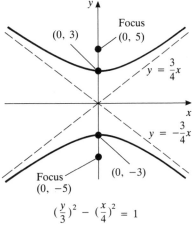

19. $\dfrac{(x-1)^2}{4} + \dfrac{(y+1)^2}{9} = 1$ is an ellipse with its

center at (1, -1) and its major axis lying on the

vertical line through the center, namely, x = 1.

We know this because the larger number is under the

y^2 type term. a = 2 and b = 3 in this problem so the

vertices must lie 3 units above and below the center

which puts them at (1, 2) and (1, - 4). The major axis

connects them. The minor axis must connect points

which are 2 units away from the center left or right;

these must be the points (-1, -1) and (3, -1). As

Sketch:

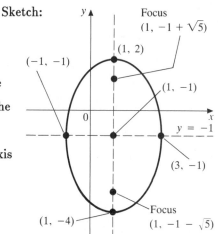

for the foci: c = $\sqrt{9 - 4}$ = $\sqrt{5}$ and so the foci must be located $\sqrt{5}$ units above and below the

center (1, -1) at (1,-1 - $\sqrt{5}$) and (1, -1+$\sqrt{5}$). The eccentricity is given by e = c/b = $\sqrt{5}/3$.

(for ellipses it's always c/ (larger of a and b))

21. $\dfrac{(x+2)^2}{25} + \dfrac{(y-5)^2}{25} = 0$ can only be satisfied by the point (- 2, 5).

23. $x^2 + 2x - y^2 + 2y = 0$

$(x^2 + 2x + 1) - (y^2 - 2y + 1) = 0 + 1 - 1$

$(x + 1)^2 - (y - 1)^2 = 0$ is a degenerate hyperbola (two straight lines)

The equations of these lines can be found by factoring the left side as the difference of 2 squares:

$[(x + 1) + (y - 1)][(x + 1) - (y - 1)] = 0$

$(x + y)(x - y + 2) = 0$ so that x + y = 0 and x - y + 2 = 0 are the equations of the lines

These equations can also be written as y = - x and y = x + 2 (they cross at the "center" (-1, 1))

Sketch:

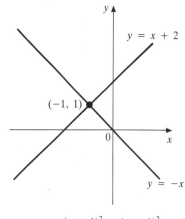

$(x + 1)^2 = (y - 1)^2$

25. $4x^2 + 4x + 3y^2 + 24y = 5$

    $4(x^2 + x + 1/4) + 3(y^2 + 8y + 16) = 5 + 1 + 48$

    $4(x + 1/2)^2 + 3(y + 4)^2 = 54$ $\Rightarrow$ $\dfrac{(x + 1/2)^2}{27/2} + \dfrac{(y + 4)^2}{18} = 1$

    which is an ellipse centered at (- 1/2, - 4)

    $a = \sqrt{27/2}$ and $b = \sqrt{18} = 3\sqrt{2}$ and b is larger so the major axis will connect points which are

    located $3\sqrt{2}$ units above and below the center at (-1/2, - 4 + $3\sqrt{2}$) and (- 1/2, - 4 - $3\sqrt{2}$).

    The minor axis is horizontal connecting the points (-1/2 - $\sqrt{27/2}$, - 4) and (-1/2 + $\sqrt{27/2}$, - 4).

    Since b > a the value of c will be $\sqrt{18 - 27/2} = \sqrt{9/2} = 3/\sqrt{2}$ which puts the foci

    this distance above and below the center at (- 1/2, - 4 - $3/\sqrt{2}$) and (- 1/2, - 4 + $3/\sqrt{2}$).

    The eccentricity e is given by c/b = $(3/\sqrt{2})/ 3\sqrt{2} = (3/\sqrt{2})\cdot(1/3\sqrt{2}) = 1/2$.

    Sketch:

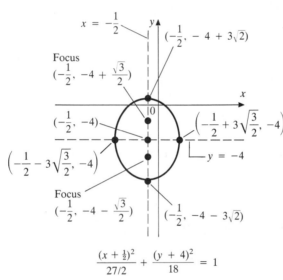

27. From the foci we see that c = 3 and the center is at the origin making the equation $\dfrac{x^2}{a^2} + \dfrac{y^2}{b^2} = 1$

    with a > b (since the foci are on the x-axis). The eccentricity is given as 0.6 and this must be

    equal to c/a so we have 3/a = 0.6 which implies that 3 = 0.6a so a is 5. Since $c^2 = a^2 - b^2$ we

    can also surmise that 9 = 25 - b^2 which means $b^2 = 16$ and b = 4. The equation can now be pieced

    together as $\dfrac{x^2}{25} + \dfrac{y^2}{16} = 1$

29. Foci and vertices are on y-axis equidistant from the origin so the equation must be of the form

    $\dfrac{y^2}{a^2} - \dfrac{x^2}{b^2} = 1$. From the points given we can see that c = 3 and a = 2. For hyperbolas it is always

    true that $c^2 = a^2 + b^2$ so 9 = 4 + b^2 and $b^2 = 5$. The equation is then: $\dfrac{y^2}{4} - \dfrac{x^2}{5} = 1$

CHAPTER SIX
Exponential and Logarithmic Functions

Chapter Objectives

In this chapter you will learn about exponential and logarithmic functions. These functions have many important applications in mathematics, computer science, the natural sciences, business and economics, and many other areas.

You learn how to evaluate, graph and apply these functions. The natural logarithm, natural exponential, common logarithm, and common exponential are the most important for most applications, so they are emphasized.

Chapter Summary

- An **exponential function** is a function of the form $f(x) = a^x$ where $a > 0$, $a \neq 1$.

- The **number e** is the number approached by the expression $\left(1 + \frac{1}{m}\right)^m$ as $m \to \infty$. To 10 decimal places, $e \approx 2.7182818285$. The function $y = e^x$ is the most important exponential function in applications.

- The function $y(t) = y(0)e^{kt}$ is the equation of **exponential growth** if $k > 0$ or **exponential decay** if $k < 0$.

- The **logarithm function** $y = \log_a x$ is the inverse of $y = a^x$. We have $y = \log_a x$ if and only if $x = a^y$.

- **Properties of Logarithm Functions**

  These properties hold for $a > 0$ (except $a = 1$) and where arguements are > 0.

  $$a^{\log_a x} = x \qquad \log_a a^x = x \qquad \log_a a = 1 \qquad \log_a 1 = 0$$

  $$\log_a xw = \log_a x + \log_a w \qquad \log_a \frac{x}{w} = \log_a x - \log_a w$$

  $$\log_a \frac{1}{x} = -\log_a x \qquad \log_a x^r = r\log_a x \qquad \log_a b = \frac{1}{\log_b a} \qquad \log_a x = \frac{\log x}{\log a} = \frac{\ln x}{\ln a}$$

- **Common Logarithms** are logarithms to the base 10.
- **Natural Logarithms** are logarithms to the base e.

Solutions to Chapter Six Problems

Problems 6.1

1.

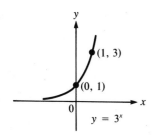

$y = 3^x$

3.

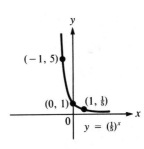

$y = (\tfrac{1}{5})^x$

5.

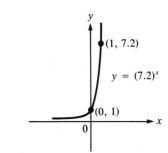

$y = (7.2)^x$

7.

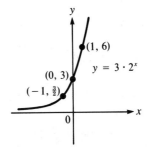

$y = 3 \cdot 2^x$

9.

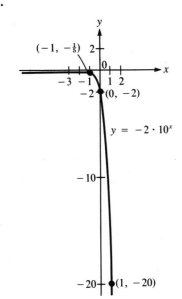

$y = -2 \cdot 10^x$

11.

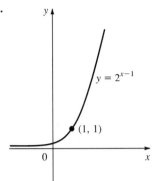

$y = 2^{x-1}$

(1, 1)

13.

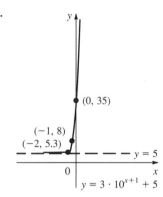

(0, 35)

(−1, 8)
(−2, 5.3)

$y = 5$

$y = 3 \cdot 10^{x+1} + 5$

15. $10^{2.2} \approx 158.4893192$

17. $4^{-1.6} \approx 0.108818820$

19. $(0.35)^{0.42} \approx 0.643440775$

21. $3^{\sqrt{2}} \approx 4.728804388$

23. $S = 5000(1 + \frac{0.06}{1})^{1 \cdot 4} = 5000(1.06)^4 = \$6312.38$ $I = \$6312.38 - \$5000 = \$1312.38$

25. $S = 5000\left(1 + \frac{.06}{12}\right)^{4 \cdot 12} = 5000(1.270489161) = \$6352.45$ $I = \$1352.45$

27. $S = 8000\left(1 + \frac{.11}{4}\right)^{4 \cdot 4} = 8000(1.5435094) = \$12,348.08$ $I = \$4348.08$

29. $S = 8000\left(1 + \frac{.11}{100}\right)^{400} = 8000(1.5523317) = \$12,418.65$ $I = \$4418.65$

31. $S = 10,000\left(1 + \frac{.085}{4}\right)^{24} = 10,000(1.656416961) = \$16,564.17$ $I = \$6,564.17$

33. The future value of the investment compounded annually; $P(1.06)^{10} = 1.7908447P$, an increase of approximately 79.08%.   For the investment compounded quarterly, the future value would be $P\left(1 + \frac{.06}{4}\right)^{40} = 1.8140184$, an increase of approximately 81.4%.

The percentage difference is 81.4% - 79.08% = 2.32%

35. The effective annual interest rate on 5% compounded daily is

$\left(1 + \frac{.05}{365}\right)^{365} - 1 = 1.0512675 - 1 = .0512675 = 5.12675\%$

The effective annual interest rate for $5\frac{1}{8}\%$ compounded semiannually is

$(1 + \frac{.05125}{2})^2 - 1 = 1.0519066 - 1 = .0519066 = 5.19066\%$

The $5\frac{1}{8}\%$ compounded semiannually gives the better investment.

37. Let x = nominal rate of interest paid per quarter. Then

$$1000 = 750(1 + x)^{32}$$
$$1.333333 = (1 + x)^{32}$$
$$\sqrt[32]{1.333333} = 1 + x$$

   x = 1.0090306 - 1 = .009036 or .90306% per quarter. Multiply by 4

   to obtain a nominal annual interest rate of 3.61224%.

39. a) The effective interest rate for account A is $\left(1 + \frac{.12}{2}\right)^2 - 1 = .1236 = 12.36\%$

   The effective rate for account B is $\left(1 + \frac{.11}{12}\right)^{12} - 1 = .1157 = 11.57\%$.

   b) The investment in A, after 5 years, would be worth $400(1.06)^{10} = \$716.34$

   The investment in B would be worth $400\left(1 + \frac{.11}{12}\right)^{60} = \$691.57$

   A is better by \$24.77

41. $f(x) = 3^x$ is graph d

43. $f(x) = \left(\frac{1}{3}\right)^x$ is graph a

Problems 6.2

1. $e^{3.15} \approx 23.3361$

3. $e^{\sqrt{3}} \approx 5.65223$

5. $e^{29.4} \approx 5.86486 \times 10^{12}$

7. $e^{-15.9} \approx 1.24371 \times 10^{-7}$

9. $e^{\pi} \approx 23.1407$ 11. f   13. a   15. c   17. d   19. i

21.

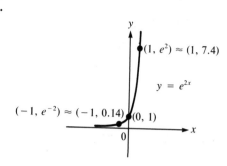

23.

6.2

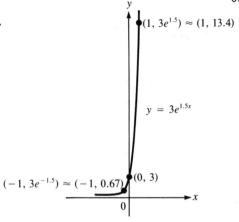

25.

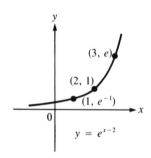

27.

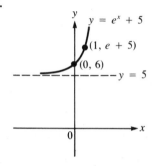

29. a) Just plug .13 into the formula in place of x. You will obtain 1.1388284

b) Plug in -.37 and obtain .6907343

31. $S = 9000e^{5(.10)} = 9000e^{.50} = 9000(1.648721271) = \$14,838.49$ Int: 14,838.49 - 9000 = \$5838.49

33. $S = 10,000e^{1.04} = 10,000(2.829217014) = \$28,292.17$ I = \$18,292.17

35. $S = 10,000e^{.72} = \$20,544.33$ I = \$10,544.33

37. The effective interest rate for 5% compounded continuously is

$e^{.05} - 1 = 1.0512711 - 1 = .0512711 = 5.12711\%$.

The effective rate for $5\frac{1}{8}\%$ compounded annually is 1.05125 - 1 = 5.125%.

Choose the continuously compounded investment.

39. You have $2 = e^{.12t}$. Just keep plugging in values for t and obtain 5.8.

In Section 6.4 you will learn to solve the problem using the following method:

$2 = e^{.12t}$ or $\ln 2 = .12t$

so $t = \frac{\ln 2}{.12} \approx 5.8$ years

41. a) The effective interest rate for account A is $\left(1 + \frac{.12}{2}\right)^2 - 1 = .1236 = 12.36\%$.

The effecive rate for account B is $e^{.11} - 1 = 11.63\%$.

b) Investment A is worth $\left(1 + \frac{.12}{2}\right)^{10} = \$716.34$

Investment B is worth $400e^{.55} = \$693.30$. A is better by \$23.04

43. Treat time $t = 0$ as the year 1986. The population of the world t years later is given by $4,845,000,000e^{.01t}$.

a) Plug in 4 (1990 is 4 years later) in place of t to obtain 5,042,728,201

b) Now $t = 14$ for 14 years later; plug in and obtain 5,573,076,555.

c) Plug in $t = 24$ and obtain 6,159,202,133

45. The initial temperature difference is 130° (180 - 50). Let D = the temperature difference (at time t minutes after noon) between the coal and the air, and one obtains $D = 130e^{-.12t}$.

a) plug in $t = 5$ to obtain d= 71.35°, meaning the temperature of the coal is 71.35° + 50 or 121.35°F

b) Plug in $t = 15$, get d= 21.49, or coal temperature is 71.49°F

c) plug in $t = 30$ and then add 50 to the result to obtain temperature 53.55°F.

47. a) $\sinh(-x) = \frac{e^{-x} - e^{x}}{2} = \frac{-(e^{x} - e^{-x})}{2} = -\sinh(x)$ so odd.

b)

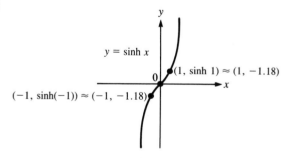

49. a) $\cosh(-x) = \dfrac{e^{-x} + e^{x}}{2} = \dfrac{e^{x} + e^{-x}}{2} = \cosh(x)$ so even.

b) $\cosh 0 = \dfrac{e^{0} + e^{-0}}{2} = \dfrac{1 + 1}{2} = \dfrac{2}{2} = 1$

If $x \neq 0$ then $\cosh(x) = \dfrac{e^{x} + e^{-x}}{2}$. Multiply numerator and denominator by

e^{2x} to obtain $\dfrac{e^{2x} + 1}{2e^{x}}$ (since $e^{x} \cdot e^{x} = e^{x+x} = e^{2x}$ and $e^{x} \cdot e^{-x} = e^{x-x} = e^{0} = 1$)

Now $e^{x} - 1 = 0$ only if $x = 0$. So if $x \neq 0$, $e^{x} - 1 \neq 0$ so $(e^{x} - 1)^{2} > 0$.
Therefore, $e^{2x} - 2e^{x} + 1 > 0$, which means $e^{2x} + 1 > 2e^{x}$.
From the above expression for $\cosh(x)$, the numerator is larger than the
denominator, (and both numerator and denominator are positive) which means
the fraction, and hence $\cosh(x)$, is greater than 1.

c)

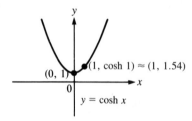

$y = \cosh x$

$(0, 1)$

$(1, \cosh 1) \approx (1, 1.54)$

Problems 6.3

1. $16^{1/2} = 4.$

3. $\left(\dfrac{1}{2}\right)^{-4} = 16$

5. $12^{0} = 1$

7. $10^{-3} = 0.001$

9. $e^{2} = e^{2}$

11. $\log_{2} 32 = 5$

13. $\log_{1/3} 9 = -2$

15. $\log_{10} 2 = 0.301029995$

17. $3^{y} = 27 \quad$ so $y = 3$

19. $\left(\dfrac{1}{2}\right)^{u} = 4 \quad$ so $\quad u = -2$

21. $\left(\frac{1}{4}\right)^s = 2$

   $s = -\frac{1}{2}$ since 2 is the reciprocal of the square root of $\frac{1}{4}$.

23. $y = 5$

25. $\log_5 \frac{1}{125} = \log_5 5^{-3} = -3$, so $y = 3(-3) = -9$

27. $\frac{1}{3^{4.5}} = 3^{-4.5}$ so $u = -4.5$

29. $y = \log_3 x$ is graph b

31. $y = \log_{(1/3)} x$ is graph c

33. $\log_e 8 = \log_e 2^3 = 3 \cdot \log_e 2 = 3(.6931) = 2.0793$ (using $\log_a x^r = r\log_a x$)

35. $\log_e 36 = \log_e 6^2 = 2 \cdot \log_e 6 = 2(1.7918) = 3.5836$

37. $\log_{10} 345 = \log_{10}(3.45 \cdot 100) = \log_{10}(3.45) + \log_{10} 100 = .5378 + 2 = 2.5378$

39. $\log_{10} 0.345 = \log_{10}(3.45 \cdot \frac{1}{10}) = \log_{10} 3.45 + \log_{10} \frac{1}{10} = .5378 - 1 = -0.4622$

41. $\log_{10}(3.45)^{1/2} = \frac{1}{2}\log_{10}(3.45) = \frac{1}{2}(.5378) = .2689$

43. $\log_{10} \frac{1}{\sqrt{3450}} = \log_{10}(3450)^{-1/2} = -\frac{1}{2}\log_{10}(3.45 \cdot 1000) = -\left(\frac{1}{2}\log_{10} 3.45 + \log_{10} 1000\right)$

   $= -\frac{1}{2}(0.5378 + 3) = -1.7689$

45. $\log_{10}(\frac{1}{34.5})^3 = \log_{10}(34.5)^{-3} = -3\log_{10}(3.45 \cdot 10) = -3[\log_{10}(3.45) + \log_{10} 10]$

   $= -3(0.5378 + 1) = -4.6134$

47. $\log_a x^4 y^3 = \log_a x^4 + \log_a y^3$

   $= 4\log_a x + 3\log_a y$

49. $(x^4 y^{3/2})^{1/2} = x^2 y^{3/4}$ so

   $\log_a x^2 y^{3/4} = \log_a x^2 + \log_a y^{3/4} = 2\log_a x + \frac{3}{4}\log_a y$

51. $\log_a \dfrac{x^{1/2} y^{1/3}}{z^{1/5}} = \log_a x^{1/2} + \log_a y^{1/3} - \log_a z^{1/5} = \tfrac{1}{2}\log_a x + \tfrac{1}{3}\log_a y - \tfrac{1}{5}\log_a z$

53. $\tfrac{4}{5}\log_a \dfrac{xy}{z} = \tfrac{4}{5}(\log_a x + \log_a y - \log_a z)$

55. $\log_a(x \cdot \tfrac{1}{x}) = \log_a 1 = 0$

57. $\log_a x + \log_a y^2 = \log_a\left(xy^2\right)$

59. $\log_a(4x \cdot 5w) = \log_a(20xw)$

61. $\log_a x + \log_a y^2 + \log_a z^3 = \log_a(xy^2z^3)$

63. $\log_a \dfrac{x^2 - 4}{x + 2} = \log_a \dfrac{(x - 2)(x + 2)}{x + 2} = \log_a(x - 2)$

65. $\log_a[(x^2 - 2y)(y + 2)] = \log_a(x^2 y + 2x^2 - 2y^2 - 4y)$

67. $\log_a(x^2 - y^2) - \log_a(x - y)^4 + \log_a(x + y)^3$

$= \log_a \dfrac{(x^2 - y^2)(x + y)^3}{(x - y)^4} = \log_a \dfrac{(x - y)(x + y)(x + y)^3}{(x - y)^4} = \log_a \dfrac{(x + y)^4}{(x - y)^3}$

Problems 6.4

1. $10^3 = 1000$ so $x = 3$ 3. $z = 20$

5. since 10^x and $\log_{10} x$ are inverses, $v = 3.4$ 7. since e^x and $\ln x$ are inverses, $x = 0.235$

9. $x = \pi$ since $\ln x$ and e^x are inverses   11. $x = 6.4$

13. $\ln(x(x + 3)) = \ln(x^2 + 3x)$ (Using $\ln ab = \ln a + \ln b$)

15. $\ln(xyz)$ 17. $\ln z^{1/2} + \ln w^{3/4} - \ln(x^2 + 1)^2 = \ln\dfrac{z^{1/2} w^{3/4}}{(x^2 + 1)^2} = \ln\dfrac{\sqrt{z}\,\sqrt[4]{w^3}}{(x^2 + 1)^2}$

19. $\ln 20 - (\ln 2 + \ln 5) = \ln 20 - \ln(2 \cdot 5) = \ln 20 - \ln 10 = \ln\dfrac{20}{10} = \ln 2$

21. $\ln\left(\frac{2}{w}\right)^3 - \ln\left(\frac{z}{3}\right)^4 = \ln\frac{8}{w^2} - \ln\frac{z^4}{81} = \ln\frac{8/w^2}{z^4/81} = \ln\ \ \frac{8}{w^2}\cdot\frac{81}{z^4} = \ln\frac{648}{w^3z^4}$

For 22-35, just recall that the argument of ln must be positive.

23. $(-1, \infty)$ 25. $(-\infty, 1)$ 27. $(-\infty, 0) \cup (0, \infty)$ 29. $(-\infty, -1) \cup (1, \infty)$

31. $(-\infty, -3) \cup (4, \infty)$ 33. $(-\infty, \infty)$ 35. $(0, \infty)$ 37. f   39. i   41. e   43. b

45. $\log x = \log 2^4 + \log 3^2 = \log 16 + \log 9 = \log(16\cdot 9) = \log 144,$ so $x = 144$

47. $\log x = \log b^a + \log a^b = \log(b^a\cdot b^a)$ so $x = b^a\cdot a^b$

49. $\log x = \log 2 + \log 5 - (\log 3 + \log 7) = \log (2\cdot 5) - \log (3\cdot 7) = \log 10 - \log 21$
    so $\log x = \log\frac{10}{21}$.   Therefore $x = \frac{10}{21}$.

51. $\ln x^{1/3} = \ln 3^2 + \ln 6 = \ln 9 + \ln 6 = \ln 6\cdot 9 = \ln 54$
    Therefore $x^{1/3} = 54$ and $x = 54^3 = 157,464$

53. $\frac{3}{2}\ln x - \ln x = \ln\frac{9}{3}$ so $\frac{1}{2}\ln x = \ln 3$

    $\ln x^{1/2} = \ln 3$ which means $x^{1/2} = 3$ or $x = 9$

55. a) $\log 684 = \log (6.84\cdot 100) = \log 6.84 + \log 100 \approx 0.8351 + 2 \approx 2.8351$
    b) $\log 6,840,000 = \log (6.84\cdot 1,000,000) = \log 6.84 + \log 1,000,000 \approx 6.8351$
    c) $\log 0.0684 = \log(6.84\cdot 0.01) = \log 6.84 + \log (0.01) \approx 0.8351 - 2 = -1.1649$
    d) $0.8351 - 7 = -6.1649$

57. a) $\log 95.2 = \log(9.52\cdot 10) = \log 9.52 + \log 10 = 0.9786 + 1 = 1.9786$
    b) $\log 95,200 = \log(9.52\cdot 10,000) = 0.9786 + 4 = 4.9786$
    c) $\log .0952 = \log(9.52\cdot\frac{1}{100}) = .9786 - 2 = -1.0214$
    d) $\log (95.2 \times 10^{-6}) = \log (9.52 \times 10 \times 10^{-6}) = \log(9.52 \times 10^{-5}) = 0.9876 - 5 = 4.0214$

59. $\log 1 - \log .00952 = 0 - \log (9.52\cdot 0.001) = 0 - [0.9786 - 3] = 2.0214$

61. $\log (29{,}700)^{1/2} = \frac{1}{2}\log(2.97 \cdot 10{,}000) = \frac{1}{2}(\log 2.97 + \log 10{,}000) = \frac{1}{2}(.4728 + 4) = 2.2364$

63. $\log (64.8)^{-1/2} = -\frac{1}{2}\log(6.48 \cdot 10) = -\frac{1}{2}(\log 6.48 + \log 10) = -\frac{1}{2}(.8351 + 1) = -0.91755$

65. $\log 68.4 + \log 29.7 + \log 95.2 = \log(6.84 \cdot 10) + \log(2.97 \cdot 10) + \log (9.52 \cdot 10)$

$= 1.8351 + 1.4728 + 1.9786 = 5.2865$

67. $\frac{1}{2}\log 6840 + \frac{1}{3}\log 95.2 - 5\log .297 = \frac{1}{2}(3.8351) + \frac{1}{3}(1.9786) - 5(.4728 - 1)$

$= 1.9176 + .6595 - 2.3640 + 5 = 5.2131$

69. Start with $y = 3\ln x$. a) If x triples , now you have 3x

$y = 3\ln 3x = 3\ln (3 \cdot x) = 3(\ln 3 + \ln x) = 3\ln 3 + 3\ln x$. So y increases

by $3\ln 3 = \ln 27$.

b) Now have $y = 3\ln \frac{x}{10} = 3 (\ln x - \ln 10) = 3\ln x - 3\ln 10$.

So y decreases by $3\ln 10 = \ln 1000$.

71. 1.757699625

73. 4.698987376

75. 6.907755279

77. -3.42651519

79. .497149872

81. $\pi \approx 3.141592654$

83. 1.561842388

85. 1.3513983465

87. $\log_3 4 = \dfrac{\log_{10} 4}{\log_{10} 3} \approx 1.261859507$ or can write as $\dfrac{\ln 4}{\ln 3}$ (This does not equal $\ln \frac{4}{3}$!!)

89. $x = \dfrac{\ln 25}{\ln 6} \approx 1.796488803$ 91. $x = \dfrac{\ln (1/3)}{\ln (1/4)} \approx 0.79248125$ 93. $x = \dfrac{\ln 39.5}{\ln 28.2} \approx 1.10091231$

95. a) use $\log_a b = \dfrac{1}{\log_b a}$ to obtain $\log_5 2 = \dfrac{1}{2.3219} = 0.4307$

b) $\log_5 4 = \log_5 2^2 = 2 \log_5 2 = 2(0.4307) = 0.8614$

c) $\log_5 2^{-3} = -3 \log_5 2 = -3(0.4307) = -1.2921$

97. a) if $x = 0.8$, then $A = \frac{0.8 - 1}{0.8 + 1} = -\frac{.2}{1.8} \approx -0.111111111$.

Plug this into formula in place of A and obtain $\ln 0.8 \approx -0.223143491$

Now $x = \frac{1.2 - 1}{1.2 + 1} \approx .09090909$. Plug in and obtain $\ln 1.2 \approx 0.182321542$

b) $\ln \frac{3}{2}$; $A = \frac{1.5 - 1}{1.5 + 1} = \frac{.5}{2.5} = .2$ Plug in and obtain $.4055$

$\ln \frac{4}{3} = \frac{4/3 - 1}{4/3 + 1} = 0.14285714$. Plug in and obtain $.2877$

$\ln 2 = \ln (\frac{3}{2} \cdot \frac{4}{3}) = \ln \frac{3}{2} + \ln \frac{4}{3} = .4055 + .6932 = .6932$

c) $\ln 3 = \ln (\frac{3}{2} \cdot 2) = \ln \frac{3}{2} + \ln 2 = .4055 + .6932 = 1.0987$

$\ln 8 = \ln 2^3 = 3 \ln 2 = 3(.6932) = 2.0796$

99. $100! \approx \sqrt{2 \cdot \pi \cdot 100} \left(\frac{100}{e}\right)^{100} \approx (25.07)(36.788)^{100}$. Let this expression $= x$.

Then $\log x \approx \log 25.07 + 100 \log 36.788 \approx 1.39909 + 156.57055 = 157.9696$.

So $x \approx 10^{157.9696} = 10^{.9696} \times 10^{157} \approx 9.32 \times 10^{157}$

$200! \approx \sqrt{1256}(73.5759)^{200} = 35.45(73.5759)$ Let this expression $= x$,

obtain $\log x = \log 35.45 + 200 \log 73.5759 = 374.89661$.

So $x = 10^{374.89661} = 10^{.89661} \times 10^{374} \approx 7.88 \times 10^{374}$

101. a) if $\sinh x = 2$, then $2 = \frac{e^x - e^{-x}}{2}$. Multiply both sides by 2, $4 = e^x - e^{-x}$.

b) Multiply both sides of $4 = e^x - e^{-x}$ by e^x to obtain $4e^x = e^x e^x - e^{-x} e^x$.

Now $e^x e^x = e^{x+x} = e^{2x}$ and $e^{-x} e^x = e^{-x+x} = e^0 = 1$.

therefore, $4e^x = e^{2x} - 1$ or $e^{2x} - 4e^x - 1 = 0$.

c) Let $w = e^x$ and the $w^2 - 4w - 1 = 0$

By the quadratic formula, $w = \frac{4 \pm \sqrt{16 + 4}}{2} = \frac{4 \pm 2\sqrt{5}}{2} = 2 \pm \sqrt{5}$.

d) e^x cannot be negative, so e^x cannot equal $2 - \sqrt{5}$, but can equal $2 + \sqrt{5}$

e) Take $\ln$ of both sides of $e^x = 2 + \sqrt{5}$ to obtain $x = \ln (2 + \sqrt{5}) \approx 1.4436$

103. $\sinh x = 4$

$e^x - e^{-x} = 8$

$e^{2x} - 8e^x - 1 = 0$

Set $w = e^x$ to obtain $w^2 - 8w - 1 = 0$.

By the quadratic formula $w = \dfrac{8 \pm \sqrt{64 + 4}}{2} = \dfrac{8 \pm \sqrt{68}}{2} = \dfrac{8 \pm 2\sqrt{17}}{2} = 4 \pm \sqrt{17}$

Since e^x must be positive, take + above, so $e^x \approx 8.1231056$

Take ln to obtain $x \approx 2.0947$

105. $e^x + e^{-x} = 4$

$e^{2x} - 4e^x + 1 = 0$. Set $e^x = w$ to obtain $w^2 - 4w + 1 = 0$

From the quadratic formula, $w = \dfrac{4 \pm \sqrt{16 - 4}}{2} = \dfrac{4 \pm \sqrt{12}}{2} \approx 3.73205$ or 0.2679492

So $e^x = $ either 3.73205 or 0.2679492. Take ln, and obtain $x = \pm 1.316957897$

Problems 6.5

1. $x = 2$ 3. $-x = 1$ so $x = -1$ 5. $x = 2^8 = 256$

7. $x = e^{-1} = \frac{1}{e}$ 9. $10^1 = 10$ so $x = 10$ 11. $10^0 = 1$ so $x = 1$

13. $x = \ln 10 \approx 2.303$ 15. no solutions since e^x cannot be negative

17. $\ln 7^x = \ln 14$ 19. $\ln 231^x = \ln 8$

    $x \ln 7 = \ln 14$ $x \ln 231 = \ln 8$

    $x = \dfrac{\ln 14}{\ln 7} \approx 1.356$ $x = \dfrac{\ln 8}{\ln 231} \approx 0.382$

21. $e^{1.6} \approx 4.953$ 23. $10^{-1.57} \approx 0.0269$

25. $x - 1 = \ln 2$ so $x = 1 + \ln 2 \approx 1.693$

27. Take ln of both sides to obtain $\frac{t}{2} = \ln 4$. Then $t = 2\ln 4 \approx 2.773$

29. Divide both sides by 10 and obtain $e^{-h/5} = 5$. Now take ln of both sides

and obtain $-\frac{h}{5} = \ln 5$. Therefore, $h = -5 \ln 5 \approx -8.047$

31. Use the addition property to obtain $\ln 3x = 1$.

Use each side as the exponent of e to obtain $e^{\ln 3x} = e^1$ or $3x = e$.

Therefore, $x = \frac{e}{3} \approx 0.906$

33. Use addition property to obtain $\ln(v \cdot \frac{1}{v}) = \ln(v - 3)$

$$\ln 1 = \ln(v - 3) \quad \text{or} \quad 0 = \ln(v - 3)$$
$$e^0 = e^{\ln(v - 3)} \quad \text{so} \quad 1 = v - 3 \quad \text{or} \quad v = 4.$$

35. $\ln (0.172w) = 0.428$

$e^{\ln (0.172w)} = e^{0.428}$

$0.172w = e^{0.428}$ so $w = \frac{e^{.428}}{.172} \approx 8.920$

37. $\ln \frac{z - 3}{z + 4} = \ln 2$ Now Property (6), page 227, says if $\log_a x = \log_a y$, then $x = y$.

$\frac{z - 3}{z + 4} = 2.$ Multiply both sides by $z + 4$ to obtain

$z - 3 = 2(z + 4)$

$z - 3 = 2z + 8$

$-11 = z$. But if this is plugged back into the original

equation, you'd take ln of negative numbers. So, no solutions.

39. $e^x + 8e^{-x} - 6 = 0$ (by first subtracting 6 from both sides). Now multiply through by e^x to obtain

$(e^x)^2 - 6e^x + 8 = 0$ (since $e^x \cdot e^{-x} = e^{x-x} = e^0 = 1$)

Can factor the result (or use the substitution $u = e^x$) to obtain $(e^x - 4)(e^x - 2) = 0$

Therefore, $e^x = 4$ or $e^x = 2$. Taking ln of each side of both of these, you obtain

$x = \ln 4 \approx 1.386$ or $x = \ln 2 \approx 0.693$

41. subtract 6, multiply through by $e^{(1/2)x}$ to obtain $e^x + e^{(1/2)x} - 6 = 0$. Again, either

use substitution or factoring. If substitution, $u = e^{(1/2)x}$ to obtain $u^2 + u - 6 = 0$

or $(u - 2)(u + 3) = 0$, which means $u = 2$ or $u = -3$. This gives $e^{(1/2)x} = -3$ (not possible)

or $e^{(1/2)x} = 2$. Take ln of both sides to get $\frac{1}{2}x = \ln 2$, so $x = 2 \ln 2 \approx 1.386$

45. $z = 48$ since b^x and $\log_b x$ are inverse functions   47. As in #45, $q = .0023$

49. $s = 3^{\log_3 16^{1/2}} = 16^{1/2} = 4$ 51. $u = e^{\log_e 100^{-1/2}} = 100^{-1/2} = \frac{1}{10}$

53. $\log_5 2x = \frac{1}{3}$, so $5^{1/3} = 2x$,   or $x = \frac{5^{1/3}}{2} \approx 0.855$

55. Rewrite and obtain $2^4 = u^4$, so $u = 2$ or $u = -2$

57. $w^{-3} = 125$, or $w = \frac{1}{5}$.

59. $2^{x-1} = 4$. Multiply both sides by 2 to obtain $2^x = 8$, or $x = 3$

61. $\log_4 x^2 = -3$

   $4^{-3} = x^2$

   $\frac{1}{64} = x^2$,   so $x = \frac{1}{8}$ ($-\frac{1}{8}$ is not a solution since argument of log is always positive).

63. $10^{-3} = x + 5$

   $x = 10^{-3} - 5 = \frac{1}{1000} - 5 = -\frac{4999}{1000} = -4.999$

65. $2 + \frac{1}{x}$ becomes $\frac{1}{x} = 3$ (use log and exponential inverse property) so $x = \frac{1}{3}$

67. Only $10^0 = 1$, so $t^2 + 2t - 8 = 0$

   $(t - 2)(t + 4) = 0$

   $t = -4$ since t negative is specified

69. $\log_4 \frac{x}{x+1} = 1$ (Using $\log_a x - \log_a y = \log_a \frac{x}{y}$).

   $4^1 = \frac{x}{x+1}$.   Multiply both sides by $x + 1$ to obtain $4x + 4 = x$.

   $3x = -4$ or $x = -\frac{4}{3}$ so no solutions since argument of log must be positive.

71. $x + 3 = 2x - 5$ so $8 = x$ 73. $10^{10^x} = 10^1$ Equating exponents, $10^x = 1$ or $x = 0$

Problems 6.6

1. If the principle P has increased by half, it is 1.5 times its original amount.
   So we want to solve the equation $1.5P = P\left(1 + \frac{.04}{12}\right)^{12t}$ (using the formula for compounding interest a finite number of times per year). Divide both sides by P to obtain $1.5 = \left(1 + \frac{.04}{12}\right)^{12t}$. Now take ln of both sides

$$\ln(1.5) = \ln\left(1 + \frac{.04}{12}\right)^{12t} = 12t\cdot\ln\left(1 + \frac{.04}{12}\right) \text{ by property of ln.}$$

   Now solve for t to obtain
$$t = \frac{\ln(1.5)}{12\ln\left(1 + \frac{.04}{12}\right)} \approx 10.1 \text{ years}$$

3. For each part we start with $3P = P\left(1 + \frac{r}{4}\right)^{4t}$ or $3 = \left(1 + \frac{r}{4}\right)^{4t}$.

   Take ln of both sides to obtain $\ln 3 = \ln\left(1 + \frac{r}{4}\right)^{4t} = 4t\cdot\ln\left(1 + \frac{r}{4}\right)$.

   Solve for t and obtain $t = \dfrac{\ln 3}{4\cdot\ln\left(1 + \frac{r}{4}\right)}$

   For part a), substitute $r = .02$ into the above and obtain 55.07 years

        b) substitute $r = .05$ and obtain 22.11 years

        c) substitute $r = .08$ and obtain 13.87 years

        d) substitute $r = .10$ and obtain 11.12 years

        e) substitute $r = .15$ and obtain 7. 46 years

5. Now use $3P = Pe^{rt}$ or $3 = e^{rt}$

   Take ln of both sides and get $\ln 3 = \ln e^{rt}$ or $\ln 3 = rt$

   Solve for t and obtain $t = \frac{\ln 3}{r}$

   For part a) $r = .02$ and $t = 54.93$ years

        b) $r = .05$ and $t = 21.97$ years

        c) $r = .08$ and $t = 13.73$ years

        d) $r = .10$ and $t = 10.99$ years

        e) $r = .15$ and $t = 7.32$ years

7. $10,000 = Pe^{0.085 \cdot 1}$ so $P = \dfrac{10,000}{e^{0.085}} = \$9185.12$

9. $10,000 = P(1 + .09)^5$ or $P = \dfrac{10,000}{(1.09)^5} = \$6499.31$

11. $1000 = 750e^{r8}$ or $e^{8r} = \dfrac{1000}{750} = \dfrac{4}{3}$

Take ln of both sides and obtain $\ln \dfrac{4}{3} = \ln e^{8r} = 8r$

So $r = \dfrac{1}{8} \cdot \ln \dfrac{4}{3} \approx .03596$ or 3.596%

13. $3P = Pe^{r15}$ or $3 = e^{15r}$. Take ln of both sides and obtain

$\ln 3 = \ln e^{15r} = 15\,r$. Therefore, $r = \dfrac{\ln 3}{15} \approx .07324$ or 7.324%

15. $P(t) = 4.845e^{.019t}$ (billion)

$8 = 4.845e^{.019t}$ or $\dfrac{8}{4.845} = e^{.019t}$. Take ln of both sides as usual

and obtain $\ln \dfrac{8}{4.845} = \ln e^{.019t} = .019t$. Therefore, $t = \dfrac{\ln \dfrac{8}{4.845}}{.019} \approx 26.39$

It would reach 8 billion in 2012 (26.39 years later)

17. First, we must find r. $248,709,873 = 226,549,448e^{r10}$ (since 1990 is

10 years after 1980). Divide both sides by 226,549,448 and obtain

$1.09878172 = e^{10r}$. Take ln of both sides to obtain $\ln 1.09878172 = 10r$

so $r = \dfrac{\ln 1.09878172}{10} \approx 0.0094202038$

The equation giving the population is $P(t) = 203,302,031e^{0.0094202038t}$

a) Plug $t = 30$ into the above equation and obtain $P(30) \approx 273,037,968$

b) Set the above equation equal to 500,000,000.

$500,000,000 = 226,549,448e^{0.0094202038t}$

$2.207023696 = e^{0.0094202038t}$.   Take ln of both sides

$\ln 2.207023696 = 0.0094202038t$ or $t = \dfrac{\ln 2.20703696}{0.0094202038} \approx 84.83$ years

So the population will reach $\dfrac{1}{2}$ billion in the year 2065 (approximately 84.83

years later).

19. $16,643,000 = 13,400,000e^{r10}$ or $1.242014925 = e^{16r}$

Take ln of both sides and obtain $\ln 1.242014925 = 16r$

or $r = \dfrac{\ln 1.242014925}{16} \approx 0.013545937$

So the population equatin is $P(t) = 13,400,000e^{0.013545937t}$

a) Plug in $t = 26$ and obtain $19,057,278$

b) Set $20,000,000 = 13,400,000e^{0.013545937t}$. Divide both sides by $13,400,000$

and obtain $1.49253731 = e^{0.013545937t}$. Take ln of both sides to obtain

$\ln 1.49253731 = 0.013545937t$ or $t = \dfrac{\ln 1.49253731}{0.013545937} \approx 29.56$

So the population will reach 20 million in 2004 (approximately 29.56 years

later).

21. From problem 20 we had that the population of New York t years after 1980

is $P(t) = 17,558,165e^{0.0024322t}$. Now we find the equation that gives

the population of Florida t years after 1980.

We have $P(t) = 9,747,197e^{kt}$. Now

$12,937,9267 = P(10) = 9,747,197e^{10k}$

$1.3273484 = e^{10k}$

$k = \dfrac{\ln 1.3273484}{10} = 0.0283183$

So the population for Florida is $P(t) = 9,747,197e^{0.0283183t}$

To determine when the population of Florida will exceed that of New York,

set the two population equations equal to each other.

$9,747,197e^{0.0283183t} = 17,558,165e^{0.0024322t}$. Divide both sides by $9,747,197$

and by $e^{0.0024322t}$ to obtain

$1.8013553 = e^{0.0258861t}$ (subtract the exponents of e)

$t = \dfrac{\ln 1.8013553}{0.0258861} \approx 22.74$ years later, in 2003.

23. $262 = 100e^{r7}$. Divide both sides by 100, get $2.62 = e^{7r}$.

Take ln of both sides, get $\ln 2.62 = 7r$ or $r = \dfrac{\ln 2.62}{7} \approx .01376$ or 13.76%

25. $116.3 = 88.7e^{10r}$ or $1.31116122 = e^{10r}$. Therefore, $r = \dfrac{\ln 1.31116122}{10} \approx .0271$ or 2.71%

27. Again use $S = Pe^{rt}$. If no gain or loss in real income, then in 1960 salary is \$4743 and in 1970, \$7564 so $7564 = 4743e^{10r}$ and $1.5947712 \approx e^{10r}$. Let x = the unknown CPI in 1970. $x = 88.7e^{10r} = 88.7e^{10r} \approx 88.7(1.5947712) \approx 141.5$

The CPI in 1970 would have to be approximately 141.5

29. The initial temperature difference is 160° and the temperature of the surrounding medium is -10°, so the equation giving the temperature of the coal is

$Q(t) = -10 + 160e^{-kt}$. $60 = Q(30) = -10 + 160e^{-30k}$. Need to solve for k.

First add 10 to both sides to obtain $70 = 160e^{-30k}$. Divide both sides by 160 to obtain $\frac{7}{16} = e^{-30k}$. Now take ln of both sides to obtain $\ln \frac{7}{16} = -30k$ or $k = \frac{\ln\frac{7}{16}}{-30} \approx 0.027555952$

So the equation giving the temperature in °C after t seconds is

$Q(t) = -10 + 160e^{-.027555952t}$.

a) Plug in t = 120 (seconds) and obtain $-10 + 160e^{120(-.027555952)} \approx -4.138°C$

b) $0 = -10 + 160e^{-.027555952t}$

$10 = 160e^{-.027555952t}$

$\frac{1}{16} = e^{-.027555952t}$ so $\ln \frac{1}{16} = -.027555952t$ or $t = \frac{\ln\frac{1}{16}}{-.027555952} \approx 100.62$ seconds

31. a) Let Q_0 be the initial amount. Then $0.60Q_0 = Q_0e^{-k100}$ (40% lost means 60% is left)

$0.60 = e^{-100k}$. Take ln of boths sides to obtain $\ln (0.60) = \ln e^{-100k} = -100k$.

Therefore, $k = \frac{\ln 0.60}{-100} = -0.005108256$. The half-life of something $= \frac{\ln 2}{k} = \frac{\ln 2}{0.005108256}$, or approximately 135.69 years.

b) $0.10Q_0 = Q_0e^{-0.005108256t}$ or $0.10 = e^{-0.005108256t}$. Take ln of both sides to obtain $\ln (0.10) = \ln e^{-0.005108256t} = -0.005108256t$ so $t = \frac{\ln 0.10}{-0.005108256} \approx 450.76$ years.

33. Use the relationship that half-life $= \frac{\ln 2}{k}$ and obtain $\frac{\ln 2}{1.5 \times 10^{-7}} \approx 4,620,981$ years

35. The pressure at height h is given by $P(h) = 1013.25e^{-kh}$. We must find k.

$845.6 = P(1500) = 1013.25e^{-1500k}$ so $0.8345 = e^{-1500k}$

$\ln 0.8345 = -1500k$ so $k = \dfrac{\ln 0.8345}{-1500} \approx -.00012058$

Therefore, the equation is $P(h) = 1013.25e^{-.00012058h}$

a) $P(4000) = 1013.25e^{4000(-.00012058)} \approx 625.53$ mbar

b) $P(10,000) \approx 303.42$ mbar

c) $P(-86) - P(4418) = 1023.8120 - 594.7825 = 429.03$ mbar

d) $P(8848) = 348.63$ mbar

e) $1 = 1023.15e^{-.00012058h}$

$\quad .00097737 = e^{-.00012058h}$ so $\ln .00097737 = -.00012058h$

$\quad\quad h = \dfrac{\ln .00097737}{-.0012058} \approx 57,396.9$ meters

37. One way to do the problem is to let $t = 0$ correspond to 12 hours after the count began. Then $Q(0) = 5969$ and $Q(t) = 5969e^{-kt}$.

$3563 = Q(12) = 5969e^{-12k}$

$.59692 = e^{-12k}$

$k = \dfrac{\ln .59692}{-12} \approx -.04299804$

$Q(t) = 5969e^{-.04299804t}$

a) Since we started at $t = 0$ corresponding to 12 hours after initial counting, $t = 12$ would correspond to when the counting started. So we want $Q(-12) = 10,000$ (the number of bacteria must be an integer).

b) One week corresponds to 7 days or 168 hours. But $t = 0$ corresponded to 12 hours so we have to have $Q(156) = 5969e^{156(-.04299804)} \approx 7$

c) $1 = 5969e^{-.04299804t}$

$\quad .000167532 = e^{-.04299804t}$

$\quad t = \dfrac{\ln .000167532}{-.042998042} \approx 202.203$. Add 12 to obtain that after 214.203 hours

(approximately 8 days, 22.2 hours) there would be no more bacteria.

39. Take $- \log\left(0.6 \text{ X } 10^{-7}\right) = 7.22$

41. Take $10^{-4.5} \approx 3.162 \text{ X } 10^{-5}$ moles per liter

43. a) $(2 - 5) = 2.5\log(\text{ratio})$

$\frac{-3}{2.5} = \log(\text{rat})$ or $-1.2 = \log(\text{rat})$

$\text{rat} = 10^{-1.2} \approx .0630957$ which means the brightness ratio is about 1 to 15.849

b) $(m - n) = 2.5\log 5 = 1.747$

c) $(2.15 - 1.4) = 2.5 \log$ (ratio of brightness of Sirius to unknown)

$$.75 = 2.5\log(\text{rat})$$

$$.3 = \log(\text{rat})$$

$$10^{.3} \approx 1.995 = \text{ratio (almost twice as bright)}$$

d) $2.5\log(45,000) \approx 11.63$ magnitudes

e) $(2.95 - 1.97) = 2.5 \log(\text{ratio of brightnesses})$

$$.98 = 2.5 \log(\text{rat})$$

$$.392 = \log(\text{rat})$$

$$\text{ratio} = 10^{.392} \approx 2.46$$

Let b = brightness of weaker

Then 2.46b = brightness of stronger

b + 2.46b = 3.46b = combined brightness

Let x be the magnitude of the two combined

Then $2.95 - x = 2.5\log\left(\frac{3.46b}{b}\right) = 2.5\log(3.46)$

$2.95 - x = 1.35$

$x = 1.60$ So their combined magnitudes is 1.60.

Review Exercises for Chapter Six

1.

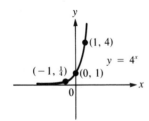

3.

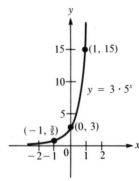

5.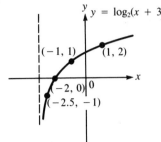

7.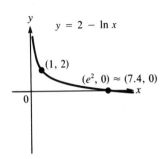

9. 5.47347392

11. 0.079913677

13. 1.45331834

15. 13.81551056

17. $y = 4$ since $2^4 = 16$

19. $x = 25$ since $25^{1/2} = 5$

21. $x = 10^{10^{-9}} \approx 1$

23. $\ln(x + 3) = 2$

$\qquad x + 3 = e^2$

$\qquad\quad x = e^2 - 3 \approx 4.389056099$

25. $\ln \dfrac{x}{x+1} = \ln 2$ or $\dfrac{x}{x+1} = 2$ $\qquad x = 2(x+1) = 2x + 2$ $\qquad$ so $x = -2$.

   Substituting this back into the original equation would require taking

   ln of -2, which is impossible. So, there are no solutions.

27. $\ln [(y+2)y] = \ln 3$ (by $\ln a + \ln b = \ln ab$) so $\ln \left(y^2 + 2y\right) = \ln 3$

   $y^2 + 2y = 3$ or $y^2 + 2y - 3 = 0$.

   This factors as $(y + 3)(y - 1) = 0$, which has solutions $y = 1$ or $y = -3$. $\qquad$ Now $y = -3$

   would require taking ln of negative numbers in original equation, so $y = 1$ is the only solution.

29. Replace x by 2x to obtain $y = 3 \ln 2x = 3(\ln 2 + \ln x) = 3 \ln 2 + 3 \ln x$.

   Therefore, y increases by $3\ln 2 = \ln 8$.

31. a) $9^{3/2} = 27$ b) $\left(\frac{1}{2}\right)^{-3} = 8$

33. $\log_3 x(x - 2) = \log_3(x^2 - 2x)$

35. $\log \dfrac{(x + 1)z^2}{x + 3}$ (The + terms all go in the numerator, the - in the denominator).

37. $\ln(x + 1)^3 - \ln(y + 4)^{1/2} + \ln(z^2 + 12)^{1/3} = \ln \dfrac{(x + 1)^3(z + 12)^{1/3}}{(y + 4)^{1/2}}$

39. $\dfrac{\ln 7}{\ln 2} = \dfrac{1.94959104}{.693147181} \approx 2.807354922$ $\left(\text{using } \log_a b = \dfrac{\ln b}{\ln a}\right)$

41. $\dfrac{\ln 37}{\ln 4} \approx 2.604726683$

43. Use $I = Prt = 5,000(.07)6 = \$2100$

45. Use $S = P\left(1 + \frac{r}{m}\right)^{mt}$

    $S = 8000\left(1 + \frac{.055}{1}\right)^{1\cdot 4} = 8000(1.238824651) = \$9910.60$ is the value

    of the investment; subtract \$8000 to obtain \$1910.60 interest.

47. $6000e^{.10\cdot 5} = 6000e^{.50} = 6000(1.648721271) = \$9892.33$ for total value,

    \$3,892.33 for interest.

49. $e^{.08} - 1 = 1.083287068 - 1 = .083287068.$ Multiply by 100 to obtain

    approximately 8.32871%

51. Want to know when $2P = P\left(1 + \frac{r}{4}\right)^{4t}$

    $2 = \left(1 + \frac{r}{4}\right)^{4t}$

    $\ln 2 = 4t \cdot \ln\left(1 + \frac{r}{4}\right)$ or $t = \dfrac{\ln 2}{4 \ln\left(1 + \frac{r}{4}\right)}$

    a) Let $r = .03$, then $t = 23.19$ years

    b) Let $r = .065$, then $t = 10.75$ years

    c) Let $r = .08$, then $t = 8.75$ years

    d) Let $r = .12$, then $t = 5.86$ years

53. The equation is $P(t) = 10,000e^{.15t}$

   After 5 years, $P(5) = 10,000e^{.75} = 10,000(2.117000017) = 21,170$ (no fractions)

   After 10 years $P(10) = 44,817$

55. Let $Q(t)$ be the temperature after 10 minutes. The temperature of the room's air is 23°C and the initial difference between the air temperature and the cake is 102°C. Then $Q(t) = 23 + 102e^{-kt}$. Find k.

   $80 = Q(10) = 23 + 102e^{-10k}$

   $57 = 102e^{10k}$

   $0.558823529 = e^{-10k}$

   $k = \dfrac{\ln 0.558823529}{-10} \approx 0.058192155$

   So $Q(t) = 23 + 102e^{-.058192155t}$

   a) $Q(20) = 23 + 102e^{20(-.058192155)} = 23 + 31.85294117 = 54.85°C$

   b) $25 = 23 + 102e^{-.058192155t}$

   $\quad 2 = 102e^{-.058192155t}$

   $\quad .019607843 = e^{-.058192155t}$

   $\quad t = \dfrac{\ln .019607843}{-.058192155} = 67.57$ minutes

57. The half-life of something is $\dfrac{\ln 2}{k}$. So we need to find k. Let t be the number of weeks. Let M be the original mass. Let $Q(t)$ be the amount left after t weeks.

   Then $Q(t) = Me^{kt}$.   Now $.80M = Q(1) = Me^{k}$ or $.80 = e^{k}$

   $k = \ln .80 = -.223143551$ and half-life $= \dfrac{\ln 2}{k} = \dfrac{-0.693147181}{-0.223143551} = 3.1063$ weeks

   or approximately 21.744 days.

CHAPTER SEVEN
The Trigonometric Functions

Chapter Objectives

In chapter seven you will be introduced to the six trigonmetric functions: the sine, cosine, tangent, cosecant, secant and cotangent functions. Each can be defined in two ways: one involving triangles and one simply involving points. The inputs to these functions are numbers which can be thought of as angles. Early in the chapter you will encounter radian measure for angles as an alternative for degree measure which you may be famliliar with. You will become adept at graphing these trigonometric functions and will gradually become proficient at manipulating them, taking advantage of many special relationships that exist among these functions (identities). Critical to your survival is knowing (i.e., memorizing) some key values of these functions for key inputs (e.g., 0, $\pi/6$, $\pi/3$, etc) and having the definitions and basic identities at your fingertips. On the practical side you will see how to find missing parts of right triangles given minimal information; on the theoretical side, you will do some work with the inverses of these six functions.

Chapter Summary

- An angle consists of two rays (an initial side and a terminal side) with a common initial point called the vertex.
- One degree (1˚) is the measure of the angle obtained by rotating the terminal side of the angle $\frac{1}{360}$ of a complete revolution in the positive (counterclockwise) direction.
- The complement of θ is 90˚ - θ; the supplement of θ is 180˚ - θ.
- 1 minute (1') = 1/60 of a degree; 1 second (") = 1/60 of a minute.
- The radian measure of an angle in standard position (vertex at the origin, initial side along the positive x-axis) is the length of the arc of the unit circle cut off by the sides of the angle.
- Conversion formulas:

$$\theta \text{ (degrees)} = \frac{180}{\pi} \cdot \theta \text{ (radians)} \qquad \theta \text{ (radians)} = \frac{\pi}{180} \cdot \theta \text{ (degrees)}$$

- If a central angle subtends (cuts off) an arc of length L in a circle of radius r, then the radian measure of θ is given by L/r.

- If an object is moving on a circle at n rpm, then the angular speed $= \omega = 2\pi n$ radians per minute.

- The linear speed $= v = 2\pi nr$ units per minute where r is the radius of the circle.

- Sine and Cosine functions:

  If the terminal side of an angle θ in standard position intersects the circle of radius r at the point (x,y), then $\cos\theta = x/r$ and $\sin\theta = y/r$. If $r = 1$ (the unit circle), then $\cos\theta = x$ and $\sin\theta = y$.

- The reference angle of θ (in standard position) is the positive acute angle, θ_r, that the terminal side of θ makes with the x-axis.

- Reference angle theorem: $\sin\theta = \pm\sin\theta_r$ and $\cos\theta = \pm\cos\theta_r$.

- Identities Involving $\sin\theta$ and $\cos\theta$ (given in radian measures)

$$\sin^2\theta + \cos^2\theta = 1$$

$\cos(\theta + 2\pi) = \cos\theta$	$\sin(\theta + 2\pi) = \sin\theta$
$\cos(-\theta) = \cos\theta$	$\sin(-\theta) = -\sin\theta$
$\cos(\theta + \pi) = -\cos\theta$	$\sin(\theta + \pi) = -\sin\theta$
$\cos(\theta - \pi) = -\cos\theta$	$\sin(\theta - \pi) = -\sin\theta$
$\cos(\pi/2 - \theta) = \sin\theta$	$\sin(\pi/2 - \theta) = \cos\theta$

- The zeros of $\sin\theta$ are $0, \pm\pi, \pm2\pi, \ldots$

- The zeros of $\cos\theta$ are $\pm\frac{\pi}{2}, \pm\frac{3\pi}{2}, \pm\frac{5\pi}{2}, \ldots$

- If x, y, and r are as in the definitions of $\sin\theta$ and $\cos\theta$,

$$\tan\theta = \frac{y}{x} \qquad \cot\theta = \frac{x}{y} \qquad \sec\theta = \frac{r}{x} \qquad \csc\theta = \frac{r}{y}$$

- $\tan\theta = \dfrac{\sin\theta}{\cos\theta} \qquad \cot\theta = \dfrac{\cos\theta}{\sin\theta} = \dfrac{1}{\tan\theta} \qquad \sec\theta = \dfrac{1}{\cos\theta} \qquad \csc\theta = \dfrac{1}{\sin\theta}$

- $\tan\theta$ and $\sec\theta$ are undefined when $\theta = \pm\frac{\pi}{2}, \pm\frac{3\pi}{2}, \pm\frac{5\pi}{2}, \ldots$

- $\cot\theta$ and $\csc\theta$ are undefined when $\theta = 0, \pm\pi, \pm2\pi, \ldots$

- Identities involving $\tan\theta$, $\cot\theta$, $\sec\theta$, and $\csc\theta$

$\tan(-\theta) = -\tan\theta$	$\cot(-\theta) = -\cot\theta$
$\sec(-\theta) = \sec\theta$	$\csc(-\theta) = -\csc\theta$
$\tan(\theta + \pi) = \tan\theta$	$\cot(\theta + \pi) = \cot\theta$
$\sec(\theta + 2\pi) = \sec\theta$	$\csc(\theta + 2\pi) = \csc\theta$
$\tan^2\theta + 1 = \sec^2\theta$	$\cot^2\theta + 1 = \csc^2\theta$

- Right Triangle Definitions: Let adj = side adjacent to angle θ in a right triangle, opp = opposite side, and hyp = the hypotenuse.

$$\sin \theta = \frac{\text{opp}}{\text{hyp}} \qquad \cos \theta = \frac{\text{adj}}{\text{hyp}} \qquad \tan \theta = \frac{\text{opp}}{\text{adj}}$$

$$\csc \theta = \frac{\text{hyp}}{\text{opp}} \qquad \sec \theta = \frac{\text{hyp}}{\text{adj}} \qquad \cot \theta = \frac{\text{adj}}{\text{opp}}$$

- The Inverse Functions

Function	Domain	Range	Also called
$\sin^{-1}x$	[-1,1]	$[-\pi/2,\ \pi/2]$	arc sin x
$\cos^{-1}x$	[-1,1]	$[0, \pi]$	arc cos x
$\tan^{-1}x$	$(-\infty, \infty)$	$(-\pi/2, \pi/2)$	arc tan x

Problems 7.1

1. 50° is in quadrant I because 0° < 50° < 90°.

3. 200° is in quadrant III because 180° < 200° < 270°.

5. 1 radian is in quadrant I because $0 < 1 < \frac{\pi}{2}$ (≈ 1.57).

7. An angle of measure 3 would terminate in quadrant II because $\frac{\pi}{2} < 3 < \pi$.

9. Quadrant IV since $\frac{3\pi}{2} < 5 < 2\pi$ (note: $\frac{3\pi}{2} \approx 4.71$ and $2\pi \approx 6.28$

11. Quadrant II since 90° < 176° < 180°

13. Quadrant II since $\frac{\pi}{2} < 2.18 < \pi$

15. 40° + 360° = 400°; 40° - 360° = -320°

17. -25° + 360° = 335°; -25° - 360° = -385°

19. $-100° + 360° = 260°$; $-100 - 360° = -460°$

21. $\frac{\pi}{3} + 2\pi = \frac{\pi}{3} + \frac{2\pi}{1} = \frac{\pi + 6\pi}{3} = \frac{7\pi}{3}$; $\frac{\pi}{3} - \frac{2\pi}{1} = \frac{\pi - 6\pi}{3} = -\frac{5\pi}{3}$

23. $\frac{4\pi}{3} + \frac{2\pi}{1} = \frac{4\pi + 6\pi}{3} = \frac{10\pi}{3}$; $\frac{4\pi}{3} - \frac{2\pi}{1} = -\frac{2\pi}{3}$

25. $\frac{9\pi}{8} + \frac{2\pi}{1} = \frac{25\pi}{8}$; $\frac{9\pi}{8} - \frac{2\pi}{1} = -\frac{7\pi}{8}$

27. comp. of $47° = 90° - 47° = 43°$

29. comp. of $66.6° = 90° - 66.6° = 23.4°$

31. comp. of $46°15' = 90° - 46°15' = 89°60' - 46°15' = 43°45'$

33. supp. of $21° = 180° - 21° = 159°$

35. supp. of $110° = 180° - 110° = 70°$

37. supp. of $48°22' = 179°60' - 48°22' = 131°38'$

39. $.35° = 0.35 \times 60' = 21'$ so $42.35° = 42°21'$

41. $.425° = 0.425 \times 60' = 25.5'$; $.5' = .5 \times 60'' = 30''$ so $121.425° = 121°25'30''$

43. $15'' = \frac{15}{60}$ of a minute $= 0.25'$; $30.25' = \frac{30.25}{60}$ of a degree $\approx 0.5042°$ so $24°30'15''$ is approximately $24.5042°$

45. $50'' = 50/60' = 0.8333'$; $10.8333' = 10.8333/60° = 0.1806°$ so $-20°10'50''$ is approximately $-20.1806°$

47. $39'' = 39/60' = 0.65'$; $17.65' = 17.65/60° = 0.2942°$ so $-48°17'39'' \approx -48.2942°$

49. $10° = 10° \times \frac{\pi}{180°} = \frac{\pi}{18}$

51. $150° = 150° \times \frac{\pi}{180°} = \frac{5\pi}{6}$

53. $540° = 540° \times \frac{\pi}{180°} = 3\pi$

55. $27° = 27° \times \frac{\pi}{180°} = 0.4712$ (radians)

57. $-71° = -71° \times \frac{\pi}{180°} = -1.2392$ (radians)

59. $\frac{\pi}{8} = \frac{\pi}{8} \times \frac{180°}{\pi} = 22.5°$

61. $-\frac{3\pi}{8} = -\frac{3\pi}{8} \times \frac{180°}{\pi} = -67.5°$

63. $\frac{5\pi}{3} = \frac{5\pi}{3} \times \frac{180°}{\pi} = 300°$

65. $\frac{7\pi}{18} = \frac{7\pi}{18} \times \frac{180°}{\pi} = 70°$

67. $-0.23 = -0.23 \times \frac{180°}{\pi} = -13.1780°$

69. $10 = 10 \times \frac{180°}{\pi} = 572.9578°$

71. $L = r\theta = (\ 1\ cm\ )\ (\pi/3) = \pi/3\ cm \approx 1.0472\ cm$

73. $L = r\theta = (\ 1\ m\ )\ (2.8) = 2.8\ m$

75. $L = r\theta = (4\ in\ )\left(\frac{7\pi}{3}\right) = \frac{28\pi}{3}\ in \approx 29.3215\ in$

77. $\theta = 75° \times \frac{\pi}{180°} \approx 1.3090$; $s = r\theta \approx (\ 2\ ft\ )\ (1.3090) = 2.618\ ft$

79. $\theta = 200° \times \frac{\pi}{180°} \approx 3.49066$; $s = r\theta = (\ 7\ m\ )\ (3.49066) = 24.435\ m$

81. since $L = r\theta$, $\theta = \frac{L}{r} = \frac{5\ cm}{3\ cm} = \frac{5}{3}$ radians $= \frac{5}{3} \times \frac{180°}{\pi} = 95.49°$

83. $\theta = \frac{L}{r} = \frac{15\ in}{2.5\ in} = 6$ radians or $6 \times \frac{180°}{\pi} = 343.77°$

85. 1 rpm = 1 revolution per minute = 2π radians per minute so...

    a) 45 rpm = 45 X 2π radians per minute = 90π radians per minute

    b) θ = angular speed $\cdot$ time = $\dfrac{90\pi \text{ radians}}{60 \text{ sec}} \cdot \dfrac{3 \text{ sec}}{1} = \dfrac{9\pi}{2}$ radians

    c) linear speed = ν = r $\cdot$ angular speed = 3.5 in $\cdot$ 90π/min = 315π in/min or 16.49 in/sec

    d) distance = rate $\cdot$ time = 16.49 in/sec $\cdot$ 3 sec = 49.5 in

87. r = 0.75 m and 40 rpm

    a) angular speed = ω = (2π X 40) radians per min = 80π rad/min

    b) ν = ωr= 80π rad/min X 0.75 m = 60π m/min

    c) distance = rate X time = (60π m/60 sec) X 8 sec = 8π m

89. r = 27 in = 2.25 ft

linear speed = $\dfrac{55 \text{ miles}}{1 \text{ hour}}$ X $\dfrac{1 \text{ hour}}{60 \text{ min}}$ X $\dfrac{5280 \text{ ft}}{1 \text{ mile}}$ = 4840 ft/min

and since linear speed = 2πnr we have n = $\dfrac{\text{linear speed}}{2\pi\text{r}}$ = $\dfrac{4840 \text{ ft/min}}{2\pi(2.25 \text{ ft})}$ $\approx$ 342 rpm

91. In one hour the minute hand sweeps out one revolution or 2π radians;

    therefore in 2 hours the angle would be $\theta = 4\pi$ radians so....

        L = rθ = (4 in) (4π) = 16π in

93. we know: rpm of wheel = 12 = n

                r = 1.5 m

                speed of paddle wheel = $\frac{1}{2}$ of speed of current

speed of paddle wheel = ν = 2πnr = $2\pi(12)(1.5)$ $\approx$ 113.0973 m/min

so speed of current is twice this or 226.1947 m/min.

Changing to km/hr requires two conversion factors:

$\dfrac{226.1947 \text{ m}}{1 \text{ min}}$ X $\dfrac{1 \text{ km}}{1000 \text{ m}}$ X $\dfrac{60 \text{ min}}{1 \text{ hr}}$ = 13.57 km/hr

95. $r = 1,169,820$ miles   and 1 revolution (2π radians) requires 16.69 days

a) In 5 days, the angle will be $\frac{5}{16.69}$ of 1 revolution (2π) or ≈ 1.8823781 rad

b) $\nu = 2\pi nr$ (we have r but must find n : 1 rev in 16.69 days is equivalent to 0.0599 rev per day and this will be our n) $= 2\pi(0.0599)(1,169,820) = 440,395.2$ miles per day

c) Jupiter's year is 11.86 times as long as our year or 4328.9 days

distance = rate X time $= 440,395.2$ mi/day X 4328.9 da $= 1.9059$ X 10^9 miles

97. $r = 3960$ miles ; θ_{comp} is the angle we need $= 49°15' = 49.25°$ X $\frac{\pi}{180°} = 0.8599$

$L = r \, \theta_{comp} = (3960 \text{ miles})(0.8599) = 3405$ miles from NYC to the North Pole

99. $\theta = 35°41' = 35.6833°$ X $\frac{\pi}{180°} = 0.6228$ radian

$L = (3960 \text{ miles})(0.6228) = 2466$ miles from Sante Fe to the equator

101. $L = 500$ and $r = 3960$ so $\theta = \frac{L}{r} = \frac{500}{3960} = 0.12626$ radian or $7.23°$

Problems 7.2

1. First thing to do is find the hypotenuse (hyp) by using the Pythagorean theorm:

$$8^2 + 15^2 = \text{hyp}^2; \quad 64 + 225 = \text{hyp}^2; \quad 289 = \text{hyp}^2; \quad \text{hyp} = 17.$$

Viewing from θ we see that opp = 15 and adj = 8; now the six values can be determined:

$$\sin \theta = \frac{\text{opp}}{\text{hyp}} = \frac{15}{17} \qquad\qquad \sec \theta = \frac{\text{hyp}}{\text{adj}} = \frac{17}{8}$$

$$\cos \theta = \frac{\text{adj}}{\text{hyp}} = \frac{8}{17} \qquad\qquad \csc \theta = \frac{\text{hyp}}{\text{opp}} = \frac{17}{15}$$

$$\tan \theta = \frac{\text{opp}}{\text{adj}} = \frac{15}{8} \qquad\qquad \cot \theta = \frac{\text{adj}}{\text{opp}} = \frac{8}{15}$$

3. $1^2 + 4^2 = \text{hyp}^2 \Rightarrow \sqrt{17} = \text{hyp};$ viewing from θ shows opp $= 4$ and adj $= 1$

$$\sin\theta = \frac{4}{\sqrt{17}}, \quad \cos\theta = \frac{1}{\sqrt{17}}, \quad \tan\theta = \frac{4}{1} = 4, \quad \sec\theta = \sqrt{17}, \quad \csc\theta = \frac{\sqrt{17}}{4}, \quad \cot\theta = \frac{1}{4}$$

5. $3^2 + \text{opp}^2 = 7^2$ so opp$^2 = 40$ which implies that opp $= \sqrt{40} = 2\sqrt{10}$; also adj $= 3$ and hyp $= 7$

$$\sin\theta = \frac{2\sqrt{10}}{7}, \quad \cos\theta = \frac{3}{7}, \quad \tan\theta = \frac{2\sqrt{10}}{3}, \quad \sec\theta = \frac{7}{3}, \quad \csc\theta = \frac{7}{2\sqrt{10}} = \frac{7\sqrt{10}}{20}, \quad \cot\theta = \frac{3}{2\sqrt{10}} = \frac{3\sqrt{10}}{20}$$

7. $2^2 + \text{opp}^2 = 3^2$ so opp$^2 = 5$ which implies that opp $= \sqrt{5};$ adj $= 2$ and hyp $= 3$ so....

$$\sin\theta = \frac{\sqrt{5}}{3}, \quad \cos\theta = \frac{2}{3}, \quad \tan\theta = \frac{\sqrt{5}}{2}, \quad \sec\theta = \frac{3}{2}, \quad \csc\theta = \frac{3}{\sqrt{5}} = \frac{3\sqrt{5}}{5}, \quad \cot\theta = \frac{2}{\sqrt{5}} = \frac{2\sqrt{5}}{5}$$

9. $(1/3)^2 + \text{opp}^2 = (1/2)^2$ so opp$^2 = 1/4 - 1/9 = 5/36$ which implies that opp $= \frac{\sqrt{5}}{6}$.

$$\sin\theta = \frac{\sqrt{5}/6}{1/2} = \frac{\sqrt{5}}{6} \cdot \frac{2}{1} = \frac{\sqrt{5}}{3}, \quad \cos\theta = \frac{1/3}{1/2} = \frac{2}{3}, \quad \tan\theta = \frac{\sqrt{5}/6}{1/3} = \frac{\sqrt{5}}{2},$$

$$\sec\theta = \frac{3}{2}(\text{ reciprocal of } \cos\theta), \quad \csc\theta = \frac{3}{\sqrt{5}} = \frac{3\sqrt{5}}{5}, \quad \cot\theta = \frac{2}{\sqrt{5}} = \frac{2\sqrt{5}}{5}$$

11. $\sin\theta = \frac{3}{5} = \frac{\text{opp}}{\text{hyp}}$

(a)   sketch a generic right triangle
label one acute angle as θ
set the side opposite θ to 4 and
the hypotenuse to 5 and find the
other side by using the Pythagorean

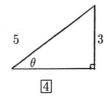

theorem: adj $= \sqrt{5^2 - 3^2} = \sqrt{16} = 4$ (any value that had to be computed will be boxed in the $\triangle$)

11. (continued)

Now, by glancing at the triangle:

(b) $\cos \theta = \dfrac{\text{adj}}{\text{hyp}} = \dfrac{4}{5}$; $\tan \theta = \dfrac{\text{opp}}{\text{adj}} = \dfrac{3}{4}$; $\csc \theta = \dfrac{\text{hyp}}{\text{opp}} = \dfrac{5}{3}$; $\sec \theta = \dfrac{\text{hyp}}{\text{adj}} = \dfrac{5}{4}$; $\cot \theta = \dfrac{4}{3}$

13. $\cot \theta = \dfrac{12}{5} = \dfrac{\text{adj}}{\text{opp}}$

(a) depicting θ in a $\triangle$:

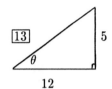

$\text{hyp} = \sqrt{5^2 + 12^2} = \sqrt{169} = 13$

(b) $\sin \theta = \dfrac{5}{13}$; $\cos \theta = \dfrac{12}{13}$; $\tan \theta = \dfrac{5}{12}$; $\csc \theta = \dfrac{13}{5}$; $\sec \theta = \dfrac{13}{12}$;

15. $\cos \theta = \dfrac{2}{3} = \dfrac{\text{adj}}{\text{hyp}}$

(a) the $\triangle$:

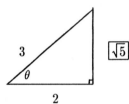

$\text{opp} = \sqrt{3^2 - 2^2} = \sqrt{5}$

(b) $\sin \theta = \dfrac{\sqrt{5}}{3}$; $\tan \theta = \dfrac{\sqrt{5}}{2}$; $\csc \theta = \dfrac{3}{\sqrt{5}}$; $\sec \theta = \dfrac{3}{2}$; $\cot \theta = \dfrac{2}{\sqrt{5}}$

17. $\cot \theta = \dfrac{4}{7} = \dfrac{\text{adj}}{\text{opp}}$

(a) the $\triangle$:

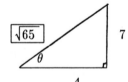

$\text{hyp} = \sqrt{4^2 + 7^2} = \sqrt{65}$

(b) $\sin \theta = \dfrac{7}{\sqrt{65}}$; $\cos \theta = \dfrac{4}{\sqrt{65}}$; $\tan \theta = \dfrac{7}{4}$; $\csc \theta = \dfrac{\sqrt{65}}{7}$; $\sec \theta = \dfrac{\sqrt{65}}{4}$

19. $\csc \theta = 1.5 = \frac{3}{2} = \frac{\text{hyp}}{\text{opp}}$

(a) the $\triangle$:
$$\text{adj} = \sqrt{3^2 - 2^2} = \sqrt{5}$$

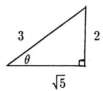

(b) $\sin \theta = \frac{2}{3}$; $\cos \theta = \frac{\sqrt{5}}{3}$; $\tan \theta = \frac{2}{\sqrt{5}}$; $\sec \theta = \frac{3}{\sqrt{5}}$; $\cot \theta = \frac{\sqrt{5}}{2}$

21. $\tan \theta = 4.5 = 4\frac{1}{2} = \frac{9}{2} = \frac{\text{opp}}{\text{adj}}$

(a) the $\triangle$:
$$\text{hyp} = \sqrt{9^2 + 2^2} = \sqrt{85}$$

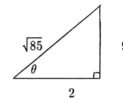

(b) $\sin \theta = \frac{9}{\sqrt{85}}$; $\cos \theta = \frac{2}{\sqrt{85}}$; $\csc \theta = \frac{\sqrt{85}}{9}$; $\sec \theta = \frac{\sqrt{85}}{2}$; $\cot \theta = \frac{2}{9}$

23. $\sin \theta = 0.3 = \frac{3}{10} = \frac{\text{opp}}{\text{hyp}}$

(a) the $\triangle$:
$$\text{adj} = \sqrt{10^2 - 3^2} = \sqrt{91}$$

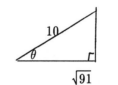

(b) $\cos \theta = \frac{\sqrt{91}}{10}$; $\tan \theta = \frac{3}{\sqrt{91}}$; $\csc \theta = \frac{10}{3}$; $\sec \theta = \frac{10}{\sqrt{91}}$; $\cot \theta = \frac{\sqrt{91}}{3}$

25. $\sec \theta = 20 = \frac{20}{1} = \frac{\text{hyp}}{\text{adj}}$

(a) the $\triangle$:
$$\text{opp} = \sqrt{20^2 - 1^2} = \sqrt{399}$$

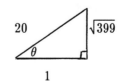

(b) $\sin \theta = \frac{\sqrt{399}}{20}$; $\cos \theta = \frac{1}{20}$; $\tan \theta = \sqrt{399}$; $\csc \theta = \frac{20}{\sqrt{399}}$; $\cot \theta = \frac{1}{\sqrt{399}}$

27. $\csc \theta = \frac{8}{5} = \frac{\text{hyp}}{\text{opp}}$

    (a) the triangle:
$$\text{adj} = \sqrt{8^2 - 5^2} = \sqrt{39}$$

    (b) $\sin \theta = \frac{5}{8}$; $\cos \theta = \frac{\sqrt{39}}{8}$; $\tan \theta = \frac{5}{\sqrt{39}}$; $\sec \theta = \frac{8}{\sqrt{39}}$; $\cot \theta = \frac{\sqrt{39}}{5}$

29. $\cos \theta = \frac{x}{2} = \frac{\text{adj}}{\text{hyp}}$ $\Rightarrow$ 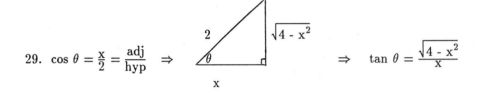 $\Rightarrow$ $\tan \theta = \frac{\sqrt{4 - x^2}}{x}$

31. $4 \cos \theta = y$ is the same as $\cos \theta = \frac{y}{4}$ $\Rightarrow$

    so $\sec \theta = \frac{4}{y}$ (which could have been gotten instantaneously using reciprocals)

33. $\sin \theta = \frac{u}{\sqrt{3}}$ $\Rightarrow$ so $\cos \theta = \frac{\sqrt{3 - u^2}}{\sqrt{3}}$

35. $\sqrt{2} \cos \theta = 7$ is equivalent to $\cos \theta = \frac{7}{\sqrt{2}}$ which is bad news for the cosine because its values must

    be in the [-1,1] range and $\frac{7}{\sqrt{2}}$ is greater than 1. No solution is possible.

37. Start with the fact that $\sin^2\theta + \cos^2\theta = 1$; divide both sides by $\cos^2\theta$ with

    the stipulation that $\theta \neq \pi/2, 3\pi/2$, etc. (so that we don't divide by 0) to get
$$\sin^2\theta/\cos^2\theta + \cos^2\theta/\cos^2\theta = 1/\cos^2\theta$$
    or $\tan^2\theta + 1 = \sec^2\theta$

Problems 7.3

1. $r = \sqrt{x^2 + y^2} = \sqrt{1^2 + 2^2} = \sqrt{5}$ so now $\cos\theta = \frac{x}{r} = \frac{1}{\sqrt{5}}$ and $\sin\theta = \frac{y}{r} = \frac{2}{\sqrt{5}}$.

3. r is still $\sqrt{5}$ but now $\cos\theta = \frac{x}{r} = \frac{-2}{\sqrt{5}}$ and $\sin\theta = \frac{y}{r} = \frac{1}{\sqrt{5}}$.

5. $r = \sqrt{(-3)^2 + (-2)^2} = \sqrt{13}$ so $\cos\theta = \frac{x}{r} = \frac{-3}{\sqrt{13}}$ and $\sin\theta = \frac{y}{r} = \frac{-2}{\sqrt{13}}$.

7. $r = \sqrt{6^2 + (-2)^2} = \sqrt{40} = 2\sqrt{10}$ so $\cos\theta = \frac{6}{2\sqrt{10}} = \frac{3}{\sqrt{10}}$ and $\sin\theta = \frac{-2}{2\sqrt{10}} = \frac{-1}{\sqrt{10}}$.

9. This one is different; we're given x and r and forced to solve for y. $r = \sqrt{x^2 + y^2}$ becomes $8 = \sqrt{(-4)^2 + y^2}$ which becomes $64 = 16 + y^2$ so $y = \pm\sqrt{48} = \pm 4\sqrt{3}$. We must choose the positive value since y is positive in QII. So now $\cos\theta = \frac{x}{r} = \frac{-4}{8} = -\frac{1}{2}$ and $\sin\theta = \frac{y}{r} = \frac{4\sqrt{3}}{8} = \frac{\sqrt{3}}{2}$.

11. (a) -180° corresponds to the point (-1, 0) as it is half a revolution clockwise.

   (b) $\sin(-180°) = y = 0$ and $\cos(-180°) = x = -1$

13. (a) 5π corresponds to (-1, 0) as it is two and a half revolutions.

   (b) $\sin 5\pi = y = 0$ and $\cos 5\pi = x = -1$

15. (a) $5\pi/2$ is 5 quarter revolutions and this is the same location as $\pi/2$ so the point is (0, 1)

   (b) $\sin 5\pi/2 = y = 1$ and $\cos 5\pi/2 = x = 0$

17. (a) -90° is a quarter revolution clockwise which leads to the point (0, -1) on the unit circle.

   (b) $\sin(-90°) = y = -1$ and $\cos(-90°) = x = 0$

19. (a) $9\pi/4$ could be thougt of as $2\frac{1}{4}\pi = 2\pi + \pi/4 =$ same location as $\pi/4$ or $(1/\sqrt{2}, 1/\sqrt{2})$

   (b) $\sin 9\pi/4 = y = 1/\sqrt{2}$ and $\cos 9\pi/4 = x = 1/\sqrt{2}$

21. (a) 390° is one revolution plus 30° so the point is $(\sqrt{3}/2, 1/2)$

   (b) $\sin 390° = y = 1/2$ and $\cos 390° = x = \sqrt{3}/2$

23. 55° is in quadrant one where ALL the trig functions are positive.

25. 247° is in quadrant three so the sine and cosine are both negative. (x and y are both negative)

27. 161° is in quadrant two where x (cosine) is negative and y (sine) is positive.

29. -200° is in quadrant two (remember to go clockwise) so sine is positive and cosine is negative.

31. 1500° (divide by 360 and see what the remainder is; it's 60.) so 1500° must terminate in QI. So both the sine and cosine of 1500° will be positive.

33. 2.1 rad is between $\pi/2$ (1.57) and π (3.14) rad so it must be in QII; sine is +'ve but cosine is -'ve

35. 8.9 radians; remember that 2π (one revolution) is approximately 6.28; subtract this from 8.9 and we're left with 2.62 as the radian measure of a coterminal angle; since 2.62 is larger than $\pi/2$ but smaller than π, it must terminate in quadrant II; the sine is positive and the cosine is negative.

37. $\frac{15\pi}{16}$ is just a nudge smaller than 1π so it's in QII; sine is positive and cosine negative.

39. 10 - 6.28 (two revolutions) = 3.72 so an angle of 3.72 radians is coterminal with an angle of 10 radians. Since 3.72 is greater than π but less than $3\pi/2$, this angle terminates in QIII and both sine and cosine are therefore negative.

41. Cosine will be positive since θ is in QI; since $\sin^2\theta + \cos^2\theta = 1$ we have $\cos^2\theta = 1 - 0.6^2$ so $\cos\theta = \sqrt{0.64} = 0.8$

43. Sine will be negative in QIII; $\sin\theta = -\sqrt{1 - (-0.75)^2} \approx -0.6614$

45. Cosine will be positive in QIV; $\cos\theta = \sqrt{1 - (-0.38)^2} \approx 0.9250$

47. If $4\sin\theta = 5$ then $\sin\theta = 5/4$ which is greater than one and no y value on the unit circle can be greater than 1 !

49. Since the sine is negative (y is negative) and cosine is positive (x is positive) it must be QIV.

51. If the sine and cosine are both positive then both x and y are positive and the first quadrant is indicated.

53. Be sure to be in degree mode; enter 37 and then tap $\boxed{\sin}$; get 0.6018 and then reenter 37 and tap $\boxed{\cos}$ to get 0.7986. $\text{Sin}^2 37° \approx 0.3622$ and $\cos^2 37° \approx 0.6378$ and these add to 1.

55. degree mode; 263 ; $\boxed{\sin}$ to get - 0.9925 . 263 ; $\boxed{\cos}$ to get - 0.1219

57. $\cos 1.5 \approx 1 - \dfrac{1.5^2}{2} + \dfrac{1.5^4}{24} - \dfrac{1.5^6}{720} = 1 - 1.125 + 0.210938 - 0.015820 = 0.070118$

   from a calculator we get $\cos 1.5 = 0.070737$; the %-error is the difference of these two values divided by the calculator value X 100% which comes out to be $\approx 0.9\%$

   $\sin 1.5 \approx 1.5 - \dfrac{1.5^3}{6} + \dfrac{1.5^5}{120} - \dfrac{1.5^7}{5040} = 1.5 - 0.5625 + 0.063281 - 0.0033 = 0.997391$

   calculator value is 0.997495 ; find difference and divide by 0.997495 and X by 100% to get 0.01%

59. 50° must be converted to radians first: $50° \times \dfrac{3.14159}{180°} \approx 0.873$

   $\cos 0.873 \approx 1 - \dfrac{0.873^2}{2} + \dfrac{0.873^4}{24} - \dfrac{0.873^6}{720} = 1 - 0.381065 + 0.024202 - 0.000615 = 0.642522$

   a calculator would give 0.6427876 for this; %-error $\approx 0.04\%$

   $\sin 0.873 \approx 0.873 - 0.110890 + 0.004226 - 0.000077 = 0.766260$ as compared to 0.7660445 from a calculator; %- error is their difference divided by the calculator value X 100% or 0.028%.

61. $\cos (-0.9) \approx 1 - \dfrac{(-0.9)^2}{2} + \dfrac{(-0.9)^4}{24} - \dfrac{(-0.9)^6}{720} = 1 - 0.405 + 0.027338 - 0.000738 = 0.6216$

   as compared to the value of 0.62161 from a calculator; the %-error = 0.002%

   $\sin (-0.9) \approx -0.9 - \dfrac{(-0.9)^3}{6} + \dfrac{(-0.9)^5}{120} - \dfrac{(-0.9)^7}{5040} = -0.9 + 0.1215 - 0.004921 + 0.00009 = -0.783326$

   and the calculator version of this answer would be -0.7833269, a percentage error of only 0.0001%

63. crd $\theta = y_1 - (-y_1) = 2y_1$ so $y_1 = \frac{1}{2}$ crd θ but we also know that $y_1 = \sin \theta/2$

   Therefore, $\sin (\theta/2) = \frac{1}{2}$ crd θ and if you multiply both sides by 2 you'll get the fact that $2 \sin (\theta/2) = $ crd θ.

65. crd $120° = 2 \sin (120°/2) = 2 \sin 60° = 2 (\sqrt{3}/2) = \sqrt{3}$

67. crd $80° = 2 \sin (80°/2) = 2 \sin 40° \approx 1.2856$ 69. crd $150° = 2 \sin 75° \approx 1.9319$

71. crd $2.35 = 2 \sin 1.175 \approx 1.8454$

73. (a) two and a half times would correspond to "π" so the point would be (-1, 0)

   (b) seven and three quarter times would correspond to 3/4 of a revolution or $3\pi/2$ radians so the point would be (0, -1)

   (c) since $\theta = L/r$ we have $\theta = 2/1$ or 2

   Then $y = \sin \theta = \sin 2 \approx 0.9093$ and $x = \cos \theta = \cos 2 \approx -0.4161$ which means that the point is (-0.4161, 0,9093)

   (d) Now $\theta = 20$ (radians) so $x = \cos 20 \approx 0.4081$ and $y = \sin 20 \approx 0.9129$ and the point is (0.4081, 0.9129)

Problems 7.4

1. if $\theta = \frac{2\pi}{3}$, $\theta_{ref} = \pi - \frac{2\pi}{3} = \frac{3\pi}{3} - \frac{2\pi}{3} = \frac{\pi}{3}$

3. if $\theta = \frac{5\pi}{3}$, $\theta_{ref} = 2\pi - \frac{5\pi}{3} = \frac{\pi}{3}$

5. if $\theta = \frac{21\pi}{10}$ (slightly larger than 2π) , $\theta_{ref} = \frac{21\pi}{10} - 2\pi = \frac{\pi}{10}$

7. if $\theta = 200°$, $\theta_{ref} = 200° - 180° = 20°$

9. if $\theta = 197°$, $\theta_{ref} = 197° - 180° = 17°$

11. $\theta_{ref} = \frac{\pi}{6}$ with θ in quadrant II so $\sin \frac{5\pi}{6} = + \sin \frac{\pi}{6} = \frac{1}{2}$ and $\cos \frac{5\pi}{6} = - \cos \frac{\pi}{6} = - \frac{\sqrt{3}}{2}$

13. $-\frac{3\pi}{2}$ is coterminal with $\frac{\pi}{2}$ so $\sin\left(-\frac{3\pi}{2}\right) = \sin\left(\frac{\pi}{2}\right) = 1$ and $\cos\left(-\frac{3\pi}{2}\right) = \cos\left(\frac{\pi}{2}\right) = 0$

15. reference angle for 210° is 30° in quadrant III: $\sin 210° = -\sin 30° = -\frac{1}{2}$
$$\cos 210° = -\cos 30° = -\frac{\sqrt{3}}{2}$$

17. -90° is coterminal with 270°;   (0,-1) on the unit circle
$$\sin(-90°) = \sin 270° = \text{the y-value} = -1$$
$$\cos(-90°) = \cos 270° = \text{the x-value} = 0$$

19. reference angle for $\frac{11\pi}{4} = 2\pi + \frac{3}{4}\pi$ (QII) is $\frac{\pi}{4}$ so $\sin\frac{11\pi}{4} = \sin\frac{\pi}{4} = \frac{\sqrt{2}}{2}$
$$\cos\frac{11\pi}{4} = -\cos\frac{\pi}{4} = -\frac{\sqrt{2}}{2}$$

21. $\frac{11\pi}{6}$ terminates in QIV and has a reference angle of $\frac{\pi}{6}$ so $\sin\frac{11\pi}{6} = -\sin\frac{\pi}{6} = -\frac{1}{2}$
and $\cos\frac{11\pi}{6} = +\cos\frac{\pi}{6} = \frac{\sqrt{3}}{2}$

23. 540° is coterminal with 180° so $\sin 540° = \sin 180° = 0$ and $\cos 540° = \cos 180° = -1$

25. 960° = 720° (2 whole revolutions) + 240° and the reference angle for 240° (Q III) is 60° so...
$$\sin 960° = -\sin 60° = -\frac{\sqrt{3}}{2} \text{ and } \cos 960° = -\cos 60° = -\frac{1}{2}$$

27. $-\frac{9\pi}{4} = -\frac{\pi}{4} - 2\pi$ so $-\frac{9\pi}{4}$ is coterminal with $-\frac{\pi}{4}$ (Q IV) for which the reference angle is $\frac{\pi}{4}$

so $\sin\left(-\frac{9\pi}{4}\right) = -\sin\frac{\pi}{4} = -\frac{\sqrt{2}}{2}$ and $\cos\left(-\frac{9\pi}{4}\right) = \cos\frac{\pi}{4} = \frac{\sqrt{2}}{2}$

29. $101\pi = 100\pi + 1\pi$ so $\sin 101\pi = \sin\pi = 0$ and $\cos 101\pi = \cos\pi = -1$

31. $\frac{61\pi}{2} = 30\pi + \frac{\pi}{2}$ so $\sin\frac{61\pi}{2} = \sin\frac{\pi}{2} = 1$ and $\cos\frac{61\pi}{2} = \cos\frac{\pi}{2} = 0$

33. Dividing 167 by 2 (longhand) gives $83\frac{1}{2}$ so view $83\frac{1}{2}\pi$ as $82\pi + 1\frac{1}{2}\pi$ and since 82π is 41 complete revolution, $\frac{167\pi}{2}$ must be coterminal with $\frac{3}{2}\pi$ where the sine is - 1 and the cosine is 0.

35. 3660° divided by 360° leaves a remainder of 60° so $\sin 3660° = \sin 60° = \sqrt{3}/2$
$$\cos 3660° = \cos 60° = 1/2$$

37. Divide by 360 to get 330 as remainder so the quadrant is IV and the reference angle is 30°
so $\sin 330° = -\sin 30° = -1/2$ and $\cos 330° = +\cos 30° = \sqrt{3}/2$

39. As θ goes from $\frac{\pi}{2}$ to $\frac{3\pi}{4}$ the point on the unit circle moves from $(0,1)$ to $(-\frac{\sqrt{2}}{2}, \frac{\sqrt{2}}{2})$ so both the x and y values are decreasing which means that both the sine and cosine are decreasing.

41. Same as $\frac{\pi}{2}$ to π so point is moving left and down so x ($\cos \theta$) is decreasing and y ($\sin \theta$) is decreasing.

43. The point on the unit circle is moving from $(-\frac{\sqrt{2}}{2}, -\frac{\sqrt{2}}{2})$ to $(0, -1)$ which means it's moving to the right (x is increasing $\Rightarrow$ cosine is increasing) and down (y is decreasing $\Rightarrow$ sine is decreasing).

45. Same as $-\frac{\pi}{2}$ to 0 (or $\frac{3\pi}{2}$ to 2π if you prefer) so the point on the unit circle is moving up and to the right which means that both sine and cosine are increasing.

47. Since 0.8134 is positive we have Q1 and QIV possibilities for α and β and the reference angle is $\cos^{-1}0.8134 \approx 0.6208$

$\alpha = 0.6208$ (the Q I possibility)
$\beta = 2\pi - 0.6208 = 5.6624$ (the Q IV possibility)

49. Since the value for cosine is negative quadrants II and III are dictated to us for α and β; the reference angle is $\cos^{-1}0.8134 \approx 0.6208$ again.
$$\alpha = \pi - 0.6208 = 2.5208 \quad \text{and } \beta = \pi + 0.6208 = 3.7624$$

51. $\cos \theta = \frac{1}{2}$ if $\theta = \pi/3$ (QI) and $\theta = 5\pi/3$ (QIV); adding 2π to each of these produces the next two smallest answers $7\pi/3$ and $11\pi/3$.

53. $\sin \pi/4 = \frac{1}{\sqrt{2}}$ so $\pi/4$ (QI), $3\pi/4$ (QII) are the first two answers. Adding 2π gives $9\pi/4$ and $11\pi/4$ and adding one last 2π to $9\pi/4$ to get $17\pi/4$.

55. $\overline{OA} = \overline{OC}$ (equal radii of same circle)

$\angle AOB = \angle COD$ (vertical angles are equal)

$\triangle DOC = \triangle BOA$ (hypotenuse angle, ie. HA)

therefore: $\overline{OD} = \overline{OB}$ (corresponding parts)

So the x coordinates will be opposites with the same absolute value.

Similarly, $\overline{DC} = \overline{BA}$ so the y values will also be opposites.

The fact that the x and y values for θ and $\theta + \pi$ are opposites accounts for the change in sign.

57. $\cos\left(\theta + \frac{\pi}{2}\right) = -\sin\theta$?

$$\cos\left(\theta + \frac{\pi}{2}\right) = \cos\left(\pi - (\pi/2 - \theta)\right) \quad \text{: not an obvious step to take}$$
$$= \cos\left(-(\pi/2 - \theta) + \pi\right) \quad \text{: just a reversal of order}$$
$$= -\cos\left(-(\pi/2 - \theta)\right) \quad \text{: identity for } \cos(*+\pi)$$
$$= -\cos(\pi/2 - \theta) \quad \text{: odd-even identity}$$
$$= -\sin\theta \quad \text{: complementary angle theorem}$$

59. want $\sin\theta < \frac{1}{2}$ with $\theta \in [0,\pi]$

This can happen in quadrants I and II. $\sin 0 = 0$ and $\sin \pi/6 = \frac{1}{2}$ so the QI interval is $[0,\pi/6]$.

In quadrant II, $\sin 5\pi/6 = \frac{1}{2}$ and $\sin \pi = 0$ so the interval there is $[5\pi/6,\pi]$.

61. $u = \sin\left(\theta - \frac{\pi}{4}\right)$ has a graph that differs

from the graph of $y = \sin\theta$ only in that it

is shifted $\frac{\pi}{4}$ (≈ 0.785) units to the RIGHT.

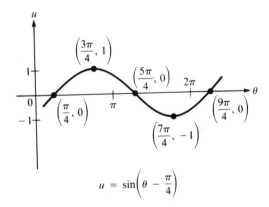

$$u = \sin\left(\theta - \frac{\pi}{4}\right)$$

63. The graph of $u = \cos\left(\theta - \frac{3\pi}{4}\right)$ is the graph
of $u = \cos\theta$ shifted $\frac{3\pi}{4}$ units to the RIGHT.
This is about 2.36

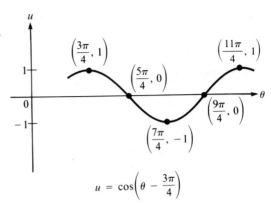

$$u = \cos\left(\theta - \frac{3\pi}{4}\right)$$

65. Shift the graph of $u = \sin\theta$
$\frac{3\pi}{2}$ (≈ 4.71) units to the LEFT

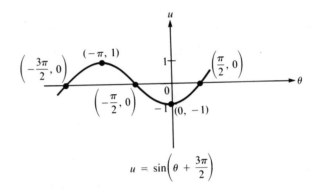

$$u = \sin\left(\theta + \frac{3\pi}{2}\right)$$

67. Lower the graph of $u = \sin\theta$ one (1) block

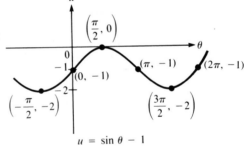

$$u = \sin\theta - 1$$

69. Lower the graph of $u = \cos\theta$ by 4 blocks

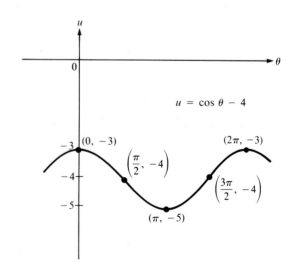

$$u = \cos\theta - 4$$

71. $\sin \theta = 1$ when $\theta = \frac{\pi}{2}$ or $\frac{\pi}{2}$ increased or decreased by a multiple of 2π

so the answer is $\frac{\pi}{2} + 2n\pi$ where n is an integer, positive or negative.

73. want $\cos \theta = \frac{\sqrt{3}}{2}$ which can happen in QI and in QIV (cosine's positive there)

so the generalized QI answer is $\frac{\pi}{6} \pm 2n\pi$

In quadrant IV, the angle itself is $2\pi - \frac{\pi}{6}$ or $\frac{11\pi}{6}$ which when generalized is $\frac{11\pi}{6} \pm 2n\pi$

Problems 7.5

1. $\theta = \frac{\pi}{4}$; $\quad \tan \theta = \frac{\sin \theta}{\cos \theta} = \frac{1/\sqrt{2}}{1/\sqrt{2}} = 1 \qquad \cot \theta = \frac{1}{\tan \theta} = \frac{1}{1} = 1$

$\qquad \sec \theta = \frac{1}{\cos \theta} = \frac{1}{1/\sqrt{2}} = \sqrt{2} \qquad \csc \theta = \frac{1}{\sin \theta} = \frac{1}{1/\sqrt{2}} = \sqrt{2}$

3. $\theta = \frac{7\pi}{6}$; $\quad \tan \theta = \frac{-1/2}{-\sqrt{3}/2} = \frac{1}{\sqrt{3}}$ or $\frac{\sqrt{3}}{3} \qquad \cot \theta = \frac{\sqrt{3}}{1} = \sqrt{3}$

$\qquad \sec \theta = \frac{1}{-\sqrt{3}/2} = -\frac{2}{\sqrt{3}}$ or $-\frac{2\sqrt{3}}{3} \qquad \csc \theta = \frac{1}{-1/2} = -2$

5. $\theta = -\frac{\pi}{2}$; $\quad \tan \theta = \frac{-1}{0}$ which is undefined $\qquad \cot \theta = \frac{0}{-1} = 0$

$\qquad \sec \theta = \frac{1}{0}$ which is undefined $\qquad \csc \theta = \frac{1}{-1} = -1$

7. $\theta = 315°$; $\quad \tan \theta = \frac{-\sqrt{2}/2}{\sqrt{2}/2} = -1 \qquad \cot \theta = \frac{1}{-1} = -1$

$\qquad \sec \theta = \frac{1}{\sqrt{2}/2} = \frac{2}{\sqrt{2}} \qquad \csc \theta = \frac{1}{-1/\sqrt{2}} = -\sqrt{2}$

9. $\theta = 300°$; $\tan \theta = \dfrac{-\sqrt{3}/2}{1/2} = -\sqrt{3}$ $\cot \theta = \dfrac{1}{-\sqrt{3}}$ or $-\dfrac{\sqrt{3}}{3}$

(ref $\angle = 60°$)

$\sec \theta = \dfrac{1}{1/2} = 2$ $\csc \theta = \dfrac{1}{-\sqrt{3}/2} = -\dfrac{2}{\sqrt{3}}$ or $-\dfrac{2\sqrt{3}}{3}$

11. $\sin \theta = \frac{1}{3}$; θ resides in QI ; the definition of sine is $\frac{y}{r}$; comparing this to $\frac{1}{3}$ we see that y = 1 and r = 3. Use the fact that $x^2 + y^2 = r^2$ to find the QI value of x: x = $\sqrt{8}$. Now sketch the triangle:

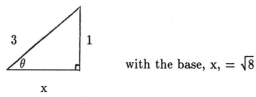

with the base, x, = $\sqrt{8}$

by glancing at the triangle for values of x, y, and r it's not difficult to find all the other values of the trigonometic functions of θ:

$\cos \theta = \frac{x}{r} = \dfrac{\sqrt{8}}{3} = \dfrac{2\sqrt{2}}{3}$ $\csc \theta = \frac{r}{y} = \dfrac{3}{1} = 3$

$\tan \theta = \frac{y}{x} = \dfrac{1}{\sqrt{8}} = \dfrac{1}{\sqrt{2}}$ $\sec \theta = \frac{r}{x} = \dfrac{3}{2\sqrt{2}}$ $\cot \theta = \frac{x}{y} = \dfrac{\sqrt{8}}{1} = 2\sqrt{2}$

13. $\tan \theta = 4$ with θ in QI; $\tan \theta = \frac{y}{x} = 4 = \dfrac{+4}{+1}$ or $\dfrac{-4}{-1}$ but only the former choices, +4 for y and +1 for x, agree with θ being in QI. $x^2 + y^2 = r^2$ tells us that r = $\sqrt{17}$.

$\sin \theta = \dfrac{4}{\sqrt{17}}$ $\sec \theta = \sqrt{17}$ $\cot \theta = \frac{1}{4}$

$\cos \theta = \dfrac{1}{\sqrt{17}}$ $\csc \theta = \dfrac{\sqrt{17}}{4}$

15. $\sec \theta = -\frac{3}{2}$ which can be written as $\dfrac{-3}{2}$ or $\dfrac{3}{-2}$ which we compare to $\frac{r}{x}$;

since r is always positive we'll choose to let r = 3 and x = -2 which forces y to be $\pm \sqrt{5}$ and we choose + because of the QII stipulation.

$\sin \theta = \frac{y}{r} = \dfrac{\sqrt{5}}{3}$ $\tan \theta = \frac{y}{x} = \dfrac{\sqrt{5}}{-2} = -\dfrac{\sqrt{5}}{2}$ $\cot \theta = -\dfrac{2}{\sqrt{5}}$

$\cos \theta = \frac{x}{r} = \dfrac{-2}{3}$ $\csc \theta = \frac{r}{y} = \dfrac{3}{\sqrt{5}}$

17. $\cot \theta = -\frac{2}{5} = \frac{-2}{5}$ or $\frac{2}{-5}$ and since $\cot \theta = \frac{x}{y}$ and we want to be in QII we'll choose
to let x = -2 and y = 5 which forces r to be $\sqrt{29}$.

$$\sin \theta = \frac{5}{\sqrt{29}} \qquad \tan \theta = -\frac{5}{2} \qquad \sec \theta = \frac{\sqrt{29}}{-2} = -\frac{\sqrt{29}}{2}$$

$$\cos \theta = \frac{-2}{\sqrt{29}} \qquad \csc \theta = \frac{\sqrt{29}}{5}$$

19. $\cos \theta = \frac{4}{5} = \frac{x}{r}$ so x = 4, r = 5, and y = + 3 will put θ in QI

$$\sin \theta = \frac{3}{5} \quad \tan \theta = \frac{3}{4} \quad \csc \theta = \frac{5}{3} \quad \sec \theta = \frac{5}{4} \quad \cot \theta = \frac{4}{3}$$

21. $\sin \theta = \frac{3}{5}$ and $\cos \theta = \frac{4}{5}$ implies that θ must be in QI (both x and y are > 0)

$$\text{since tangent} = \frac{\text{sine}}{\text{cosine}} \quad \text{we get } \tan \theta = \frac{3/5}{4/5} = \frac{3}{4}$$

using the reciprocal relationships we get: $\csc \theta = \frac{5}{3}$, $\sec \theta = \frac{5}{4}$, $\cot \theta = \frac{4}{3}$

23. $\sin \theta = -\frac{3}{5}$ and $\cos \theta = \frac{4}{5}$ imply that y = -3 and x = 4 (quadrant IV)

$$\tan \theta = \frac{-3/5}{4/5} = -\frac{3}{4} \text{ , } \csc \theta = -\frac{5}{3} \text{ , } \sec \theta = \frac{5}{4} \text{ , } \cot \theta = -\frac{4}{3}$$

25. $\tan \theta = -\frac{8}{15}$ (Q's II or IV) and $\sec \theta = \frac{17}{15}$ (Q's I or IV) so it must be QIV
with x = 15, y = -8 and r= 17 (by comparing the given values to the definitions)
by reciprocal relationships: $\cot \theta = -\frac{15}{8}$ and $\cos \theta = \frac{15}{17}$
by x, y, r definitions: $\sin \theta = -\frac{8}{17}$ and $\csc \theta = -\frac{17}{8}$

27. $\tan \theta = \frac{8}{15}$ and $\sec \theta = -\frac{17}{15}$ could only occur signwise in QIII with x = - 15 and
y = - 8 and r = 17   so...
$\cos \theta = -\frac{15}{17}$ and $\cot \theta = \frac{15}{8}$ can be deduced using reciprocals;

$\sin \theta = -\frac{8}{17}$ and $\csc \theta = -\frac{17}{8}$ can be found using the x, y, r definitions.

29. $\csc \theta = \frac{25}{24}$ and $\cos \theta = -\frac{7}{25}$ csc (same sign as sin) is positive and cos is negative only in QII.

csc is y over r so y = 25 and r = 24; cos is x over r so x must be -7

by reciprocals: $\sin \theta = \frac{24}{25}$ and $\sec \theta = -\frac{25}{7}$.

by x, y, r definitions $\tan \theta = \frac{y}{x} = -\frac{24}{7}$ and $\cot \theta = -\frac{7}{24}$

31. cosine positive and tangent negative only in quadrant IV

$\sec \theta$ = reciprocal of $\cos \theta$ = 6.9832 (use $\boxed{1/x}$ key)

$\cot \theta = $ " " $\tan \theta = -0.1447$

$\sin \theta = -\sqrt{1 - \cos^2\theta} = -0.9897$

$\csc \theta$ = reciprocal of $\sin \theta$ = -1.0104

33. csc (follows sin) is negative and sec (follows cos) is positive only in QIV

$\sin \theta$ = reciprocal of $\csc \theta$ = -0.2397

$\cos \theta$ = reciprocal of $\sec \theta$ = 0.9709

$\tan \theta = \sin \theta / \cos \theta$ = -0.2397/0.9709 = -0.2469

$\cot \theta$ = reciprocal of $\tan \theta$ = -4.0505

35. sec 3:   set mode to radians (rad)

keystrokes: $\boxed{3}$ $\boxed{\cos}$ $\boxed{1/x}$ and you'll see -1.0101 (approximately)

37. cot 0.9:   set mode to rad

strike $\boxed{\cdot}$ $\boxed{9}$ $\boxed{\tan}$ $\boxed{1/x}$ to get 0.7936

39. csc 130°:   set mode to degrees (deg)

strike $\boxed{1}$ $\boxed{3}$ $\boxed{0}$ $\boxed{\sin}$ $\boxed{1/x}$ to obtain 1.3054

41. tan 260°:   be in deg mode and strike $\boxed{2}$ $\boxed{6}$ $\boxed{0}$ $\boxed{\tan}$ to get $\approx$ 5.6713

43. $\csc \left(\frac{1}{3.29}\right)$: set calc to rad and enter $\boxed{3}$ $\boxed{\cdot}$ $\boxed{2}$ $\boxed{9}$ $\boxed{1/x}$ $\boxed{\sin}$ $\boxed{1/x}$ to get 3.3412

45. $\tan (\pi - \theta) = - \tan \theta$?

$$\tan (\pi - \theta) = \frac{\sin (\pi - \theta)}{\cos (\pi - \theta)} = \frac{\sin \theta}{- \cos \theta} = - \tan \theta \quad \text{(used (12) and (13) from 6.3 text)}$$

47. $\sec (\pi - \theta) = - \sec \theta$?

$$\sec (\pi - \theta) = \frac{1}{\cos (\pi - \theta)} = \frac{1}{- \cos \theta} = - \sec \theta$$

49. $\csc (\pi - \theta) = \csc \theta$?

$$\csc (\pi - \theta) = \frac{1}{\sin (\pi - \theta)} = \frac{1}{\sin \theta} \text{ (from 6.3)} = \csc \theta$$

51. $\tan (\pi/2 - \theta) = \cot \theta$?

$$\tan (\pi/2 - \theta) = \frac{\sin (\pi/2 - \theta)}{\cos (\pi/2 - \theta)} = \frac{\cos \theta}{\sin \theta} = \cot \theta$$

53. $\cot (\pi/2 - \theta) = \tan \theta$?

$$\cot (\pi/2 - \theta) = \frac{\cos (\pi/2 - \theta)}{\sin (\pi/2 - \theta)} = \frac{\sin \theta}{\cos \theta} = \tan \theta$$

55. a little trickery: do you agree that $\theta + \pi/2 = \pi - (\pi/2 - \theta)$?

then $\tan (\theta + \pi/2) = \tan [\pi - (\pi/2 - \theta)] = - \tan (\pi/2 - \theta) = - \cot \theta$

57. $\sec \theta \le 2$ is equivalent to $\frac{1}{\cos \theta} \le 2$; since $\theta \in [0, \pi/2)$, QI, we know that

the cosine is positive so we can multiply by $\cos \theta$ to get:

$1 \le 2 \cos \theta$ or $\cos \theta \ge 1/2$ and this occurs for θ in $[0, \pi/3]$ since $\cos 0 = 1$ and

$\cos \pi/3 = 1/2$ and cosine decreases as θ goes from 0 to $\pi/2$.

59. $\cot \theta > \sqrt{3}$ is equivalent to $\frac{1}{\tan \theta} > \sqrt{3}$ or $\tan \theta < \frac{1}{\sqrt{3}}$

now $\tan 0 = 0$ and $\tan \pi/6 = \frac{1}{\sqrt{3}}$ and tan increases in QI so answer is $(0, \pi/6)$ for θ.

Problems 7.6

1. $\csc \frac{7\pi}{4}$: in quadrant IV (csc like sin is -'ve there) and the reference angle is $\frac{\pi}{4}$

$$= - \csc \frac{\pi}{4} = - \frac{1}{\sin \pi/4} = - \frac{1}{\sqrt{2}/2} = - \frac{2}{\sqrt{2}} = - \sqrt{2}$$

3. $\sec 11\pi$: 11π is coterminal with π

$$= \frac{1}{\cos \pi} = \frac{1}{-1} = - 1$$

5. $\cot 11\pi = \cot \pi = \frac{\cos \pi}{\sin \pi} = \frac{-1}{0}$ which is undefined.

7. $\cot \frac{19\pi}{6}$ (first subtract 2π or $\frac{12\pi}{6}$ to get a smaller coterminal angle of $\frac{7\pi}{6}$ in QIII)

$$= \cot \frac{7\pi}{6} = \frac{1}{\tan \frac{7\pi}{6}} = \frac{1}{1/\sqrt{3}} = \sqrt{3}$$

9. $\tan 750° = \tan 30°$ after subtracting away $720°$, two complete revolutions.

$$= \frac{1}{\sqrt{3}}$$

11. $\tan \theta = 1$ is possible in quadrants I and III ; $\theta_r = \frac{\pi}{4}$

QI solutions: $\frac{\pi}{4} \pm n\pi$ (period of tan is π)

QIII solutions are included in the QI generalized solutions

13. $\csc \theta = 1$ is equivalent to $\sin \theta = 1$ and this occurs only at $\frac{\pi}{2}$ or in general at $\frac{\pi}{2} \pm 2n\pi$

15. To sketch $u = \tan \left(\theta - \frac{\pi}{6} \right)$ sketch $u = \tan \theta$.

Then slide the graph (asymptotes and all)

$\frac{\pi}{6}$ ($\approx$ a half) of a unit to the right.

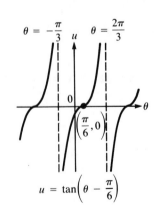

17. To sketch $u = \sec \theta - 1$ lower the sketch of $u = \sec \theta$ (asymptotes whenever $\cos \theta = 0$) 1 block.

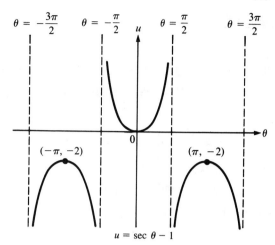

$u = \sec \theta - 1$

19. Since the period of cotangent is π, the graph of $u = \cot (\theta - \pi)$ is the same as the graph of $\cot \theta$. (asymptotes whenever $\cos \theta = 0$)

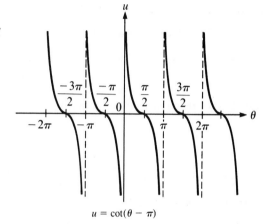

$u = \cot(\theta - \pi)$

Problems 7.7

1. $\cos^{-1} \dfrac{\sqrt{3}}{2}$: what angle in $[\,0,\,\pi]$ has a cosine value of $\dfrac{\sqrt{3}}{2}$?
$$= \dfrac{\pi}{6}$$

3. $\cos^{-1}\left(-\dfrac{1}{2}\right)$: must be a second quadrant angle with a reference angle whose cosine is $\dfrac{1}{2}$
: so reference angle must be $\dfrac{\pi}{3}$ and the second quadrant angle must be $\dfrac{2\pi}{3}$.

5. $\tan^{-1} \dfrac{1}{\sqrt{3}} = \dfrac{\pi}{6}$ because $\tan \dfrac{\pi}{6} = \dfrac{1}{\sqrt{3}}$

7. $\tan^{-1}\left(-1\right) : \text{QIV} = -\pi/4$

9. $\tan^{-1}(\sin 30\pi) = \tan^{-1} 0 = 0$ since $\tan 0 = 0$

11. $\cos [\sin^{-1} (-3/5)] = ?$

let $x = \sin^{-1} (-3/5)$ which means that x must lie in QIV. Draw a $\triangle$ for x_r where

$$\sin x_r = \frac{3}{5} = \frac{opp}{hyp}$$

4 (found using Pythaqgorean theorem)

so $\cos [x] = \cos x_r = \frac{4}{5}$ (+'ve since QIV)

13. $\sin \left(\tan^{-1} 3/5 \right) = ?$

let $x = \tan^{-1} 3/5$ which means that

$\tan x = \frac{3}{5} = \frac{opp}{adj}$ so the triangle looks like this

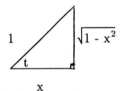

so $\sin x$ (the original problem) $= \dfrac{3}{\sqrt{34}}$

15. $\tan (\sin^{-1} 5) = ?$

let $x = \sin^{-1} 5$ which means that $\sin x = 5$ and this can't be because the range of the sine function (the domain of the inverse sine function) is $[-1,1]$

17. $\sin (\cos^{-1} x) = ?$

let $t = \cos^{-1} x$ which implies that $\cos t = x = \frac{x}{1} = \frac{adj}{hyp}$ from which we can sketch a triangle and find the "opp" side:

$opp = \sqrt{1^2 - x^2} = \sqrt{1 - x^2}$

so now we can see that $\sin t = \dfrac{\sqrt{1 - x^2}}{1} = \sqrt{1 - x^2}$ (if $0 \le x \le 1$)

or

$-\sqrt{1 - x^2}$ (if $-1 \le x \le 0$)

19. $\sec\left(\sin^{-1}\frac{3}{5}\right) = ?$

let $x = \sin^{-1}\frac{3}{5}$ so that $\sin x = \frac{3}{5} = \frac{\text{opp}}{\text{hyp}}$ and our drawing is

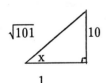

so $\sec x = \frac{\text{hyp}}{\text{adj}} = \frac{5}{4}$

21. $\sin\left(\cos^{-1}\frac{9}{7}\right)$ can't exist because $\frac{9}{7}$ is greater than 1 and is not in the domain of the inverse cosine function.

23. $\cos(\cos^{-1}10^{-55}) = 10^{-55}$ because in general $f(f^{-1}(x)) = x$ and 10^{-55} is positive; so its inverse cosine value would be in quadrant I (the simple case)

25. $\cos^{-1}\left(\cos\frac{5\pi}{6}\right) = \frac{5\pi}{6}$ (as long as $\frac{5\pi}{6}$ is in the range of the inverse cosine and it is)

27. $\tan^{-1}(\tan 100\pi)$: doesnt just equal 100π because 100π is not in the range of

the inverse tangent function

$= \tan^{-1}0 = 0$ (since 100π is coterminal with 0 and $\tan 0 = 0$)

29. $\sec(\tan^{-1}10) = ?$

let $x = \tan^{-1}10$ which implies that $\tan x = 10 = \frac{10}{1} = \frac{\text{opp}}{\text{adj}}$ so we have

as our triangle from which we see that $\sec x = \sqrt{101}$

31. $\csc\left(\cos^{-1}\frac{4}{7}\right) = ?$

let $x = \cos^{-1}\frac{4}{7}$ so $\cos x = \frac{4}{7}$ and we can picture

The unknown opposite side can be calculated as $\sqrt{49-16} = \sqrt{33}$

from which we see that $\csc x = \frac{\text{hyp}}{\text{opp}} = \frac{7}{\sqrt{33}}$.

33. $\cot(\tan^{-1}(-2)) = $ the reciprocal of $-2 = -\frac{1}{2}$

35. $\cos\left(\sin^{-1}(-5/6)\right)$: the inverse sine value must be in QIV because $-5/6 < 0$

let $x = \sin^{-1}(-5/6)$ so that $\sin x = -5/6$ with x in QIV

call the reference angle x_r and assume $\sin x_r = \dfrac{5}{6} = \dfrac{\text{opp}}{\text{hyp}}$

the triangle for x_r would be 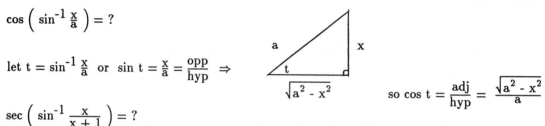 so $\cos x = +\cos x_r = \dfrac{\text{adj}}{\text{hyp}} = \dfrac{\sqrt{11}}{6}$

37. $\cos\left(\sin^{-1}\dfrac{x}{a}\right) = ?$

let $t = \sin^{-1}\dfrac{x}{a}$ or $\sin t = \dfrac{x}{a} = \dfrac{\text{opp}}{\text{hyp}} \Rightarrow$ 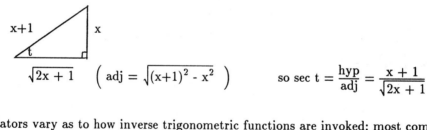 so $\cos t = \dfrac{\text{adj}}{\text{hyp}} = \dfrac{\sqrt{a^2 - x^2}}{a}$

39. $\sec\left(\sin^{-1}\dfrac{x}{x+1}\right) = ?$

let $t = \sin^{-1}\dfrac{x}{x+1}$ which means that $\sin t = \dfrac{x}{x+1} = \dfrac{\text{opp}}{\text{hyp}}$ which leads to..

so $\sec t = \dfrac{\text{hyp}}{\text{adj}} = \dfrac{x+1}{\sqrt{2x+1}}$

$\left(\text{adj} = \sqrt{(x+1)^2 - x^2}\right)$

41. note: calculators vary as to how inverse trigonometric functions are invoked; most common keystrokes are $\boxed{\text{inv}}$ and perhaps $\boxed{\text{sin}}$ preceded by the number which you want the inverse sine of. while in degree mode:

strike $\boxed{\cdot}\ \boxed{3}\ \boxed{1}\ \boxed{\text{inv}}\ \boxed{\text{cos}}\ \boxed{=}$ to find out that

$\cos^{-1} 0.31 \approx 71.9408°$ or $71°\ 56'$ ($56'$ from $.9408 * 60$)

while in radian mode:

≈ 1.2556 (radians) using the same keystrokes

43. $\sin^{-1} -0.42 \approx -24.8346°$ or -0.4334 radians $\left(\boxed{+/-}\text{ struck after entering } 0.42\right)$

45. $\tan^{-1} 2.7 \approx 69.6769°$ or 1.2161 depending on mode selection

47. $\tan^{-1}\dfrac{1}{8} = \tan^{-1} 0.125 \approx 7.1250°$ or 0.1244 radians

49. $\sin^{-1} \pi/4 = \sin^{-1} 0.785398 \approx 51.7575°$ or 0.9033 radians

51. $\sin \theta = 0.2$ and θ is to be in QII:

first find θ_r which is $\sin^{-1} 0.2$ which is $\approx 11.5370°$ or 0.2014 rad

now we place the reference angle in QII by subtracting it from 180° or π:

in degrees: $\theta \approx 180° - 11.5370° = 168.4630°$

in radians: $\theta \approx \pi (3.14159) - 0.2014 = 2.9402$

53. $\sin \theta = -0.2$ and QIV is stipulated. The calculator will give the answer directly as -11.5370° or -0.2014 radians.

55. $\cos \theta = -\frac{2}{3}$ with QIII a must for θ. Plan is to use $\cos^{-1}$ function with just the value 2/3 and then "place it" in QIII:

$$\theta_r = \cos^{-1} 0.66666.. \approx 48.1897° \text{ or } .84107 \text{ rad}$$

"placing" θ_r in QIII requires adding it to 180° (π) so we

obtain the final values 228.1897° or 3.9827 radians

57. $\tan \theta = -\frac{1}{2}$; QIV

$\theta_r = \tan^{-1}\left(-\frac{1}{2}\right) = -26.5651°$ or - 0.4636 radians

subtracting these from 360° and 2π respectively gives us 333.435° and 5.8196 rad

59. $\cos \theta = -\frac{4}{17}$; QIII so $\theta_r = \cos^{-1} \frac{4}{17} \approx 76.3910°$ or 1.3333

adding these to 180° and π produce QIII solutions 256.3910° or 4.4749

61. for $\cos \theta = \frac{1}{2}$ to be true θ would have to be in quadrant I or IV and the reference angle here is $\pi/3$. Adding $\pi/3$ to 4π produces the QI answer of $13\pi/3$; subtracting $\pi/3$ from 6π give the QIV version of $17\pi/3$.

63. $\tan \theta$ can equal 7 in either quadrant I or III; the reference number is $\tan^{-1} 7$ which is approximately 1.4289 (rad mode)

$10\pi + 1.4289 \approx 32.8448$ is the first quadrant answer

$11\pi + 1.4289 \approx 35.9864$ is the third quadrant possibility

65. consider the left side sin ($\cos^{-1}$ x):

let t = $\cos^{-1}$ x which means that cos t = x = $\frac{x}{1}$ = $\frac{adj}{hyp}$ so a triangle with x as an acute angle would look like:

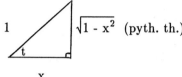

from which we can see that sin t = $\frac{\sqrt{1-x^2}}{1}$ = $\sqrt{1-x^2}$.

Now consider the right side which is cos ($\sin^{-1}$ x):

let s = $\sin^{-1}$ x so that sin s = x = $\frac{x}{1}$ = $\frac{opp}{hyp}$ $\Rightarrow$

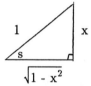

so now we can see that cos s (the right side) = $\frac{adj}{hyp}$ = $\sqrt{1-x^2}$ which is the same as what we got for the left side so we're done!

67. let the values for x decrease "without bound" (ie $\rightarrow$ - ∞) and let t = $\tan^{-1}$ x; this means that tan t = x = $\frac{x}{1}$ = $\frac{opp}{adj}$ and for negative values of x we would know that t must be in the fourth quadrant. Trying to sketch a right triangle that captures this information would produce:

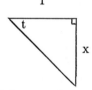

and if you think about it and let x go down and down and down ($\rightarrow$ - ∞) t would be drawn downward also and would get closer and closer to the -$\pi/2$ position. If t is approaching - $\pi/2$ then so is $\tan^{-1}$ x because they were defined to be the same thing.

69. $\sin^{-1}$ (- x) = - $\sin^{-1}$ x ?   let t = $\sin^{-1}$(- x)

then sin t = - x which implies that - sin t = x

which implies that sin (- t) = x   (odd-even identity)

which implies that $\sin^{-1}$x = - t

which implies that - $\sin^{-1}$x = t

and t originally was standing for $\sin^{-1}$(-x) so we've done it!

71. $\tan^{-1}(-x) = -\tan^{-1}x$??

      let $s = \tan^{-1}(-x) \Rightarrow \tan s = -x$

                      $\Rightarrow -\tan s = x$

                      $\Rightarrow \tan(-s) = x$

                      $\Rightarrow \tan^{-1}x = -s$

                      $\Rightarrow -\tan^{-1}x = s$ which was standing for $\tan^{-1}(-x)$

Problems 7.8

1. $\sin 34° = \dfrac{\text{opp}}{\text{hyp}} = \dfrac{2}{y}$

      $0.5592 = \dfrac{2}{y}$

      $0.5592y = 2$

      $y = \dfrac{2}{0.5592} \approx 3.5765$

    finding α:

      $\alpha = 90° - 34°$

      $\alpha = 56°$

finding x:

$$x^2 + 2^2 = y^2$$
$$x^2 + 4 = (3.5765)^2$$
$$x^2 + 4 = 12.7914$$
$$x^2 = 8.7914$$
$$x = \sqrt{8.7914}$$
$$x \approx 2.9650$$

3. $\tan 28.28° = \dfrac{\text{opp}}{\text{adj}} = \dfrac{x}{5}$

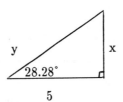

so $\tan 28.28° = \dfrac{x}{5}$

        $0.5381 = \dfrac{x}{5}$

      $(0.5381)(5) = x$

      so $x = 2.6903$

finding y: $x^2 + 5^2 = y^2$

$$(2.6903)^2 + 25 = y^2$$
$$32.2379 = y^2$$

so $y \approx 5.6778$

$\alpha = 90° - 28.28° = 61.72°$

5. $\sin 55.83° = \dfrac{y}{35.4}$

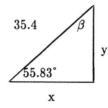

$y = (35.4) \sin 55.83$ °

$= (35.4)(0.82737)$

$= 29.2902$

finding x: $x^2 + y^2 = 35.4^2$

$x^2 + 857.9171 = 1253.16$

$x^2 = 395.2429$

$x \approx 19.8824$

finding β:

$\beta = 90° - 55.83°$

$= 34.17°$

7. finding θ:

$\tan \theta = \dfrac{\text{opp}}{\text{adj}} = \dfrac{30}{50} = 0.6$

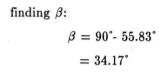

so $\theta = \tan^{-1} 0.6 = 30.96°$

now α:

$\alpha = 90° - 30.96° = 59.04°$

now x:

$30^2 + 50^2 = x^2$

$3400 = x^2$

$x \approx 58.3095$

9. finding α: (have adj and hyp)

$\cos \alpha = \dfrac{\text{adj}}{\text{hyp}} = \dfrac{180}{550} = 0.32727$

$\alpha = \cos^{-1} 0.32727 \approx 70.90°$

finding β:

$\beta = 90° - 70.90° = 19.10°$

finding z:

$180^2 + z^2 = 550^2$

$z^2 = 270100$

$z = 519.7115$

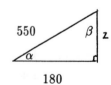

11. let the height of the tree = x ft
    we know angle A and we know its
    adjacent side while the tree is the
    opposite side; use the TANGENT:

    $$\tan 57.3333° = \frac{opp}{adj} = \frac{x}{45}$$

    $$1.559653 = \frac{x}{45}$$

    so x = 45(1.559653) = 70.2 ft

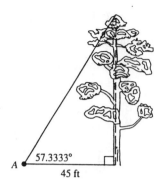

13. let distance across pond = x (solid line)
    x is opposite the 61° angle and the 82.5 m
    is the adjacent side so use the tangent:

    $$\tan 61° = \frac{x}{82.5}$$

    $$1.80405 = \frac{x}{82.5}$$

    so x ≈ 148.8 m

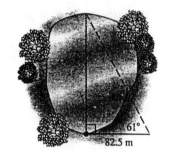

15. let d = distance from A to the top of the tree
    which forms the hypotenuse of the triangle.
    now use cosine because we seek the hypotenuse
    and we know the adjacent (45 ft)

    $$\cos 57.3333° = \frac{adj}{hyp} = \frac{45}{d}$$

    $$0.53975 = \frac{45}{d}$$

    $$d = \frac{45}{0.53975} \approx 83.4 \text{ ft}$$

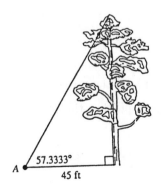

17. $\overline{CB}$ = 38 m - 3 m = 35 m (this is the opposite side)

$\overline{AB}$ (the adjacent) is sought

$\angle A = \angle DCA = 23.5°$ (alternate interior angles)

so $\tan \angle A = \dfrac{opp}{adj} = \dfrac{CB}{AB}$; ie $\tan 23.5° = \dfrac{35}{AB}$

$0.43481 = \dfrac{35}{AB}$ so $AB = \dfrac{35}{0.43481} \approx 80.5$ m

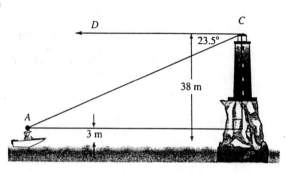

19. By checking out the figure PB must be 420 - 50 = 370 ft (opp)

we want the line of sight distance, ie AP, (hyp)

$\angle A = 6.3333°$

so $\sin 6.3333° = \dfrac{370}{AP}$ or $0.11031 = \dfrac{370}{AP}$

which is the same as $AP = \dfrac{370}{0.11031} = 3354.1$ ft

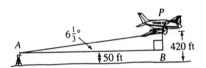

21. let x = the length of the shadow

$\tan 52° = \dfrac{opp}{adj} = \dfrac{85}{x}$

$1.27994 = \dfrac{85}{x}$ so $x = \dfrac{85}{1.27994} \approx 66.4$ ft

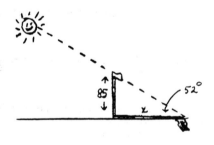

23. let t = the height of the tower

(we want the opp; know the adj)

$\tan 67.3833° = \dfrac{x}{125}$; $2.39448 = \dfrac{x}{125}$

so $x = 125(2.39448) = 299.3$ m

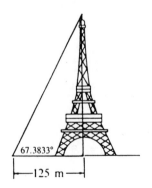

25. We'll work in △ACD and △BCD which are right triangles.

∠ A = 48° by alternate interior angles

∠ DBC = 54.17° by "addition" and alternate angles again

In each triangle we know the opp and can find the adj; then subtract to find the desired distance:

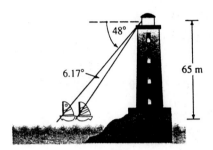

in △ACD: $\tan 48° = \frac{65}{AC}$ in △ BCD: $\tan 54.17° = \frac{65}{BC}$

$1.11061 = \frac{65}{AC}$ $1.38500 = \frac{65}{BC}$

AC = 58.5 m   BC = 46.9 m

so AB = AC - BC = 11.6 m

27. Drop a perpendicular segment from the vertex down to the base and work in either right triangle:

$\sin 14° = \frac{opp}{hyp}$ becomes $0.24192 = \frac{x}{23}$

so x ≈ 5.564 cm which makes the base, 2x,

about 11.1 cm long

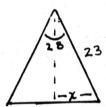

29. Let x be this angle

$\tan x = \frac{opp}{adj} = \frac{5}{8} = 0.625$

so $x = \tan^{-1} 0.625 = 32.0°$

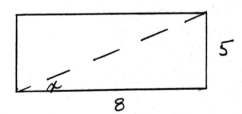

31. let P_1 designate the position of plane 1

and let P_2 represent plane 2.

$\triangle$'s ABP_1 and ABP_2 are right $\triangle$'s of interest.

Note: the angles of depression are vertexed at

the top of the tower; the angles in these triangles

are alternate interior angles to those (they're equal).

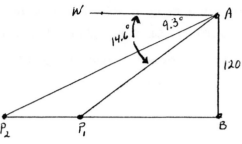

in $\triangle ABP_1$: $\tan 14.6° = \dfrac{120}{BP_1}$; in $\triangle ABP_2$: $\tan 9.3° = \dfrac{120}{BP_2}$

$$BP_1 = \frac{120}{0.26048} = 460.7 \text{ ft} \qquad BP_2 = \frac{120}{0.16376} = 732.8 \text{ ft}$$

so $P_1P_2 = (732.8 - 460.7) \text{ ft} = 272.1 \text{ ft}$

33. M = edge of moat; P = princess

K = knight (12 ft back); O = point under princess

In the drawing we need to find MP, agree?

If we could find MO or OP we could then

track down MP. To do this we'll have to

involve two right triangles and several

unknowns.

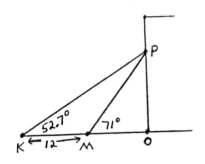

in $\triangle MOP$:   and in $\triangle KOP$:

$$\tan 71° = \frac{OP}{MO} \qquad\qquad \tan 52.7° = \frac{OP}{12 + MO}$$

the relationship on the left tells us that $OP = MO \tan 71°$

substituting this in the other relationship gives

$$\tan 52.7° = \frac{MO(\tan 71°)}{12 + MO} \quad : \text{now cross multiply}$$

$(12 + MO) \tan 52.7° = MO(\tan 71°)$

$(12 + MO) \, 1.3127 = MO \, (2.9042)$

$15.7524 + 1.3127 \, MO = 2.9042 \, MO$

$15.7524 = 1.5915 \, MO$ so $MO = 9.90$ ft

now for the ladder....

in $\triangle MOP$ $\cos 71° = \dfrac{MO}{MP}$ so $\cos 71° = \dfrac{9.91}{MP}$ which means

that $MP = \dfrac{9.91}{\cos 71°} = \dfrac{9.91}{0.3256} = 30.4$ ft

35. In the drawing at the right, let C = the center of the earth

S = point on surface of the earth beneath the plane

A = the plane

H = the point corresponding to the horizon

We seek HA!

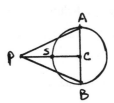

facts: SC = HC = radius of earth = 7926.4 mi

AS = 6.6 mi

∠ H is a right angle (tangent to a circle

is perpendicular to a radius drawn to the point of tangency)

pythagorean theorem applies--we know two sides, CH and CA

$$CH^2 + HA^2 = CA^2$$

$$HA^2 = (3963.2 + 6.6)^2 - (3963.2)^2$$

so HA ≈ 229.3 mi.

37. In the drawing: P = the probe

A and B = endpoints of a diameter

C = the center of Jupiter

S = point on Jupiter's surface

facts:

AB = 88,000 mi (so AC = 44,000 mi)

∠ APB = 57.7° (so ∠ APC = 28.85°)

we seek PS

Plan: find PC and subtract SC from it.

In right △PCA $\tan \angle APC = \dfrac{opp}{adj}$ which becomes $\tan 28.85° = \dfrac{44,000}{PC}$

so $PC = \dfrac{44,000}{0.55089} = 79,870.5$ mi; now we can say that PS = PC - SC = 79870.5 - 44000

$$= 35,870.5 \text{ mi}$$

39. By the definition given in the problem we know that $grade = \dfrac{\text{vertical change}}{\text{length}}$

Substituting the given values (no trig involved!) into this set-up gives:

$$0.22 = \frac{\text{vertical change}}{12}$$

so vertical change = 12(0.22) = 2.6 m

41. In the drawing: P = port

    A = point where ship changed course

    B = point 41 km from A

    facts: ∠ A = 27.25° ; AB = 41 ; PA = 55

    we seek BP right?

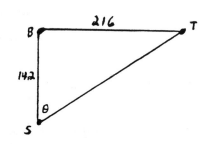

    Plan: we need right triangles so run a perpendicular

    segment from B over to PA and label the point R where

    they meet. We can find AR and BR in the bottom △ and

    then we'll know what PR is as well. Then we can use the Pythagorean theorem to find BP

    RA = 41 cos 27.25° ≈ 36.45; BR = 41 sin 27.25° ≈ 18.77; PR ≈ 18.55 by subtraction

$$\text{Therefore BP} = \sqrt{BR^2 + PR^2} \approx \sqrt{18.77^2 + 18.55^2} \approx 26.4 \text{ km}$$

43. Let S denote the first "ship" mentioned;

    let T denote the second ship.

    We need angle BST or θ

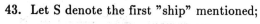

$$\tan \theta = \frac{\text{opp}}{\text{adj}} = \frac{216}{142} = 1.5211$$

    so θ = tan⁻¹ 1.5211 = 56.7° and the bearing is N 56.7° E (east of north)

45. steps: 1) find θ using the right triangle ASB

    2) add θ to 63°

    3) convert to a "bearing"

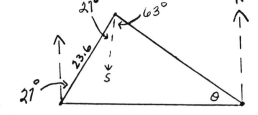

    1) $\tan \theta = \frac{23.6}{37.2} = 0.6344$ so θ = tan⁻¹ 0.6344 = 32.4°

    2) angle from north (west of) = 63° + 32.4° = 95.4°

    3) bearings needn't exceed 90° so N 95.4° W is the same as S 84.6° W

47. In the triangle at the right we seek θ

$$\tan \theta = \frac{\text{opp}}{\text{adj}} = \frac{\text{rise}}{\text{half of span}} = \frac{3}{3.5} = 0.85714$$

    so θ ≈ 40.6°

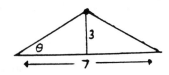

49. pitch $= \frac{\text{rise}}{\text{span}}$

let 2x = the span (or you could use any number)

then tan 28.67° $= \frac{\text{rise}}{\text{x}}$ so rise = x tan 28.67° = 0.54674 x

now when we do the pitch ratio, the x's should cancel:

$$\text{pitch} = \frac{\text{rise}}{\text{span}} = \frac{0.54674 \text{ x}}{2\text{x}} = 0.3$$

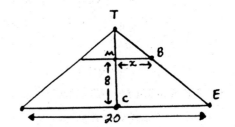

51. plan: 1) use pitch to find rise

   2) use similar triangles to find x

   3) double x to find length of collar beam

1) pitch $= \frac{\text{rise}}{\text{span}} = \frac{\text{rise}}{20} = 0.5$ so rise = 10

2) if collar beam is placed 8 feet above top of plate, there must be 2 ft left above it;

△'s TMB and TCE are similar so $\frac{2}{x} = \frac{10 \text{ (the rise)}}{10 \text{ (half the span)}}$ so x = 2

3) doubling 2 gives 4 ft

53. an easy one:

B A must be 40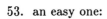

tan 26° $= \frac{\text{opp}}{\text{adj}} = \frac{\text{AB}}{40}$

so AB = 40 tan 26° = 40 (0.48773) ≈ 19.5 ft

Review for Chapter 7

1. 114° is betwee 90° and 180° so it must be in quadrant II.

3. 5.924 radians.... let's see; $\pi \approx 3.14$ radians and $3\pi/2 \approx 4.71$ and $2\pi \approx 6.28$
   so 5.924 is between $3\pi/2$ and 2π which puts it in quadrant IV.

5. 287° is between 270° and 360°, that is, it's in quadrant IV.

7. supplement = 180° - 43°42' = 179°60' - 43°42' = 136°18'

9. $43" = \frac{43}{60}$ ' = 0.71667 ' ; 28.71667 ' $= \frac{28.71667}{60}$ ° = 0.4786° so we have 127.4786°

11. (a) $165° = 165° \times \frac{\pi}{180°} = \frac{11\pi}{12}$ or 2.8798 (radians)

    (b) $-67\frac{1}{2}° = -\frac{135°}{2} \times \frac{\pi}{180°} = -\frac{3\pi}{8}$ or - 1.1781

    (c) $112° = 112° \times \frac{\pi}{180°} = 1.9548$

13. (a) 35 rpm = 35 rpm $\times \frac{2\pi \text{ radians}}{1 \text{ revolution (r)}} = 70\pi$ radians per min

    (b) linear speed = angular speed X radius = $70\pi(3m) = 210\pi$ meters per min

    (c) distance = rate X time = 210π m/min X $\frac{12}{60}$ min = 42π meters (≈ 131.9m)

15. We'll use the fact that $s = r\theta$ with r = 3960 mi and θ = the complement of
    40°42' (since we want the angle to the north pole and not the equator).
    Also remember that θ must be in radians for $s = r\theta$ to work!

    So $\theta = 49°18' = 49.3° = 49.3° \times \frac{3.14 \text{ radians}}{180°} = 0.8604$ (radians)

    Therefore, s (the distance we're looking for) = 3960 mi X 0.8604
    $$= 3407 \text{ mi}$$

17. $\sin 30° = \frac{1}{2}$; $\cos 30° = \frac{\sqrt{3}}{2}$; $\tan 60° = \frac{1}{\sqrt{3}}$; $\csc 60° = 2$; $\sec 30° = \frac{2}{\sqrt{3}}$;

$\cot 30° = \sqrt{3}$ Hint: memorize at least the sine and cosine of the main

angles 0, 30°, 45°, 60°, 90° ; use identities from later

in the chapter to deduce the tangent, cosecant, etc.

19. $\sin 0 = 0$; $\cos 0 = 1$; $\tan 0 = 0$; $\csc 0$ and $\cot 0$ are undefined ; $\sec 0 = 1$

21. Note: fourth quadrant so cosine and its reciprocal, the secant, are positive!

$\sin\left(-\frac{\pi}{6}\right) = -\sin\frac{\pi}{6} = -\frac{1}{2}$; $\cos\left(-\frac{\pi}{6}\right) = +\frac{\sqrt{3}}{2}$; $\tan\left(-\frac{\pi}{6}\right) = -\frac{1}{\sqrt{3}}$

$\csc\left(-\frac{\pi}{6}\right) = -2$; $\sec\left(-\frac{\pi}{6}\right) = \frac{2}{\sqrt{3}}$; $\cot\left(-\frac{\pi}{6}\right) = -\sqrt{3}$

23. 21π is coterminal with π; the point on the unit circle corresponding to π is (- 1, 0). So....

$\sin 21\pi = 0$; $\cos 21\pi = -1$; $\tan 21\pi = 0$; $\csc 21\pi$ is undefined; $\sec 21\pi = -1$; $\cot 21\pi$ is

also undefined.

25. The point on the unit circle corresponding to 270° is (0, - 1) and so...

$\sin 270° = -1$; $\cos 270° = 0$; $\tan 270°$ is undefined ; $\csc 270° = -1$; $\sec 270°$ is undefined ;

$\cot 270° = 0$

27. 315° is in quadrant 4 (cos and sec will be +'ve) and has a reference angle of 45°.

$\sin 315° = -\sin 45° = -\frac{1}{\sqrt{2}}$; $\cos 315° = \cos 45° = \frac{1}{\sqrt{2}}$; $\tan 315° = -\tan 45° = -1$

$\csc 315° = -\csc 45° = -\sqrt{2}$; $\sec 315° = \sec 45° = \sqrt{2}$; $\cot 315° = -\cot 45° = -1$

29. For the sin, cos, and tan values enter the angle (be sure to be in the proper mode) and hit the

proper function key:

$\sin 3.85 = -0.6506$; $\cos 3.85 = -0.7594$; $\tan 3.85 = 0.8568$

For the csc, sec, and cot values find the reciprocals respectively of the above numbers using the

$\boxed{1/x}$ key:

$\csc 3.85 = -1.5370$; $\sec 3.85 = -1.3168$; $\tan 3.85 = 1.1672$

31. sin 217° = - 0.6018 ; cos 217° = - 0.7986 ; tan 217° = 0.7536 ; csc 217° = - 1.6616

      sec 217° = -1.2521 ; cot 217° = 1.3270

33. sin (- 1.76) = - 0.9822 ; cos (- 1.76) = - 0.1881 ; tan (- 1.76) = 5.2221

    csc (- 1.76) = - 1.0182 ; sec (- 1.76) = - 5.3170 ; cot (- 1.76) = 0.1915

35. First convert to degrees: 71°28' 9" = 71.4692°

     sin 71.4692° = 0.9482 ; cos 71.4692° = 0.3178 ; tan 71.4692° = 2.9833

     csc 71.4692° = 1.0547 ; sec 71.4692° = 3.1465 ; cot 71.4692° = 0.3352

37. $\tan \theta = 4$ with θ in quadrant I

     Plan: Draw a "helper" triangle from ratio given; deduce third side using the Pythagorean theorem;

     deduce remaining trig functions.

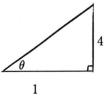

$$\tan \theta = \frac{4}{1} = \frac{\text{opp}}{\text{adj}}$$

The hypotenuse would be $\sqrt{4^2 + 1^2} = \sqrt{17}$ and now all the other functions are available:

$$\sin \theta = \frac{\text{opp}}{\text{hyp}} = \frac{4}{\sqrt{17}} \; ; \; \cos \theta = \frac{1}{\sqrt{17}} \; ; \; \csc \theta = \frac{\sqrt{17}}{4} \; ; \; \sec \theta = \sqrt{17} \; ; \; \cot \theta = \frac{1}{4}$$

39. $\cos \theta = -\frac{1}{2}$ with θ in quadrant III (only the tan and cot will be +'ve)

     Ignore the negative sign for triangle drawing purposes:

$$\cos \text{"}\theta\text{"} = \frac{1}{2} = \frac{\text{adj}}{\text{hyp}} \quad \text{leads to} \dots$$

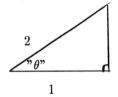

The opposite side is found to be $\sqrt{3}$ and the other trig functions are (now remember that the true

     θ is in QIII) :

$$\sin \theta = -\frac{\sqrt{3}}{2} \; ; \quad \tan \theta = \sqrt{3} \; ; \quad \csc \theta = -\frac{2}{\sqrt{3}} \; ; \quad \sec \theta = -2 \; ; \quad \cot \theta = \frac{1}{\sqrt{3}}$$

41. if $\sec \theta = 3$ then $\cos \theta = \frac{1}{3}$ and θ is given in QIV (only the cosine and secant will be positive)

$\cos\ ”\theta” = \frac{1}{3} = \frac{\text{adj}}{\text{hyp}}$ leads to . . .

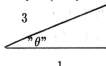

The remaining opposite side is found to be $\sqrt{8} = 2\sqrt{2}$ by the pythagorean theorem. The remaining trig functions (all negative) are:

$$\sin \theta = -\frac{\sqrt{8}}{3} = -\frac{2\sqrt{2}}{3} \ ; \ \tan \theta = -2\sqrt{2} \ ; \ \csc \theta = -\frac{3}{2\sqrt{2}} \ ; \ \cot \theta = -\frac{1}{2\sqrt{2}}$$

43. if $\cot \theta = -6$ then $\tan \theta = -\frac{1}{6}$ and θ is given in quadrant II

$\tan\ ”\theta” = \frac{1}{6} = \frac{\text{opp}}{\text{adj}}$ leads to . . .

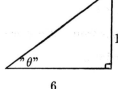

from which we see that the hypotenuse is $\sqrt{37}$.

When finding the other trig functions remember θ IS IN QUADRANT II !

$$\sin \theta = \frac{1}{\sqrt{37}} \ ; \ \cos \theta = -\frac{6}{\sqrt{37}} \ ; \ \csc \theta = \sqrt{37} \ ; \ \sec \theta = -\frac{\sqrt{37}}{6}$$

45. since $\cos x = \sin (\pi/2 - x)$ and we want $\sin \alpha = 0.2190$ knowing that

$\cos 1.35 = 0.2190$ we must have $\alpha = \pi/2 - 1.35 \approx 0.2208$

(the first quadrant solution; x replaced by 1.35).

Since the sine is positive also in QII the other answer, β, is $\pi - 0.2208 = 2.9208$.

47. The first occurrences of θ for $\cos \theta = \frac{\sqrt{3}}{2}$ occur in quadrants I and IV

The reference angle is $\pi/6$ and the answers are $\pi/6$ and $11\pi/6$.

The next answers will differ from each of these by $+2\pi$: $13\pi/6$ and $23\pi/6$.

49. If $\sin \theta = 3/5$ and $\cos \theta = -4/5$ θ must be in quadrant II

By the reciprocal relationships $\csc \theta = 5/3$ and $\sec \theta = -5/4$

We'll use a triangle to help find tan and cot (which must be negative)

49. (continued)

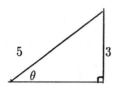

from $\sin \theta = \frac{3}{5}$

The adjacent side is 4 by inspection or the pythagorean theorem.

Therefore $\tan \theta = -3/4$ and $\cot \theta = -4/3$

51. It's the cosine curve with a phase shift of $\pi/3$ to the right:

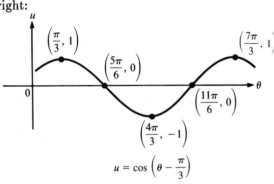

53. If $\tan \theta = \frac{x}{3}$ we can draw

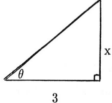

and the missing hypotenuse can be found to be $\sqrt{x^2 + 9}$ which means that $\sin \theta$ must be $\frac{\text{opp}}{\text{hyp}}$

$$= \frac{x}{\sqrt{x^2 + 9}}$$

55. $\sin^{-1}$ answers must lie in the range $-\pi/2$ to $\pi/2$ so $\sin^{-1} \frac{1}{2} = \frac{\pi}{6}$ (ask yourself: what angle has a sine of $\frac{1}{2}$?)

57. What angle, between $-\pi/2$ and $\pi/2$ inclusive, has a tangent of 1? $\frac{\pi}{4}$ does.

59. let $x = \sin^{-1} \frac{5}{8}$ so that $\sin x = \frac{5}{8}$ so that x can be depicted in the triangle:

so the missing adjacent side is $\sqrt{39}$; now we can find $\cos (\sin^{-1} \frac{5}{8})$ because

that's the same as finding $\cos x$ which from the triangle is $\frac{\sqrt{39}}{8}$ (adj over hyp)

61. Once again, draw a triangle from the "inner" expression: let $x = \tan^{-1}\frac{1}{3}$

visually . . .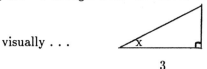

so the missing hypotenuse is $\sqrt{10}$ and the sine of x must be $\dfrac{1}{\sqrt{10}}$

63. $\cot \sin^{-1}\left(-\frac{3}{8}\right)$ is a little trickier because now the inner angle must be in the fourth quadrant and there the cot is negative. We can now ignore the negative sign on the $-\frac{3}{8}$ in the drawing of our triangle:

Let $x = \sin^{-1}\frac{3}{8}$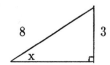

so the missing adjacent side is found to be $\sqrt{55}$ and the cot x answer (which

be negative) is $-\sqrt{55}/3$ (adj over opp)

65. Let $y = \cos^{-1}\frac{x}{4}$ so that we can draw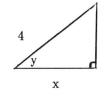

and the missing opposite side is found to be $\sqrt{16-x^2}$ and the

sin of y (that is, the original problem) can now be seen to be $\dfrac{\sqrt{16-x^2}}{4}$

67. θ must equal 50° because the acute angles of a right triangle are complementary

$$\sin 40° = \frac{\text{opp}}{\text{hyp}} = \frac{3}{y} \quad \text{which implies that } y = \frac{3}{\sin 40°} = 4.6672$$

$$\tan 40° = \frac{\text{opp}}{\text{adj}} = \frac{3}{x} \quad \text{which implies that } x = \frac{3}{\cos 40°} = 3.5753$$

69. The angle at the top left must be 90° - 17.17° or 72.83°

$$\sin 17.17° = \frac{y}{10} \quad \text{so } y = 10 \,(\sin 17.17°) = 2.9521$$

$$\cos 17.17° = \frac{x}{10} \quad \text{so } x = 10(\cos 17.17°) = 9.5543$$

71. AB is opposite the known angle ; 38 m is adjacent to the known angle

The tangent function involves the opposite and the adjacent:

$$\tan 48.43° = \frac{\text{opp}}{\text{adj}} = \frac{AB}{38} \quad \text{so } AB = 38 \tan 48.43° = 42.85 \text{ m}$$

73. In the figure at the right, T = top of billboard

B = bottom of billboard

E = end of shadow

facts: △EBT is a right triangle

∠ E = 36°

TB = 42 ft

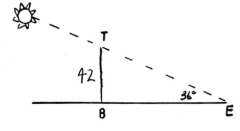

we seek BE: $\tan 36° = \dfrac{\text{opp}}{\text{adj}} = \dfrac{42}{x}$ so $x = \dfrac{42}{\tan 36°} \approx 57.81$ ft

75. C = center of earth

H = point called horizon by person in plane

S = point on earth beneath plane

P = the plane

facts: CS = CH = half of diameter = 3963.2 mi

SP = 30,000 ft ≈ 5.68 mi (so CP ≈ 3968.88 mi)

we seek PH which can be found using the pythagorean theorem:

$CH^2 + HP^2 = CP^2$ so that $HP^2 = CP^2 - CH^2 = 3968.88^2 - 3963.2^2 = 45,054.2$

so HP = the square root of this number which is ≈ 212.26 mi

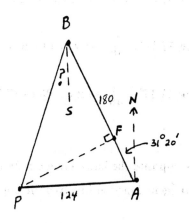

77. P = the port

A = point at which ship changed course

B = point from which it wishes to return to port

S = a point due south of B

N = a point due north of A

facts: PA = 124 ; AB = 180 ; ∠ NAB = 31°20'

(this means that ∠ PAB = 58°40')

In the process of doing #76 there would be a segment drawn perpendicular to AB from P.

Its length would have been found to be 105.915 ; Also, FB would   have been found to be 115.52

Plan: If we can find ∠ FBP we can find the bearing back to port by subtracting

off ∠ SBF which has the same measure as ∠ PAF (alt int angles)

in right triangle PFB tan ∠FBP = $\dfrac{\text{opp}}{\text{adj}}$ = $\dfrac{105.92}{115.52}$ = 0.9168

so ∠ FBP is the inverse tangent of 0.9168 which turns out to be 42°31"

Subtract ∠ SBF which is 31°20" from this and you have 11°11" left which

is ∠ SBP which is slightly west of south so the final answer is S 11°11" W

CHAPTER EIGHT
Trigonometric Identities and Graphs

Chapter Objectives

Goals for this chapter include mastery of proving identities which involve the trigonometric functions of double and half angles as well as sums and differences of angles. You will be confronted with many formulas each with a certain purpose (for example, expressing a sum of two different functions, say sine and cosine, as a product). You will learn how to find **ALL** solutions to trigonometric equations and lastly you will see how the trigonometric functions can be used as models for physical phenomenon such as sound, electricity and predator-prey situations.

Chapter Summary

- Trigonometric identities that follow from the definitions

$$\tan x = \frac{\sin x}{\cos x} \qquad \cot x = \frac{\cos x}{\sin x} = \frac{1}{\tan x} \qquad \sec x = \frac{1}{\cos x} \qquad \csc x = \frac{1}{\sin x}$$

- Circular or Pythagorean Identities

$$\sin^2 x + \cos^2 x = 1 \qquad 1 + \tan^2 x = \sec^2 x \qquad 1 + \cot^2 x = \csc^2 x$$

- Even-Odd Identities

$$\sin(-x) = -\sin x \qquad \csc(-x) = -\csc x$$

$$\cos(-x) = \cos x \qquad \sec(-x) = \sec x$$

$$\tan(-x) = -\tan x \qquad \cot(-x) = -\cot x$$

- Reduction Identities

$$\sin(x + 2\pi) = \sin x \qquad \cos(x + 2\pi) = \cos x \qquad \tan(x + \pi) = \tan x$$

$$\sin\left(\frac{\pi}{2} - x\right) = \cos x \qquad \csc\left(\frac{\pi}{2} - x\right) = \sec x$$

$$\cos\left(\frac{\pi}{2} - x\right) = \sin x \qquad \sec\left(\frac{\pi}{2} - x\right) = \csc x$$

- Reduction Identities (continued)

$$\tan\left(\frac{\pi}{2} - x\right) = \cot x \qquad \cot\left(\frac{\pi}{2} - x\right) = \tan x$$

$$\sin(x + \pi) = -\sin x \qquad \sin(\pi - x) = \sin x$$

$$\cos(x + \pi) = -\cos x \qquad \cos(\pi - x) = -\cos x$$

$$\sin\left(\frac{\pi}{2} + x\right) = \cos x \qquad \cos\left(\frac{\pi}{2} + x\right) = -\sin x$$

- Sum and Difference Identities

$$\sin(\alpha + \beta) = \sin\alpha\cos\beta + \sin\beta\cos\alpha \qquad \sin(\alpha - \beta) = \sin\alpha\cos\beta - \sin\beta\cos\alpha$$
$$\cos(\alpha + \beta) = \cos\alpha\cos\beta - \sin\alpha\sin\beta \qquad \cos(\alpha - \beta) = \cos\alpha\cos\beta + \sin\alpha\sin\beta$$

$$\tan(\alpha + \beta) = \frac{\tan\alpha + \tan\beta}{1 - \tan\alpha\tan\beta} \qquad\qquad \tan(\alpha - \beta) = \frac{\tan\alpha - \tan\beta}{1 + \tan\alpha\tan\beta}$$

- Double-Angle Identities

$$\sin 2x = 2\sin x\cos x$$
$$\cos 2x = \cos^2 x - \sin^2 x \text{ or } 2\cos^2 x - 1 \text{ or } 1 - 2\sin^2 x$$
$$\tan 2x = \frac{2\tan x}{1 - \tan^2 x}$$

- Half-Angle Identities

$$\sin\frac{x}{2} = \pm\sqrt{\frac{1 - \cos x}{2}} \qquad \cos\frac{x}{2} = \pm\sqrt{\frac{1 + \cos x}{2}}$$

$$\tan\frac{x}{2} = \pm\sqrt{\frac{1 - \cos x}{1 + \cos x}} = \frac{1 - \cos x}{\sin x} = \frac{\sin x}{1 + \cos x}$$

- Product-to-Sum Identities

$$\sin\alpha\sin\beta = \frac{1}{2}\left[\cos(\alpha - \beta) - \cos(\alpha + \beta)\right]$$

$$\sin\alpha\cos\beta = \frac{1}{2}\left[\sin(\alpha + \beta) + \sin(\alpha - \beta)\right]$$

$$\cos\alpha\sin\beta = \frac{1}{2}\left[\sin(\alpha + \beta) - \sin(\alpha - \beta)\right]$$

$$\cos\alpha\cos\beta = \frac{1}{2}\left[\cos(\alpha + \beta) + \cos(\alpha - \beta)\right]$$

- Sum-to-Product Identities

$$\sin \alpha + \sin \beta = 2 \sin \frac{\alpha + \beta}{2} \cos \frac{\alpha - \beta}{2}$$

$$\sin \alpha - \sin \beta = 2 \cos \frac{\alpha + \beta}{2} \sin \frac{\alpha - \beta}{2}$$

$$\cos \alpha + \cos \beta = 2 \cos \frac{\alpha + \beta}{2} \cos \frac{\alpha - \beta}{2}$$

$$\cos \alpha - \cos \beta = 2 \sin \frac{\alpha + \beta}{2} \sin \frac{\beta - \alpha}{2} \; ; \; (\text{ note the order of subtraction in this one!})$$

- Harmonic Identities

$$a \cos \omega t + b \sin \omega t = \sqrt{a^2 + b^2} \cos (\omega t - \delta) \; \text{ where } \tan \delta = \frac{b}{a}$$

$$a \cos \omega t + b \sin \omega t = \sqrt{a^2 + b^2} \sin (\omega t + \phi) \; \text{ where } \tan \phi = \frac{a}{b}$$

- In the graphs of $y = A \cos (\omega t - \delta)$ and $y = A \sin (\omega t - \delta)$, with $\omega > 0$:

$|A|$ is the amplitude, $\frac{2\pi}{\omega}$ is the period, and $\frac{|\delta|}{\omega}$ is the phase shift which is to the right if $\delta > 0$ and to the left if $\delta < 0$

Solutions to Chapter Eight Problems

Problems 8.1

1. $\cos x \tan x = \frac{\cos x}{1} \cdot \frac{\sin x}{\cos x} = \sin x$

3. $\cos \theta \cot \theta = \frac{\cos \theta}{1} \cdot \frac{\cos \theta}{\sin \theta} = \frac{\cos^2 \theta}{\sin \theta}$

5. $\frac{1 + \sec t}{1 - \tan t} = \frac{1 + \frac{1}{\cos t}}{1 - \frac{\sin t}{\cos t}}$: (multiply by $\frac{\cos t}{\cos t}$) $= \frac{\cos t + 1}{\cos t - \sin t}$

7. $\frac{\tan u + \sec u}{\tan u - \sec u} = \frac{\frac{\sin u}{\cos u} + \frac{1}{\cos u}}{\frac{\sin u}{\cos u} - \frac{1}{\cos u}} = \frac{\sin u + 1}{\sin u - 1}$

9. $\dfrac{1 + \tan s}{1 - \tan s} = \dfrac{1 + \frac{\sin s}{\cos s}}{1 - \frac{\sin s}{\cos s}} = \dfrac{\cos s + \sin s}{\cos s - \sin s}$

11. $\sin^2 x = 1 - \cos^2 x$ since $\sin^2 x + \cos^2 x = 1$

13. $\dfrac{1 + \sec x}{1 - \sec x} = \dfrac{1 + \frac{1}{\cos x}}{1 - \frac{1}{\cos x}} = \dfrac{\cos x + 1}{\cos x - 1}$

15. $(1 - \sin x)^2 + 2\sin x = (1 - 2\sin x + \sin^2 x) + 2\sin x = 1 + \sin^2 x$
$$= 1 + (1 - \cos^2 x) = 2 - \cos^2 x$$

17. $\cos x \cot x = \dfrac{\cos x}{1} \cdot \dfrac{\cos x}{\sin x} = \dfrac{\cos^2 x}{\sin x} = \dfrac{1 - \sin^2 x}{\sin x}$

19. $\tan^2 x = \dfrac{\sin^2 x}{\cos^2 x} = \dfrac{\sin^2 x}{1 - \sin^2 x}$

21. want to show that $\sin \theta \csc \theta = 1$ is an identity

start with left side (the most complicated one) and use substitutions until desired form is reached

$\sin \theta \csc \theta = \dfrac{\sin \theta}{1} \cdot \dfrac{1}{\sin \theta} = 1$

23. $\tan u \cot u = 1$? $\qquad\qquad \tan u \cot u = \dfrac{\tan u}{1} \cdot \dfrac{1}{\tan u} = 1$

25. $\dfrac{\sec \beta}{\csc \beta} = \tan \beta$?
$$\dfrac{\sec \beta}{\csc \beta} = \dfrac{\frac{1}{\cos \beta}}{\frac{1}{\sin \beta}} = \dfrac{1}{\cos \beta} \cdot \dfrac{\sin \beta}{1} = \dfrac{\sin \beta}{\cos \beta} = \tan \beta$$

27. $\cos \gamma (\tan \gamma + \cot \gamma) = \csc \gamma$?

$\cos \gamma (\tan \gamma + \cot \gamma) = \cos \gamma \tan \gamma + \cos \gamma \cot \gamma$
$$= \cos \gamma \cdot \dfrac{\sin \gamma}{\cos \gamma} + \cos \gamma \cdot \dfrac{\cos \gamma}{\sin \gamma}$$
$$= \dfrac{\sin \gamma}{1} + \dfrac{\cos^2 \gamma}{\sin \gamma} \quad \text{: add by "cross multiplying"}$$
$$= \dfrac{\sin^2 \gamma + \cos^2 \gamma}{\sin \gamma} = \dfrac{1}{\sin \gamma} = \csc \gamma$$

29. $(\cos x - \sin x)^2 = 1 - 2\cos x \sin x$?

$(\cos x - \sin x)^2 = \cos^2 x - 2\cos x \sin x + \sin^2 x$

$$= (\sin^2 x + \cos^2 x) - 2\cos x \sin x$$

$$= 1 - 2\cos x \sin x$$

31. $\tan \theta + \cot \theta = \sec \theta \csc \theta$?

$\tan \theta + \cot \theta = \dfrac{\sin \theta}{\cos \theta} + \dfrac{\cos \theta}{\sin \theta} = \dfrac{\sin^2 \theta + \cos^2 \theta}{\cos \theta \sin \theta} = \dfrac{1}{\cos \theta \sin \theta} = \dfrac{1}{\cos \theta} \cdot \dfrac{1}{\sin \theta} = \sec \theta \csc \theta$

33. $1 - 2\cos^2 \beta = 2\sin^2 \beta - 1$?

$1 - 2\cos^2 \beta = (\sin^2 \beta + \cos^2 \beta) - 2\cos^2 \beta$

$= \sin^2 \beta - \cos^2 \beta \;$: want only sines so solve $\sin^2 \beta + \cos^2 \beta = 1$ for $\cos^2 \beta$ and substitute it in

$= \sin^2 \beta - (1 - \sin^2 \beta) \; = \sin^2 \beta - 1 + \sin^2 \beta = 2\sin^2 \beta - 1$

35. $\dfrac{\tan x - \sin x}{\tan x + \sin x} = \dfrac{\sec x - 1}{\sec x + 1}$?

$\dfrac{\tan x - \sin x}{\tan x + \sin x} = \dfrac{\dfrac{\sin x}{\cos x} - \sin x}{\dfrac{\sin x}{\cos x} + \sin x} \;$ (by factoring) $= \dfrac{\sin x \left(\dfrac{1}{\cos x} - 1\right)}{\sin x \left(\dfrac{1}{\cos x} + 1\right)} = \dfrac{\sec x - 1}{\sec x + 1}$

37. $\cos x \sec x - \sin^2 x = \cos^2 x$?

$\cos x \sec x - \sin^2 x = \cos x \cdot \dfrac{1}{\cos x} - \sin^2 x = 1 - \sin^2 x = \cos^2 x$ where the last equality is a version of the basic pythagorean identity $\sin^2 * + \cos^2 * = 1$

39. $(1 - \cos^2 \theta)(1 + \cot^2 \theta) = 1$?

$(1 - \cos^2 \theta)(1 + \cot^2 \theta) = (\sin^2 \theta)(\csc^2 \theta)$

$$= (\sin \theta \cdot \csc \theta)^2$$

$$= (\sin \theta \cdot \dfrac{1}{\sin \theta})^2 = 1^2 = 1$$

41. $\dfrac{\sec \beta - \cos \beta}{\sin^2 \beta} = \sec \beta$?

$\dfrac{\sec \beta - \cos \beta}{\sin^2 \beta} = \dfrac{\dfrac{1}{\cos \beta} - \dfrac{\cos \beta}{1}}{\sin^2 \beta} = \dfrac{\dfrac{1 - \cos^2 \beta}{\cos \beta}}{\dfrac{\sin^2 \beta}{1}} = \dfrac{\sin^2 \beta}{\cos \beta} \cdot \dfrac{1}{\sin^2 \beta} = \dfrac{1}{\cos \beta} = \sec \beta$

43. $\dfrac{2}{\sin x + 1} - \dfrac{2}{\sin x - 1} = 4\sec^2 x$?

$$\frac{2}{\sin x + 1} - \frac{2}{\sin x - 1} = \frac{2(\sin x - 1) - 2(\sin x + 1)}{(\sin x + 1)(\sin x - 1)}$$

$$= \frac{2\sin x - 2 - 2\sin x - 2}{\sin^2 x - 1}$$

$$= \frac{-4}{-(1 - \sin^2 x)} = \frac{-4}{-\cos^2 x} = \frac{4}{\cos^2 x} = 4 \cdot \frac{1}{\cos^2 x} = 4\sec^2 x$$

45. $\dfrac{\tan \alpha}{\sec \alpha + 1} = \dfrac{1}{\cot \alpha + \csc \alpha}$?

$$\frac{\tan \alpha}{\sec \alpha + 1} = \frac{\frac{\sin \alpha}{\cos \alpha}}{\frac{1}{\cos \alpha} + 1} \quad : \quad \text{to get rid of complex fraction multiply by } \frac{\cos \alpha}{\cos \alpha}$$

$$= \frac{\sin \alpha}{1 + \cos \alpha} \quad : \quad \text{notice that final result is to have a numerator of 1 so now divide}$$

numerator and denominator by $\sin \alpha$

$$= \frac{1}{\frac{1}{\sin \alpha} + \frac{\cos \alpha}{\sin \alpha}} = \frac{1}{\csc \alpha + \cot \alpha}$$

47. $\dfrac{1}{1 - \cos \gamma} + \dfrac{1}{1 + \cos \gamma} = 2\csc^2 \gamma$?

$$\frac{1}{1 - \cos \gamma} + \frac{1}{1 + \cos \gamma} = \frac{1 + \cos \gamma + 1 - \cos \gamma}{1 - \cos^2 \gamma} \; : \; \text{adding by "cross multiplying"}$$

$$= \frac{2}{\sin^2 \gamma} = 2\csc^2 \gamma$$

49. $\cos^4 \theta + \sin^2 \theta = \cos^2 \theta + \sin^4 \theta$?

$\cos^4 \theta + \sin^2 \theta \; : \;$ play around with $\cos^4 \theta$ to get more $\sin^2 \theta$'s into the problem

$$= (\cos^2 \theta)(\cos^2 \theta) + \sin^2 \theta$$

$$= \cos^2 \theta(1 - \sin^2 \theta) + \sin^2 \theta$$

49. (continued)

$= \cos^2\theta - \cos^2\theta \sin^2\theta + \sin^2\theta$: factor $\sin^2\theta$ from last two terms

$= \cos^2\theta - \sin^2\theta (\cos^2\theta - 1)$: use variation of pythagorean

$= \cos^2\theta - \sin^2\theta(-\sin^2\theta)$

$= \cos^2\theta + \sin^4\theta$

51. $\dfrac{1}{\tan \beta + \cot \beta} = \sin \beta \cos \beta$?

$$\frac{1}{\tan \beta + \cot \beta} = \frac{1}{\dfrac{\sin \beta}{\cos \beta} + \dfrac{\cos \beta}{\sin \beta}} = \frac{1}{\dfrac{\sin^2\beta + \cos^2\beta}{\cos \beta \sin \beta}} = \frac{\cos \beta \sin \beta}{\sin^2\beta + \cos^2\beta} = \frac{\cos \beta \sin \beta}{1}$$

53. $\dfrac{1 - \sin v}{1 + \sin v} = (\tan v - \sec v)^2$? : new trick- multiply by $\dfrac{1 - \sin v}{1 - \sin v}$ to get to a variation of the pythagorean identity

$$\frac{1 - \sin v}{1 + \sin v} = \frac{1 - \sin v}{1 + \sin v} \cdot \frac{1 - \sin v}{1 - \sin v}$$

$$= \frac{1 - 2\sin v + \sin^2 v}{1 - \sin^2 v}$$

$$= \frac{1 - 2\sin v + \sin^2 v}{\cos^2 v} \quad : \text{now work a little on the right side to get a "meeting point"}$$

$$(\tan v - \sec v)^2 = \tan^2 v - 2\tan v \sec v + \sec^2 v$$

$$= \frac{\sin^2 v}{\cos^2 v} - 2 \cdot \frac{\sin v}{\cos v} \cdot \frac{1}{\cos v} + \frac{1}{\cos^2 v} \quad : \text{common denominator!}$$

$$= \frac{\sin^2 v - 2\sin v + 1}{\cos^2 v} \quad : \text{and this matches our previous stopping point}$$

55. $\dfrac{\sec \theta \sin \theta}{\tan \theta + \cot \theta} = \sin^2\theta$?

$$\frac{\sec \theta \sin \theta}{\tan \theta + \cot \theta} = \frac{\dfrac{1}{\cos \theta} \cdot \sin \theta}{\dfrac{\sin \theta}{\cos \theta} + \dfrac{\cos \theta}{\sin \theta}} = \frac{\dfrac{\sin \theta}{\cos \theta}}{\dfrac{\sin^2\theta + \cos^2\theta}{\cos \theta \sin \theta}} = \frac{\sin \theta}{\cos \theta} \cdot \frac{\cos \theta \sin \theta}{\sin^2\theta + \cos^2\theta} = \frac{\sin^2\theta}{1} = \sin^2\theta$$

57. $(1 - \sin^2 v)(1 + \tan^2 v) = 1$?

$(1 - \sin^2 v)(1 + \tan^2 v) = (\cos^2 v)(\sec^2 v) = (\cos^2 v)\left(\dfrac{1}{\cos^2 v}\right) = 1$: pythagoreans!

59. $\dfrac{\cot \alpha}{1 + \sin\alpha} = \dfrac{1 - \sin \alpha}{\sin \alpha \cos \alpha}$?

$$\frac{\cot \alpha}{1 + \sin \alpha} = \frac{\cot \alpha}{1 + \sin \alpha} \cdot \frac{1 - \sin \alpha}{1 - \sin \alpha} = \frac{\cot \alpha(1 - \sin \alpha)}{1 - \sin^2\alpha} = \frac{\cos \alpha(1 - \sin \alpha)}{\sin \alpha \,(\cos^2\alpha)} = \frac{1 - \sin \alpha}{\sin \alpha \cos \alpha}$$

61. $\dfrac{\cos \gamma - \sin\gamma + 1}{\cos \gamma + \sin \gamma - 1} = \dfrac{\cos \gamma + 1}{\sin \gamma}$?? : multiply by $\dfrac{\cos \gamma + \sin \gamma + 1}{\cos \gamma + \sin \gamma + 1}$ and hope!

$$\frac{\cos \gamma - \sin \gamma + 1}{\cos \gamma + \sin \gamma - 1} = \frac{\cos \gamma - \sin \gamma + 1}{\cos \gamma + \sin \gamma - 1} \cdot \frac{\cos \gamma + \sin \gamma + 1}{\cos \gamma + \sin \gamma + 1}$$

$$= \frac{\cos^2\gamma + \cos \gamma \sin \gamma + \cos \gamma - \sin \gamma \cos \gamma - \sin^2\gamma - \sin \gamma + \cos \gamma + \sin \gamma + 1}{\cos^2\gamma + \cos \gamma \sin \gamma + \cos \gamma + \sin \gamma \cos \gamma + \sin^2\gamma + \sin \gamma - \cos \gamma - \sin \gamma - 1}$$

$$= \frac{\cos^2\gamma + 2\cos \gamma + (\,1 - \sin^2\gamma)}{(\sin^2\gamma + \cos^2\gamma) + 2\sin \gamma \cos \gamma - 1}$$

$$= \frac{2 \cos^2\gamma + 2 \cos \gamma}{2\sin \gamma \cos \gamma} : \text{ two applications of } \sin^2* + \cos^2* = 1$$

$$= \frac{2\cos \gamma(\,\cos \gamma + 1\,)}{2 \sin \gamma \cos \gamma} = \frac{\cos \gamma + 1}{\sin \gamma}$$

63. $\dfrac{1 + \sin x + \cos x}{1 + \sin x - \cos x} = \csc x + \cot x$?

$$\frac{1 + \sin x + \cos x}{1 + \sin x - \cos x} = \frac{1 + \sin x + \cos x}{1 + \sin x - \cos x} \cdot \frac{1+ \sin x + \cos x}{1 + \sin x + \cos x}$$

$$= \frac{1+ 2\sin x + \sin^2x + 2\cos x + 2\sin x \cos x + \cos^2x}{1 + 2\sin x + \sin^2x - \cos^2x}$$

$$= \frac{2 + 2\sin x + 2\cos x + 2\sin x \cos x}{\sin^2x + 2\sin x + \sin^2x}$$

$$= \frac{2(1 + \sin x) + 2\cos x(1 + \sin x)}{2\sin x(\,\sin x + 1)}$$

$$= \frac{2(1 + \sin x)[\,1 + \cos x]}{2(1 + \sin x) \sin x} = \frac{1 + \cos x}{\sin x} = \frac{1}{\sin x} + \frac{\cos x}{\sin x} = \csc x + \cot x$$

65. $\dfrac{\sin^3\theta - \cos^3\theta}{1 + \sin\theta\cos\theta} = \sin\theta - \cos\theta$?

$$\frac{\sin^3\theta - \cos^3\theta}{1 + \sin\theta\cos\theta} = \frac{(\sin\theta - \cos\theta)(\sin^2\theta + \sin\theta\cos\theta + \cos^2\theta)}{1 + \sin\theta\cos\theta}$$

$$= \frac{(\sin\theta - \cos\theta)(1 + \sin\theta\cos\theta)}{(1 + \sin\theta\cos\theta)} = \sin\theta - \cos\theta$$

67. $\sqrt{\dfrac{1 + \cos\alpha}{1 - \cos\alpha}} = \dfrac{1 + \cos\alpha}{|\sin\alpha|}$?

$$\sqrt{\frac{1 + \cos\alpha}{1 - \cos\alpha}} = \sqrt{\frac{1 + \cos\alpha}{1 - \cos\alpha}\cdot\frac{1 + \cos\alpha}{1 + \cos\alpha}} = \sqrt{\frac{(1 + \cos\alpha)^2}{1 - \cos^2\alpha}} = \frac{\sqrt{(1 + \cos\alpha)^2}}{\sqrt{\sin^2\alpha}}$$

$$= \frac{1 + \cos\alpha}{|\sin\alpha|} \quad \text{note: absolute value signs not necessary around } 1 + \cos\alpha \text{ because it is never negative.}$$

69. $(a\cos\gamma + b\sin\gamma)^2 + (b\cos\gamma - a\sin\gamma)^2 = a^2 + b^2$?

original left side $= a^2\cos^2\gamma + 2ab\cos\gamma\sin\gamma + b^2\sin^2\gamma + b^2\cos^2\gamma - 2ab\cos\gamma\sin\gamma + a^2\sin^2\gamma$

$$= a^2(\cos^2\gamma + \sin^2\gamma) + b^2(\sin^2\gamma + \cos^2\gamma)$$

$$= a^2(1) + b^2(1) = a^2 + b^2$$

71. $a^2\csc^2 x - a^2 = a^2\cot^2 x$?

$a^2\csc^2 x - a^2 = a^2(\csc^2 x - 1)$

$$= a^2(\cot^2 x)$$

73. $(a\cot u + a)^2 + (a\cot u - a)^2 = 2a^2\csc^2 u$?

$(a\cot u + a)^2 + (a\cot u - a)^2 = a^2\cot^2 u + 2a^2\cot u + a^2 + a^2\cot^2 u - 2a^2\cot u + a^2$

$$= 2a^2\cot^2 u + 2a^2$$

$$= 2a^2(\cot^2 u + 1)$$

$$= 2a^2\csc^2 u$$

75. $\ln|\csc\theta| = -\ln|\sin\theta|$?

$\ln|\csc\theta| = \ln|1/\sin\theta| = \ln 1 - \ln|\sin\theta| = 0 - \ln|\sin\theta| = -\ln|\sin\theta|$

77. $\ln|\cot\beta| = \ln|\cos\beta| - \ln|\sin\beta|$?

$\ln|\cot\beta| = \ln|\cos\beta/\sin\beta| = \ln|\cos\beta| - \ln|\sin\beta|$

79. $\ln |\sin x| + \ln |\csc x| = 0$?

$$\ln |\sin x| + \ln |\csc x| = \ln |\sin x| + \ln |1/\sin x|$$
$$= \ln |\sin x| + \ln 1 - \ln |\sin x|$$
$$= \ln 1 = 0$$

81. $\sin x = \sqrt{\sin^2 x}$ is not true all the time because $\sqrt{}$ means the principal square root which is never negative. Pick any angle not in QI or QII such as $3\pi/2$:

$$\sin 3\pi/2 = \text{-}1 \text{ but } \sqrt{\sin^2 3\pi/2} = \sqrt{(\text{-}1)^2} = \sqrt{1} = 1$$

83. $3\pi/2$ again: $\sin 3\pi/2 = \text{-}1$ but $\sqrt{1 - \cos^2 3\pi/2} = \sqrt{1 - 0^2} = \sqrt{1} = 1$

85. pick an angle where cotangent is negative and a "nice number"; how about $3\pi/4$?

$$\cot 3\pi/4 = \text{-}1 \text{ but } \sqrt{\csc^2 3\pi/4 - 1} = \sqrt{2 - 1} = \sqrt{1} = 1$$

87. almost anything should work! try $\pi/3$

$$\sqrt{\sin^2 \pi/3 + \cos^2 \pi/3} = \sqrt{3/4 + 1/4} = \sqrt{1} = 1 \text{ but } \sin \pi/3 + \cos \pi/3 \text{ comes out to be}$$
$$\sqrt{3}/2 + 1/2 \text{ or } (\sqrt{3} + 1)/2 \text{ which is not 1!}$$

89. let $x = 0$ $\ln \dfrac{1}{\cos 0} = \ln \dfrac{1}{1} = \ln 1 = 0$ but the other side turns out to be $\dfrac{1}{\ln \cos 0} = \dfrac{1}{\ln 1} = \dfrac{1}{0}$ which is undefined.

91. $\sin (\sec x) = 1$? try $x = 0$: $\sin (\sec 0) = \sin (1) \approx 0.8415$ which isn't 1.

93. $\sin x = \sqrt{1 - \cos^2 x}$ should be true whenever $\sin x \geq 0$ which is for $0 \leq x \leq \pi$ or $2\pi \leq x \leq 3\pi$ or (even) $\pi \leq x \leq$ (even $+ 1) \pi$

95. $\sqrt{a^2 - x^2} = \sqrt{a^2 - (a \sin \theta)^2} = \sqrt{a^2 - a^2 \sin^2 \theta} = \sqrt{a^2(1 - \sin^2 \theta} = \sqrt{a^2 \cos^2 \theta}$
$= a \cos \theta$ since $\cos \theta \geq 0$ on the interval given

97. $\sqrt{4 + x^2} = \sqrt{4 + (2 \tan \theta)^2} = \sqrt{4 + 4 \tan^2 \theta} = \sqrt{4(1 + \tan^2 \theta)} = \sqrt{4 \sec^2 \theta} = 2 \sec \theta$

99. $\dfrac{2}{\sqrt{9-x^2}} = \dfrac{2}{\sqrt{9-(3\sin\theta)^2}} = \dfrac{2}{\sqrt{9-9\sin^2\theta}} = \dfrac{2}{\sqrt{9(1-\sin^2\theta)}} = \dfrac{2}{\sqrt{9\cos^2\theta}} = \dfrac{2}{3\cos\theta}$ or $\dfrac{2}{3}\sec\theta$

101. $\sqrt{x^2-a^2} = \sqrt{(a\sec\theta)^2-a^2} = \sqrt{a^2(\sec^2\theta-1)} = \sqrt{a^2\tan^2\theta} = a\tan\theta$

103. $\dfrac{1}{(x^2-4)^{3/2}} = \dfrac{1}{(4\sec^2\theta-4)^{3/2}} = \dfrac{1}{[4(\sec^2\theta-1)]^{3/2}} = \dfrac{1}{[4\tan^2\theta]^{3/2}} = \dfrac{1}{8\tan^3\theta}$

105. $\dfrac{x^2}{\sqrt{3-x^2}} = \dfrac{3\cos^2\theta}{\sqrt{3-3\cos^2\theta}} = \dfrac{3\cos^2\theta}{\sqrt{3(1-\cos^2\theta)}} = \dfrac{3\cos^2\theta}{\sqrt{3\sin^2\theta}} = \dfrac{3\cos^2\theta}{\sqrt{3}\sin\theta} = \dfrac{\sqrt{3}\cos^2\theta}{\sin\theta}$

107. $\dfrac{x^n}{(a^2-x^2)^{n/2}} = \dfrac{a^n\sin^n\theta}{(a^2-a^2\sin^2\theta)^{n/2}} = \dfrac{a^n\sin^n\theta}{[a^2(1-\sin^2\theta)]^{n/2}} = \dfrac{a^n\sin^n\theta}{[a^2\cos^2\theta]^{n/2}} = \dfrac{a^n\sin^n\theta}{a^n\cos^n\theta} = \tan^n\theta$

109. $\dfrac{\sqrt{x^2-a^2}}{x^n} = \dfrac{\sqrt{a^2\sec^2\theta-a^2}}{a^n\sec^n\theta} = \dfrac{\sqrt{a^2(\sec^2\theta-1)}}{a^n\sec^n\theta} = \dfrac{\sqrt{a^2\tan^2\theta}}{a^n\sec^n\theta} = \dfrac{a\tan\theta}{a^n\sec^n\theta} = \dfrac{\tan\theta}{a^{n-1}\sec^n\theta}$

Problems 8.2

1. $\cos\left(\dfrac{7\pi}{12}\right) = \cos\left(\dfrac{\pi}{3} + \dfrac{\pi}{4}\right) = \cos\pi/3\,\cos\pi/4 - \sin\pi/3\,\sin\pi/4$

$$= \left(\tfrac{1}{2}\right)\left(\tfrac{\sqrt{2}}{2}\right) - \left(\tfrac{\sqrt{3}}{2}\right)\left(\tfrac{\sqrt{2}}{2}\right) = \dfrac{\sqrt{2}-\sqrt{6}}{4}$$

3. $\tan\left(\dfrac{7\pi}{12}\right) = \tan\left(\dfrac{\pi}{3} + \dfrac{\pi}{4}\right) = \dfrac{\tan\pi/3 + \tan\pi/4}{1 - \tan\pi/3\,\tan\pi/4} = \dfrac{\sqrt{3}+1}{1-(\sqrt{3})(1)} = \dfrac{\sqrt{3}+1}{1-\sqrt{3}}\cdot\dfrac{1+\sqrt{3}}{1+\sqrt{3}} =$

$$= \dfrac{1+2\sqrt{3}+3}{1-3} = \dfrac{4+2\sqrt{3}}{-2} = -2-\sqrt{3}$$

5. $\tan 15° = \tan (45 - 30)° = \dfrac{\tan 45° - \tan 30°}{1 + \tan 45° \tan 30°} = \dfrac{1 - \dfrac{1}{\sqrt{3}}}{1 + 1 \cdot \dfrac{1}{\sqrt{3}}}$: multiply by $\dfrac{\sqrt{3}}{\sqrt{3}}$ to get

$= \dfrac{\sqrt{3} - 1}{\sqrt{3} + 1}$ (if rationalized, $2 - \sqrt{3}$)

7. $\sin \left(\dfrac{13\pi}{12} \right) = \sin \left(\dfrac{3\pi}{4} + \dfrac{\pi}{3} \right) = \sin 3\pi/4 \cos \pi/3 + \sin \pi/3 \cos 3\pi/4$

$= \sqrt{2}/2 \cdot 1/2 + \sqrt{3}/2 \cdot (-\sqrt{2}/2)$

$= \sqrt{2}/4 - \sqrt{6}/4 = (\sqrt{2} - \sqrt{6})/4$

NOTE: the textbook gives the formula $\sin (\alpha + \beta) = \sin \alpha \cos \beta + \cos \alpha \sin \beta$ and I seem to consistently use $\sin (\alpha + \beta) = \sin \alpha \cos \beta + \sin \beta \cos \alpha$. There is no difference due to the commutative property of multiplication. I hope this slight variation does not cause any confusion.

9. $\sin 345° = - \sin 15°$ ($345°$ terminates in QIV where the sine function is negative)

$= - \sin (45 - 30)° = - (\sin 45 \cos 30 - \sin 30 \cos 45) = - (\sqrt{2}/2 \cdot \sqrt{3}/2 - 1/2 \cdot \sqrt{2}/2)$

$= (-\sqrt{6} + \sqrt{2})/ 4$

11. $\cos 22° \cos 47° - \sin 22° \sin 47°$: cosine of a sum formula "in reverse": $= \cos (22+47)° = \cos 69°$

13. $\sin 1.5 \cos 2.3 - \cos 1.5 \sin 2.3 = \sin (1.5 - 2.3) = \sin (-0.8)$ or $- \sin 0.8$ using the odd even identity

15. $\cos 3.4 \cos 2.6 + \sin 3.4 \sin 2.6 = \cos (3.4 - 2.6) = \cos (0.8)$ (cosine of a difference formula)

17. $\dfrac{\tan 1 + \tan \frac{1}{2}}{1 - \tan 1 \tan \frac{1}{2}} = \tan \left(1 + \tfrac{1}{2} \right) = \tan (3/2)$

19. $\sin x \cos 2x + \cos x \sin 2x = \sin (x + 2x) = \sin 3x$

21. $\dfrac{\tan \alpha - \tan 4\alpha}{1 + \tan \alpha \tan 4\alpha} = \tan (\alpha - 4\alpha) = \tan (-3\alpha)$ or $-\tan 3\alpha$

23. since $\cos \alpha = \frac{4}{5}$ and α terminates in QI we can deduce values for $\sin \alpha$ and $\tan \alpha$ which we will need shortly:

from $\sin^2 * + \cos^2 * = 1$ we find that $\sin \alpha = +\sqrt{1 - (\tfrac{4}{5})^2} = \sqrt{\dfrac{25 - 16}{25}} = \dfrac{3}{5}$

(sine is positive in QI)

23. (continued)

from $\tan * = \frac{\sin *}{\cos *}$ we see that $\tan \alpha = \frac{3/5}{4/5} = \frac{3}{4}$ (also positive since QI)

similarly since $\sin \beta = \frac{12}{13}$, $\cos \beta$ turns out to be $\frac{5}{13}$ and $\tan \beta = \frac{12}{5}$, both positive since β is also a first quadrant angle.

(a) $\sin (\alpha + \beta) = \sin \alpha \cos \beta + \sin \beta \cos \alpha$

$$= \left(\frac{3}{5}\right)\left(\frac{5}{13}\right) + \left(\frac{12}{13}\right)\left(\frac{4}{5}\right) = \frac{15}{65} + \frac{48}{65} = \frac{63}{65}$$

$\sin (\alpha - \beta) = \sin \alpha \cos \beta - \sin \beta \cos \alpha$ (see directly above for values)

$$= \frac{15}{65} - \frac{48}{65} = -\frac{33}{65}$$

$\cos (\alpha + \beta) = \cos \alpha \cos \beta - \sin \alpha \sin \beta$

$$= \left(\frac{4}{5}\right)\left(\frac{5}{13}\right) - \left(\frac{3}{5}\right)\left(\frac{12}{13}\right) = \frac{20}{65} - \frac{36}{65} = -\frac{16}{65}$$

$\cos (\alpha - \beta) = \cos \alpha \cos \beta + \sin \alpha \sin \beta$ (very similar to above)

$$= \frac{20}{65} + \frac{36}{65} = \frac{56}{65}$$

$\tan (\alpha + \beta) = \frac{\tan \alpha + \tan \beta}{1 - \tan \alpha \tan \beta} = \frac{3/4 + 12/5}{1 - (3/4)(12/5)}$: multiply by $\frac{20}{20}$

$$= \frac{15 + 48}{20 - 36} = -\frac{63}{16}$$

$\tan (\alpha - \beta) = \frac{\tan \alpha - \tan \beta}{1 + \tan \alpha \tan \beta}$ which by changing signs in the last step

$$= \frac{15 - 48}{20 + 36} = -\frac{33}{56}$$

(b) Since $\sin (\alpha + \beta)$ is positive, but it's cosine and tangent are negative, $\alpha + \beta$ must terminate in quadrant II. Since $\cos (\alpha - \beta)$ is positive but it's sine and tangent are negative, it must terminate in QIV.

25. $\sin \alpha = -\frac{3}{5}$ along with QIV being specified allows us to deduce that $\cos \alpha = \frac{4}{5}$ and $\tan \alpha = -\frac{3}{4}$.

$\cos \beta = -\frac{12}{13}$ with β in QII tells us that $\sin \beta = \frac{5}{13}$ and $\tan \beta = -\frac{5}{12}$.

25. (continued)

(a) $\sin(\alpha + \beta) = \sin \alpha \cos \beta + \sin \beta \cos \alpha = \left(-\frac{3}{5}\right)\left(-\frac{12}{13}\right) + \left(\frac{5}{13}\right)\left(\frac{4}{5}\right) = \frac{36}{65} + \frac{20}{65} = \frac{56}{65}$

for $\sin (\alpha - \beta)$ the sign of the second term will change giving $\frac{36}{65} - \frac{20}{65} = \frac{16}{65}$

$\cos (\alpha + \beta) = \cos \alpha \cos \beta - \sin \alpha \sin \beta = \left(\frac{4}{5}\right)\left(-\frac{12}{13}\right) - \left(-\frac{3}{5}\right)\left(\frac{5}{13}\right) = -\frac{33}{65}$

$\cos (\alpha - \beta) = $ (by changing the middle sign above) $= -\frac{48}{65} - \frac{15}{65} = -\frac{63}{65}$

$\tan (\alpha + \beta) = \dfrac{-\frac{3}{4} + \left(-\frac{5}{12}\right)}{1 - \left(-\frac{3}{4}\right)\left(-\frac{5}{12}\right)} \cdot \dfrac{\frac{48}{1}}{\frac{48}{1}} = \dfrac{-36 - 20}{48 - 15} = -\frac{56}{33}$

by changing signs in numerator and denomiator $\tan (\alpha - \beta) = \dfrac{-36 + 20}{48 + 15} = -\frac{16}{63}$

(b) $\alpha + \beta$ and $\alpha - \beta$ are second quadrant angles because their sine is positive while their cosines and tangents are negative.

27. An alternative to using the pythagorean identities for obtaining values of the different trigonometric function is to use a "helper triangle" along with the right triangle definitions of sin, cos, and tan. Care still must be taken to see that each value has the correct sign for its quadrant: given fact: $\sin \alpha = \frac{5}{13} = \frac{\text{opp}}{\text{hyp}}$ leads to this "helper triangle"

the pythagorean theorem (or familiarity) tells us
the adjacent side is 12 so now we can rattle off the needed values: $\cos \alpha = \frac{\text{adj}}{\text{hyp}} = \frac{12}{13}$
and $\tan \alpha = \frac{\text{opp}}{\text{adj}} = \frac{5}{12}$ by looking at the triangle.

$\tan \beta = \frac{4}{3} = \frac{\text{opp}}{\text{adj}}$ leads us to draw

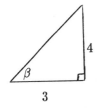

and the missing hypotenuse is 5

27. (continued)

so $\sin \beta = -\frac{4}{5}$ (QIII remember?) and $\cos \beta = -\frac{3}{5}$

(a) $\sin(\alpha + \beta) = \left(\frac{5}{13}\right)\left(-\frac{3}{5}\right) + \left(-\frac{4}{5}\right)\left(\frac{12}{13}\right) = -\frac{15}{65} - \frac{48}{65} = -\frac{63}{65}$

$\sin(\alpha - \beta) = -\frac{15}{65} + \frac{48}{65} = \frac{33}{65}$

$\cos(\alpha + \beta) = \left(\frac{12}{13}\right)\left(-\frac{3}{5}\right) - \left(\frac{5}{13}\right)\left(-\frac{4}{5}\right) = \frac{-36}{65} + \frac{20}{65} = -\frac{16}{65}$

$\cos(\alpha - \beta) = \frac{-36}{65} - \frac{20}{65} = -\frac{56}{65}$

$\tan(\alpha + \beta) = \dfrac{\frac{5}{12} + \frac{4}{3}}{1 - \frac{5}{12} \cdot \frac{4}{3}} = \text{(multiply by } \frac{36}{36}) \; \frac{15 + 48}{36 - 20} = \frac{63}{16}$

$\tan(\alpha - \beta) = \frac{15 - 48}{36 + 20} = -\frac{33}{56}$

(b) since sine and cosine of $\alpha + \beta$ were positive but the tangent is negative $\alpha + \beta$ must be in QIII

since cosine and tangent of $\alpha - \beta$ were negative but the sine is positive, $\alpha - \beta$ is in QII

29. $\sin(x + 2\pi) = \sin x$?

$\sin(x + 2\pi) = \sin x \cos 2\pi + \sin 2\pi \cos x$

$= (\sin x)(1) + (0)(\cos x) = \sin x$

31. $\tan(x + \pi) = \tan x$?

$\tan(x + \pi) = \dfrac{\tan x + \tan \pi}{1 - \tan x \tan \pi} = \dfrac{\tan x + 0}{1 - (\tan x)(0)} = \tan x / 1 = \tan x$

33. $\cos(x + \pi) = -\cos x$?

$\cos(x + \pi) = \cos x \cos \pi - \sin x \sin \pi$

$= (\cos x)(-1) - (\sin x)(0) = -\cos x$

35. $\tan(\pi - x) = -\tan x$?

$\tan(\pi - x) = \dfrac{\tan \pi - \tan x}{1 + \tan \pi \tan x} = \dfrac{0 - \tan x}{1 + (0)\tan x} = -\tan x / 1 = -\tan x$

37. $\cos 2x = \cos^2 x - \sin^2 x$?

    $\cos 2x = \cos(x + x) = \cos x \cos x - \sin x \sin x = \cos^2 x - \sin^2 x$

39. $\cos 2x = 2\cos^2 x - 1$?

    $\cos 2x = \cos^2 - \sin^2 x = \cos^2 x - (1 - \cos^2 x) = \cos^2 x - 1 + \cos^2 x = 2\cos^2 x - 1$

41. $\cos(\pi/2 + x) = -\sin x$?

    $\cos(\pi/2 + x) = \cos \pi/2 \cos x - \sin \pi/2 \sin x$

                $= (0)\cos x - (1)\sin x = -\sin x$

43. $\cos(x + \pi/4) = 1/\sqrt{2}\,(\cos x - \sin x)$?

    $\cos(x + \pi/4) = \cos x \cos \pi/4 - \sin x \sin \pi/4$

               $= \cos x\,(1/\sqrt{2}) - \sin x\,(1/\sqrt{2}) = 1/\sqrt{2}\,(\cos x - \sin x)$

45. $\cos(3\pi/2 - x) = -\sin x$?

    $\cos(3\pi/2 - x) = \cos 3\pi/2 \cos x + \sin 3\pi/2 \sin x = (0)\cos x + (-1)\sin x = -\sin x$

47. $\tan(x - \pi/4) = \dfrac{\tan x - 1}{\tan x + 1}$?

    $\tan(x - \pi/4) = \dfrac{\tan x - \tan \pi/4}{1 + \tan x \tan \pi/4} = \dfrac{\tan x - 1}{1 + \tan x}$

49. $\cos^2 x - \sin^2 y = \cos(x + y)\cos(x - y)$? (start with right side this time!!)

    $\cos(x + y)\cos(x - y) = (\cos x \cos y - \sin x \sin y)(\cos x \cos y + \sin x \sin y)$

                        $= \cos^2 x \cos^2 y - \sin^2 x \sin^2 y$

                   : nothing factors so we've got to add and subtract a

                   contrived term to effect factoring by grouping

                        $= \cos^2 x \cos^2 y + \cos^2 x \sin^2 y - \cos^2 x \sin^2 y - \sin^2 x \sin^2 y$

                        $= \cos^2 x\,(\cos^2 y + \sin^2 y) - \sin^2 y\,(\cos^2 x + \sin^2 x)$

                        $= \cos^2 x - \sin^2 y$

51. $\cos(x + y) - \cos(x - y) = -2\sin x \sin y$?

    $\cos(x + y) - \cos(x - y) = \cos x \cos y - \sin x \sin y - (\cos x \cos y + \sin x \sin y)$

                       $= \cos x \cos y - \sin x \sin y - \cos x \cos y - \sin x \sin y$

                       $= -2\sin x \sin y$

53. $\sin(n\pi + x) = (-1)^n \sin x$?

$\sin(n\pi + x) = \sin n\pi \cos x + \sin x \cos n\pi$

$\qquad = (0)\cos x + (\sin x)(-1)^n$: if n is even, $\cos n\pi = 1$, else -1

$\qquad = (-1)^n \sin x$

55. $\sin(\cos^{-1}\frac{1}{2} + \sin^{-1} 0) = \sin(\pi/3 + 0) = \sin \pi/3 = \sqrt{3}/2$

57. $\cos(\cos^{-1}\frac{1}{2} + \sin^{-1}\frac{\sqrt{2}}{2}) = \cos(\pi/3 + \pi/4) = \cos \pi/3 \cos \pi/4 - \sin \pi/3 \sin \pi/4$

$$= \frac{1}{2}\cdot\frac{\sqrt{2}}{2} - \frac{\sqrt{3}}{2}\cdot\frac{\sqrt{2}}{2} = \frac{\sqrt{2} - \sqrt{6}}{2}$$

59. $\tan(\sin^{-1}\frac{1}{2} + \cos^{-1}\frac{1}{2}) = \tan(\pi/6 + \pi/3) = \tan \pi/2$ which is undefined!

61. let $x = \cos^{-1}\frac{3}{4}$; then $\cos x = \frac{3}{4} = \frac{adj}{hyp}$ so "helper triangle" looks like this:

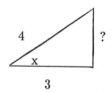

missing opp $= \sqrt{7}$

let $y = \sin^{-1}\frac{4}{5}$; then $\sin y = \frac{4}{5} = \frac{opp}{hyp}$; its triangle:

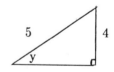

missing adj $= 3$

now, $\tan(\cos^{-1}\frac{3}{4} + \sin^{-1}\frac{4}{5}) = \tan(x+y) = \frac{\tan x + \tan y}{1 - \tan x \tan y}$

: use the triangles!

$$= \frac{\frac{\sqrt{7}}{3} + \frac{4}{3}}{1 - \frac{\sqrt{7}}{3}\cdot\frac{4}{3}} = (\text{ multiply by } \frac{9}{9}) \quad \frac{3\sqrt{7} + 12}{9 - 4\sqrt{7}}$$

63. let $x = \tan^{-1} 3$; then $\tan x = 3 = \frac{3}{1} = \frac{opp}{adj}$ which leads to

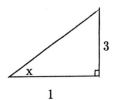

missing hyp $= \sqrt{10}$

let $y = \tan^{-1} \frac{1}{3}$; then $\tan y = \frac{1}{3}$ which leads to

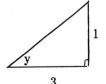

hyp $= \sqrt{10}$ again

$\sin (\tan^{-1} 3 - \tan^{-1} \frac{1}{3}) = \sin (x - y) = \sin x \cos y - \sin y \cos x$

$$= \frac{3}{\sqrt{10}} \cdot \frac{3}{\sqrt{10}} - \frac{1}{\sqrt{10}} \cdot \frac{1}{\sqrt{10}}$$

$$= 9/10 - 1/10 = 8/10 = 4/5$$

65. let $a = \tan^{-1} x$; $\tan a = \frac{x}{1}$:

let $b = \tan^{-1} (x + 2)$; $\tan b = \frac{x + 2}{1}$:

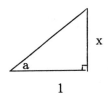 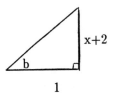

in the first $\triangle$ hyp $= \sqrt{x^2 + 1}$, in the second hyp $= \sqrt{x^2 + 4x + 5}$

$\cos (a - b) = \cos a \cos b + \sin a \sin b$

$$= \frac{1}{\sqrt{1 + x^2}} \cdot \frac{1}{\sqrt{x^2 + 4x + 5}} + \frac{x}{\sqrt{1 + x^2}} \cdot \frac{x + 2}{\sqrt{x^2 + 4x + 5}}$$

$$= \frac{x^2 + 2x + 1}{\sqrt{(1 + x^2)(x^2 + 4x + 5)}}$$

67. let a $= \sin^{-1} 2x$ let b $= \cos^{-1} x$

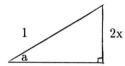   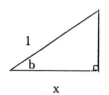

on the left, adj $= \sqrt{1 - 4x^2}$; on the right, opp $= \sqrt{1 - x^2}$

$\sin (a + b) = \sin a \cos b + \sin b \cos a$

$$= (2x)(x) + \sqrt{1 - x^2} \sqrt{1 - 4x^2} = 2x^2 + \sqrt{1 - 5x^2 + 4x^4}$$

69. let a $= \cos^{-1} x$ let b $= \sin^{-1} x$

                                          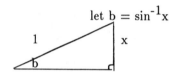

on the left, opp $= \sqrt{1 - x^2}$; on the right, adj $= \sqrt{1 - x^2}$

$$\sin (a + b) = \sin a \cos b + \sin b \cos a = \sqrt{1 - x^2} \sqrt{1 - x^2} + x \cdot x = (1 - x^2) + x^2 = 1$$

71. $\cos 4x = \cos (2x + 2x)$

$\qquad = \cos 2x \cos 2x - \sin 2x \sin 2x$

$\qquad = (\cos^2 x - \sin^2 x)^2 - (2 \sin x \cos x)^2$

$\qquad = \cos^4 x - 2 \cos^2 x \sin^2 x + \sin^4 x - 4 \sin^2 x \cos^2 x$

$\qquad = \cos^4 x - 6 \cos^2 x \sin^2 x + \sin^4 x$

73. $\tan (\alpha - \beta) = \tan (\alpha + (- \beta)) = \dfrac{\tan \alpha + \tan (- \beta)}{1 - \tan \alpha \tan (- \beta)} = \dfrac{\tan \alpha - \tan \beta}{1 + \tan \alpha \tan \beta}$

notes: the first equality is the definition of subtraction;

$\qquad$ the next is identity (9);

$\qquad$ and the last is the "odd-even" identity that tan (-x) $= -$ tan x.

75.

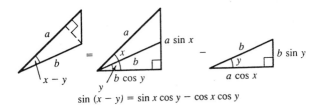

$$\sin(x - y) = \sin x \cos y - \cos x \cos y$$

First, look at the middle triangle, the one that is a right triangle consisting of a "lower" triangle and an "upper" triangle. This proof is based on the principle that the area of the "upper" triangle could be expressed as the area of the entire middle triangle minus the area of the "lower" triangle. Agree? If so, the area of the "upper" triangle which is shown on the left side of the equal signs would be the product of $\frac{1}{2}$ * its base * its height. We can use a as the base but we'll have to determine the height (the length of that dotted segment perpendicular to a) using some trigonometry:

$$\sin(x \text{ - } y) = \frac{\text{opp}}{\text{hyp}} = \frac{\text{height}}{\text{b}} \quad \text{so height} = \text{b} \sin(x \text{ - } y).$$

Therefore, the area of this triangle is $\frac{1}{2}$ * a * b sin (x - y) or more simply $\frac{ab \sin(x \text{ - } y)}{2}$.

Now, on the right side of the equal signs, the area of the large composite triangle is also $\frac{1}{2}$ * its base * its height $= \frac{1}{2}$ * b cos y * a sin x or more simply $\frac{ab \sin x \cos y}{2}$.

The area of the triangle after the "-" sign is $\frac{1}{2}$ * a cos x * b sin y or $\frac{ab \cos x \sin y}{2}$ by the same reasoning. Putting all these area computations in place as the pictures suggest gives us the true statement:

$$\frac{ab \sin(x \text{ - } y)}{2} = \frac{ab \sin x \cos y}{2} \text{ - } \frac{ab \cos x \sin y}{2}$$

and if you multiply both sides by $\frac{2}{ab}$ you will see that sin (x - y) = sin x cos y - cos x sin y.

Problems 8.3

1. $\sin 15° = ?$ 15 is $\frac{1}{2}$ of 30 ; use formula $\sin \frac{x}{2} = \pm \sqrt{\frac{1 - \cos x}{2}}$ with $x = 30°$

$\sin 15° = \sin \frac{30°}{2} = \pm \sqrt{\frac{1 - \cos 30°}{2}} = \pm \sqrt{\frac{1 - \sqrt{3}/2}{2}}$: we must choose only the "+" sign because 15°

is clearly a 1st quadrant angle.

to simplify this a little, multiply inside the radical by $\frac{2}{2}$ to obtain $\sqrt{\frac{2 - \sqrt{3}}{4}}$ or $\frac{\sqrt{2 - \sqrt{3}}}{2}$

3. $\tan 15° = \tan \frac{30°}{2}$: using $\frac{1 - \cos x}{\sin x}$ version of formula for $\tan \frac{x}{2}$

$$= \frac{1 - \cos 30°}{\sin 30°} = \frac{1 - \frac{\sqrt{3}}{2}}{\frac{1}{2}} = 2 - \sqrt{3} \text{ (simply multiplied by } \frac{2}{2} \text{)}$$

5. $\cos \frac{\pi}{24}$: $\frac{\pi}{24}$ is half of $\frac{\pi}{12}$ which is half of $\frac{\pi}{6}$ so we'll have to "nest" applications of the

$\cos \frac{x}{2} = \pm \sqrt{\frac{1 + \cos x}{2}}$ formula:

$$\cos \frac{\pi}{24} = + \sqrt{\frac{1 + \cos \pi/12}{2}} = \sqrt{\frac{1 + \sqrt{\frac{1 + \cos \pi/6}{2}}}{2}} = \sqrt{\frac{1 + \sqrt{\frac{1 + \sqrt{3}/2}{2}}}{2}} =$$

$$= \sqrt{\frac{1 + \frac{\sqrt{2 + \sqrt{3}}}{2}}{2}} = \frac{\sqrt{2 + \sqrt{2 + \sqrt{3}}}}{2} \quad \text{(whew!!)}$$

7. $\frac{3\pi}{8}$ is half of $\frac{3\pi}{4}$ so $\sin \frac{3\pi}{8} = \sqrt{\frac{1 - \cos 3\pi/4}{2}} = \sqrt{\frac{1 - \left(-\frac{\sqrt{2}}{2}\right)}{2}} = \sqrt{\frac{1 + \frac{\sqrt{2}}{2}}{2}} = \frac{\sqrt{2 + \sqrt{2}}}{2}$

notes: $3\pi/4$ is a second quadrant angle and cosine is negative there; the last simplification was

multiplication by 2/2 under the big $\sqrt{}$

9. $\frac{11\pi}{12}$ is half of $\frac{11\pi}{6}$; the former is in QII, the latter in QIV

$$\tan \frac{11\pi}{12} = \frac{1 - \cos 11\pi/6}{\sin 11\pi/6} = \frac{1 - \frac{\sqrt{3}}{2}}{\left(-\frac{1}{2}\right)} = -2 + \sqrt{3}$$

11. $\sec \pi/8 = \dfrac{1}{\cos \pi/8} = \dfrac{1}{\sqrt{\dfrac{1 + \cos \pi/4}{2}}} = \sqrt{\dfrac{2}{1 + \dfrac{\sqrt{2}}{2}}} = \dfrac{2}{\sqrt{2 + \sqrt{2}}}$

13. $\cos \theta = \frac{4}{5} = \dfrac{\text{opp}}{\text{hyp}}$ so the triangle

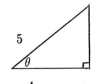

will help

The missing opposite side is 3.

$\sin 2\theta = 2 \sin \theta \cos \theta = 2 \ (3/5) \ (4/5) = 24/25$

$\cos 2\theta = \cos^2\theta - \sin^2\theta = 16/25 - 9/25 = 7/25$

$\tan 2\theta = \sin 2\theta/\cos 2\theta = 24/25 \div 7/25 = 24/7$

$$\sin \frac{\theta}{2} = +\sqrt{\frac{1 - \cos \theta}{2}} = \sqrt{\frac{1 - \frac{4}{5}}{2}} = \sqrt{\frac{1}{10}} \text{ or } \frac{\sqrt{10}}{10}$$

$$\cos \frac{\theta}{2} = +\sqrt{\frac{1 + \cos \theta}{2}} = \sqrt{\frac{1 + \frac{4}{5}}{2}} = \sqrt{\frac{9}{10}} \text{ or } \frac{3\sqrt{10}}{10}$$

$$\tan \frac{\theta}{2} = \sin \frac{\theta}{2} \div \cos \frac{\theta}{2} = \frac{\sqrt{10}}{10} \cdot \frac{10}{3\sqrt{10}} = \frac{1}{3}$$

15. $\tan \theta = \frac{5}{12} = \dfrac{\text{opp}}{\text{adj}}$ with θ in QIII

triangle:

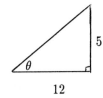

The missing hypotenuse is 13

Now we can rattle off

$\sin 2\theta = 2 \ (-5/13) \ (-12/13) = 120/169$ (note negative signs for QIII)

$\cos 2\theta = 144/169 - 25/169 = 119/169$

15. (continued)

$\tan 2\theta = 120/169 \div 119/169 = 120/119$

$\sin \frac{\theta}{2}$: θ is more than 180° but less than 270° so $\theta/2$ is more than 90° but less than 135° so QII

$$= +\sqrt{\frac{1 - (-12/13)}{2}} = \sqrt{\frac{25}{26}} = \frac{5}{\sqrt{26}} \text{ or } \frac{5\sqrt{26}}{26}$$

$$\cos \frac{\theta}{2} = -\sqrt{\frac{1 + (-12/13)}{2}} = -\sqrt{\frac{1}{26}} = -\frac{\sqrt{26}}{26}$$

$$\tan \frac{\theta}{2} = \text{sine / cosine} = \frac{5}{\sqrt{26}} \cdot -\frac{26}{\sqrt{26}} = -5$$

17. $\sin \theta = -\frac{3}{7}$ with θ in QIII

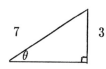

$\text{adj} = \sqrt{40} = 2\sqrt{10}$

$$\sin 2\theta = 2\left(-\frac{3}{7}\right)\left(-\frac{2\sqrt{10}}{7}\right) = \frac{12\sqrt{10}}{49}$$

$$\cos 2\theta = \left(-\frac{2\sqrt{10}}{7}\right)^2 - \left(-\frac{3}{7}\right)^2 = \frac{40}{49} - \frac{9}{49} = \frac{31}{49} \qquad \tan 2\theta = \frac{12\sqrt{10}}{49} \cdot \frac{49}{31} = \frac{12\sqrt{10}}{31}$$

$$\sin \frac{\theta}{2} = \sqrt{\frac{1 - (-2\sqrt{10}/7)}{2}} = \sqrt{\frac{(7 + 2\sqrt{10})/7}{2}} = \sqrt{\frac{7 + 2\sqrt{10}}{14}} \quad (\text{+'ve as half of a QIII angle is a QII angle})$$

$$\cos \frac{\theta}{2} = \sqrt{\frac{1 + (-2\sqrt{10}/7)}{2}} = \sqrt{\frac{(7 - 2\sqrt{10})/7}{2}} = \sqrt{\frac{7 - 2\sqrt{10}}{14}}$$

$$\tan \frac{\theta}{2} = \frac{1 - \cos \theta}{\sin \theta} = \frac{1 - \left(-\frac{2\sqrt{10}}{7}\right)}{-\frac{3}{7}} = \frac{-7 - 2\sqrt{10}}{3} \quad \text{after multiplying by } \frac{(-7)}{(-7)}$$

19. $\cos \theta = -\frac{2}{3}$ with θ in QII :

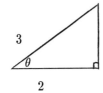

missing opp $= \sqrt{5}$

$\sin 2\theta = 2 \left(\frac{\sqrt{5}}{3} \right)\left(-\frac{2}{3}\right) = -\frac{4\sqrt{5}}{9}$ $\cos 2\theta = \frac{4}{9} - \frac{5}{9} = -\frac{1}{9}$

$\tan 2\theta = -\frac{4\sqrt{5}}{9} \div -\frac{1}{9} = 4\sqrt{5}$

$\sin \frac{\theta}{2} = \sqrt{\frac{1 - (-2/3)}{2}} = \sqrt{\frac{5}{3} \cdot \frac{1}{2}} = \sqrt{\frac{5}{6}} = \frac{\sqrt{30}}{6}$

(positive since half of a QII angle would be a QI angle)

$\cos \frac{\theta}{2} = \sqrt{\frac{1 + (-2/3)}{2}} = \sqrt{\frac{1}{6}} = \frac{\sqrt{6}}{6}$

$\tan \frac{\theta}{2} =$ the quotient of the two items directly above $= \sqrt{5}$

21. $\sec \theta = 4$; θ in QI (same as $\cos \theta = \frac{1}{4}$) :

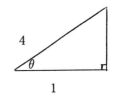

missing opposite side is $\sqrt{15}$

$\sin 2\theta = 2 \left(\frac{\sqrt{15}}{4} \right)\left(\frac{1}{4}\right) = \frac{\sqrt{15}}{8}$

$\cos 2\theta = \frac{1}{16} - \frac{15}{16} = -\frac{14}{16} = -\frac{7}{8}$ $\tan 2\theta = \frac{\sqrt{15}}{8} \div -\frac{7}{8} = -\frac{\sqrt{15}}{7}$

$\sin \frac{\theta}{2} = \sqrt{\frac{1 - 1/4}{2}} = \sqrt{\frac{3}{8}} = \frac{\sqrt{6}}{4}$ $\cos \frac{\theta}{2} = \sqrt{\frac{1 + 1/4}{2}} = \sqrt{\frac{5}{8}} = \frac{\sqrt{10}}{4}$

$\tan \frac{\theta}{2} = \sqrt{\frac{3}{8}} \div \sqrt{\frac{5}{8}} = \sqrt{\frac{3}{5}} = \frac{\sqrt{15}}{5}$

23. $\cos 3x = \cos(2x + x)$

$= \cos 2x \cos x - \sin 2x \sin x$

$= (2\cos^2 x - 1)(\cos x) - (2\sin x \cos x)\sin x$

$= 2\cos^3 x - \cos x - 2\sin^2 x \cos x$

$= 2\cos^3 x - \cos x - 2(1 - \cos^2 x)\cos x$

$= 2\cos^3 x - \cos x - 2\cos x + 2\cos^3 x = 4\cos^3 x - 3\cos x$

25. $\sin 4x = \sin 2(2x)$

$= 2\sin 2x \cos 2x$

$= 2(2\sin x \cos x)(1 - 2\sin^2 x)$

$= 4\sin x \cos x - 8\sin^3 x \cos x$ (other answers possible!)

27. $\sin^3\theta \cos^3\theta = \sin\theta \sin^2\theta \cos\theta \cos^2\theta$

$= \sin\theta \left(\dfrac{1 - \cos 2\theta}{2}\right) \cos\theta \left(\dfrac{1 + \cos 2\theta}{2}\right)$

$= \dfrac{\sin\theta \cos\theta}{4}(1 - \cos^2 2\theta) = \dfrac{\sin\theta \cos\theta}{4}\left(1 - \left(\dfrac{1 + \cos 4\theta}{2}\right)\right)$

$= \dfrac{\sin\theta \cos\theta}{4}\left(\dfrac{1 - \cos 4\theta}{2}\right) = \dfrac{1}{8}\sin\theta \cos\theta(1 - \cos 4\theta)$

29. $\cos 2x = \cos(x + x) = \cos x \cos x - \sin x \sin x = \cos^2 x - \sin^2 x$

31. $\cos 2x = \cos^2 x - \sin^2 x = (1 - \sin^2 x) - \sin^2 x = 1 - 2\sin^2 x$

33. $\csc 2x = \dfrac{\sec x \csc x}{2}$?

$\csc 2x = \dfrac{1}{\sin 2x} = \dfrac{1}{2\sin x \cos x} = \dfrac{1}{2} \cdot \dfrac{1}{\sin x} \cdot \dfrac{1}{\cos x} = \dfrac{1}{2}\csc x \sec x$ (same !)

35. $\sin 2x = \dfrac{2\tan x}{\sec^2 x}$? $\dfrac{2\tan x}{\sec^2 x} = \dfrac{2 \cdot \frac{\sin x}{\cos x}}{\frac{1}{\cos^2 x}} = \dfrac{2\sin x}{\cos x} \cdot \dfrac{\cos^2 x}{1} = 2\sin x \cos x = \sin 2x$

37. $\tan x = \dfrac{\sin 2x}{1 + \cos 2x}$? $\dfrac{\sin 2x}{1 + \cos 2x} = \dfrac{2 \sin x \cos x}{1 + (2\cos^2 x - 1)} = \dfrac{2 \sin x \cos x}{2 \cos^2 x} = \dfrac{\sin x}{\cos x} = \tan x$

39. $\tan 2x = \dfrac{\sin 4x}{1 + \cos 4x}$? $\dfrac{\sin 4x}{1 + \cos 4x} = \dfrac{2 \sin 2x \cos 2x}{1 + (2\cos^2 2x - 1)} = \dfrac{2 \sin 2x \cos 2x}{2 \cos^2 2x} = \dfrac{\sin 2x}{\cos 2x} = \tan 2x$

41. $\cot 2x = \dfrac{\cot^2 x - 1}{2 \cot x}$?

$\cot 2x = \dfrac{1}{\tan 2x} = \dfrac{1}{\dfrac{2 \tan x}{1 - \tan^2 x}} = \dfrac{1 - \tan^2 x}{2 \tan x} = \dfrac{1 - \dfrac{1}{\cot^2 x}}{2 \cdot \dfrac{1}{\cot x}} = \dfrac{\cot^2 x - 1}{\cot^2 x} \cdot \dfrac{\cot x}{2} = \dfrac{\cot^2 x - 1}{2 \cot x}$

43. $\csc \dfrac{x}{2} = \pm \dfrac{\sqrt{2(1 + \cos x)}}{\sin x}$??

$\csc \dfrac{x}{2} = \dfrac{1}{\sin \dfrac{x}{2}} = \dfrac{1}{\pm \sqrt{\dfrac{1 - \cos x}{2}}} = \pm \sqrt{\dfrac{2}{1 - \cos x}} = \pm \sqrt{\dfrac{2}{1 - \cos x} \cdot \dfrac{1 + \cos x}{1 + \cos x}}$

$= \pm \sqrt{\dfrac{2(1 + \cos x)}{1 - \cos^2 x}} = \pm \sqrt{\dfrac{2(1 + \cos x)}{\sin^2 x}} = \pm \dfrac{\sqrt{2(1 + \cos x)}}{\sin x}$

45. $\csc 2x = \dfrac{1}{2}(\tan x + \cot x)$?

$\csc 2x = \dfrac{1}{\sin 2x} = \dfrac{1}{2 \sin x \cos x}$

(Now work with other side)

$\dfrac{1}{2}(\tan x + \cot x) = \dfrac{1}{2}\left(\dfrac{\sin x}{\cos x} + \dfrac{\cos x}{\sin x}\right) = \dfrac{1}{2}\left(\dfrac{\sin^2 x + \cos^2 x}{\cos x \sin x}\right) = \dfrac{1}{2 \sin x \cos x}$ They match !

47. $\tan x = \dfrac{1}{2}\sin 2x (1 + \tan^2 x)$?

$\dfrac{1}{2}\sin 2x (1 + \tan^2 x) = \dfrac{1}{2}(2 \sin x \cos x)(\sec^2 x) = \sin x \cos x \cdot \dfrac{1}{\cos^2 x} = \dfrac{\sin x}{\cos x} = \tan x$

49. $\cot x = \dfrac{\cos^2 \frac{x}{2} - \sin^2 \frac{x}{2}}{\sin x}$??

(the numerator is a variation of the cosine of a double angle formula)

$$\frac{\cos^2 \frac{x}{2} - \sin^2 \frac{x}{2}}{\sin x} = \frac{\cos 2(\frac{x}{2})}{\sin x} = \frac{\cos x}{\sin x} = \cot x$$

51. $\sec x = \dfrac{1 + \tan^2 \frac{x}{2}}{1 - \tan^2 \frac{x}{2}}$?(attack the right side with the half angle formula for tan that uses only cos)

$$\frac{1 + \tan^2 \frac{x}{2}}{1 - \tan^2 \frac{x}{2}} = \frac{1 + \frac{1 - \cos x}{1 + \cos x}}{1 - \frac{1 - \cos x}{1 + \cos x}} = \frac{(1 + \cos x) + (1 - \cos x)}{(1 + \cos x) - (1 - \cos x)} \quad \text{: (by Xing by the LCD)}$$

$$= \frac{2}{2 \cos x} = \frac{1}{\cos x} = \sec x$$

53. area = (1/2 of the base) X height

let x = 1/2 of the base and y = height

work in the right triangle on the left: $\cos 70° = \frac{x}{5}$ so $x = 5 \cos 70°$ (≈ 1.7101)

$\sin 70° = \frac{y}{5}$ so $y = 5 \sin 70°(\approx 4.6985)$ area = $(5 \cos 70°)(5\sin 70°) \approx 8.0349$

55. equation (13) is $\tan \frac{x}{2} = \pm \sqrt{\dfrac{1 - \cos x}{1 + \cos x}}$ squaring both sides gives

$$\tan^2 \frac{x}{2} = \frac{1 - \cos x}{1 + \cos x}$$

$$= \frac{1 - \cos x}{1 + \cos x} \cdot \frac{1 - \cos x}{1 - \cos x}$$

$$= \frac{(1- \cos x)^2}{\sin^2 x} \quad \text{and now take square roots}$$

$$\tan \frac{x}{2} = \frac{1 - \cos x}{\sin x} \quad \text{Now multiply by 1 + cos x over itself :}$$

$$= \frac{1 - \cos x}{\sin x} \cdot \frac{1 + \cos x}{1 + \cos x} = \frac{\sin^2 x}{\sin x \, (1 + \cos x)} = \frac{\sin x}{1 + \cos x}$$

57.

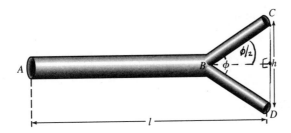

We are trying to find the sum $\overline{AB} + \overline{BC}$: (assume $\triangle BCD$ is isosceles)

Consider the right triangle relationships that are available in right triangle BEC above:

since $\text{sine} = \frac{\text{opp}}{\text{hyp}}$ we have $\sin \frac{\phi}{2} = \frac{h/2}{\overline{BC}}$; this tells us that $\overline{BC} = \dfrac{h/2}{\sin \frac{\phi}{2}}$

secondly, if we can find $\overline{BE}$, we can subtract it from l and we'll then have $\overline{AB}$

we can find $\overline{BE}$ using the cosine function: $\cos \frac{\phi}{2} = \frac{\overline{BE}}{\overline{BC}}$; so $\overline{BE} = \overline{BC} \cos \frac{\phi}{2} = \dfrac{h/2}{\sin \frac{\phi}{2}} \cdot \cos \frac{\phi}{2}$

which implies that $\overline{AB} = 1 - \dfrac{h/2 \cos \frac{\phi}{2}}{\sin \frac{\phi}{2}}$;   putting all this together lets us see that

$$\overline{AB} + \overline{BC} = \left(1 - \dfrac{h/2 \cos \frac{\phi}{2}}{\sin \frac{\phi}{2}}\right) + \dfrac{h/2}{\sin \frac{\phi}{2}} = 1 + h/2 \left(\dfrac{1 - \cos \frac{\phi}{2}}{\sin \frac{\phi}{2}}\right) = 1 + h/2 \tan\left(\frac{\phi/2}{2}\right)$$

by factoring and identity (14) with $x = \frac{\phi}{2}$

$$= 1 + \frac{h}{2} \tan \frac{\phi}{4}$$

Problems 8.4

1. using (2) from the text on page 503:

$\sin 30° \cos 50° = \frac{1}{2} [\sin (30 + 50)° + \sin (30 - 50)°]$

$\qquad\qquad\quad = \frac{1}{2} [\sin 80° + \sin (- 20°)] = \frac{1}{2} [\sin 80° - \sin 20°]$

3. using (2) :

$\sin 54° \cos 46° = \frac{1}{2} [\sin (54 + 46)° + \sin (54 - 46)°]$

$\qquad\qquad\quad = \frac{1}{2} [\sin 100° + \sin 8°]$

5. using (3) :

$$\cos \frac{\pi}{4} \sin \frac{\pi}{7} = \frac{1}{2} [\sin (\pi/4 + \pi/7) - \sin (\pi/4 - \pi/7)]$$

$$= \frac{1}{2} [\sin 11\pi/28 - \sin 3\pi/28]$$

7. using (4) :

$$\cos \frac{2}{3} \cos \frac{3}{2} = \frac{1}{2} [\cos (2/3 + 3/2) + \cos (2/3 - 3/2)] = \frac{1}{2} [\cos 13/6 + \cos (- 5/6)]$$

$$= \frac{1}{2} [\cos 13/6 + \cos 5/6]$$

9. using (3) :

$$\cos 3\alpha \sin 4\alpha = \frac{1}{2} [\sin (3\alpha + 4\alpha) - \sin (3\alpha - 4\alpha)] = \frac{1}{2} [\sin 7\alpha - \sin (-\alpha)]$$

$$= \frac{1}{2} [\sin 7\alpha + \sin \alpha]$$

11. using (2) :

$$3 \sin (x + 1) \cos (x - 1) = 3 \cdot \frac{1}{2} \left(\sin [(x + 1) + (x - 1)] + \sin [(x + 1) - (x - 1)] \right)$$

$$= \frac{3}{2} \left(\sin 2x + \sin 2 \right)$$

13. use (4) on the first two factors :   remember: cos (-t) = cos t !

$$\cos t \cos 2t \cos 4t = \frac{1}{2} [\cos 3t + \cos t] \cos 4t$$

$$= \frac{1}{2} \cos 3t \cos 4t + \frac{1}{2} \cos t \cos 4t$$

$$= \frac{1}{2} \left(\frac{1}{2} [\cos 7t + \cos t] \right) + \frac{1}{2} \left(\frac{1}{2} [\cos 5t + \cos 3t] \right)$$

$$= \frac{1}{4} (\cos 7t + \cos t + \cos 5t + \cos 3t)$$

15. using (7) from the text page 504:

$$\cos 15° + \cos 35° = 2 \cos \frac{15° + 35°}{2} \cos \frac{15° - 35°}{2} = 2 \cos 25° \cos (-10°) = 2 \cos 25° \cos 10°$$

17. using (7): $\quad \cos 3 + \cos 5 = 2 \cos \dfrac{3+5}{2} \cos \dfrac{3-5}{2} = 2 \cos 4 \cos (-1) = 2 \cos 4 \cos 1$

19. using (6):

$$\sin \tfrac{\pi}{8} - \sin \tfrac{\pi}{12} = 2 \cos \dfrac{\pi/8 + \pi/12}{2} \sin \dfrac{\pi/8 - \pi/12}{2} = 2 \cos 5\pi/48 \sin \pi/48$$

21. using (6): $\quad \sin 5x - \sin 2x = 2 \cos \dfrac{5x+2x}{2} \sin \dfrac{5x-2x}{2} = 2 \cos \tfrac{7x}{2} \sin \tfrac{3x}{2}$

23. using (6):

$$\sin (-2\theta) - \sin (-5\theta) = 2 \cos \dfrac{(-2\theta) + (-5\theta)}{2} \sin \dfrac{(-2\theta) - (-5\theta)}{2} = 2 \cos \tfrac{7\theta}{2} \sin \tfrac{3\theta}{2}$$

25. $2 \cos t + 3 \sin t = A \cos (t - \delta)$ where $A = \sqrt{2^2 + 3^2} = \sqrt{13}$

and δ is such that: $\cos \delta = \dfrac{2}{\sqrt{13}} \approx 0.5547$ and $\sin \delta = \dfrac{3}{\sqrt{13}} \approx 0.8321$

Both are positive so δ must be in QI and we have $\delta = \sin^{-1} 0.8321 \approx 0.9828$

so this first answer is $\sqrt{13} \cos (t - 0.9828)$

It is also possible for $2 \cos t + 3 \sin t = A \sin (t + \phi)$

Same A as above and ϕ is such that: $\cos \phi = \dfrac{3}{\sqrt{13}} \approx 0.8321$ and $\sin \phi = \dfrac{2}{\sqrt{13}} \approx 0.5547$

ϕ is also in quadrant I and its value is $\sin^{-1} 0.5547 \approx 0.5880$

so the second answer is $\sqrt{13} \sin (t + 0.5880)$

27. $- 2 \cos t - 3 \sin t$ is to be converted: $A = \sqrt{4 + 9} = \sqrt{13}$

δ such that $\cos \delta = \dfrac{-2}{\sqrt{13}} \approx - 0.5547$ and $\sin \delta = \dfrac{-3}{\sqrt{13}} \approx - 0.8321$ (QIII for δ)

so the reference angle for $\delta = \cos^{-1} 0.5547 = 0.9828$ and adding this to π will place it in the third quadrant. So $\delta = 4.1244$ and the first answer is $\sqrt{13} \cos (t - 4.1244)$.

Now, ϕ such that $\sin \phi \approx - 0.5547$ and $\cos \phi \approx - 0.8321$ so $\phi_{ref} \approx 0.5879$ and the real ϕ in QIII is 3.7296. Second answer: $\sqrt{13} \sin (t + 3.7296)$

29. $4 \cos 2\theta + 3 \sin 2\theta$ is to be converted: $A = \sqrt{16 + 9} = 5$

δ such that $\cos \delta = \tfrac{4}{5} = 0.8$ and $\sin \delta = \tfrac{3}{5} = 0.6$ so $\delta \approx 0.6435$ (Q I !)

First answer: $5 \cos (2\theta - 0.6435)$

ϕ such that $\cos \phi = 0.6$ and $\sin \phi = 0.8$ so $\phi \approx 0.9273$ and the second answer

is: $5 \sin (2\theta + 0.9273)$

31. $-5 \cos \frac{\alpha}{2} + 12 \sin \frac{\alpha}{2}$ is to be converted:

$A = \sqrt{25 + 144} = 13$

δ such that $\cos \delta = \frac{-5}{13} = -0.3846$ and $\sin \delta = \frac{12}{13} = 0.9231$ (Q II)

a reference angle is inverse cosine of 0.3846 or 1.176

subtracting this from π will place δ in Q II and give 1.9656

first answer: $13 \cos \left(\frac{\alpha}{2} - 1.9656 \right)$

ϕ must be such that $\cos \phi = 0.9231$ and $\sin \phi = -0.3846$ (Q IV)

reference angle $=$ inverse cosine of 0.9231 or 0.3947 and this must be

subtracted from 2π to give a fourth quadrant ϕ of 5.8885 and we now have

the second answer: $13 \sin \left(\frac{\alpha}{2} + 5.8885 \right)$

33. $0.5 \cos 4\beta - 0.25 \sin 4\beta = \frac{1}{2} \cos 4\beta - \frac{1}{4} \sin 4\beta$ is to be converted:

$A = \sqrt{1/4 + 1/16} = \sqrt{5/16} = \sqrt{5}/4$

δ such that $\cos \delta = \dfrac{1/2}{\sqrt{5}/4} \approx 0.8944$ and $\sin \delta = \dfrac{-1/4}{\sqrt{5}/4} \approx -0.4472$ (Q IV)

reference angle $= \cos^{-1} 0.8944 = 0.4636$ with δ in QIV being 5.8195

first answer: $\dfrac{\sqrt{5}}{4} \cos (4\beta - 5.8195)$

Now to find ϕ such that $\cos \phi = -0.4472$ and $\sin \phi = 0.8944$ (Q II)

the reference angle would be $\sin^{-1} 0.8944$ or 1.1071 which must be subtracted from π to give

$\phi = 2.0344$; the second answer is $\dfrac{\sqrt{5}}{4} \sin (4\beta + 2.0344)$

35. $-7 \cos \frac{\pi}{2}x - 3 \sin \frac{\pi}{2}x$ is to be converted:

$A = \sqrt{49 + 9} = \sqrt{58}$

δ such that $\cos \delta = \dfrac{-7}{\sqrt{58}} \approx -0.9191$ and $\sin \delta = \dfrac{-3}{\sqrt{58}} \approx -0.3939$ (Q III)

$\delta_{\text{ref}} = \cos^{-1} 0.9191 \approx 0.4049$ so add π to get $\delta = 3.5465$

first answer: $\sqrt{58} \cos \left(\frac{\pi}{2}x - 3.5465 \right)$

ϕ such that $\cos \phi = -0.3939$ and $\sin \phi = -0.9191$ (still Q III)

reference angle is inverse cosine of 0.3939 or 1.1659 which when added to

π yields 4.3075 so other answer is $\sqrt{58} \sin \left(\frac{\pi}{2}x + 4.3075 \right)$

37. $\sin (\alpha + \beta) = \sin \alpha \cos \beta + \sin \beta \cos \alpha$

+ $\underline{\sin (\alpha - \beta) = \sin \alpha \cos \beta - \sin \beta \cos \alpha}$

$\sin (\alpha + \beta) + \sin (\alpha - \beta) = 2 \sin \alpha \cos \beta$: now divide both sides by 2 to get:

$$\tfrac{1}{2} [\sin (\alpha + \beta) + \sin (\alpha - \beta)] = \sin \alpha \cos \beta$$

39. $\cos (\alpha + \beta) = \cos \alpha \cos \beta - \sin \alpha \sin \beta$

+ $\underline{\cos (\alpha - \beta) = \cos \alpha \cos \beta + \sin \alpha \sin \beta}$

$\cos (\alpha + \beta) + \cos (\alpha - \beta) = 2 \cos \alpha \cos \beta$ which yields upon division by 2

$$\tfrac{1}{2} [\cos (\alpha + \beta) + \cos (\alpha - \beta)] = \cos \alpha \cos \beta$$

41. identity (4) is that $\cos A \cos B = \tfrac{1}{2} [\cos (A + B) + \cos (A - B)]$ with A and B used rather than α and β for clarity's sake.

to prove (7) let $\alpha = A + B$ and $\beta = A - B$

then $\alpha + \beta = 2A$ so that $A = \dfrac{\alpha + \beta}{2}$

also $\alpha - \beta = 2B$ so that $B = \dfrac{\alpha - \beta}{2}$

making these substitutions above:

$$\cos \frac{\alpha + \beta}{2} \cos \frac{\alpha - \beta}{2} = \tfrac{1}{2} [\cos \alpha + \cos \beta]$$

multiplying both sides by 2 gives:

$$2 \cos \frac{\alpha + \beta}{2} \cos \frac{\alpha - \beta}{2} = \cos \alpha + \cos \beta \quad \text{which is (7)}$$

43. Here is the context for this identity:

$$A = \sqrt{a^2 + b^2}, \quad \cos \phi = b/A \quad \text{and} \quad \sin \phi = a/A$$

We wish to show that $a \cos \omega t + b \sin \omega t = A \sin (\omega t + \phi)$.   Start with the right side:

$A \sin (\omega t + \phi) = A (\sin \omega t \cos \phi + \sin \phi \cos \omega t)$

$= A \left((\sin \omega t) \cdot b/A + a/A \cdot \cos \omega t \right)$

$= (\sin \omega t) \cdot b + a \cos \omega t$

45. $\dfrac{\cos 3x + \cos x}{2 \cos 2x} = \cos x$? Using (7) on the numerator gives $\dfrac{2 \cos 2x \cos x}{2 \cos 2x}$ which equals cos x

47. $\dfrac{\cos 10x - \cos 6x}{\sin 10x - \sin 6x} = -\tan 8x$??

using (8) on the numerator and (6) on the denominator gives

$$\frac{2 \sin 8x \sin (-2x)}{2 \cos 8x \sin 2x} = \frac{- 2 \sin 8x \sin 2x}{2 \cos 8x \sin 2x} = \frac{- \sin 8x}{\cos 8x} = -\tan 8x$$

49. $\tan \dfrac{\alpha + \beta}{2} = \dfrac{\sin \alpha + \sin \beta}{\cos \alpha + \cos \beta}$

work on the right side using (5) on the top and (7) on the bottom to get

$$\frac{2 \sin \dfrac{\alpha + \beta}{2} \cos \dfrac{\alpha - \beta}{2}}{2 \cos \dfrac{\alpha + \beta}{2} \cos \dfrac{\alpha - \beta}{2}} \text{ which simplifies to } \frac{\sin \dfrac{\alpha + \beta}{2}}{\cos \dfrac{\alpha + \beta}{2}} \text{ or } \tan \dfrac{\alpha + \beta}{2}$$

51. $\dfrac{\cos 2\theta + \cos 4\theta}{\sin 2\theta + \sin 4\theta} = \cot 3\theta$?

on the left use (7) on the top and (5) on the denominator to get

$$\frac{2 \cos 3\theta \cos (-\theta)}{2 \sin 3\theta \cos (-\theta)} \text{ which equals } \frac{\cos 3\theta}{\sin 3\theta} \text{ which is the same as } \cot 3\theta$$

53. $\dfrac{\sin \alpha + \sin \beta}{\sin \alpha - \sin \beta} = \dfrac{\tan \left(\dfrac{\alpha + \beta}{2} \right)}{\tan \left(\dfrac{\alpha - \beta}{2} \right)}$??

on the left use (5) on the numerator and (6) on the denominator to get

$$\frac{2 \sin \dfrac{\alpha + \beta}{2} \cos \dfrac{\alpha - \beta}{2}}{2 \cos \dfrac{\alpha + \beta}{2} \sin \dfrac{\alpha - \beta}{2}} = \tan \dfrac{\alpha + \beta}{2} \cdot \cot \dfrac{\alpha - \beta}{2} = \tan \dfrac{\alpha + \beta}{2} \cdot \frac{1}{\tan \dfrac{\alpha - \beta}{2}}$$

55. $\sin \theta + \sin (\theta + \pi) = 2 \sin \left(\dfrac{2\theta + \pi}{2} \right) \cos \left(\dfrac{\theta - (\theta + \pi)}{2} \right) = 2 \sin \left(\theta + \dfrac{\pi}{2} \right) \cos \left(-\dfrac{\pi}{2} \right)$

$$= 2 \sin (\theta + \pi/2) \cdot 0 = 0$$

57. (b) $\sin 10° \sin 50° = \frac{1}{2}[\cos 40° - \cos 60°] = \frac{1}{2}[\cos 40° - \frac{1}{2}] = \frac{1}{2}\cos 40° - \frac{1}{4}$

substituting this result into the original gives:

$\left(\frac{1}{2}\cos 40° - \frac{1}{4}\right)\sin 70° = \frac{1}{2}\cos 40° \sin 70° - \frac{1}{4}\sin 70°$

$= \frac{1}{2}\left(\frac{1}{2}(\sin 110° + \sin 30°)\right) - \frac{1}{4}\sin 70°$

$= \frac{1}{4}\sin 110° + \frac{1}{4}\sin 30° - \frac{1}{4}\sin 70°$

but by our knowledge of reference angles we know that $\sin 110° = \sin 70°$ so the first and last terms cancel and the result is $\frac{1}{4}\left(\frac{1}{2}\right) = \frac{1}{8}$.

59. Let $y_1 = 2\sin 528\pi t$ be the first wave and $y_2 = 2\sin 264\pi t$ be the second. The wave produced when both are struck together would be $y = y_1 + y_2 = 2\sin 528\pi t + 2\sin 264\pi t$

$= 2(\sin 528\pi t + \sin 264\pi t)$

$= 2\left(2\sin\frac{528\pi t + 264\pi t}{2}\cos\frac{528\pi t - 264\pi t}{2}\right)$

$= 4\sin 396\pi t \cos 132\pi t$

<u>Problems 8.5</u>

Whenever k is used in answers k can be 0, ±1, ±2, ±3, ±4, . . .

1. $\cos x = 0$ if $x = \frac{\pi}{2}$, $\frac{3\pi}{2}$, $\frac{5\pi}{2}$,... or $-\frac{\pi}{2}$, $-\frac{3\pi}{2}$, $-\frac{5\pi}{2}$ ... or simply $\frac{\pi}{2} + k\pi$ since each item in either list is π more or less than the preceding item.

3. $\cos \theta = -\frac{1}{2}$ will have solutions in quadrants II and III

   The reference angle is $\pi/3$ because $\cos \pi/3 = \frac{1}{2}$

   The second quadrant answers will be $(\pi - \pi/3) + 2k\pi$ or $2\pi/3 + 2k\pi$

   The third quadrant answers will be $(\pi + \pi/3) + 2k\pi$ or $4\pi/3 + 2k\pi$

5. $\sqrt{3} \tan 4\alpha + 1 = 0$ if $\tan 4\alpha = -1/\sqrt{3}$ which will have second and 4th quadrant solutions:

   the reference angle for 4α is $\pi/6$ since $\tan \pi/6 = 1/\sqrt{3}$; The second (and fourth) quadrant answers for 4α are $5\pi/6 + k\pi$ and so answers for α are $5\pi/24 + (k/4) \pi$

7. $\sin x - \cos x = 0$

   $\sin x = \cos x$ and now divide both sides by $\cos x$ (assuming $\cos x \neq 0$)

   $\frac{\sin x}{\cos x} = 1$ or equivalently $\tan x = 1$ which has solutions in Q's I and III

   The reference angle is $\pi/4$ since $\tan \pi/4$ is 1.

   The solutions are then given by $\pi/4 + k\pi$ (when k is odd, you get those third quadrant solutions)

9. $\csc \omega = -2$ is the same as $\sin \omega = -1/2$ which has solutions in Q's III and IV.

   The reference angle is $\pi/6$ ($\sin \pi/6 = 1/2$)

   The third quadrant solutions are $7\pi/6 + 2k\pi$

   The fourth quadrant solutions are $11\pi/6 + 2k\pi$ (or $-\pi/6 + 2k\pi$)

11. $\cot 3x = 0$ whenever $\cos 3x = 0$ (and $\sin 3x \neq 0$) since $\cot * = \cos */\sin *$

    Solutions for 3x are $\frac{\pi}{2} + 2k\pi$ and $\frac{3\pi}{2} + 2k\pi$ which, if you stop to think about it, can be combined into the single expression $\frac{\pi}{2} + n\pi$. So at this point we have $3x = \frac{\pi}{2} + n\pi$. Dividing by 3 gives solutions for x as $\frac{\pi}{6} + \frac{n\pi}{3}$

13. $\sin \theta = \frac{2}{3}$ will have answers in quadrants I and II

The reference angle is $\sin^{-1} 0.6667 \approx 0.7297$

Quadrant I solutions: $0.7297 + 2k\pi$

Quadrant II solutions: $(\pi - 0.7297) + 2k\pi$ or $2.4119 + 2k\pi$

15. $\cos \alpha = 0.55$ has quadrant I and quadrant IV solutions.

The reference angle is $\cos^{-1} 0.55 \approx 0.9884$ In QI: $0.9884 + 2k\pi$

In QIV: $(2\pi - 0.9884) + 2k\pi$ or $5.2948 + 2k\pi$ (or $-.9884 + 2k\pi$)

17. $\sec \theta - \csc \theta = 0$ is equivalent to $\sec \theta = \csc \theta$ or $\dfrac{1}{\cos \theta} = \dfrac{1}{\sin \theta}$

or $\dfrac{\sin \theta}{\cos \theta} = 1$ or finally $\tan \theta = 1$ (as long as $\cos \theta \neq 0$ and $\sin \theta \neq 0$)

solutions for θ: $\frac{\pi}{4} + k\pi$

19. assume $\tan x + \cot x = 0$

then $\tan x = - \cot x$

or $\tan x = - \dfrac{1}{\tan x}$ but this would imply that $\tan^2 x = - 1$ which can't be.

so no solutions! (complex numbers are not in the range of $\tan x$)

21. $4 \cos^2 \theta = 3$ is equivalent to $\cos^2 \theta = \frac{3}{4}$ or $\cos \theta = \pm \dfrac{\sqrt{3}}{2}$

First possibility:

$\cos \theta = \sqrt{3}/2$ has solutions in Q's I and IV with a reference angle of $\frac{\pi}{6}$

The solutions are $\pi/6$ and $2\pi - \pi/6$ or $\pi/6$ and $11\pi/6$

Second possibility:

$\cos \theta = - \sqrt{3}/2$ has solutions in Q's II and III with the same reference angle, $\frac{\pi}{6}$

These answers are $5\pi/6$ and $7\pi/6$.

23. $\sin^3 x + \sin x = 0$ factors into $(\sin x)(\sin^2 x + 1) = 0$

$\sin x = 0$ or $\sin^2 x + 1 = 0$ (no hope on this latter one, right?)

Solving $\sin x = 0$:

$\sin x = 0$ if $x = 0$ or π or 2π

25. $\sin^3 \beta + 2 \sin \beta = 0$ factors into $(\sin \beta)(\sin^2 \beta + 2) = 0$ so...

$\qquad$ $\sin \beta = 0$ or $\sin^2 \beta + 2 = 0$ (no way could this last one ever be true!)

$\quad$ Solving $\sin \beta = 0$:

$\qquad$ $\sin \beta = 0$ if $\beta = 0$ or π or 2π (the reference angle is zero)

27. $2 \cos 3x = 1$ means that $\cos 3x = \frac{1}{2}$ so values for $3x$ may lie in either quadrant I or IV. Since $\cos \pi/3 = \frac{1}{2}$, answers for $3x$ could be $\pi/3$ plus any multiple of 2π or $5\pi/3$ plus any multiple of 2π. In this problem, we're supposed to find all answers for x in the interval $[0,2\pi]$; We'll find these by setting $3x = \pi/3 + 2\pi$, 4π etc and by setting $3x = 5\pi/3 + 2\pi$, 4π, etc, and solving for x until x exceeds 2π: so......

$\qquad$ if $3x = \pi/3$ then $x = \pi/9$

$\qquad$ if $3x = \pi/3 + 2\pi = 7\pi/3$ then $x = 7\pi/9$

$\qquad$ if $3x = \pi/3 + 4\pi = 13\pi/3$, then $x = 13\pi/9$

$\qquad$ if $3x = \pi/3 + 6\pi$, x will be larger than 2π so don't bother

$\qquad$ if $3x = 5\pi/3$ then $x = 5\pi/9$

$\qquad$ if $3x = 5\pi/3 + 2\pi = 11\pi/3$ then $x = 11\pi/9$

$\qquad$ if $3x = 5\pi/3 + 4\pi = 17\pi/3$ then $x = 17\pi/9$ (just under 2π!)

29. $2 \sin^2 x - \cos x - 1 = 0$ is the same as $2(1 - \cos^2 x) - \cos x - 1 = 0$ or

$\qquad$ $- 2 \cos^2 x - \cos x + 1 = 0$ or $2 \cos^2 x + \cos x - 1 = 0$

$\qquad$ which factors into $(2 \cos x - 1)(\cos x + 1) = 0$

$\quad$ Solving $2 \cos x - 1 = 0$:

$\qquad$ $2 \cos x = 1$ so $\cos x = \frac{1}{2}$ which mean x is in quadrant I or IV with reference angle $\pi/3$;

$\qquad$ answers: $\pi/3$ and $5\pi/3$

$\quad$ Solving $\cos x + 1 = 0$: $\qquad$ $\cos x = - 1$ only at $x = \pi$

31. $2 \cos^2 x + 3 \cos x + 1 = 0$ factors into $(2 \cos x + 1)(\cos x + 1) = 0$

$\quad$ Solving $2 \cos x + 1 = 0$:

$\qquad$ $\cos x = - 1/2$ has solutions in Q's II and III with a reference angle of $\pi/3$

$\qquad$ In QII we have $\pi - \pi/3$ or $2\pi/3$

$\qquad$ In QIII it'll be $\pi + \pi/3$ or $4\pi/3$

$\quad$ Solving $\cos x + 1 = 0$: $\cos x = -1$ has but one solution, π (reference angle of 0)

33. $2 \cos^2\theta = \cos\theta$ or $2\cos^2\theta - \cos\theta = 0$ or $(\cos\theta)(2\cos\theta - 1) = 0$

Solving $\cos\theta = 0$: θ can be $\pi/2$ or $3\pi/2$

Solving $2\cos\theta - 1 = 0$:

$\cos\theta = \frac{1}{2}$ will have solutions in Q's I and IV; reference angle of $\pi/3$

in QI θ it's $\pi/3$ itself; in QIV, θ is $2\pi - \pi/3$ or $5\pi/3$

35. $6\cos^2\theta + 7\cos\theta - 5 = 0$ is same as $(3\cos\theta + 5)(2\cos\theta - 1) = 0$

Solving $3\cos\theta + 5 = 0$:

$\cos\theta = -5/3 = -1.6666666$ has no solutions since neither the sine nor cosine is ever < -1

Solving $2\cos\theta - 1 = 0$:

$\cos\theta = \frac{1}{2}$; θ in QI or QIV again ; reference angle is $\pi/3$

θ is $\pi/3$ itself or $(2\pi - \pi/3)$ or $5\pi/3$

37. $\sin^2 u - 5\sin u + 6 = 0$ is equivalent to $(\sin u - 3)(\sin u - 2) = 0$

Neither $\sin u - 3 = 0$ nor $\sin u - 2 = 0$ has any solutions because $|\sin u| \le 1$

39. $\sin^2 x - \frac{1}{2}\cos x = 0.76$ is same as $2\sin^2 x - \cos x = 1.52$ which is the same as

$2(1 - \cos^2 x) - \cos x = 1.52$ or $2 - 2\cos^2 x - \cos x - 1.52 = 0$ or

$-2\cos^2 x - \cos x + 0.48 = 0$ or $2\cos^2 x + \cos x - 0.48 = 0$

Awkward to factor so use quadratic formula to solve for cos x:

$$\cos x = \frac{-1 \pm \sqrt{1 - 4\cdot 2 \cdot (-0.48)}}{2\cdot 2} = \frac{-1 \pm \sqrt{1 + 3.84}}{4} = \frac{-1 \pm 2.2}{4} = -0.8 \text{ or } 0.3$$

Solving $\cos x = -0.8$:

Solutions in Q's II and III ; ref ang $= \cos^{-1} 0.8 = 0.6435$

$x = \pi - 0.6435 = 2.4981$ or $x = \pi + 0.6435 = 3.7851$

Solving $\cos x = 0.3$:

Solutions in Q's I and IV: reference angle $= \cos^{-1} 0.3 = 1.2661$

$x = 1.2661$ itself (QI) or $2\pi - 1.266 = 5.0171$ (QIV)

41. $2\cos^2\theta + 3\cos\theta + 4 = 0$ doesn't seem to factor; try quadratic formula...

$$\cos\theta = \frac{-3 \pm \sqrt{9 - 4(2)(4)}}{4} = \frac{-3 \pm \sqrt{9 - 32}}{4}$$ and at this point you can see that the discriminant

is negative so no real solutions for $\cos\theta$ and there fore NO solutions for θ.

43. $\tan^2\theta + \sec\theta - 4 = 0$ is equivalent to $(\sec^2\theta - 1) + \sec\theta - 4 = 0$ or

$\sec^2\theta + \sec\theta - 5 = 0$ and the quadratic formula is necessary.

$$\sec\theta = \frac{-1 \pm \sqrt{1 - 4(1)(-5)}}{2} = \frac{-1 \pm \sqrt{21}}{2} \approx 1.7913 \text{ or } -2.7913$$

Solving $\sec\theta = 1.7913$: (take reciprocals of both sides to get on familiar ground)

$\cos\theta = 0.5583$ which has QI and QIV solutions; the reference angle is $\cos^{-1} 0.5583$ or 0.9785

In QI we have 0.9785 ; in QIV it's $2\pi - 0.9785$ or 5.3047

Solving $\sec\theta = -2.7913$:

$\cos\theta = -0.3583$ will have θ in quadrants II and III.

The reference angle is $\cos^{-1} 0.3583 \approx 1.2044$

In QII we have $\pi - 1.2044$ or 1.9372

In QIII it's $\pi + 1.2044$ or 4.3460

45. $2\sec^2 x - 5\tan x - 3 = 0$ or $2(\tan^2 x + 1) - 5\tan x - 3 = 0$ which can be written

as $2\tan^2 x - 5\tan x - 1 = 0$ which requires the quadratic formula:

$$\tan x = \frac{5 \pm \sqrt{25 - 4(2)(-1)}}{4} = \frac{5 \pm \sqrt{33}}{4} \approx 2.6861 \text{ or } -0.1861$$

Solving $\tan x = 2.6861$ (Q I or Q III) ref $\angle = \tan^{-1} 2.6861 \approx 1.2144$

In QI we have $x = 1.2144$; in QIII we have $x = \pi + 1.2144$ or 4.3560

Solving $\tan x = -0.1861$ (QII or QIV) ref $\angle = \tan^{-1} 0.1861 = 0.1840$

In QII $x = \pi - 0.184$ or 2.9576 ; in QIV it's $x = 2\pi - 0.1840$ or 6.0992

47. $\sin x = \sin\frac{x}{2}$ (view $\sin x$ as $\sin 2\left(\frac{x}{2}\right)$ and use the $\sin 2\theta$ formula on the left)

$2\sin\frac{x}{2}\cos\frac{x}{2} = \sin\frac{x}{2}$ or $2\sin\frac{x}{2}\cos\frac{x}{2} - \sin\frac{x}{2} = 0$ which can be factored as

$\sin\frac{x}{2}\left(2\cos\frac{x}{2} - 1\right) = 0$

Solving $\sin\frac{x}{2} = 0$:

$\sin\frac{x}{2} = 0$ if $\frac{x}{2} = 0$, π, or 2π implying $x = 0$, 2π, or 4π (out of bounds)

47. (continued)

Solving $2 \cos \frac{x}{2} - 1 = 0$:

$\cos \frac{x}{2} = \frac{1}{2}$ if $\frac{x}{2} = \frac{\pi}{3}$ or $\frac{5\pi}{3}$ which means x can $= \frac{2\pi}{3}$ or $\frac{10\pi}{3}$ (no good)

So the answers in $[0, 2\pi]$ are 0, $\frac{2\pi}{3}$, and 2π

49. $\tan^2\theta + \sec^2\theta = 3$ is the same as $\tan^2\theta + (\tan^2\theta + 1) = 3$ or $2\tan^2\theta = 2$
or $\tan^2\theta = 1$ so we can have $\tan\theta = \pm 1$. All quadrants are available for
θ and the reference angle is the inverse tangent of 1 which is $\pi/4$. Placing
$\pi/4$ in all four quadrants gives these answers: $\pi/4$, $3\pi/4$, $5\pi/4$, $7\pi/4$

51. $4\tan 2\beta + 5\cot \beta = 0$ (if you try using a double angle formula for $\tan 2\beta$ and $1/\tan\beta$ for
$\cot \beta$ and then combining fractions it appears that there are no solutions; however trial and
error shows that $\pi/2$ and $3\pi/2$ do work. For example: $4\tan 2(\pi/2) + 5\cot \pi/2 = 4\tan \pi$
$+ 5\cot \pi/2 = 0 + 0 = 0$. The solutions are $\pi/2$ and $3\pi/2$.

53. $\sin \theta \cos \theta = c$; What's the largest c can be?
Since $\sin 2\theta = 2\sin \theta \cos \theta$ we can replace $\sin \theta \cos \theta$ by $\frac{\sin 2\theta}{2}$ and say that we must have
$\frac{\sin 2\theta}{2} = c$ or $\sin 2\theta = 2c$. Now the largest that the sin of any angle can be is 1 so the largest
value for $\sin 2\theta$ is 1 which means the largest 2c can be is 1 so the largest c can be is $1/2$.

55. $\frac{\sin \theta_1}{\sin \theta_2} = \frac{v_1}{v_2}$ becomes $\frac{\sin 30°}{\sin \theta_2} = \frac{300{,}000}{220{,}400}$ so $\sin \theta_2 = \frac{(0.5)(220{,}400)}{300{,}000}$

or $\sin \theta_2 = 0.3673$ which means that $\theta_2 = \sin^{-1} 0.3673$ or $21°33'$

57. We are given that $\cos \theta = \left(\frac{r_2}{r_1}\right)^4$ and that $r_1 = 2r_2$. Substituting the latter gives

$$\cos \theta = \left(\frac{r_2}{2r_2}\right)^4 = \left(\frac{1}{2}\right)^4 = \frac{1}{16} = 0.0625$$

so $\theta = \cos^{-1} 0.0625 = 1.15083$ radians or $86°25'$

59. Stated mathematically, the requirement is that we must have $10 \sin \frac{\pi}{4}t \geq 7$.

(For simplicity, we can try to find the intervals in $[0,2\pi]$ for which the inequality $\sin \theta \geq 0.7$ is true and then divide then divide the length of this inverval by 2π to get a percentage.)

The values of θ for which $\sin \theta = 0.7$ are located in quadrants I and II; using a calculators inverse sine capabilities gives us a reference angle of 0.7754, the quadrant I value and leads to 2.3662 ($\pi - 0.7754$) as the QII solutions. This means that for any θ greater than or equal 0.7754 but less than or equal to 2.3662, $\sin \theta$ will be greater than or equal to 0.7. The length of this interval is 1.5908.

The function will repeat this behavior on every interval of 2π so the percentage of time that the voltage is at least 7 can be found by comparing 1.5908 to 2π by division:

The computation: $\frac{1.5908}{2\pi}$ X $100\% = 25.32\%$

61. $P(t) = 10,000 + 2000 \cos \frac{\pi}{3}t$

a) $P(0) = 10,000 + 2000 \cos 0 = 10,000 + 2000 = 12,000$ since $\cos 0 = 1$.

b) Replace $\frac{\pi}{3}t$ with x:

try to find for how much of 2π is $8500 \leq 10000 + 2000 \cos x \leq 11,000$

$- 1500 \leq 2000 \cos x \leq 1000$

$- 0.75 \leq \cos x \leq 0.5$

The cosine function starts out at height 1 ($\cos 0 = 1$) and decreases to -1 ($\cos \pi = -1$)

The cosine is at the value 0.5 when $x = \pi/3$ or approximately 1.0472

The cosine is at the value - 0.75 when $x = 2.4189$ in quadrant II

For all x values between 1.0472 and 2.4189 the double inequality above is true and there will be an interval of identical length stretching from quandrant III to IV that also contains solutions. The total length of the intervals is 2 X 1.3717 or 2.7434 and if this is divided by 2π and converted to a percent by multiplying by 100, you get 43.66%.

Problems 8.6

1. Choose the cosine function because when t = 0, the graph is at its maximum. Choose A = 1 because the maximal height of the graph is 1. Observe that the graph completes one complete cycle as t goes from 0 to π so this means that the period (T) is equal to π. We can use the relationship $\omega = \frac{2\pi}{T}$ to solve for ω: $\omega = \frac{2\pi}{\pi} = 2$ and so the final result is y = 1 cos 2t or just y = cos 2t.

3. Choose sine because the graph contains the origin; choose A = 2 because it's the maximal height; observe that the graph goes through one cycle as t goes from 0 to 2.
$$\text{so } T = 2 \text{ and } \omega = \frac{2\pi}{T} = \frac{2\pi}{2} = \pi$$
   The equation must therefore be y = 2 sin πt.

5. Choose sine because the graph goes through the origin. Notice that the graph heads down immediately; this means we need a (-1) factor! Choose A = 3 because of the maximal height reached and observe that t has to go from 0 to $\frac{\pi}{2}$ for the graph to do one cycle so T = $\frac{\pi}{2}$ and $\omega = \frac{2\pi}{T} = \frac{2\pi}{\pi/2} = 4$ and so the final result is y = - 3 sin 4t.

7. The graph goes through the origin so it must be a sine; highest points are at a height of 10 so A = 10 and the graph completes a cycle as t goes from 0 to $\frac{2\pi}{5}$ so T = $\frac{2\pi}{5}$ and $\omega = \frac{2\pi}{T} = \frac{2\pi}{2\pi/5} = 5$ which makes the equation come out to be y = 10 sin 5t.

9. When t = 0, the graph is at its lowest point so it must be a cosine and there must be a (-1) factor. The maximum height is 1 so A = 1; we can see one cycle of the graph as t goes from - 4π to 4π so the period T must be 8π and $\omega = \frac{2\pi}{8\pi} = \frac{1}{4}$ and so the equation is y = - cos $\frac{1}{4}$t

11. It will have to be a sine as the graph passes through the origin; A = 2 from seeing the maximum height; a cycle of the graph is completed as t goes from 0 to $2\pi^2$ so T = $2\pi^2$ and $\omega = \frac{2\pi}{2\pi^2} = \frac{1}{\pi}$ and so the equation is y = 2 sin $\frac{1}{\pi}$t.

13. $y = 2 \sin t$ has an amplitude of 2 (the coefficient of the sine function) and a period of

$2\pi \div$ (the coefficient of t) $= 2\pi \div 1 = 2\pi$. The phase shift is found by setting $t = 0$ and solving for t (in this case trivial!)

Sketch: Fit a sine curve with amplitude 2 into the interval

[phase shift, phase shift + period] or [0, 2π]

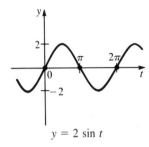

$y = 2 \sin t$

15. $y = - 2 \sin t$ (reflect graph from #1 about the x axis) or use same basic fact list as in #1 about

amplitude, period etc. only start the graph off downward because of the negative sign.

Sketch:

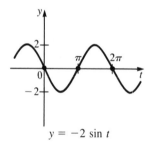

$y = -2 \sin t$

17. $y = 2 \sin 2t$ has an amplitude of 2, a period of $2\pi \div 2 = \pi$ and a phase shift found by setting

$2t = 0$ and getting $t = 0$ (ie., no phase shift)

Sketch: fit a sine curve of amplitude 2 into the interval $[0, \pi]$

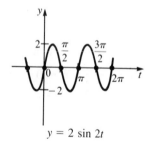

$y = 2 \sin 2t$

19. $y = \frac{1}{2} \sin \frac{1}{2}t$ has an amplitude of $\frac{1}{2}$, a period of $2\pi \div \frac{1}{2} = 4\pi$ and no phase shift

Sketch: horizontal dotted lines at $y = \frac{1}{2}$ and at $-\frac{1}{2}$ and a sine curve stretched over the interval

$[0, 4\pi]$

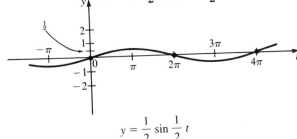

$$y = \frac{1}{2} \sin \frac{1}{2} t$$

21. $y = -\sin \pi t$ has an amplitude of 1 (it'll start downward), a period of $2\pi \div \pi = 2$ and 0 phase

shift (the solution of $\pi t = 0$ is $t = 0$).

Sketch: initially downward sine curve squeezed into the interval $[0, 2]$

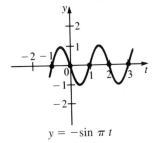

$$y = -\sin \pi t$$

23. $y = \cos (t - \pi)$ has an amplitude of 1, a period of $2\pi \div 1 = 2\pi$ and a phase shift of $+ \pi$. To

sketch it draw a cosine curve over the interval $[\pi, 3\pi]$:

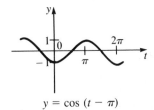

$$y = \cos (t - \pi)$$

25. $y = \cos \left(\frac{t}{2} + \frac{\pi}{4}\right)$ has an amplitude of 1, a period of $2\pi \div \frac{1}{2} = 4\pi$ and a phase shift found by

solving $\frac{t}{2} + \frac{\pi}{4} = 0$; $\frac{t}{2} = -\frac{\pi}{4}$; $t = -\frac{\pi}{2}$. The sketch will be a cosine curve (amp $= 1$) sketched over

the interval $[-\frac{\pi}{2}, -\frac{\pi}{2} + 4\pi]$:

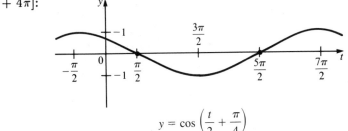

$$y = \cos \left(\frac{t}{2} + \frac{\pi}{4}\right)$$

27. We can choose A = 2 because of the obvious height of the hightest points on the graph.

We can see the period, T, as follows: the graph goes through one complete cycle as t goes from 0 to π; therefore T = π and now we can find ω using the formula $\omega = \frac{2\pi}{T}$: we get $\omega = \frac{2\pi}{\pi} = 2$.

Now for δ: from the graph you can see a "standard" sine curve starting at the point $(\frac{\pi}{4},0)$; this means that the phase shift can be taken to be $\frac{\pi}{4}$ and we can use the relationship

$$\text{phase shift} = \frac{|\delta|}{\omega}$$

to find δ.

Putting in our values gives the equation $\frac{\pi}{4} = \frac{|\delta|}{2}$ which implies that $|\delta| = \frac{\pi}{2}$ so δ can be either $\frac{\pi}{2}$ or - $\frac{\pi}{2}$; we must choose the positive one because we can see that the phase shift is to the right.

29. A = 3 from observing the height of the hightest points; the graph goes through half a cycle as t goes from - $\frac{2\pi}{3}$ to $\frac{4\pi}{3}$ so $\frac{6\pi}{3}$ or 2π is half a period so T, a full period, must be 4π. The value of ω can be found by evaluating the formula $\omega = \frac{2\pi}{T}$ with T = 4π to get $\omega = \frac{1}{2}$. One possibility for the actual phase shift is $\frac{7\pi}{3}$ (a sine curve can be seen "starting" at ($\frac{7\pi}{3}$, 0). Now we must substitute our known values into "phase shift = $\frac{|\delta|}{\omega}$" to get $\frac{7\pi}{3} = \frac{|\delta|}{1/2}$ which implies that $|\delta| = \frac{7\pi}{6}$ so we'll choose $\delta = \frac{7\pi}{6}$ to ensure the phase shift to the right.

31. A = $\frac{1}{3}$ simply by seeing how high the graph gets; one cycle is completed as t goes from $\pi - \pi^2$ to $\pi + \pi^2$ so the difference is $2\pi^2$ which is the period T. This means that $\omega = \frac{2\pi}{T} = \frac{2\pi}{2\pi^2} = \frac{1}{\pi}$. The phase shift can be observed to be π (to the right) so the relationship

$$\text{phase shift} = \frac{|\delta|}{\omega} \quad \text{becomes}$$

$$\pi = \frac{|\delta|}{1/\pi} \quad \text{which implies that } |\delta| = 1 \quad \text{(cross multiply)}$$

and we will choose the positive solution, 1, as our value for δ because it was a phase shift to the right.

Note: the answers to problems 27, 29, and 31 are not unique as one may choose the phase shift to the left rather than the one to the right or to use a negative rather than a positive value for A.

PROBLEMS 8.7

1. $y = \sin t + 2$ will have as its graph the graph of $y = \sin t$ shifted UP two blocks: (horizontal

    dotted lines at $y = -1 + 2$ or $y = 1$ and at $y = 1 + 2$ or 3)

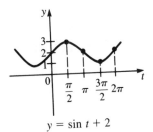

$$y = \sin t + 2$$

3. $y = 2 \cos \left(t - \frac{\pi}{6} \right) + 1$ will have as its graph the graph of $y = 2 \cos \left(t - \frac{\pi}{6} \right)$ raised up one block:

    amplitude is two so after the raising of one block we'll have the horizontal dotted lines at -1 and

    at 3 (normally with an amplitude of 2 they would be at - 2 and 2). The period is $2\pi \div 1$ or 2π

    and the phase shift is $\pi/6$ to the right. To sketch, fit a cosine curve (it starts at its high

    point) between the dotted lines over the interval $[\ \pi/6,\ \pi/6 + 2\pi]$:

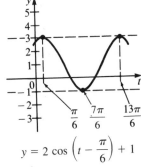

$$y = 2 \cos \left(t - \frac{\pi}{6} \right) + 1$$

5. $y = \sin t + \cos t = \sqrt{2} \sin \left(t + \frac{\pi}{4} \right)$ if you convert into A sin $(\omega t + \phi)$ form. Now it is clear that

    the amplitude is $\sqrt{2}\ (\approx 1.41)$, the period is 2π, and the phase shift is - $\pi/4$. The sketch is a sine

    curve of proper amplitude over the interval $[-\pi/4,\ -\pi/4 + 2\pi]$ or $[-\pi/4,\ 7\pi/4]$:

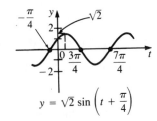

$$y = \sqrt{2} \sin \left(t + \frac{\pi}{4} \right)$$

7. $y = \cos t - \sin t$ is equivalent to $\sqrt{2} \cos \left(t + \pi/4 \right)$ (or you could use the sine) and the information is identical to that in # 5 except that now a cosine curve would be fitted into the interval $[-\pi/4, 7\pi/4]$:

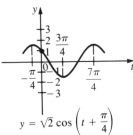

$$y = \sqrt{2} \cos \left(t + \frac{\pi}{4} \right)$$

9. $y = -2 \sin t - 3 \cos t$, if converted into a single sine function yields $y = \sqrt{13} \sin (t + 4.12)$ here $\sin \phi = -3/\sqrt{13}$ and $\cos \phi = -2/\sqrt{13}$ so ϕ had to be in QIII; the reference angle was 0.98 which was added to π. The amplitude is $\sqrt{13}$ or ≈ 3.61, the period is 2π, and the phase shift is -4.1   Sketch:

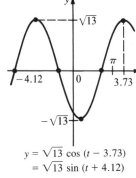

$$y = \sqrt{13} \cos (t - 3.73)$$
$$= \sqrt{13} \sin (t + 4.12)$$

11. $y = -2 \sin t + 3 \cos t$ becomes $\sqrt{13} \sin (t + 2.16)$ when $\cos \phi = -2/\sqrt{13}$ and $\sin \phi = 3/\sqrt{13}$ and ϕ is found in QII (reference angle was 0.98). The amplitude is $\sqrt{13}$, the period is 2π, and the phase shift is -2.16. To sketch: sine curve of amplitude 3.6 over the interval $[-2.16, -2.16 + 2\pi]$

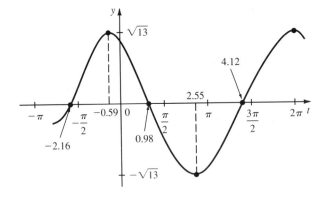

13. $y = 4 \cos 2t - 3 \sin 2t$ is equivalent to $5 \sin (2t + 2.21)$ where $\sin \phi = 4/5$ and $\cos \phi = -3/5$ (QII) and a reference angle of 0.93 was used. The amplitude is 5, the period is $2\pi \div 2 = \pi$, and the phase shift is found by solving $2t + 2.21 = 0$; $2t = -2.21$; $t = -1.10$ The sketch will be a sine curve of amplitude 5 squeezed over the interval $[-1.10, -1.10 + \pi]$ or $[-1.10, 2.04]$

13. (continued)

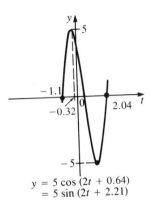

$$y = 5 \cos (2t + 0.64)$$
$$= 5 \sin (2t + 2.21)$$

15. The graph of $y = -4 \cos 2t + 3 \sin 2t$ is the reflection of the graph of #13 about the x-axis:

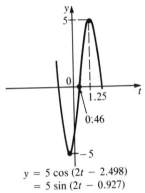

$$y = 5 \cos (2t - 2.498)$$
$$= 5 \sin (2t - 0.927)$$

17. $y = -\cos \frac{\pi}{2}t + 4 \sin \frac{\pi}{2}t$ can be rewritten $\sqrt{17} \cos \left(\frac{\pi}{2}t - 1.82\right)$ where we used the A cos ($\omega t - \delta$)

model with $A = \sqrt{(-1)^2 + 4^2} = \sqrt{17}$ and $\cos \delta = \frac{-1}{\sqrt{17}}$ and $\sin \delta = \frac{4}{\sqrt{17}}$ which lead to δ being in

quadrant 2 and a reference angle of 1.32 ; ie $\delta = \pi - 1.32 = 1.82$

The sketch is a cosine curve with an amplitude of $\sqrt{17}$, a period of 4 ($2\pi \div \frac{\pi}{2}$), and a phase shift

found by setting $\pi/2 \ t - 1.82 = 0$; $t = 3.64/\pi \approx 1.16$ (to the right). One period of the graph

could begin above 1.16 and end above 1.16 + 4:

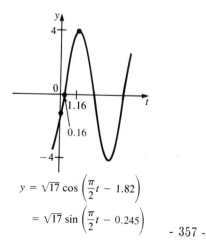

$$y = \sqrt{17} \cos \left(\frac{\pi}{2}t - 1.82\right)$$

$$= \sqrt{17} \sin \left(\frac{\pi}{2}t - 0.245\right)$$

19. $y = 12 \cos \frac{\pi}{4}t + 5 \sin \frac{\pi}{4}t$ can be expressed as $13 \sin \left(\frac{\pi}{4}t + 1.18 \right)$

where $A = \sqrt{12^2 + 5^2} = 13$, $\sin \phi = \frac{12}{13}$ and $\cos \phi = \frac{5}{13}$ which implies $\phi \approx 1.18$ and so we sketch a sine curve with an amplitude of 13, a period of $2\pi \div \frac{\pi}{4} = 8$ and a phase shift of -1.5 (solve the equation $\frac{\pi}{4}t + 1.18 = 0$)

Sketch : a sine curve strecthed out over the interval [-1.5, -1.5 + 8] with an amplitude of 13

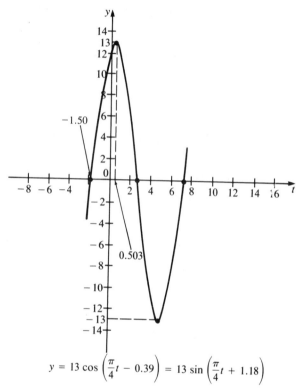

$$y = 13 \cos \left(\frac{\pi}{4}t - 0.39 \right) = 13 \sin \left(\frac{\pi}{4}t + 1.18 \right)$$

21. $y = \sin t - \cos 2t$ can be sketched by the "addition of ordinates" method (or "adding y-values" or "adding heights"). We'll sketch the graph of sin t and the graph of - cos 2t on the same axes and then try to "add" the two graphs together.

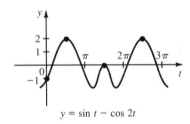

$y = \sin t - \cos 2t$

Some general hints on adding graphs:

   1) Look for places where one graph crosses the x-axis. Wherever the other graph is will be a point on the resultant graph, the "sum" graph.

   2) Look for places where one graph is just as far above the x-axis as the other one is below the x-axis. These locations will be the x-intercepts of the resultant graph.

   3) Closely examine peaks and low points of each of the graphs (eg, if two peaks occur at the same x value surely the resultant graph will peak at that same x value.

   4) Look for places where the curves intersect. Double this height (or depth) and you have a point on the resultant graph.

23. $y = - \sin t - \cos 2t$

   - sin t is a downward heading sine on the interval $[0, 2\pi]$

   - cos 2t is a cosine curve which starts low, at (0,-1), and has a shortened period of π.

Sketch:

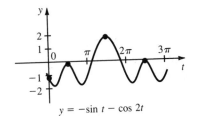

$$y = -\sin t - \cos 2t$$

25. $y = \cos \frac{\pi}{2}t - \sin \pi t$

   $\cos \frac{\pi}{2}t$ is a cosine curve of amplitude 1 with a period of $2\pi \div \frac{\pi}{2}$ or 4

   - sin πt is an initially downward heading sine curve with a period of 2

Note: the resultant graph has a period which is the least common multiple of the individual periods (assuming the same phase shift). In this problem the period of the resultant graph is 4.

Sketch:

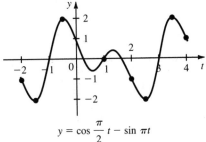

$$y = \cos \frac{\pi}{2} t - \sin \pi t$$

Problems 8.8

1. $E = 20 \sin 2\pi t$ volts and $R = 10$ ohms

   a) frequency $= \frac{\omega}{2\pi} = \frac{2\pi}{2\pi} = 1$ hz (cycle per second)

   b) current $= I = \frac{E}{R} = \frac{20 \sin 2\pi t}{10} = 2 \sin 2\pi t$ which has amplitude 2 so 2 amps

3. $E = 50 \sin \frac{\pi}{2}t$ and $R = 1$ ohm

   a) frequency $= \frac{\pi/2}{2\pi} = \frac{1}{4}$ cycle per second

   b) max current $= 50$ amp because E will be divided by 1 and the 50 remains

5. a) set $y(t) = 0$: $e^{t/4}(1000 \cos 2t - 100 \sin 2t) = 0$

   since $e^{t/4}$ can't be 0 it's up to $1000 \cos 2t - 100 \sin 2t$ to equal 0:

   $$\text{so } 1000 \cos 2t = 100 \sin 2t$$
   $$10 = \frac{\sin 2t}{\cos 2t}$$

   $$10 = \tan 2t \text{ so } 2t = \tan^{-1}10 \approx 1.47$$
   $$\text{which means that } t \approx 0.735 \text{ years}$$

   b) find $x(0.735) \approx e^{0.735/4}(200 \cos 1.47 + 500 \sin 1.47)$
   $$\approx 1.202 \, (\, 200 \cdot 0.1006 + 500 \cdot 0.9949 \,)$$
   $$\approx 622$$

7. a) $y(t) = e^{t/10}(400 \cos \frac{\pi}{2}t - 600 \sin \frac{\pi}{2}t) = 0$

   $$400 \cos \frac{\pi}{2}t = 600 \sin \frac{\pi}{2}t$$

   $$0.66667 \approx \tan \frac{\pi}{2}t \text{ so } \frac{\pi}{2}t = \tan^{-1}0.66666... = 0.5880$$

   $$\text{so } t \approx \frac{2 \, (0.5880)}{\pi} \approx 0.3743$$

7. (continued)

   b) $x(0.3743) = e^{0.03743}(1000 \cos 0.5880 + 1500 \sin 0.5880)$

   $\approx 1.0381 \ (\ 832 + 832 \ )$ or 1727

9. $E = 50 \sin 200\pi t$

   a) $\dfrac{200\pi}{2\pi} = 100$ cycles per second

   b) $E = 50 \sin (200 \ \pi \cdot 0.002) = 50 \sin .4\pi \approx 50 \ (0.9511) = 47.553$ volts

11. $A = 5$ and $f = 1056$ cycles per second

   $f = \dfrac{\omega}{2\pi}$ becomes $1056 = \dfrac{\omega}{2\pi}$ so $\omega = 2112\pi$

   the equation is $y = 5 \sin (2112\pi t)$

13. $A = \dfrac{1}{5}$ and 1 cycle is accomplished as t goes from 0 to 200 so the period is 200

   Since $\omega = \dfrac{2\pi}{T}$ we have $\omega = \dfrac{2\pi}{200} = \dfrac{\pi}{100}$ and the equation is $y = \dfrac{1}{5} \sin \dfrac{\pi}{100} t$

Chapter 8 Review

1. $\tan x \sec x = \dfrac{\sin x}{\cos x} \cdot \dfrac{1}{\cos x} = \dfrac{\sin x}{\cos^2 x}$

3. $\dfrac{1 - \sec x}{1 + \sec x} \csc^2 x = \dfrac{1 - \dfrac{1}{\cos x}}{1 + \dfrac{1}{\cos x}} \cdot \dfrac{1}{\sin^2 x} = \dfrac{\cos x - 1}{\cos x + 1} \cdot \dfrac{1}{1 - \cos^2 x} =$

   $\dfrac{\cos x - 1}{\cos x + 1} \cdot \dfrac{1}{(1 + \cos x)(1 - \cos x)} = - \dfrac{1}{(\cos x + 1)^2}$

5. $\dfrac{\csc \alpha}{\sec \alpha} = \cot \alpha \ ?$ $\dfrac{\csc \alpha}{\sec \alpha} = \dfrac{1/\sin \alpha}{1/\cos \alpha} = \dfrac{\cos \alpha}{\sin \alpha} = \cot \alpha$

7. $\sin \theta \csc \theta - \cos^2\theta = \sin^2\theta$?

$\sin \theta \csc \theta - \cos^2\theta = 1 - \cos^2\theta = \sin^2\theta$

9. wherever the secant function is positive, ie, quadrants I & IV, ie for

$$0 \le x < \pi/2 \quad \text{and} \quad 3\pi/2 < x \le 2\pi$$

11. $\sqrt{16 + x^2} = \sqrt{16 + 16\tan^2\theta} = \sqrt{16(1 + \tan^2\theta} = \sqrt{16\sec^2\theta} = 4\sec \theta$

13. $\tan \dfrac{5\pi}{12} = \tan \dfrac{1}{2}\left(\dfrac{5\pi}{6}\right) = \dfrac{\sin \frac{5\pi}{6}}{1 + \cos \frac{5\pi}{6}} = \dfrac{\frac{1}{2}}{1 + \left(-\frac{\sqrt{3}}{2}\right)} = \dfrac{1}{2 - \sqrt{3}} = 2 + \sqrt{3}$ (rationalized)

15. $\tan \dfrac{3\pi}{8} = \tan \dfrac{1}{2}\left(\dfrac{3\pi}{4}\right) = \dfrac{\sin \frac{3\pi}{4}}{1 + \cos \frac{3\pi}{4}} = \dfrac{\frac{1}{\sqrt{2}}}{1 + \left(-\frac{1}{\sqrt{2}}\right)} = \dfrac{1}{\sqrt{2} - 1}$ or $\sqrt{2} + 1$

17. $\sin 37°\cos 41° - \cos 37°\sin 41° = \sin (37 - 41)° = \sin (-4°) = - \sin 4°$

19. $\cos^2\alpha - \sin^2\beta = \cos (\alpha - \beta) \cos (\alpha + \beta)$? (start with right side)

$\cos (\alpha - \beta) \cos (\alpha + \beta) = (\cos \alpha \cos \beta + \sin \alpha \sin \beta)(\cos \alpha \cos \beta - \sin \alpha \sin \beta)$

$\qquad = \cos^2\alpha \cos^2\beta - \sin^2\alpha \sin^2\beta$

$\qquad = \cos^2\alpha \cos^2\beta - (1 - \cos^2\alpha)(1 - \cos^2\beta)$

$\qquad = \cos^2\alpha \cos^2\beta - 1 + \cos^2\beta + \cos^2\alpha - \cos^2\alpha \cos^2\beta$

$\qquad = \cos^2\alpha + \cos^2\beta - 1$

$\qquad = \cos^2\alpha + (- \sin^2\beta)$

21. $\cos \alpha = -\dfrac{2}{3}$ with α in QIII; triangle:

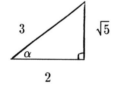

$\cot \beta = -2 = -\dfrac{2}{1}$ with β in QII ; triangle:

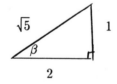

21. (continued)

(a) $\sin(\alpha+\beta) = \sin\alpha\cos\beta + \sin\beta\cos\alpha = \left(-\frac{\sqrt{5}}{3}\right)\left(-\frac{2}{\sqrt{5}}\right) + \left(\frac{1}{\sqrt{5}}\right)\left(-\frac{2}{3}\right)$

$$= \frac{2\sqrt{5}-2}{3\sqrt{5}} \text{ or } \frac{2}{3} - \frac{2}{3\sqrt{5}} \text{ or } \frac{10-2\sqrt{5}}{15}$$

$\sin(\alpha-\beta)$: (change middle sign above)

$$= \frac{10+2\sqrt{5}}{15} \text{ or any of its variations;}$$

$\cos(\alpha+\beta) = \cos\alpha\cos\beta - \sin\alpha\sin\beta = \left(-\frac{2}{3}\right)\left(-\frac{2}{\sqrt{5}}\right) - \left(-\frac{\sqrt{5}}{3}\right)\left(\frac{1}{\sqrt{5}}\right)$

$$= \frac{4+\sqrt{5}}{3\sqrt{5}}$$

$\cos(\alpha-\beta)$: just change the above middle sign

$$= \frac{4-\sqrt{5}}{3\sqrt{5}}$$

$\tan(\alpha+\beta) = \frac{\sin(\alpha+\beta)}{\cos(\alpha+\beta)} = \frac{2\sqrt{5}-2}{4+\sqrt{5}}$ (cancelling the like denominators)

$\tan(\alpha-\beta)$: change middle signs in numerator and denominator above

$$= \frac{2\sqrt{5}+2}{4-\sqrt{5}}$$

(b) for $\alpha+\beta$: sine and cosine and tangent are positive; therefore QI.

for $\alpha-\beta$: sine and cosine and tangent are positive; therefore QI also.

23. $\tan(\sin^{-1}(-2/3) - \cos^{-1}3/7) = ?$

Let $A = \sin^{-1}(-2/3)$ and $B = \cos^{-1}3/7$ Note: A in QIV ; B in QI by convention

The triangles:

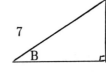

3   2   7

A   B

missing adj $= \sqrt{5}$ 3   missing opp $= 2\sqrt{10}$

23. (continued)

The original problem could now be stated as tan (A - B) which expands to

$$\frac{\tan A - \tan B}{1 + \tan A \tan B} \qquad (\text{ now check out the triangles })$$

which equals $\dfrac{\dfrac{-2}{\sqrt{5}} - \dfrac{2\sqrt{10}}{3}}{1 + \left(\dfrac{-2}{\sqrt{5}}\right)\left(\dfrac{2\sqrt{10}}{3}\right)}$ (tan A is negative because QIV)

Multiplying the numerator and denominator of this fraction by $3\sqrt{5}$ gives

$$\frac{-6 - 10\sqrt{2}}{3\sqrt{5} - 4\sqrt{10}}$$

25. $\sin \theta = \frac{3}{5}$ with θ in Q I

use and the missing adj is 4

Now all the questions can be answered:

$$\sin 2\theta = 2 \sin \theta \cos \theta = 2 \left(\frac{3}{5}\right)\left(\frac{4}{5}\right) = \frac{24}{25}$$

$$\cos 2\theta = \cos^2\theta - \sin^2\theta = \frac{16}{25} - \frac{9}{25} = \frac{7}{25}$$

$$\tan 2\theta = \sin 2\theta \div \cos 2\theta = \frac{24}{25} \div \frac{7}{25} = \frac{24}{7}$$

$$\sin \frac{\theta}{2} = \pm \sqrt{\frac{1 - \cos \theta}{2}} = + \sqrt{\frac{1 - 4/5}{2}} = \frac{1}{\sqrt{10}}$$

$$\cos \frac{\theta}{2} = \pm \sqrt{\frac{1 + \cos \theta}{2}} = + \sqrt{\frac{1 + 4/5}{2}} = \sqrt{\frac{9}{10}} = \frac{3}{\sqrt{10}}$$

$$\tan \frac{\theta}{2} = \sin \frac{\theta}{2} \div \cos \frac{\theta}{2} = \frac{1}{\sqrt{10}} \div \frac{3}{\sqrt{10}} = \frac{1}{3}$$

27. $\cos \theta = -\frac{3}{4}$ with θ in QIII

use [triangle with hypotenuse 4, base 3, angle θ]

3   and the missing opp is $\sqrt{7}$

$$\sin 2\theta = 2\left(-\frac{\sqrt{7}}{4}\right)\left(-\frac{3}{4}\right) = \frac{3\sqrt{7}}{8} \qquad \cos 2\theta = \left(-\frac{3}{4}\right)^2 - \left(-\frac{\sqrt{7}}{4}\right)^2 = \frac{9}{16} - \frac{7}{16} = \frac{1}{8}$$

$$\tan 2\theta = \frac{3\sqrt{7}}{8} \div \frac{1}{8} = 3\sqrt{7}$$

$$\sin \frac{\theta}{2} = + \sqrt{\frac{1 - \left(-\frac{3}{4}\right)}{2}} = \sqrt{\frac{7/4}{2}} = \sqrt{\frac{7}{8}} \qquad \cos \frac{\theta}{2} = -\sqrt{\frac{1 + \left(\frac{-3}{4}\right)}{2}} = -\sqrt{\frac{1}{8}}$$

$$\tan \frac{\theta}{2} = \sqrt{\frac{7}{8}} \div -\sqrt{\frac{1}{8}} = -\sqrt{7}$$

29. $\dfrac{2 \cot \theta}{\cot^2\theta - 1} = \tan 2\theta$??

$$\tan 2\theta = \frac{2 \tan \theta}{1 - \tan^2\theta} = \frac{2 \cdot \frac{1}{\cot \theta}}{1 - \frac{1}{\cot^2\theta}} = \frac{2 \cot \theta}{\cot^2\theta - 1} \quad \text{(last step was accomplished by}$$

multiplying by $\cot^2\theta / \cot^2\theta$)

31. $\sin 55° \cos 15°$ (using (2) from p. 503) $= \frac{1}{2} [\sin 70° + \sin 40°]$

33. $\sin 2x \sin 5x$ (using (1) from p. 503) $= \frac{1}{2} [\cos (-3x) - \cos 7x]$ or $\frac{1}{2} (\cos 3x - \cos 7x)$

35. $\sin 20° + \sin 50°$ (using (5) from p. 504) $= 2 \sin 35° \cos (-15°)$ or $2 \sin 35° \cos 15°$

37. $5 \cos t - 3 \sin t$ is to be converted: $A = \sqrt{25 + 9} = \sqrt{34}$

    going into $A \cos (\omega t - \delta)$ form:

    $\cos \delta = \frac{5}{\sqrt{34}} \approx 0.8575$ while $\sin \delta = \frac{-3}{\sqrt{34}} \approx -0.5145$ (QIV is indicated)

    the reference angle is $\cos^{-1} 0.8575 = 0.5404$ which when placed in QIV becomes $2\pi - 0.5404$ or

    5.7428

    Answer: $\sqrt{34} \cos (t - 5.7428)$ or (add 2π) $\sqrt{34} \cos (t + 0.5404)$

37. (continued)

going into A sin ($\omega t + \phi$) form:

$\cos \phi = -0.5145$ and $\sin \phi = 0.8575$ (QII) so $\phi \approx \pi - \sin^{-1} 0.8575$

$= \pi - 1.0304$ or 2.1112 Answer: $\sqrt{34} \sin (t + 2.1112)$

39. $- 6 \cos 3\theta - 8 \sin 3\theta$ is to be converted: $A = \sqrt{36 + 64} = 10$

into A cos ($\omega t - \delta$) form:

$\cos \delta = -6/10 = - 0.6$ and $\sin \delta = -8/10 = - 0.8$ so it's QIII for δ

the reference angle is inv cos of 0.6 or 0.9273; added to π: 4.0689

Answer: $10 \cos (3\theta - 4.0689)$

into A sin ($\omega t + \phi$) form:

$\cos \phi = - 0.8$ and $\sin \phi = - 0.6 \Rightarrow \phi = 3.7851$ (also in QIII)

Answer: $10 \sin (3\theta + 3.7851)$

41. $\dfrac{\sin \alpha - \sin \beta}{\sin \alpha + \sin \beta} = \dfrac{\tan \left(\dfrac{\alpha - \beta}{2} \right)}{\tan \left(\dfrac{\alpha + \beta}{2} \right)}$??

the left side $= \dfrac{2 \cos \dfrac{\alpha + \beta}{2} \sin \dfrac{\alpha - \beta}{2}}{2 \sin \dfrac{\alpha + \beta}{2} \cos \dfrac{\alpha - \beta}{2}} = \dfrac{\sin \dfrac{\alpha - \beta}{2}}{\cos \dfrac{\alpha - \beta}{2}} = \cot \dfrac{\alpha + \beta}{2} \tan \dfrac{\alpha - \beta}{2}$

$= \dfrac{1}{\tan \dfrac{\alpha + \beta}{2}} \tan \dfrac{\alpha - \beta}{2} = $ the right side of original

43. $\sin \theta = - \dfrac{\sqrt{3}}{2}$ has solutions in quadrants three and four.

The reference angle is $\pi/3$ since $\sin \pi/3 = \dfrac{\sqrt{3}}{2}$. Placing $\pi/3$ in quadrants III and IV involves

adding it to π and subtracting it from 2π to give the answers $4\pi/3$ and $5\pi/3$.

45. $\cot 3\alpha = - 1$ is the same as $\tan 3\alpha = \dfrac{1}{-1} = - 1$ and this occurs if 3α terminates in QII or IV.

The reference number will be $\dfrac{\pi}{4}$ and values for 3α will be allowed as long as they're less than

6π because in the act of solving for α we'll be dividing by 3:

$3\alpha = 3\pi/4, 7\pi/4, 11\pi/4, 15\pi/4, 19\pi/4,$ or $23\pi/4$ (add π and 2π to 1st two answers)

so... $\alpha = \pi/12, 7\pi/12, 11\pi/12, 5\pi/4, 19\pi/12$ or $23\pi/12$

47. $\sin \theta = 3/4 = 0.75$ allows for θ to be in QI or QII

The reference angle is inv sin of 0.75 or 0.8481 which is the QI answer.

In QII it's π - 0.8481 or 2.2935

49. $2 \sin x + \tan x = 0$ is the same as $2 \sin x + \frac{\sin x}{\cos x} = 0$ and if we assume that $\cos x \neq 0$ we can
multiply through by it:
$$2 \sin x \cos x + \sin x = 0$$
$$\sin x \ (2 \cos x + 1) = 0$$

Solving $\sin x = 0$:

occurs at $x = 0$, $x = \pi$, and $x = 2\pi$

Solving $2 \cos x + 1 = 0$:

$\cos x = $ -1/2 has solutions in QII and QIII

the reference angle is $\pi/3$ because $\cos \pi/3 = 1/2$

In QII we would have $x = \pi$ - $\pi/3$ or $2\pi/3$

In QIII it's $x = \pi + \pi/3$ or $4\pi/3$

Note: none of these answers violate our original assumption

51. $\cos^2\theta$ - $3 \cos \theta$ - $4 = 0$ factors into $(\cos \theta$ - $4)(\cos \theta + 1) = 0$

$\cos \theta$ - $4 = 0$ requires $\cos \theta = 4$ which is impossible.

$\cos \theta + 1 = 0$ implies $\cos \theta = $ - 1 and this occurs if $\theta = \pi$.

53. $\cos x = \cos \frac{x}{2}$ is the same as $\cos 2\left(\frac{x}{2}\right) = \cos \frac{x}{2}$ which allows a double angle formula to be used:
$$2 \cos^2 \frac{x}{2} - 1 = \cos \frac{x}{2} \quad \text{or} \quad 2 \cos^2 \frac{x}{2} - \cos \frac{x}{2} - 1 = 0$$

which factors into $(2 \cos \frac{x}{2} + 1)(\cos \frac{x}{2}$ - $1) = 0$

$2 \cos \frac{x}{2} + 1 = 0$ is the same as $\cos \frac{x}{2} = $ - $\frac{1}{2}$ (answers for $\frac{x}{2}$ in Q's II and III)

The reference angle is $\pi/3$ since $\cos \pi/3 = \frac{1}{2}$ so the answers for $\frac{x}{2}$ are $2\pi/3$ and $4\pi/3$. Doubling
these gives answers for x of $4\pi/3$ and $8\pi/3$ but $8\pi/3$ is bigger than 2π and need not be listed.

From the other factor:

$\cos \frac{x}{2}$ - $1 = 0$ if $\cos \frac{x}{2} = 1$ which occors at $\frac{x}{2} = 0$ or 2π which would force x to be either 0 or 4π
with 4π also being too big to list.

Answers for x: 0 and $4\pi/3$

55. sketch y = 2 sin 3t

   Amplitude: 2 (Horizontal dotted guidelines at y = 2 and at y = -2)

   Period: (normal period for sin) ÷ (coefficient of t) = $2\pi/3$

   Phase shift: set 3t = 0 ; solve for t ; get 0 (no phase shift)

   Sketch: fit a sine curve with amplitude 2 into the interval [0, $2\pi/3$]

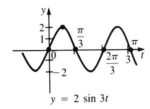

$y = 2 \sin 3t$

57. $y = 3 \cos\left(2t - \dfrac{\pi}{6}\right)$ Amplitude: 3   Period: $2\pi/2 = \pi$

   Phase shift: set $2t - \dfrac{\pi}{6} = 0$; $2t = \dfrac{\pi}{6}$; $t = \dfrac{\pi}{12}$ (≈ 0.26)

   Sketch: Fit a cosine curve with amplitude 3 into the interval [$\pi/12$, $\pi/12 + \pi$]
        or approximately the interval [0.26, 3.4]

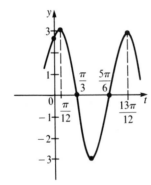

$y = 3 \cos\left(2t - \dfrac{\pi}{6}\right)$

59. $4 \cos\dfrac{t}{2} + 3 \sin\dfrac{t}{2} = 5 \cos\left(\dfrac{t}{2} - 0.64\right)$ if converted into the A cos ($\omega t - \delta$) form

   where $A = \sqrt{4^2 + 3^2} = 5$ and cos $\delta = 4/5$ and sin $\delta = 3/5$

   Now:

   Amplitude: 5   period: $2\pi \div 1/2$ (the coefficient of t) = 4π (≈ 12.56)

   Phase shift: set $\dfrac{t}{2} - 0.64 = 0$; $\dfrac{t}{2} = 0.64$; t = 1.28 (a starting point)

59. (continued)

Sketch: a cosine curve with amplitude 5 strectched out over the interval

[1.28, 1.28 + 4π] or [1.28, 13.84]

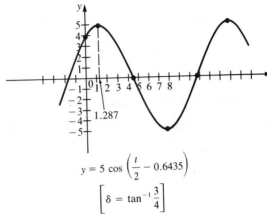

$$y = 5 \cos\left(\frac{t}{2} - 0.6435\right)$$

$$\left[\delta = \tan^{-1}\frac{3}{4} \right]$$

61. $y = 3 \sin \frac{\pi}{2}t + 2 \cos 2\pi t$ by addition of ordinates:

graph $3 \sin \frac{\pi}{2}t$: amp = 3 ; period = $2\pi \div \frac{\pi}{2} = 4$; phase shift = 0

graph $2 \cos 2\pi t$: amp = 2 ; period = $2\pi \div 2\pi = 1$; phase shift = 0

Sketch:

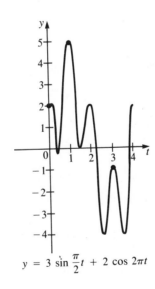

$y = 3 \sin \frac{\pi}{2}t + 2 \cos 2\pi t$

CHAPTER NINE
Further Applications of Trigonometry

Chapter Objectives

In chapter nine you will see even more applications of trigonometry. By virtue of two laws, the law of sines and the law of cosines, you will be able to solve non-right (oblique) triangles. Real world uses including surveying and navigation will be detailed. You will become acquainted with **vectors** and will see the variety of problems, mostly from physics, which can be solved using vector analysis. Lastly you will see an alternative to rectangular (x, y) coordinates called polar coordinates. You will graph polar equations and use the polar form of complex numbers in the finding of their powers and roots.

SUMMARY

- A, B, C denote angles and a, b, c denote sides of a triangle

- Law of Sines $\quad \dfrac{\sin A}{a} = \dfrac{\sin B}{b} = \dfrac{\sin C}{c}$

- Law of Cosines $\quad a^2 = b^2 + c^2 - 2bc \cos A$

$$\cos A = \frac{b^2 + c^2 - a^2}{2bc}$$

- Heron's Formula

  Area of triangle ABC $= \sqrt{s(s - a)(s - b)(s - c)} \quad$ where $s = \frac{1}{2}(a + b + c)$

- A directed line segment from P to Q, denoted by $\overline{PQ}$, is the straight line segment that extends from P to Q.

- A vector **v** is a collection of directed line segments.

  If one representation starts at (0,0) and extends to (a,b), then the vector can be denoted by **v** = (a,b).

- The magnitude of **v** = (a,b) is $|\mathbf{v}| = \sqrt{a^2 + b^2}$

- The direction of **v** is the angle θ in standard position that the vector makes with the positive x axis. By definition, $0 \leq \theta < 2\pi \quad (\, 0° \leq \theta < 360° \,)$

- Vector Algebra addition: $(a_1, b_1) + (a_2, b_2) = (a_1 + a_2,\ b_1 + b_2)$

  scalar multiplication: $\alpha(a, b) = (\alpha a, \alpha b)$

- In R^2, let $i = (1,0)$ and $j = (0,1)$. Then $v = (a,\ b)$ can be written $ai + bj$

- A unit vector u in R^2 is a vector that satisfies $|\ u\ | = 1$. In R^2, a unit vector can be written

  $u = (\ \cos\theta\ )\ i + (\ \sin\theta\ )\ j$ where θ is the direction of u.

- Let $u = (a_1,\ b_1)$ and $v = (a_2,\ b_2)$. The dot product of u and v, written $u \cdot v$ is given by

  $u \cdot v = a_1 a_2 + b_1 b_2$

- Properties of the Dot Product in R^2. For any vectors u, v, w, and scalar α,

  (i) $u \cdot v = v \cdot u$ (ii) $(\ u + v\ ) \cdot w = u \cdot w + v \cdot w$ (iii) $(\alpha u) \cdot v = \alpha(u \cdot v)$

  (iv) $u \cdot u \geq 0$ and $u \cdot u = 0$ if and only if $u = 0$ (v) $|u|^2 = u \cdot u$

- The angle ϕ between two vectors u and v in R^2 is the unique number in $[0, \pi]$ that satisfies

  $\cos\phi = \dfrac{u \cdot v}{|u|\ |v|}$

- Two vectors in R^2 are parallel if the angle between them is 0 or π.

- Two vectors in R^2 are orthogonal if the angle between them is $\dfrac{\pi}{2}$.

- Let u and v be two nonzero vectors in R^2. Then the projection of u on v is a vector, denoted by

  $\text{Proj}_v u = \dfrac{u \cdot v}{|v|^2}\ v$

- The vector $\dfrac{u \cdot v}{|v|}$ is called the component of u in the direction v.

- Polar coordinates

$P = (r,\ \theta)$ where r is the distance from the origin to the point and θ is the angle from the polar axis to OP. We generally have $r > 0$ and $0 \leq \theta < 2\pi$.

- Conversion Formulas

  From polar to rectangular coordinates   From rectangular to polar coordinates

  $x = r \cos \theta \qquad y = r \sin \theta$ $r = \sqrt{x^2 + y^2}$ and $\tan \theta = \frac{y}{x}, \quad x \neq 0$

  also: $(-r, \theta) = (r, \theta + \pi)$

- If $z = a + bi$ is a complex number, then the **absolute value** of $z = |z| = \sqrt{a^2 + b^2}$.
- The **argument** of $z = a + bi$, denoted by arg z, is the angle θ between Oz and the positive real axis. $\tan \theta = \frac{b}{a}$ and $-\pi < \arg z \leq \pi$.
- Polar form of a complex number:

  $z = r(\cos \theta + i \sin\theta) \qquad$ where $r = |z|$ and $\theta = \arg z$

- Complex algebra

  $z\bar{z} = |z|^2 \quad$ and $\bar{z} = r(\cos \theta - i \sin \theta)$

  $z_1 z_2 = |z_1| |z_2| [\cos (\arg z_1 + \arg z_2) + i \sin (\arg z_1 + \arg z_2)]$

  $\dfrac{z_1}{z_2} = \dfrac{|z_1|}{|z_2|} [\cos (\arg z_1 - \arg z_2) + i \sin (\arg z_1 - \arg z_2)$

- De Moivre's Theorem

  $[r(\cos \theta + i \sin \theta)]^n = r^n(\cos n\theta + i \sin n\theta)$

- nth roots of a complex number z

  are $w_1, w_2, \ldots, w_n$ where

  $w_k = r^{1/n} [\cos \left(\dfrac{\theta + 2k\pi}{n}\right) + i \sin \left(\dfrac{\theta + 2k\pi}{n}\right)]$, $k = 0, 1, 2, \ldots, n - 1$

  $r = |z|$ and $\theta = \arg z$

<u>Problems 9.1</u>

1. $C = (180 - 80 - 35)° = 65°$

   using the law of sines $\dfrac{a}{\sin A} = \dfrac{b}{\sin B} = \dfrac{c}{\sin C}$ we have $\dfrac{a}{\sin 35°} = \dfrac{b}{\sin 80°} = \dfrac{6}{\sin 65°}$

   or $\dfrac{a}{0.5736} = \dfrac{b}{0.9848} = \dfrac{6}{0.9063}$ from the rightmost proportion we get $0.9063b = 5.9088$

   so $b = 6.5197$

   from the outer proportion we get $0.9063a = 3.4416$

   so $a = 3.7974$

3. $C = (180 - 15 - 100)° = 65°$

   $\dfrac{a}{\sin 100°} = \dfrac{170}{\sin 15°} = \dfrac{c}{\sin 65°}$ gives the facts that $a = \dfrac{170 \sin 100°}{\sin 15°} = 646.85$

   $c = \dfrac{170 \sin 65°}{\sin 15°} = 595.29$

5. $\dfrac{13}{\sin 40°} = \dfrac{b}{\sin B} = \dfrac{6}{\sin C}$; from the outer members of this extended proportion we see that

   $\sin C = \dfrac{6 \sin 40°}{13} = 0.2967$: (now use INV sin)

   $C = 17.25°$: now we can find B

   $B = (180 - 40 - 17.25)° = 122.75°$

   $b = \dfrac{13 \sin 122.75°}{\sin 40°} = 17$

   (note: C could also theoretically be $162.25°$ $(180 - 17.25)$ but then A being $40°$ makes this impossible)

7. $\dfrac{10}{\sin 110°} = \dfrac{b}{\sin B} = \dfrac{6}{\sin C}$ $\Rightarrow$ $\sin C = \dfrac{6 \sin 110°}{10} = 0.5638$ and so $C = $ inv sin $0.5638 \approx 34.3°$

   Note: C could also be considered a second quadrant angle with a reference angle of $34.3°$. This
   gives us another possibility for C, namely, $(180 - 34.3)°$ or $145.7°$ but an angle this large
   can't be in a triangle that is known to have an angle of measure $110°$ in it!

   If $C = 34.3°$ then B must equal $(180 - 110 - 34.3)° = 35.7°$

   In this case b can now be found from the leftmost proportion above: $\dfrac{10}{\sin 110°} = \dfrac{b}{\sin 35.7°}$

   which implies than $b = \dfrac{10 \sin 35.7°}{\sin 110°} \approx 6.210$

9.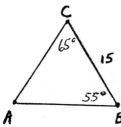

$A = (180 - 55 - 65)° = 60°$

$$\frac{15}{\sin 60°} = \frac{b}{\sin 55°} = \frac{c}{\sin 65°}$$

$$b = \frac{15 \sin 55°}{\sin 60°} = 14.19 \; ; \quad c = \frac{15 \sin 65°}{\sin 60°} = 15.70$$

11.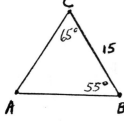

$B = (180 - 10 - 105)° = 65°$

$$\frac{a}{\sin 10°} = \frac{b}{\sin 65°} = \frac{150}{\sin 105°}$$

$$a = \frac{150 \sin 10°}{\sin 105°} = 26.97 \; ; \quad b = \frac{150 \sin 65°}{\sin 105°} = 140.74$$

13.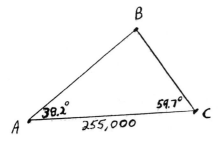

$B = (180 - 38.2 - 59.7)° = 82.1°$

$$\frac{a}{\sin 38.2°} = \frac{255,000}{\sin 82.1°} = \frac{c}{\sin 59.7°}$$

$$a = \frac{255,000 \sin 38.2°}{\sin 82.1°} = 159,205$$

$$c = \frac{255,000 \sin 59.7°}{\sin 82.1°} = 222,275$$

15.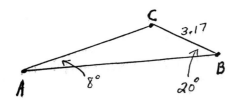

$C = (180 - 8 - 20)° = 152°$

$$\frac{3.17}{\sin 8°} = \frac{b}{\sin 20°} = \frac{c}{\sin 152°}$$

$$b = \frac{3.17 \sin 20°}{\sin 8°} = 7.79 \; ; \quad c = \frac{3.17 \sin 152°}{\sin 8°} = 10.69$$

17. $\dfrac{6}{\sin 110°} = \dfrac{b}{\sin B} = \dfrac{5}{\sin C}$ if solved for sin C gives $\sin C = \dfrac{5 \sin 110°}{6} = 0.7831$ so C could either be $\sin^{-1} 0.7831$ which is about 51.5° or its supplement 128.5°. The supplement has to be ruled out because the 180° sum for angles of a triangle would be violated . Therefore, there is exactly one triangle with sides a = 6, c = 5, and A = 110°. For the record, B = 180°- 110° - 51.5° = 18.5° and $b = \dfrac{6 \sin 18.5°}{\sin 110°} = 2.026$

19. $\dfrac{6}{\sin 58°} = \dfrac{10}{\sin B} = \dfrac{c}{\sin C} \quad \Rightarrow \quad \sin B = \dfrac{10 \sin 58°}{6} = 1.4134$ which is impossible so NO TRIANGLE

21. $\dfrac{30}{\sin A} = \dfrac{40}{\sin 27°} = \dfrac{c}{\sin C} \quad \Rightarrow \quad \sin A = \dfrac{30 \sin 27°}{40} = 0.3405$ so either $A = 19.9°$ or $160.1°$ (too big)

   now we can find C: $C = (180 - 27 - 19.9)° = 133.1°$

   now we can find c: $c = \dfrac{40 \sin 133.1°}{\sin 27°} = 64.33$

23. $\dfrac{30}{\sin A} = \dfrac{20}{\sin 27°} = \dfrac{c}{\sin C} \quad \Rightarrow \quad \sin A = \dfrac{30 \sin 27°}{20} = 0.68099$ so A will either equal $42.9°$ or $137.1°$

   if $A = 42.9°$, then $C = (180 - 27 - 42.9)° = 110.1°$; finding c: $\dfrac{20}{\sin 27°} = \dfrac{c}{\sin 110.1°} \rightarrow c \approx 41.4$

   if $A = 137.1°$, then $C = (180 - 27 - 137.1)° = 15.9°$; finding c: $\dfrac{20}{\sin 27°} = \dfrac{c}{\sin 15.9°} \rightarrow c \approx 12.1$

25. $\dfrac{a}{\sin A} = \dfrac{2}{\sin B} = \dfrac{3}{\sin 102°} \quad \Rightarrow \quad \sin B = \dfrac{2 \sin 102°}{3} = 0.6521 \quad \Rightarrow \quad B = 40.7°$ or $139.3°$ (too big)

   now finding A: $A = (180 - 102 - 40.7)° = 37.3°$

   now for a: $a = \dfrac{2 \sin 37.3°}{\sin 40.7°} = 1.86$

27. $\dfrac{100}{\sin 46.8°} = \dfrac{b}{\sin B} = \dfrac{150}{\sin C} \quad \Rightarrow \quad \sin C = \dfrac{150 \sin 46.8°}{100} = 1.0935 \quad$ NO WAY!

29. The law of sines says $\dfrac{10}{\sin 62°} = \dfrac{x}{\sin B}$

   a) for there to be no triangle we would have to have $\sin B > 1$ and from the above proportion we
   see that $\sin B = \dfrac{x \sin 62°}{10} = 0.08829\, x$

   so $\sin B > 1 \quad \Rightarrow \quad 0.08829\, x > 1$ and this will occur if $x > 11.3257$

   b) by part (ii) of the theorem on page 460 we have one triangle if $10 = x \sin 62°$ or $10 > x$
   with our x being their c, that is, if $x = 11.3257$ or $x < 10$.

29. (continued)

    c) by part (iii) of the same theorem: two triangles if x sin 62° < 10 < x

        solving the left side: x sin 62° < 10

$$0.88295 \, x < 10$$

$$x < 11.3257$$

    blending this with the right side gives us   10 < x < 11.3257

31. We know that the area of a triangle is given by $\dfrac{\text{base} \cdot \text{height}}{2}$ and if we use 3 as the base, we will need the height perpendicular to it which we can find using the sine function:

$$\sin 27° = \frac{\text{height}}{2} \;\Rightarrow\; \text{height} = 2 \sin 27° \approx 0.908$$

so area $= \dfrac{bh}{2} \approx \dfrac{(3)(0.908)}{2} = 1.362$ square units

33. We can't compute the height to the side 6 until we get the side opposite the 28° angle and we can't get that until we get the third angle which equals (180 - 81 - 28)° = 71°.
Now, by the law of sines we have $\dfrac{x}{\sin 28°} = \dfrac{6}{\sin 71°}$
so $x = \dfrac{6 \sin 28°}{\sin 71°} \approx 2.979$ (this is the side opposite 28°)

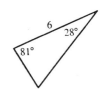

    Now we can calculate the height to the side that is 6:

$$\sin 81° = \frac{\text{height}}{2.979} \;\Rightarrow\; \text{height} = 2.979 \sin 81° \approx 2.942$$

So at last we can compute the area as $\dfrac{bh}{2} \approx \dfrac{(6)(2.942)}{2} = 8.827$ sq. units

35. Third angle is (180 - 67 - 31)° = 82°

    Use law of sines to compute length of side opposite 31°:

$$\frac{x}{\sin 31°} = \frac{7}{\sin 82°} \;\Rightarrow\; x = \frac{7 \sin 31°}{\sin 82°} \approx 3.6407$$

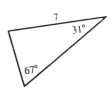

    Now for the height to side 7: $\sin 82° = \dfrac{\text{height}}{3.6407}$ and so

    height $\approx 3.6407 \sin 82° \approx 3.6053$

    Area $= \dfrac{bh}{2} \approx \dfrac{7(3.6053)}{2} = 12.6$ sq. units

37. If C is a right angle, sin C = sin 90° = 1 and the law of sines becomes $\frac{a}{\sin A} = \frac{b}{\sin B} = c$

to prove that $\frac{a}{\sin A} = c$: in the right triangle sin A $= \frac{opp}{hyp} = \frac{a}{c}$ which if solved for c yields

exactly what we wanted to prove (cross multiply and divide by sin A).

Now, to prove that $\frac{b}{\sin B} = c$, use the right triangle approach again to see that sin B $= \frac{opp}{hyp} = \frac{b}{c}$

which similarly can be solved for c giving c $= \frac{b}{\sin B}$ which completes what we had to show.

39. (i) from the law of sines we know that $\frac{a}{\sin A} = \frac{b}{\sin B}$

        cross multiplying yields b sin A = a sin B

        since sin B $\leq$ 1 for any B we have b sin A $\leq$ a · 1 which is the same as a $\geq$ b sin A

  (ii) now start with a sin B = b sin A as above and utilize the fact that sin A $\leq$ 1 to get

                 a sin B $\leq$ b   or b $\geq$ a sin B

 (iii) same idea but start with b sin C = c sin B

Problems 9.2

1. Let's name the unknown side a and the   using the law of cosines:

    angles A, B, and C as shown

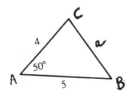

$$a^2 = 4^2 + 5^2 - 2 \cdot 4 \cdot 5 \cos 50°$$
$$= 16 + 25 - 40(0.6428)$$
$$= 15.288$$

so   a $\approx$ 3.91

now by the law of sines: $\frac{3.91}{\sin 50°} = \frac{4}{\sin B}$ which implies that sin B $= \frac{4 \sin 50°}{3.91} = 0.784$

so B $= \sin^{-1} 0.784 \approx 51.6°$ (or its supplement 128.4° --- but this is too big, right?)

if B = 51.6° then C = (180 - 50 - 51.6)° = 78.4°

3.

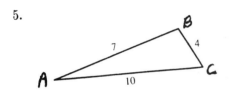

$a^2 = 180^2 + 160^2 - 2(180)(160) \cos 10°$

$= 32,400 + 25,600 - 57,600(0.9848)$

$= 1275.52 \quad$ so a ≈ 35.7

the law of sines gives us $\dfrac{35.7}{\sin 10°} = \dfrac{180}{\sin B}$ so $\sin B = \dfrac{180 \sin 10°}{35.7} = 0.8755$

so $B = \sin^{-1} 0.8755 \approx 61.1°$ (or its supplement $118.9°$)

If B were $61.1°$, then C would equal $180 - 10 - 61.1 = 108.9°$ and this would make C the largest angle in the triangle. But then C, the largest angle, should have the largest side opposite it. Since in the diagram 160, not the largest side, is opposite C, we have a contradiction if we insist on B being $61.1°$. Therefore, the correct answer for B must be $118.9°$ and this makes $C = 180 - 10 - 118.9 = 51.1°$

5.

$$\text{use } \cos A = \frac{b^2 + c^2 - a^2}{2bc} \text{ to get}$$

$$\cos A = \frac{100 + 49 - 16}{2(7)(10)} = 0.95$$

so $A = \cos^{-1} 0.95 \approx 18.2°$

The law of sines can now be used to set up a proportion to track down another angle:

$\dfrac{a}{\sin A} = \dfrac{b}{\sin B}$ becomes $\dfrac{4}{\sin 18.2°} = \dfrac{10}{\sin B}$ which implies that $\sin B = \dfrac{10 \sin 18.2°}{4} \approx 0.7808$

so one of the possibilities for B is $\sin^{-1} 0.7808 \approx 51.3°$ (the other possibility is $128.7°$)
By reasoning exactly like that used in problem 3 (side b is the longest side so B must be the largest angle) we must conclude that $B \approx 128.7°$ and then $C \approx (180 - 18.2 - 128.7)° = 33.1°$

7. Two sides of a triangle must always add to a result greater than the third side. This is the quick way to tell that this problem can't be solved. ($147 + 185 = 322$ which is not greater than 353). If you don't notice this you could try setting up the solution for cos A but you would get a value greater than 1 or less than -1 for cos A and the range of the cosine is $-1 \le \cos x \le 1$.

9.

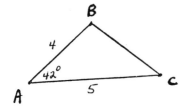

$a^2 = b^2 + c^2 - 2bc \cos A$ becomes

$a^2 = 25 + 16 - 40 \cos 42° = 11.274$

so $a \approx 3.36$ and now we can use the Law of Sines:

$$\frac{a}{\sin A} = \frac{c}{\sin C} \quad \text{becomes} \quad \frac{3.36}{\sin 42°} = \frac{4}{\sin C} \quad \text{so}$$

$$\sin C = \frac{4 \sin 42°}{3.36} = 0.7966 \Rightarrow C = 52.8° \text{ (or } 127.2°)$$

and using the 180° quota for triangles $B = 85.2°$ (or 10.8°)

11.

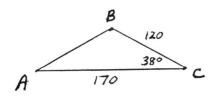

$c^2 = 120^2 + 170^2 - 2(120)(170) \cos 38°$

$= 11{,}149 \quad \text{so} \quad c \approx 105.59$

now using the law of sines we get:

$$\frac{105.59}{\sin 38°} = \frac{120}{\sin A} \quad \text{so } \sin A = 0.6997$$

so $A \approx 44.4°$

which forces B to be 97.6°

13.

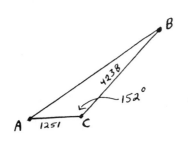

$c^2 = 1251^2 + 4238^2 - 2(1251)(4238) \cos 152°$

$= 1{,}565{,}001 + 17{,}960{,}644 + 9{,}362{,}313.6$

$= 28{,}887{,}958.6 \quad \text{so } c \approx 5{,}374.8$

from the law of sines:

$$\frac{5374.8}{\sin 152°} = \frac{1251}{\sin B} \quad \text{so } \sin B = \frac{1251 \sin 152°}{5374.8} = 0.1093$$

Therefore $B = 6.3°$ and $A = 21.7°$

15.

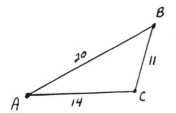

$$\cos A = \frac{b^2 + c^2 - a^2}{2bc} = \frac{196 + 400 - 121}{2(14)(20)} = 0.8482$$

so $A = \cos^{-1} 0.8482 \approx 32°$

switching now to the law of sines: $\frac{a}{\sin A} = \frac{b}{\sin B}$

becomes $\frac{11}{\sin 32°} = \frac{14}{\sin B}$ so $\sin B = \frac{14 \sin 32°}{11}$

which equals 0.6744 so $B \approx 42.4°$ which dictates that $C \approx 105.6°$

17.

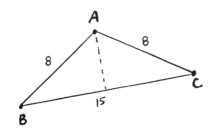

Since the triangle is isosceles, drop a perpendicular from the vertex down to the base and use right triangle trigonometry (the base is bisected, right?)

$\cos B = \frac{adj}{hyp} = \frac{7.5}{8} = 0.9375$ so $B \approx 20.4°$

Base angles of an isosceles triangle are equal so we know that $C \approx 20.4°$ also. That leaves 139.2° for A.

19. $s = \frac{1}{2}(a + b + c) = \frac{1}{2}(5 + 12 + 13) = 15$

area $= \sqrt{s(s - a)(s - b)(s - c)} = \sqrt{15(15 - 5)(15 - 12)(15 - 13)} = \sqrt{15 \cdot 10 \cdot 3 \cdot 2} = \sqrt{900} = 30$

21. Since the two smaller sides, 8 and 13, just sum to 21 rather than exceed it, these can't be the sides of a triangle and there is no area to be found.

23. $s = \frac{1}{2}(0.71 + 0.94 + 1.23) = 1.44$

area $= \sqrt{(1.44)(0.73)(0.5)(0.21)} = \sqrt{0.110376} = 0.33$

25. $s = \frac{1}{2}(3127 + 4183 + 2801) = 5055.5;$ area $= \sqrt{5055.5(1928.5)(872.5)(2254.5)} \approx 4,379,249.8$

27. we know that $a^2 = b^2 + c^2 - 2bc \cos A$

$$b^2 = a^2 + c^2 - 2ac \cos B$$

and $\qquad c^2 = a^2 + b^2 - 2ab \cos C \quad$ right??

The left side of the desired result contains $a^2 + b^2 + c^2$ so let's add the three equations above together:

$$a^2 + b^2 + c^2 = (b^2 + c^2 - 2bc \cos A) + (a^2 + c^2 - 2ac \cos B) + (a^2 + b^2 - 2ab \cos C)$$

$$= (2a^2 + 2b^2 + 2c^2) - 2bc \cos A - 2ac \cos B - 2ab \cos C$$

: subtract $(2a^2 + 2b^2 + 2c^2)$ from both sides

$$-a^2 - b^2 - c^2 = -2bc \cos A - 2ac \cos B - 2ab \cos C$$

: multiply both sides by (-1) and do a little rearranging...

$$a^2 + b^2 + c^2 = 2(ab \cos C + bc \cos A + ac \cos B)$$

29. use the result from problem 27 directly above as a starting point:

$$a^2 + b^2 + c^2 = 2ab \cos C + 2bc \cos A + 2ac \cos B$$

: divide both sides by 2abc to force the desired result

$$\frac{a^2 + b^2 + c^2}{2abc} = \frac{2ab \cos C + 2bc \cos A + 2ac \cos B}{2abc}$$

$$= \frac{\cos C}{c} + \frac{\cos A}{a} + \frac{\cos B}{b} \quad \text{by doing the individual divisions}$$

31. let $x =$ the length of any one of the sides

then $s = \frac{1}{2}(x + x + x) = \frac{3x}{2}$ and the expression for the area would be $\sqrt{\frac{3x}{2}\left(\frac{3x}{2} - x\right)^3}$

this simplifies to $\sqrt{\frac{3x^4}{16}} = \frac{x^2 \sqrt{3}}{4}$ and the area is to be 25 so we have the equation

$$25 = \frac{x^2 \sqrt{3}}{4}$$

$$100 = x^2 \cdot \sqrt{3}$$

$$57.735 \approx x^2$$

$$7.6 \approx x \ (\text{ the length of one side })$$

33. Show that $\dfrac{a - b}{c} = \dfrac{\sin\left(\dfrac{A - B}{2}\right)}{\cos\left(\dfrac{C}{2}\right)}$

Let's assume that $\angle A$ is greater than $\angle B$ and then as a consequence $a > b$.

Place the triangle with C at the origin and B on the positive x axis. Select point D down the x-axis at a distance of b from the origin. Then DB = a - b right?

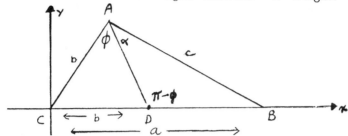

By the law of sines $\dfrac{\sin\alpha}{a - b} = \dfrac{\sin(\pi - \phi)}{c}$ or equivalently $\dfrac{a - b}{c} = \dfrac{\sin\alpha}{\sin(\pi - \phi)}$

so at this point we have formed the left side of the statement we're trying to prove.

Now we have to work on that right side: $\alpha + \phi = A$ and $\phi = \alpha + B$ if solved simultaneously for α give $\alpha = \dfrac{A - B}{2}$ which fits nicely into the top of the right side and makes it match the top of the right side of what we're proving. The last work involves $\sin(\pi - \phi)$:

$\sin(\pi - \phi) = \sin(\pi/2 + \pi/2 - \phi) = \cos(\pi/2 - \phi)$

But $C + 2\phi = \pi$ because $\triangle ACD$ is isosceles and the angles of a triangle add to π radians.

so $C = \pi - 2\phi$ which means that $\dfrac{C}{2} = \pi/2 - \phi$ and therefore $\cos(\pi/2 - \phi) = \cos\dfrac{C}{2}$

and we have our conclusion since $\cos(\pi/2 - \phi)$ is equal to $\sin(\pi - \phi)$.

35. Consider the simpler diagram on the right first; it's easy to see that the Pythagorean theorem justifies the fact that $AB^2 = AP^2 + PB^2$ and we will try to find alternate names for each of these terms by checking out the more complicated triangle on the left:

AB: since this side is opposite angle C in what must have been the original triangle, c is certainly a valid replacement name for AB.

AP: $\overline{AP}$ was formed apparently by dropping a perpendicular from A to $\overline{CB}$; in right triangle APC we have $\sin\theta = \dfrac{\text{opp}}{\text{hyp}} = \dfrac{AP}{b}$ which implies that $AP = b\sin\theta$.

PB: by simple subtraction we can see that PB = CB - CP; it's also true that CB is opposite angle A so its alternate name is a; right triangle trigonometry tells us that $CP = b\cos\theta$. Substituting these two results yields $PB = a - b\cos\theta$.

Making all these three substitutions into $AB^2 = AP^2 + PB^2$ yields

$c^2 = (b\sin\theta)^2 + (a - b\cos\theta)^2 = b^2\sin^2\theta + a^2 - 2ab\cos\theta + b^2\cos^2\theta$

$= a^2 + b^2(\sin^2\theta + \cos^2\theta) - 2ab\cos\theta = a^2 + b^2 - 2ab\cos\theta$ which is the law of cosines

1. $\overline{AB}$ could also be called c and by the law of cosines:

$$c^2 = a^2 + b^2 - 2ab \cos C$$

$$= 87.8^2 + 75.2^2 - 2(87.8)(75.2) \cos 38.77°$$

$$= 3068.30$$

so   c ≈ 55.4 ft

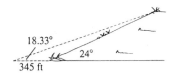

3. We seek $\overline{BC}$ or a in the drawing at the right.

$\overline{AB}$ or c is given as 345 ft

∠ ABC is 156° because of the fact that it is

the supplement to 24°.

∠ ACB = 5.67° because a triangle contains 180°

Use the law of sines: $\dfrac{a}{\sin A} = \dfrac{c}{\sin C}$

$$\frac{a}{\sin 18.33°} = \frac{345}{\sin 5.67°}$$

$$a = \frac{345 \sin 18.33°}{\sin 5.67°} \approx 1098 \text{ ft}$$

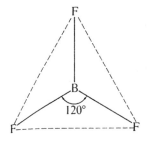

5. In the triangle at the right we seek the value of x

knowing each of the other two sides is 1.29 angstroms

and the angle included between the two sides is 120°.

(2 sides + included angle ⇒ law of cosines)

$$x^2 = 1.29^2 + 1.29^2 - 2(1.29)(1.29) \cos 120°$$

$$= 1.6641 + 1.6641 + 1.6641 = 4.9923$$

so x ≈ 2.23 angstroms

7. Assign $\overline{HS}$ a convenient number, say 1, since were only

interested in the ratio of $\overline{HH}$ to $\overline{HS}$. Now find $\overline{HH}$ using

the law of cosines: (call $\overline{HH}$ h for short)

$$h^2 = 1^2 + 1^2 - 2(1)(1) \cos 92°$$

$$= 2.0698$$

so h ≈ 1.439 and the ratio is 1.439 to 1

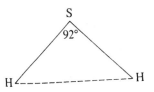

9. Make a simplified drawing and show all given information.

Then start deducing measures of other angles:

$\angle$ CAD $= 59.8° - 44.58° = 15.22°$

$\angle$ EAD $= 120.2°$ (supp. to $59.8°$)

$\angle$ EDA $= 13.38°$

We'll have to start working in $\triangle$EAD since we know one of its sides. If we could find $\overline{AD}$ we might then be able to track down $\overline{BD}$ or $\overline{CD}$ or $\overline{BC}$:

use law of sines since only one side is known:

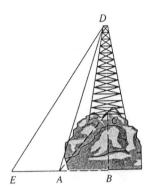

$$\frac{\overline{AD}}{\sin 46.42°} = \frac{55}{\sin 13.38°} \quad \text{so } \overline{AD} = 172.2 \text{ m}$$

Idea: Now that we know $\overline{AD}$ and $\angle$ BAD we could find $\overline{AB}$ using simple right triangle relationship

$$\cos 59.8° = \frac{\overline{AB}}{172.2} \quad \text{so } \overline{AB} = 86.6 \text{ m} \quad \text{and now the door is opened to work in } \triangle\text{'s ABC and}$$
ABD to find $\overline{BC}$ and $\overline{BD}$ and afterwards deduce $\overline{DC}$.

in $\triangle$ABC, $\overline{BC}$ (the hill) can be found as follows:

$$\tan 44.58° = \frac{\overline{BC}}{86.6} \quad \text{so } \overline{BC} = 85.3 \text{ m}$$

(end of part a)

in $\triangle$ABD $\tan 59.8° = \frac{\overline{BD}}{86.6}$ so $\overline{BD}$ (hill and tower) $= 148.8$ m

so finally, the tower itself, $\overline{CD}$, $= 148.8 - 85.3 = 63.5$ m

11. Plan: use right $\triangle$ADB to find $\angle$A; then find $\angle$C; then use law of sines in $\triangle$ACB to find $\overline{AC}$

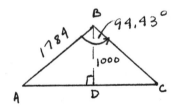

$$\sin A = \frac{1000}{1784} = 0.5605 \quad \text{so } A = \sin^{-1}0.5605 = 34.09°$$

$$C = (180 - 34.09 - 94.43)° = 51.48°$$

$$\frac{\overline{AC}}{\sin 94.43°} = \frac{1784}{\sin 51.48°} \quad \text{so } \overline{AC} = 2274 \text{ km}$$

13. The angle of 5.5° is the offset of the tower from the vertical. The sun's angle can be shown at the end of the shadow (the first ray not blocked by the tower makes an angle of 52° with the ground).

so ∠ LTP = 84.5° and ∠ L = 52° and $\overline{PT}$ = 55.6 m

need ∠ LPT before you can use law of sines: ∠ P = (180 - 52 - 84.5)° = 43.5°

$$\frac{\overline{LT}}{\sin 43.5°} = \frac{55.6}{\sin 52°}$$ so $\overline{LT}$ = 48.6 m (the shadow)

15. Three sides are known and ∠ TGR (if it's acute) is sought. Plan: use law of cosines to find angle TGR:

$$\cos \angle TGR = \frac{r^2 + t^2 - g^2}{2rt} = \frac{49 + 112.36 - 144}{2(7)(10.6)}$$

$$= 0.11698$$

so ∠ TGR = $\cos^{-1}$ 0.11698 = 83.28° = 83°17'

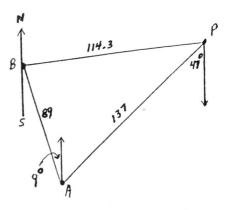

17. The answer to #16 is 114.25 km (by law of cosines with ∠ A = 56°) To return to port the ship would have to travel up BP in the drawing. Its bearing will depend on ∠ NBP: ∠ SBA = 9° because it's an alternate interior angle to the angle which corresponds to the second bearing. ∠ ABP can be found from the law of sines:

$$\frac{137}{\sin \angle ABP} = \frac{114.25}{\sin 56°} \Rightarrow \sin \angle ABP = 0.9941$$

so ∠ ABP = 83.774°

Since ∠ SBA + ∠ ABP + ∠ NBP = 180° ∠ NBP = (180 - 9 - 83.774)° = 87.226° = 87°14'
and the direction will be east of north or N 87°14' E.

19. The ship is now at C and must travel down CP to return to port. The correct answer to #18 for PC is 139.3 m which necessitated finding ∠ PBC to be 36.226° (law of sines). Now we can use the law of sines in △ BPC to find ∠ BCP which we will subtract from 51° (∠ SCB) to determine ∠ SCP which will be the bearing (west of south) .

$$\frac{114.3}{\sin \angle BCP} = \frac{139.3}{\sin 36.226°} \Rightarrow \sin \angle BCP = 0.4848$$

so ∠ BCP = 28.995° which leaves (51 - 28.995)° or 22.005° or 22° for ∠ SCP and this is the bearing in the direction west of south or S 22° W

21. Let x = the length of the third side.
By the law of cosines:

$$x^2 = 85^2 + 63^2 - 2(85)(63) \cos 104°$$
$$= 7225 + 3969 - 2(85)(63)(-0.2419)$$
$$= 13784.98$$
so x = 117.4 ft

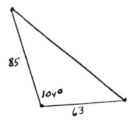

We can use Heron's formula to find the area: Area = $\sqrt{s(s - a)(s - b)(s - c)}$
For this particular triangle we can let a = 63, b = 85, and c = 117.4 so s (half their sum) = 132.7
Therefore, area = $\sqrt{132.7(132.7 - 63)(132.7 - 85)(132.7 - 117.4)}$ ≈ 2,598.1 sq. ft.

23. The distance from Earth to the sun is still 93 million miles and the distance from Venus to the sun is 67 million miles. We are asked to find ∠ E in the drawing; use law of cosines since we know all three sides:

$$\cos \angle E = \frac{s^2 + v^2 - e^2}{2sv}$$

$$= \frac{100^2 + 93^2 - 67^2}{2(100)(93)} = 0.7613$$

so ∠ E = $\cos^{-1} 0.7613 = 40.4°$

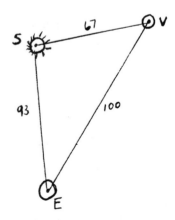

25. F = first lighthouse; S = second lighthouse; B = the boat (ship)

∠ F = 98° and ∠ S = 79° from given information so ∠ B = 3°

knowing three angles and one side permits using the law of sines:

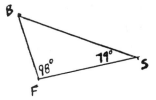

$$\frac{f}{\sin 98°} = \frac{58}{\sin 3°} = \frac{s}{\sin 79°}$$

solving for f yields $f = \frac{58 \sin 98°}{\sin 3°} = 1097.4$ km (boat from second lighthouse)

solving for s gives $s = \frac{58 \sin 79°}{\sin 3°} = 1087.9$ km (boat from first lighthouse, the closer one)

27. A = the airplane; P and Q are the markers; X marks the spot directly beneath the airplane.

$\overline{AX}$ is the altitude we seek; to find this we'll need $\overline{AQ}$.

∠ AQP = 109° so ∠ QAP = 14° (supplement and 180° △ fact)

now, by the law of sines:

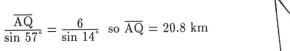

$$\frac{\overline{AQ}}{\sin 57°} = \frac{6}{\sin 14°} \text{ so } \overline{AQ} = 20.8 \text{ km}$$

now, by right △ trig in △ AXQ: $\sin 71° = \frac{\text{opp}}{\text{hyp}} = \frac{\overline{AX}}{20.8}$

so the altitude, $\overline{AX}$, = 20.8(sin 71°) = 19.7 km

29. Let S = ship at noon; R = radar; A = ship at 12:20;

B = ship at 1 PM (hope you did the previous problem!)

From #28 we know that RA = 64.84 km, ∠ BAR = 136.67°

SA = 5.67 (1/3 of 17) ; AB = 11.33 (2/3 of 17)

RB can be found using the law of cosines and then ∠ B,

the bearing, can be found using the law of sines.

$$\overline{RB}^2 = 64.84^2 + 11.33^2 - 2(64.84)(11.33) \cos 136.67° = 5401$$

so $\overline{RB} = 73.49$ km

now using the law of sines: $\frac{64.84}{\sin B} = \frac{73.49}{\sin 136.67°}$ so sin B = 0.6550 so ∠ B = 37.26°

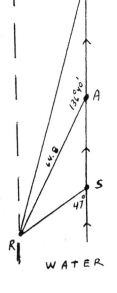

The direction is clearly west of south so the final bearing is S 37°16' W

31. In the drawing let L = the lighthouse, H = the ship, and B = the base

    To find the bearing we need to determine ∠ NHB

    on the left side of the figure. A way we can do

    this is to find ∠ NHL (on the left) and ∠ BHL

    and subtract them.

    ∠ NHL = ∠ SLH (alt int ∠'s)

          = 36°51' (a given bearing)

    ∠ BHL is tougher; first find HB (law of cosines)

       and then you can use law of sines to get to ∠ BHL:

    $HB^2 = 286^2 + 406^2 - 2\ (286)(406)\cos 135°47'$

    [Note: the 135°47' is ∠ HLB (180 - ∠ NLB - ∠SLH)]

         = 81,796 + 164,836 + 166,442.5

         = 413,074.5 so HB = 642.7 km

    Now switch to law of sines:

    $$\frac{406}{\sin \angle LHB} = \frac{642.7}{\sin 135°47'} \Rightarrow \sin \angle LHB = 0.4405 \Rightarrow \angle LHB = 26.138° \text{ or } 26°8'$$

    which means that ∠ NHB = ∠ NHL - ∠ LHB = 36°51' - 26°8' = 10°43' in the

    direction east of north or N 10°43' E

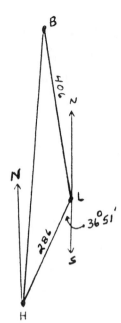

33. Let O = her original position; F = point of her first change of direction; C = second change

    We must find $\overline{OC}$ (f) and the bearing she must take.

    ∠ F = 49.433° (alt. int. ∠'s and subtraction)

    Now we can use law of cosines to find f:

    $f^2 = 5^2 + 8^2 - 2(5)(8)\cos 49.433°$ (in 1000's)

        = 36.973   so f = 6.081 (1000) or 6081 m $(\overline{OC})$

    Now for the bearing of O from C:

       It's obviously west of south right?

       Put in an auxiliary line pointing south (S)

    The angle we need to write the bearing is ∠ OCS

    We'll find it by finding ∠ OCF and ∠ FCS and subtracting them:

    by the law of sines $\frac{6081}{\sin 49.433°} = \frac{8000}{\sin \angle OCF}$ so ∠ OCF = 87.97° or its supplement 92.03°

    the supplement is the correct one as could be verified by the law of cosines. ∠ SCF = 32°7'

    by alternate interior angles so our bearing is 92°2' - 32°7' or S 59°55' W

Problems 9.4

1. Magnitude $= \sqrt{a^2 + b^2} = \sqrt{1^2 + 3^2} = \sqrt{10}$

   direction (1st quadrant formula) $= \tan^{-1}\frac{b}{a} = \tan^{-1}\frac{3}{1} \approx 71.6°$ or 1.25 radians

   i,j form: $1i + 3j$

3. Magnitude $= \sqrt{1^2 + (-3)^2} = \sqrt{10}$

   direction (4th quadrant formula) $= 2\pi + \tan^{-1}\frac{b}{a} = 2\pi + \tan^{-1}(-3) \approx 360° + (-71.6°) = 288.4°$

   other form: $i - 3j$

5. $a = 5 - 2 = 3$ and $b = 9 - 4 = 5$

   Magnitude $= \sqrt{3^2 + 5^2} = \sqrt{34}$

   direction (1st quadrant) $= \tan^{-1}\frac{5}{3} \approx 59°$ or 1.03 radians

   i,j form: $3i + 5j$

7. $a = -3 - 2 = -5$ and $b = 5 - 1 = 4$

   Magnitude $= \sqrt{(-5)^2 + 4^2} = \sqrt{41}$

   direction (2nd quadrant formula) $= \pi + \tan^{-1}\left(\frac{4}{-5}\right) \approx 180° + (-38.7°) = 141.3°$ or 2.467 radians

   i,j form: $-5i + 4j$

9. $a = -3 - (-3) = 0$ and $b = 2 - (-2) = 4$

   Magnitude $= \sqrt{0^2 + 4^2} = 4$

   direction $= \frac{\pi}{2}$ by observation (the tangent would be undefined at this angle)

   i,j form: $0i + 4j$

11. If $v = (2, 5)$ and vector PQ is to be a representation

    of v and P is the point $(1, -2)$, the coordinates of Q can

    be found by "adding" v to P: $Q = (1 + 2, -2 + 5) = (3, 3)$

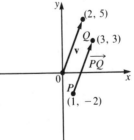

13. $Q = v + P = (7 + (-3), -3 + 7) = (4, 4)$

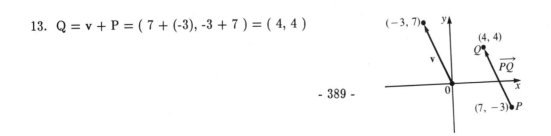

15. $Q = P + v = (-7 + 5, -2 + (-3)) = (-2, -5)$

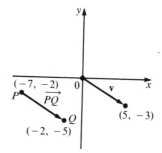

17. $v = (4, 4)$ has magnitude $\sqrt{a^2 + b^2} = \sqrt{4^2 + 4^2} = \sqrt{32} = 4\sqrt{2}$

v is clearly in the first quadrant so its direction, θ, is $\tan^{-1}\frac{b}{a} = \tan^{-1}\frac{4}{4} = \tan^{-1} 1 = \frac{\pi}{4}$.

19. $v = (4, -4)$ has magnitude $\sqrt{(4)^2 + (-4)^2} = \sqrt{32} = 4\sqrt{2}$ and is in the fourth quadrant. Therefore,

$$\theta = 2\pi + \tan^{-1}\frac{(-4)}{(4)} = 2\pi + \tan^{-1}(-1) = 2\pi + \left(-\frac{\pi}{4}\right) = \frac{7\pi}{4}$$

21. $v = (\sqrt{3}, 1)$ has magnitude $\sqrt{\sqrt{3}^2 + 1^2} = \sqrt{3 + 1} = 2$

since it is in the first quadrant its direction is $\theta = \tan^{-1}\frac{1}{\sqrt{3}}$ or $\tan^{-1}\frac{\sqrt{3}}{3} = \frac{\pi}{6}$ (from memory, right?)

23. $v = (-1, \sqrt{3})$ has magnitude 2 computed nearly identically to problem 21 above. Since v is in the

second quadrant $\theta = \pi + \tan^{-1}\frac{b}{a} = \pi + \tan^{-1}\frac{\sqrt{3}}{(-1)} = \pi + \tan^{-1}(-\sqrt{3}) = \pi + \left(-\frac{\pi}{3}\right) = \frac{2\pi}{3}$

25. $v = (-1, -\sqrt{3})$ has magnitude of 2 again (see problem 21) and is in quadrant III. Its direction

is given by $\theta = \pi + \tan^{-1}\frac{b}{a} = \pi + \tan^{-1}\frac{(-\sqrt{3})}{(-1)} = \pi + \tan^{-1}\sqrt{3} = \pi + \frac{\pi}{3} = \frac{4\pi}{3}$

27. $v = (-5, 8)$ has magnitude $\sqrt{(-5)^2 + 8^2} = \sqrt{25 + 64} = \sqrt{89}$ and is located in quadrant II.

$\theta = \pi + \tan^{-1}\frac{b}{a} = \pi + \tan^{-1}\frac{8}{(-5)} = \pi + \tan^{-1}(-1.6) \approx 180° + (-58°) = 122°$ or 2.13 radians

29. $a = x_Q - x_P = 1 - 1 = 0$
    $b = y_Q - y_P = 3 - 2 = 1$
    so $v = ai + bj = j$

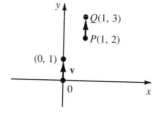

31. $a = -1 - 5 = -6$ and $b = 3 - 2 = 1$

$v = -6i + j$

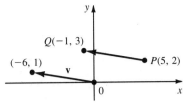

33. $a = -2 - 7 = -9$ and $b = 4 - (-1) = 5$

$v = -9i + 5j$

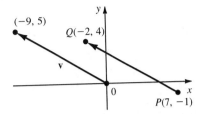

35. $a = -8 - (-3) = -5$ and $b = -3 - (-8) = 5$

$v = -5i + 5j$

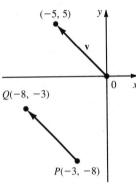

37. a) $3u = 3(2, 3) = (6, 9)$

b) $u + v = (2, 3) + (-5, 4) = (-3, 7)$

c) $v - u = (-5, 4) - (2, 3) = (-5, 4) + (-2, -3) = (-7, 1)$

d) $2u - 7v = 2(2, 3) - 7(-5, 4) = (4, 6) + (35, -28) = (39, -22)$

39. If a vector has a magnitude of 1, then it is a unit vector. The vector **i** is the same as $(1,0)$ and so its magnitude must be $\sqrt{1^2 + 0^2} = \sqrt{1} = 1$ and so it must be a unit vector. A similar argument can be given for **j**.

41. First thing to do is show that $(a/\sqrt{a^2 + b^2})\,i + (b/\sqrt{a^2 + b^2})\,j$ is a unit vector.

By the formula for magnitude we can calculate its length to be

$$\sqrt{a^2/(a^2 + b^2) + b^2/(a^2 + b^2)} = \sqrt{\frac{a^2 + b^2}{a^2 + b^2}} = \sqrt{1} = 1 \text{ so it definitely is a unit vector.}$$

Now we must show that **u** has the same direction as **v**. The algebraic signs of a and $a/\sqrt{a^2 + b^2}$ have to be the same as do the signs of b and $b/\sqrt{a^2 + b^2}$ so the general quadrant headings of both of these vectors must be the same. The direction of **v** is then uniquely determined by the value of

$\tan^{-1}\frac{b}{a}$; the direction of **u** is determined by the value of $\tan^{-1}\left(\dfrac{b/\sqrt{a^2 + b^2}}{a/\sqrt{a^2 + b^2}}\right)$ which boils down to $\tan^{-1}\frac{b}{a}$ which gives it the same direction as **v**.

43. $a = 1$ and $b = -1$

$(a/\sqrt{a^2+b^2})\, i + (b/\sqrt{a^2+b^2})\, j$ becomes $(1/\sqrt{2})i + (-1/\sqrt{2})j$.

45. $a = 3$ and $b = -4$ so $\sqrt{a^2+b^2} = \sqrt{3^2+(-4)^2} = 5$ so the unit vector in the direction of $(3, -4)$ must be $(3/5, -4/5)$.

47. $a = a$ and $b = a$ so $\sqrt{a^2+b^2} = \sqrt{a^2+a^2} = \sqrt{2a^2} = \sqrt{2}\,|a|$

so the unit vector in (a,a)'s direction $= a/(\sqrt{2}\,|a|)i + a/(\sqrt{2}|a|)j$

49. $\sin\theta = \dfrac{b}{\sqrt{a^2+b^2}} = \dfrac{-3}{\sqrt{2^2+(-3)^2}} = -\dfrac{3}{\sqrt{13}}$

$\cos\theta = \dfrac{a}{\sqrt{a^2+b^2}} = \dfrac{2}{\sqrt{13}}$

51. First let's just find a vector that is in the opposite direction; then, we'll convert it into a unit vector. If $v = i + j$ then the vector $w = -i + -j$ would be in the opposite direction of v. Now convert w into a unit vector by dividing it by its magnitude which is $\sqrt{(-1)^2+(-1)^2} = \sqrt{2}$ to get $u = (-1/\sqrt{2})\,i + (-1/\sqrt{2})\,j$.

53. Opposite direction vector $= (3, -4)$ which has a maginitude of 5; therefore $u = (3/5, -4/5)$.

55. Opposite direction vector $= 3i + 4j$ which has length 5 so $u = 3/5\, i + 4/5\, j$.

57. a) $u + v = (2, -3) + (-1, 2) = (1, -1)$ which has magnitude $\sqrt{2}$ so the unit vector in the direction of $u + v = (1/\sqrt{2}, -1/\sqrt{2})$

b) $2u - 3v = 2(2,-3) - 3(-1, 2) = (4,-6) + (3,-6) = (7,-12)$ which has a magnitude of $\sqrt{193}$ and so the answer is $(7/\sqrt{193}, -12/\sqrt{193})$

c) $3u + 8v = 3(2,-3) + 8(-1,2) = (-2, 7)$ which has magnitude $\sqrt{53}$ so the unit vector desired is $(-2/\sqrt{53}, 7/\sqrt{53})$

59. Let $R =$ the point (a, b). Consider the line passing through the origin and R. The formula for the slope of a line is given by $m = \dfrac{y_2 - y_1}{x_2 - x_1}$. If we let $(x_1, y_1) =$ the origin and $(x_2, y_2) = R = (a, b)$ then m (slope of line containing one representation of vector (a, b)) $= \dfrac{b - 0}{a - 0} = \dfrac{b}{a}$.

59. (continued)

Now consider the line containing vector PQ and let $(x_2, y_2) = Q = (c + a, d + b)$ and let (x_1, y_1) = P = (c, d). The slope of this line is $m = \dfrac{(d + b) - d}{(c + a) - c} = \dfrac{b}{a}$ as well. Therefore, the lines containing the two vectors in question are parallel and so the vectors have the same direction.

61. Because of the unit circle definitions of the sine and cosine functions we can state that a unit vector **u** in the direction $\pi/3$ would be $\cos \pi/3\ \mathbf{i} + \sin \pi/3\ \mathbf{j} = \frac{1}{2}\mathbf{i} + \frac{\sqrt{3}}{2}\mathbf{j}$.

Then the desired vector **v** = (its magnitude) **u** = 8 **u** = 4 **i** + 4$\sqrt{3}$ **j**.

63. A unit vector is θ's direction would be $\mathbf{u} = \cos \pi/2\ \mathbf{i} + \sin \pi/2\ \mathbf{j} = \mathbf{j}$. Now multiply this by the stipulated magnitude 4 and you will get **v** = 4**j**.

65. A unit vector in θ's direction would be $\mathbf{u} = \cos 2\pi/3\ \mathbf{i} + \sin 2\pi/3\ \mathbf{j} = -\frac{1}{2}\mathbf{i} + \frac{\sqrt{3}}{2}\mathbf{j}$. Multiplying this by the given magnitude for **v** gives the solution - 3 **i** + 3$\sqrt{3}$ **j**.

67. A unit vector in θ's direction would be $\mathbf{u} = \cos 11\pi/6\ \mathbf{i} + \sin 11\pi/6\ \mathbf{j} = \frac{\sqrt{3}}{2}\mathbf{i} - \frac{1}{2}\mathbf{j}$. Multiplying this by the magnitude 6 yields **v** = 3$\sqrt{3}$ **i** - 3 **j** .

Problems 9.5

1. In the drawing at the right, AD represents the weight (which is always perpendicular to the ground).

AD = 122

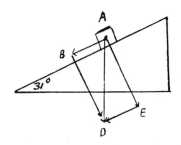

facts: $\angle$ DAE and $\angle$ BAD are complementary (they comprise a right angle)

        $\angle$ DCA and $\angle$ CAD (ie $\angle$ BAD) are complementary (acute angles of a right triangle)

    Therefore, since complements of the same angle are equal, $\angle$ DAE = $\angle$ DCA = 31°

a) we seek AB which is the same as ED which can be found by analyzing $\triangle$ DAE:

    $\sin \angle \text{DAE} = \dfrac{\text{opp}}{\text{hyp}}$ becomes $\sin 31° = \dfrac{\text{ED}}{122}$ so ED = 122 sin 31° = 122 (0.5150) = 62.835 lb

1. (continued)

   b) the force of the block against ($\perp$ to) the ramp is AE in the drawing. Now we'll use the cosine:

   $\cos \angle DAE = \dfrac{adj}{hyp}$ becomes $\cos 31° = \dfrac{AE}{122}$ so $AE = 122 \cos 31° = 122(0.8572) = 104.574$ lb

3. In the simplified drawing at the right we seek θ.

   OD = weight = 120 lb; CD must equal given force of 85 lb

   In the right triangle of forces $\sin \theta = \dfrac{opp}{hyp} = \dfrac{85}{120} = 0.7083$

   so $\theta = \sin^{-1} 0.7083 = 45.1°$

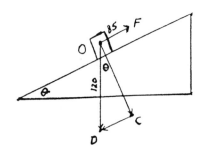

5. The tension in the rope must be the opposite of the force labeled NG in the diagram at the right.

   OG = 160 lb; $\angle NOG = 58°$

   $\sin 58° = \dfrac{opp}{hyp} = \dfrac{NG}{160}$ so $NG = 160(\sin 58°) = 135.7$ lb

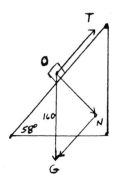

7. Now the force perpendicular to the plane, ON, is 437 N and the force parallel to the plane, NG, must be equal to the tension of 620 N. We seek θ:

   $\tan \theta = \dfrac{opp}{adj} = \dfrac{620}{437} = 1.4188$

   $\theta = \tan^{-1} 1.4188 = 54.8°$

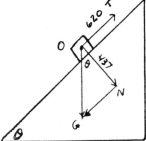

9. In the drawing at the right the horizontal and vertical components of the force, OF, (directed through the handle) are represented by x and y respectively.

   $\sin 42° = \dfrac{opp}{hyp} = \dfrac{y}{26}$ so $y = 26 \sin 42° = 17.397$ lb

   $\cos 42° = \dfrac{adj}{hyp} = \dfrac{x}{26}$ so $x = 26 \cos 42° = 19.322$ lb

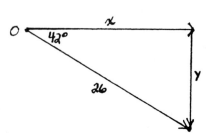

11. Let f stand for the magnitude of the force the girl exerts down through the handle and let θ represent the angle the handle makes with the horizontal.

By the Pythagorean theorm:

$$f^2 = 15^2 + 19.5^2$$

so $f = \sqrt{225 + 380.25} \approx 24.6$ lb

Now for θ:

$$\tan \theta = \frac{\text{opp}}{\text{adj}} = \frac{15}{19.5} = 0.7692$$

so $\theta = \tan^{-1} 0.7692 = 37.6°$

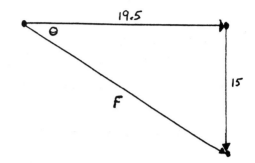

13. A grade of 3.2% means that the road rises 3.2 miles for every 100 miles of (imaginary level) road. This enables us to find θ, the angle of inclination of the hill with the horizontal as follows: $\tan \theta = \frac{\text{opp}}{\text{hyp}} = \frac{3.2}{100} = 0.032$ so $\theta = \tan^{-1} 0.032 \approx 1.833°$

In the drawing at the right, w stands for the weight of the truck. The up-the-hill force of 2850 N must be matching the down-the-hill component of the weight w.

$$\sin 1.833° = \frac{\text{opp}}{\text{hyp}} = \frac{2850}{w} \text{ so } w = \frac{2850}{\sin 1.833°} = 89,100$$

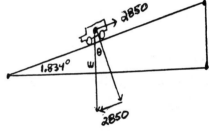

But we were asked for the mass (m) of the truck so now Newton's law is used: F = ma becomes 89,100 = m(9.8) and so m = $\frac{89,100}{9.8} = 9092$ kg

15. The drawing is all important in doing these kinds of problems! Groundspeed is what the plane ends up doing as a consequence of its airspeed capabilities and the wind added "vectorally". To make the drawing:

1.) start at the origin (the x-y axes serve as a north, south, east, west system) and depict the airspeed vector

2.) at the tip of the airspeed vector put N,S,E,W compass directions and depict the wind vector.

3.) connect the original origin with the tip of this second vector and you will be drawing the groundspeed vector.

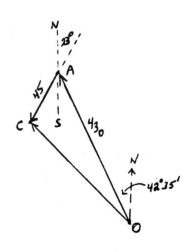

Analysis: $\angle$ OAS = 42.583° by alternate interior angles so $\angle$ OAC = 65.583° and we now know

two sides and included angle so use law of cosines:

$$OC^2 = 45^2 + 430^2 - 2(45)(430) \cos 65.583° = 170,927$$

so OC = 413.4 mph and this is the groundspeed.

To get the true course, we need $\angle$ AOC which we can then add to the 42°35'

Use law of sines: $\dfrac{45}{\sin \angle \text{AOC}} = \dfrac{413.4}{\sin 65.583°}$

$$\sin \angle \text{AOC} = 0.0991 \text{ so } \angle \text{AOC} = 5.688° = 5° 41'$$

Therefore, the true course is 42°35' + 5°41' west of north or N48°16'W

17. Since groundspeed is the sum of airspeed and wind,

airspeed must equal groundspeed plus the opposite

of the wind vector, 45 mph IN the direction of N23°E.

In the drawing OA is the desired groundspeed,

AC is the opposite of the wind,

and OC is airspeed.

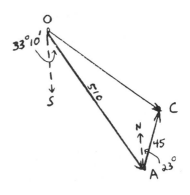

Analysis:

$\angle$ OAC = 56.17° $OC^2 = 45^2 + 510^2 - 2(45)(510) \cos 56.17° = 236,585$

so OC = 486.4 mph and this is the airspeed

To find the course at takeoff (heading) we need angle AOC (which we'll add t͓ ᴖᴖ .0')

The law of sines gives us $\dfrac{45}{\sin \angle \text{AOC}} = \dfrac{486.4}{\sin 56.17°}$

$$\sin \angle \text{AOC} = 0.07684 \text{ so } \angle \text{AOC} = 4.407° \text{ or } 4°24'$$

Therefore the course at takeoff must be 33°10' + 4°24' east of south or S37°34'E

19. Goundspeed + (opposite of wind) = airspeed idea again (#17)

In the drawing at the right OC = sought airspeed,

OA = 340 (the desired groundspeed) and AC = "- wind"

We need to figure out how big $\angle$ CAO is so we can use the

law of cosines. Do you see that $\angle$ CAS is 97°40' (it's the

supplement of $\angle$ NAC) and $\angle$ OAS = 5° (alt. int. $\angle$'s)?

$\angle$ CAO is the sum of these so $\angle$CAO = 102.67°. Now the

law of cosines can be applied:

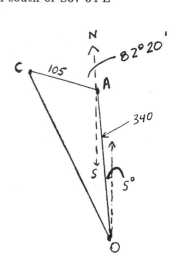

$$OC^2 = 340^2 + 105^2 - 2(340)(105) \cos 102.67°$$

$$= 142286 \text{ so } OC = 377.2 \text{ km/hr}$$

Now for the course:

$$\frac{105}{\sin \angle AOC} = \frac{377.2}{\sin 102.67°} \text{ so } \sin \angle AOC = 0.2716 \text{ so } \angle AOC = 15°46' \text{ which means}$$

that the course must be 5° + 15°46' west of north or N20°46'W

21. To reach a point directly across the river in 5

minutes, the boat must maintain a velocity of

$\frac{1465 \text{ft}}{5 \text{ min}} = 293$ ft/min EAST (this is equivalent

to groundspeed in the previous problems

and the current is equivalent to the wind).

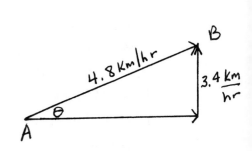

Similar to the logic in #19 we could say that:

desired velocity = velocity at necessary bearing + current

or equivalently

velocity at necessary bearing = desired velocity + (- current)

In the drawing above OA = desired velocity, AC = the opposite

of the current, and OC is the velocity at the necessary bearing.

Do you realize △OAC is a right triangle? Using the Pythagorean theorem

we have $OC^2 = 293^2 + 255^2$ so OC = 388.4 ft/min.

Using the tangent function we can find ∠AOC:

$$\tan \angle AOC = \frac{\text{opp}}{\text{adj}} = \frac{255}{293} = 0.8703 \text{ so } \angle AOC = 41° \text{ (north of east- not standard so we}$$

use its complement, 49°) in the

bearing N49°E

23. Must change $80 \frac{\text{m}}{\text{min}}$ to $\frac{\text{km}}{\text{hr}}$ so units agree:

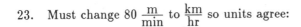

$$80 \frac{\text{m}}{\text{min}} \times \frac{60 \text{ min}}{1 \text{ hr}} \times \frac{1 \text{ km}}{1000 \text{ m}} = 4.8 \text{ km/hr}$$

Using right triangle trigonometry: $\sin \theta = \frac{\text{opp}}{\text{hyp}} = \frac{3.4}{4.8} = 0.7083$

So $\theta = \sin^{-1} 0.7083 \approx 45.1°$ or 45°6'

25. Consider the trapezoid drawn at the right.

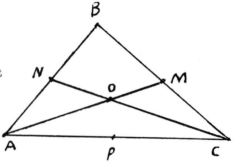

$\overrightarrow{EF} = \overrightarrow{EA} + \overrightarrow{AB} + \overrightarrow{BF}$ and

$\overrightarrow{EF} = \overrightarrow{EC} + \overrightarrow{CD} + \overrightarrow{DF}$ are both true

(they're paths from E to F)

Therefore adding these two equations gives

$2(\overrightarrow{EF}) = \overrightarrow{EA} + \overrightarrow{AB} + \overrightarrow{BF} + \overrightarrow{EC} + \overrightarrow{CD} + \overrightarrow{DF}$

But $\overrightarrow{EA} = -\overrightarrow{EC}$ (so $\overrightarrow{EA} + \overrightarrow{EC}$ "cancel") and $\overrightarrow{BF} = -\overrightarrow{DF}$ (so $\overrightarrow{BF}$ and $\overrightarrow{DF}$ also cancel to $\overrightarrow{0}$)

which means that $2(\overrightarrow{EF}) = \overrightarrow{AB} + \overrightarrow{CD}$ which implies that $\overrightarrow{EF} = \dfrac{\overrightarrow{AB} + \overrightarrow{CD}}{2}$, the desired result.

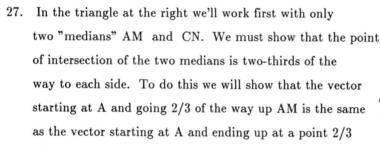

27. In the triangle at the right we'll work first with only two "medians" AM and CN. We must show that the point of intersection of the two medians is two-thirds of the way to each side. To do this we will show that the vector starting at A and going 2/3 of the way up AM is the same as the vector starting at A and ending up at a point 2/3 of the way up CN. If this is true, the common endpoint of these two vectors is the point of intersection of the two medians.

Here's what we know in terms of vectors:
$$\overrightarrow{AN} = \overrightarrow{NB} = \tfrac{1}{2}\overrightarrow{AB}$$
$$\overrightarrow{BM} = \overrightarrow{MC} = \tfrac{1}{2}\overrightarrow{BC}$$
$$\overrightarrow{AB} + \overrightarrow{BC} = \overrightarrow{AC} \quad \text{so} \quad \overrightarrow{BC} = \overrightarrow{AC} - \overrightarrow{AB}$$

The vector 2/3 of way from A to M $= 2/3\ \overrightarrow{AM} = 2/3\ (\overrightarrow{AB} + \overrightarrow{BM}) = 2/3\ (\overrightarrow{AB} + 1/2\ \overrightarrow{BC})$

$= 2/3\ (\overrightarrow{AB} + 1/2\ (\overrightarrow{AC} - \overrightarrow{AB})) = 2/3\ (\overrightarrow{AB} + 1/2\ \overrightarrow{AC} - 1/2\ \overrightarrow{AB}) = 2/3(1/2\ \overrightarrow{AB} + 1/2\ \overrightarrow{AC})$

$= 1/3\ \overrightarrow{AB} + 1/3\ \overrightarrow{AC}$

The vector starting at A and ending at a point 2/3 of the way up CN could be described as

$\overrightarrow{AC} + 2/3\ \overrightarrow{CN} = \overrightarrow{AC} + 2/3\ (\overrightarrow{CA} + \overrightarrow{AN}) = \overrightarrow{AC} + 2/3\ \overrightarrow{CA} + 2/3\ \overrightarrow{AN}$

$= \overrightarrow{AC} - 2/3\ \overrightarrow{AC} + 2/3\ (1/2\ \overrightarrow{AB}) = 1/3\ \overrightarrow{AC} + 1/3\ \overrightarrow{AB}$ which is the same as above.

So the point 2/3 of the way up AM coincides with the point 2/3 of the way up CN. Now you could bring in the third median, say BP and go through the same procedure showing that the vector starting at A and ending up at the point 2/3 of the way down BP is also $1/3\ \overrightarrow{AC} + 1/3\ \overrightarrow{AB}$ and then the proof would be complete.

Problems 9.6

1. The dot product of two vectors is the product of the coefficients of the two " i " components added
    to the product of the two " j " coefficients. So the dot product of $u = i + j = (1, 1)$ and
    $v = i - j = (1, - 1)$ would be $u \cdot v = (1)(1) + (1)(-1) = 1 - 1 = 0$.
    The cosine of the angle between two vectors is given by the formula $\cos \phi = \frac{u \cdot v}{|u| \, |v|}$.
    Here we have $|u| = \sqrt{1^2 + 1^2} = \sqrt{2}$ and $|v| = \sqrt{1^2 + (-1)^2} = \sqrt{2}$.
    Putting all this together gives us $\cos \phi = \frac{0}{\sqrt{2}\sqrt{2}} = 0$.

3. $u = - 5i = (- 5, 0)$ and $v = 18j = (0, 18)$ so $u \cdot v = (-5)(0) + (0)(18) = 0$ and as above the
    $\cos \phi$ will have to be 0 because a 0 (from the dot product) will be in the numerator again.

5. $u = 2 i + 5 j = (2, 5)$ and $v = 5 i + 2j = (5, 2)$ so the dot product is $2(5) + 5(2) = 20$.
    $| u | = \sqrt{2^2 + 5^2} = \sqrt{29}$ with an identical result for $| v |$; therefore, $\cos \phi = \frac{20}{\sqrt{29}^2} = \frac{20}{29}$.

7. $u = - 3 i + 4j = (- 3, 4)$ and $v = - 2 i - 7 j$ ($| u | = 5$ and $| v | = \sqrt{53}$)
    $u \cdot v = -3(-2) + 4(-7) = 6 - 28 = - 22$; $\cos \phi = \frac{- 22}{5\sqrt{53}}$.

9. $u = (11, - 8)$ and $v = (4, - 7)$ with respective magnitudes $\sqrt{185}$ and $\sqrt{65}$. $u \cdot v = 11(4) + -8(-7) =$
    $44 + 56 = 100$ while $\cos \phi = \frac{100}{\sqrt{185}\sqrt{65}} = \frac{100}{5\sqrt{481}} = \frac{20}{\sqrt{481}}$.

11. Two vectors are orthogonal if their dot product is zero! Here, $u = (\alpha, \beta)$ and $v = (\beta, - \alpha)$ so
    the dot product would be $u \cdot v = \alpha\beta + \beta(-\alpha) = 0$. Therefore, u and v are orthogonal.

In exercises 13 to 20 we will be checking to see if the dot product is 0 (then $\cos \phi$ would be zero
and the vectors orthogonal), or if $\cos \phi = \pm 1$ (the vectors are parallel), or if $\cos \phi$ equals any other
real number besides 0,1, or -1(vectors neither orthogonal nor parallel).

13. $u = (3, 5)$ and $v = (-6, -10)$; can just see that v is a multiple of u which means that they are
    parallel. If that got past you then you could see that $\cos \phi = \frac{3(-6) + 5(-10)}{\sqrt{34}\sqrt{136}} = \frac{- 68}{68} = - 1$ which is
    one of the signals for <u>parallel</u>.

15. $u = (2,3)$ and $v = (6, 4)$ so $u \cdot v = 2(6) + 3(4) = 24$ so these vectors are not orthogonal.
    $\cos \phi = \frac{24}{\sqrt{13}\sqrt{52}} \neq 1$ so they are not parallel either.

17. $u = (7, 0)$ and $v = (0, -23)$. The dot product is clearly 0 so these vectors are orthogonal!

19. $u = (1, 1)$ and $v = (\alpha, \alpha)$. The dot product is 2α so we must calculate $\cos \phi$:

$$\cos \phi = \frac{u \cdot v}{|u|\,|v|} = \frac{2\alpha}{\sqrt{2}\sqrt{\alpha^2 + \alpha^2}} = \frac{2\alpha}{2|\alpha|} = \text{either 1 or - 1 so in either case, as long as } \alpha \neq 0, \text{ the}$$

vectors are parallel.

21. a) For u and v to be orthogonal, we must have $u \cdot v = 0$. Here it would mean that $3(1) + 4\alpha = 0$ which implies that $\alpha = -\frac{3}{4}$.

b) For u and v to be parallel they must be multiples of each other; since the i-component of u is 3 times the i - component of v, the same must be true of the j - components: 4 must be 3 times α which means that $\alpha = \frac{4}{3}$.

c) If $\phi = \pi/4$ then we must have $\cos \pi/4 = \frac{u \cdot v}{|u|\,|v|}$ which becomes $\frac{\sqrt{2}}{2} = \frac{3 + 4\alpha}{5\sqrt{1 + \alpha^2}}$.

Squaring both sides is where to start: $\frac{1}{2} = \frac{9 + 24\alpha + 16\alpha^2}{25(1 + \alpha^2)}$: cross multiply now to get

$25 + 25\alpha^2 = 18 + 48\alpha + 32\alpha^2$ or $0 = 7\alpha^2 + 48\alpha - 7 = (7\alpha - 1)(\alpha + 7)$ so $\alpha = \frac{1}{7}$ or -7

A quick sketch of u and the two possibilities for v will show you that $\alpha = \frac{1}{7}$ is the one that makes the angle between the two vectors 45°.

d) If $\phi = \pi/3$ the equation would become $\frac{1}{2} = \frac{3 + 4\alpha}{5\sqrt{1 + \alpha^2}}$ or $\frac{1}{4} = \frac{9 + 24\alpha + 16\alpha^2}{25 + 25\alpha^2}$ which

can turn into $39\alpha^2 + 96\alpha + 11 = 0$ which when solved by the quadratic formula yields

$\frac{-96 \pm \sqrt{7500}}{2(39)} = \frac{-96 \pm 50\sqrt{3}}{2(39)} = \frac{-48 \pm 25\sqrt{3}}{39}$ and the choice of signs that seems

reasonable based on a rough sketch would be $\frac{-48 + 25\sqrt{3}}{39}$ as it is only slightly negative which makes the angle appear to be in the 60° range.

23. For u and v to have opposite directions it would have to be true that $\cos \phi = -1$. This would require that $-1 = \frac{3 + 4\alpha}{5\sqrt{1 + \alpha^2}}$ which leads to $1 = \frac{9 + 24\alpha + 16\alpha^2}{25 + 25\alpha^2}$ which simplifies to

$9\alpha^2 - 24\alpha + 25 = 0$ after cross-multiplying and collecting terms on the left hand side. If you try to solve this quadratic equation using the quadratic formula, you will get a negative discriminant which is the signal for NO REAL ZEROS which means there is no α such that the vectors in question could be in oppposite directions.

25. The vector projection of u onto v $= \text{Proj}_v \ u = \dfrac{u \cdot v}{|v|^2} \ v = \dfrac{3(1) + 0(1)}{(\sqrt{2})^2}(\ i + j \ ) = \frac{3}{2} \ i + \frac{3}{2} \ j \ .$

27. $\text{Proj}_v \ u = \dfrac{2(1) + 1(-2)}{(\sqrt{5})^2} \ (\ i - 2j \ ) = 0 \ i + 0 \ j \ = 0.$

29. $\text{Proj}_v \ u = \dfrac{1(2) + 1(-3)}{(\sqrt{13})^2} \ (\ 2 \ i \ - \ 3 \ j \ ) = - \frac{2}{13} i + \frac{3}{13} j \ .$

31. $\text{Proj}_v \ u = \dfrac{4(2) + 5(4)}{(\sqrt{20})^2} \ (\ 2 \ i + 4 \ j \ ) = \frac{7}{5} \ (\ 2 \ i + 4 \ j \ ) = \frac{14}{5} \ i + \frac{28}{5} \ j \ .$

33. $\text{Proj}_v \ u = \dfrac{-4(2) + 5(-4)}{(\sqrt{20})^2} \ (2 \ i - 4 \ j) = - \frac{14}{5} \ i + \frac{28}{5} \ j \ .$

35. $\text{Proj}_v \ u = \dfrac{\alpha(1) + \beta(1)}{(\sqrt{2})^2} \ (i + j) \ = \frac{\alpha + \beta}{2} \ i + \frac{\alpha + \beta}{2} \ j \ .$

37. $\text{Proj}_v \ u = \dfrac{\alpha - \beta}{2} \ (i + j) = \frac{\alpha - \beta}{2} \ i \ + \frac{\alpha - \beta}{2} \ j \ .$

39. If $u = a_1 i + b_1 j$ and $v = a_2 i + b_2 j$ and we want v and $\text{Proj}_v \ u$ to have the same direction (i.e., we don't want the projection to be in the direction of $- v$), we must make sure that ϕ is acute which means that $\cos \phi$ would have to be positive. But $\cos \phi = \dfrac{u \cdot v}{|u| \ |v|}$ and its denominator is certainly positive because magnitudes of vectors are positive so it boils down that $u \cdot v$ must be positive, that is, we must have $a_1 a_2 + b_1 b_2 > 0$ for v and the projection of u onto it to be in the same direction.

41. $\overrightarrow{PQ} = 3i + 4j \ ; \ \overrightarrow{RS} = -1i + 5j$ a) $\text{Proj}_{\overrightarrow{PQ}} \overrightarrow{RS} = \dfrac{\overrightarrow{RS} \cdot \overrightarrow{PQ}}{|\overrightarrow{PQ}|^2} \ \overrightarrow{PQ} = \dfrac{-3 + 20}{5^2} (3i + 4j) = \frac{51}{25}i + \frac{68}{25}j \ .$

   b) $\text{Proj}_{\overrightarrow{RS}} \overrightarrow{PQ} = \dfrac{\overrightarrow{PQ} \cdot \overrightarrow{RS}}{|\overrightarrow{RS}|^2} \ \overrightarrow{RS} \ = \dfrac{17}{(\sqrt{26})^2} (- i + 5j \ ) = - \frac{17}{26}i + \frac{85}{26}j.$

43. If both u and v are unit vectors then it must be true that $u = \cos \theta_1 i + \sin \theta_1 j$ and that $v = \cos \theta_2 i + \sin \theta_2 j$ for some θ_1 and θ_2. Since the angle between u and v is 0 we can state that $\theta_1 = \theta_2$ and this implies that u and v are the same vectors.

45. Since u and v are parallel, the unit vectors $\frac{u}{|u|}$ and $\frac{v}{|v|}$ are also parallel. If the angle between these two unit vectors is 0, problem 43 says that $\frac{u}{|u|} = \frac{v}{|v|}$. If the angle is π, then problem 44 says $\frac{v}{|v|} = - \frac{u}{|u|}$. So in either case we have $\frac{v}{|v|} = \pm \frac{u}{|u|}$ which implies that $v = \pm \frac{|v|}{|u|} \ u$. Letting $a = \pm \frac{|v|}{|u|}$ gives the desired result.

In problems 47-53 we must: 1) find a unit vector in the dircection of **F** and then multiply it by the magnitude of **F** to actually obtain the needed force vector **F**; 2) find the direction vector **D** for the motion of the object by subtracting P's coordinates from Q's; 3) dot **F** with **D** to find the work done.

47. 1) A unit vector in the direction $\pi/2$ would be **j** so **F** = 2**j** since $|$ **F** $|$ = 2; **D** = (1 - 5)**i** + (1 - 7)**j**
   = - 4**i** - 6**j** 3) **F**· **D** = 0(-4) + 2(-6) = - 12 N·m

49. 1) A unit vector in the direction of $\pi/6$ would be $\cos \pi/6 \, \mathbf{i} + \sin \pi/6 \, \mathbf{j} = \frac{\sqrt{3}}{2}\mathbf{i} + \frac{1}{2}\mathbf{j}$
   and so **F** $= 4\left(\frac{\sqrt{3}}{2}\mathbf{i} + \frac{1}{2}\mathbf{j}\right) = 2\sqrt{3}\mathbf{i} + 2\mathbf{j}$ 2) **D** = [3 - (-1)] **i** + (4 - 2)**j** = 4**i** + 2**j**
   3) work = **F** · **D** $= 4(2\sqrt{3}) + 2(2) = 8\sqrt{3} + 4$ N·m

51. 1) A unit vector in the direction of $3\pi/4$ would be $\cos 3\pi/4 \, \mathbf{i} + \sin 3\pi/4 \, \mathbf{j} = -\frac{\sqrt{2}}{2}\mathbf{i} + \frac{\sqrt{2}}{2}\mathbf{j}$ and
   so **F** $= 3(-\frac{\sqrt{2}}{2}\mathbf{i} + \frac{\sqrt{2}}{2}\mathbf{j}) = -\frac{3\sqrt{2}}{2}\mathbf{i} + \frac{3\sqrt{2}}{2}\mathbf{j}$
   2) **D** = (1 - 2)**i** + (2 - 1)**j** = -1**i** + **j**
   3) **W** = **F**· **D** $= -1(-\frac{3\sqrt{2}}{2}) + 1(\frac{3\sqrt{2}}{2}) = \frac{6\sqrt{2}}{2} = 3\sqrt{2}$ N·m

53. 1) A unit vector in the direction of 2**i** + 3**j** can be found by dividing this vector by its magnitude,
   $\sqrt{13}$, to get $\frac{2}{\sqrt{13}}\mathbf{i} + \frac{3}{\sqrt{13}}\mathbf{j}$. Multiply this by **F**'s magnitude, 4, to get **F** $= \frac{8}{\sqrt{13}}\mathbf{i} + \frac{12}{\sqrt{13}}\mathbf{j}$
   2) **D** = (-1 - 2)**i** + (3 - 0)**j** = -3**i** + 3**j**
   3) work = **F**· **D** $= \left(\frac{8}{\sqrt{13}}\right)(-3) + \left(\frac{12}{\sqrt{13}}\right)(3) = \frac{12}{\sqrt{13}}$ N·m

55. The first thing to do is make a simplified drawing
   of the situation. In the drawing at the right, H_1, H_2,
   V_1, and V_2 represent the magnitudes of the horizontal
   and vertical components of the forces exerted on the
   barge by tugs 1 and 2 respectively. Since the barge is
   moving horizontally, the vertical components of the two forces must add to zero. This can only
   happen if the vetical components are equal in magnitude but opposite in direction. We will take
   advantage of the equal vertical magnitudes but first we must to express them:
   for tug 1: $\sin 20° = \frac{V_1}{500} \Rightarrow V_1 = 500 \sin 20°$;

   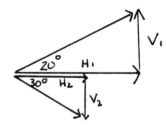

   for tug 2: $\sin 30° = \frac{V_2}{x} \Rightarrow V_2 = x \sin 30°$. Since these two magnitudes must be equal we
   can set $500 \sin 20° = x \sin 30°$ which implies that $x = \frac{500 \sin 20°}{\sin 30°} \approx 342$ N.

57. See diagram for problem 55.

Finding tug 1's horizontal component: $\cos 20° = \frac{H_1}{500} \Rightarrow H_1 = 500 \cos 20°$

So the horizontal force exerted by tug 1 on the barge is $F_1 = 500 \cos 20°$ i and the distance vector D = 750 i and so the work done by tug 1 is $F_1 \cdot D = 500 (\cos 20°)(750) \approx 352,384.7$ N·m.

Now for tug 2's horizontal component: $\cos 30° = \frac{H_2}{342} \Rightarrow H_2 = 342 \cos 30°$ and the dot product of the vector H_2 and the same distance vector D gives the work done by tug 2 as $342(\cos 30°)(750) \approx 222,136$ N·m.

59. In the diagram at the right vector F_w represents the weight of the block and d represents the distance the block slides. Since weight is a downward force, $F_w = -122j$. F_d will represent the component of F_w that is trying to push the block down the ramp. $\angle$ BAC is 31° (angle CAD must be 59° because it is congruent [by corresponding angles]

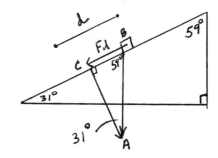

to the angle at the top right of the ramp which is 59° because it is complementary to the 31° angle at the bottom of the ramp; since angle BAC is complementary to angle CAD (59°), it's measure must be 31°. And so by right triangle trigonometry we have

a) $\sin 31° = \frac{|F_d|}{-122}$ which implies that $|F_d| = -122 \sin 31° \approx -62.83$ lb.

b)   work = force · distance = -62.83(25) = -1570.87 ft·lb.

61. The lawn mower is moving horizontally so only the horizontal component of her force is really doing work. The horizontal component would be $26 \cos 42° \approx 19.32$ lb and this force moves over a distance of 80 ft so the work done is f·d $\approx 19.32(80) = 1,545.7$ ft·lb.

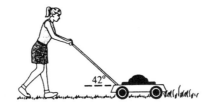

63. In the drawing at the right, C represents the car,

$F_w = -3174j$ represents the weight of the car,

and F_d = the force that tends to make the car roll

downhill. If θ represents the angle the road makes

with the horizontal, then we have $\tan \theta = \frac{4}{100} = \frac{1}{25}$

($\sin \theta = \frac{1}{\sqrt{626}}$ and $\cos \theta = \frac{25}{\sqrt{626}}$).

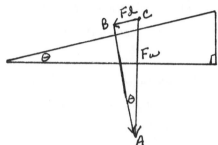

a) Angle CAB is congruent to θ (see discussion of similar situation in problem 61) so we can use right triangle trigonometry in right triangle ABC:

$$\sin \theta = \frac{\text{opp}}{\text{hyp}} = \frac{|F_d|}{-3174} \text{ so } |F_d| = -3174 \sin \theta = -3174 \left(\frac{1}{\sqrt{626}}\right) \approx -126.9 \text{ lb}$$

b) The work done by gravity on the car is force · distance = -126.9(300) = -38,058 ft lb so the work that the car does (or can do) is 38,058 ft lb.

Problems 9.7

1. $(2, 0)$ which means that $r = 2$ and $\theta = 0$ (radians)

   into rectangular:

   $\quad x = r \cos \theta = 2 \cos 0 = 2(1) = 2$

   $\quad y = r \sin \theta = 2 \sin 0 = 2(0) = 0$

   so $(2, 0)_{\text{polar}} = (2, 0)_{\text{rect}}$

3. $(-5, 0)$ is equivalent to $r = -5$ and $\theta = 0$

   $\quad x = r \cos \theta = -5(1) = -5$

   $\quad y = r \sin \theta = -5(0) = 0$

   so $(-5, 0)_{\text{polar}} = (-5, 0)_{\text{rect}}$

5. $\left(6, \dfrac{7\pi}{6}\right)$ means $r = 6$, $\theta = \dfrac{7\pi}{6}$ so...

   $\quad x = r \cos \theta = 6 \cos \dfrac{7\pi}{6} = 6\left(-\dfrac{\sqrt{3}}{2}\right) = -3\sqrt{3}$

   $\quad y = r \sin \theta = 6 \sin \dfrac{7\pi}{6} = 6\left(-\dfrac{1}{2}\right) = -3$

   so the answer is $(-3\sqrt{3}, -3)$

7. $(5, \pi) \Rightarrow r = 5$ and $\theta = \pi$ so we have

   $\quad x = 5 \cos \pi = 5(-1) = -5$

   $\quad y = 5 \sin \pi = 5(0) = 0$

   The answer in rectangular coordinates is $(-5, 0)$

9. $(-5, -\pi) \Rightarrow r = -5$ and $\theta = -\pi$ (same position as π)

   $\quad x = (-5) \cos(-\pi) = (-5) \cos \pi = (-5)(-1) = 5$

   $\quad y = (-5) \sin(-\pi) = (-5) \sin \pi = (-5)(0) = 0$

   and we have $(5, 0)$ as the rectangular coordinates.

11. $(3, \frac{11\pi}{4}) \Rightarrow r = 3$ and $\theta = \frac{11\pi}{4}$ so...

$$x = 3 \cos \frac{11\pi}{4} = 3\left(-\frac{\sqrt{2}}{2}\right) = -\frac{3\sqrt{2}}{2} \qquad y = 3 \sin \frac{11\pi}{4} = 3\left(\frac{\sqrt{2}}{2}\right) = \frac{3\sqrt{2}}{2}$$

and the x-y coordinates are $\left(-\frac{3\sqrt{2}}{2}, \frac{3\sqrt{2}}{2}\right)$

13. $(-3, -\frac{\pi}{4}) \Rightarrow r = -3$ and $\theta = -\frac{\pi}{4}$

$$x = -3 \cos\left(-\frac{\pi}{4}\right) = -3\left(\frac{\sqrt{2}}{2}\right) = -\frac{3\sqrt{2}}{2} \qquad y = -3 \sin\left(-\frac{\pi}{4}\right) = -3\left(-\frac{\sqrt{2}}{2}\right) = \frac{3\sqrt{2}}{2}$$

the x-y coordinates are therefore $\left(-\frac{3\sqrt{2}}{2}, \frac{3\sqrt{2}}{2}\right)$

15. $(1, \frac{3\pi}{2}) \Rightarrow r = 1$ and $\theta = \frac{3\pi}{2}$ so we have

$$x = (1) \cos \frac{3\pi}{2} = 1(0) = 0 \qquad y = (1) \sin \frac{3\pi}{2} = 1(-1) = -1$$

and so the rectangular coordinates are $(0, -1)$

17. $(-1, \frac{\pi}{2}) \Rightarrow r = -1$ and $\theta = \frac{\pi}{2}$

$$x = (-1) \cos \frac{\pi}{2} = (-1)(0) = 0 \qquad y = (-1) \sin \frac{\pi}{2} = (-1)(1) = -1$$

so we have $(0, -1)$ as the x-y coordinates.

19. $(2, 0)$ now means that $x = 2$ and $y = 0$ (between QIV and QI)

$$r = \sqrt{x^2 + y^2} = \sqrt{2^2 + 0^2} = \sqrt{4} = 2 \text{ and } \theta = \tan^{-1}\frac{y}{x} = \tan^{-1}\frac{0}{2} = \tan^{-1} 0 = 0$$

and so the polar coordinates are $(2, 0)$

21. $(-3, 0) \Rightarrow x = -3$ and $y = 0$ (between QII and QIII)

$$r = \sqrt{(-3)^2 + 0^2} = \sqrt{9} = 3 \qquad \theta = \pi + \tan^{-1}\frac{0}{-3} = \pi + 0 = \pi \text{ (knew QII beforehand)}$$

so the polar coordinates are $(3, \pi)$

23. $(-2, 2) \Rightarrow x = -2$ and $y = 2$ (QII)

$$r = \sqrt{(-2)^2 + (2)^2} = \sqrt{8} = 2\sqrt{2}$$

$$\theta = \pi + \tan^{-1} \frac{2}{(-2)} = \pi + \tan^{-1} (-1) = \pi + (-\tfrac{\pi}{4}) = \tfrac{3\pi}{4}$$

polar coordinates: $\left(2\sqrt{2}, \tfrac{3\pi}{4} \right)$

25. $(-2,-2)$ so $x = -2$ and $y = -2$ (QIII)

$$r = \sqrt{(-2)^2 + (-2)^2} = 2\sqrt{2} \quad \text{and} \quad \theta = \pi + \tan^{-1} \frac{(-2)}{(-2)} = \pi + \tan^{-1} (1) = \pi + \left(\tfrac{\pi}{4} \right) = \tfrac{5\pi}{4}$$

polar coordinates: $\left(2\sqrt{2}, \tfrac{5\pi}{4} \right)$

27. $(0, -1)$ so $x = 0$ and $y = -1$ (between QIII and QIV)

$$r = \sqrt{0^2 + (-1)^2} = 1 \qquad\qquad \theta = \pi + \tan^{-1} \frac{(-1)}{0} = ?$$

confused?? tangent is undefined in this way if $\theta = \tfrac{3\pi}{2}$ (see the plot of $(0, -1)$)

polar coordinates: $\left(1, \tfrac{3\pi}{2} \right)$

29. $(-2, 2\sqrt{3}) \Rightarrow x = -2$ and $y = 2\sqrt{3}$ (QII)

$$r = \sqrt{(-2)^2 + (2\sqrt{3})^2} = \sqrt{4 + 4\cdot3} = \sqrt{16} = 4$$

$$\theta = \pi + \tan^{-1} \frac{2\sqrt{3}}{(-2)} = \pi + \tan^{-1} (-\sqrt{3}) = \pi + (-\tfrac{\pi}{3}) = \tfrac{2\pi}{3}$$

polar coordinates: $\left(4, \tfrac{2\pi}{3} \right)$

31. $(-2, -2\sqrt{3}) \Rightarrow x = -2$ and $y = -2\sqrt{3}$ (Q III)
    $r = 4$ by a calculation similar to that in # 29; $\theta = \pi + \tan^{-1} \frac{(-2\sqrt{3})}{(-2)} = \pi + \tan^{-1} \sqrt{3} = \pi + \tfrac{\pi}{3} = \tfrac{4\pi}{3}$
    polar coordinates: $\left(4, \tfrac{4\pi}{3} \right)$

33. $(2\sqrt{3}, -2) \Rightarrow x = 2\sqrt{3}$ and $y = -2$ (Q IV)

   $r = 4$ (see # 29)

   $$\theta = 2\pi + \tan^{-1}\frac{(-2)}{2\sqrt{3}} = 2\pi + \tan^{-1}\left(-\frac{1}{\sqrt{3}}\right) = 2\pi + \left(-\frac{\pi}{6}\right) = \frac{11\pi}{6}$$

   polar coordinates: $\left(4, \frac{11\pi}{6}\right)$

35. $(-2\sqrt{3}, -2) \Rightarrow x = -2\sqrt{3}$ and $y = -2$ (Q III)

   $r = 4$ (again!)

   $$\theta = \pi + \tan^{-1}\frac{(-2)}{(-2\sqrt{3})} = \pi + \tan^{-1}\frac{1}{\sqrt{3}} = \pi + \frac{\pi}{6} = \frac{7\pi}{6}$$

   polar coordinates: $\left(4, \frac{7\pi}{6}\right)$

Problems 9.8

1. Any point with $r = 5$ would be on the circle of radius 5
   centered at the pole (origin). If $r = 5$ were converted
   to an x-y equation it would yield $\sqrt{x^2 + y^2} = 5$
   or $x^2 + y^2 = 25$ which is the circle described above.

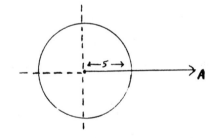

3. If $r = -4$ it is still a circle centered at the pole with
   a radius of $r = |-4| = 4$. All symmetries are present.

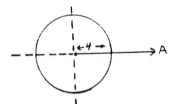

5. Any point on the graph of $\theta = -\pi/6$ (r can be anything)
   would have to lie on the straight line which makes an angle
   of $-\pi/6$ with the polar axis ($\pi/6$ clockwise). If you want
   to see the x-y version use fact that $\tan\theta = y/x$ to get
   $y = (-1/\sqrt{3})\,x$.

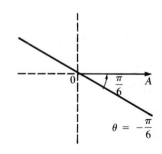

7. $r = 5 \sin \theta$ (let's change to rectangular)

   if we multiply both sides by r we get $r^2 = 5(r \sin \theta)$

   which is equivalent to $x^2 + y^2 = 5y$

   and if you complete the square you get $x^2 + (y - 5/2)^2 = 25/4$

   which is a circle centered at $(0, 5/2)$ with a radius of $5/2$

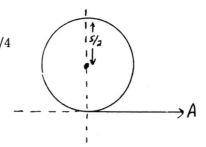

9. $r = -5 \sin \theta$ (similar to #7)

   $r^2 = -5 (r \sin \theta)$ leads to $x^2 + y^2 = -5y$

   or $x^2 + (y^2 + 5y) = 0$ and after completing the square it

   looks like this: $x^2 + (y + 5/2)^2 = 25/4$, a circle centered

   at $(0, -5/2)$ with a radius of $5/2$.

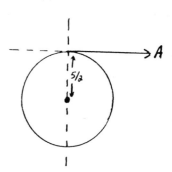

11. $r = 5 \cos \theta + 5 \sin \theta$ is equivalent to $r^2 = 5 (r \cos \theta) + 5 (r \sin \theta)$

   which in good old x's and y's is $x^2 + y^2 = 5x + 5y$ which would like

   its square completed. Some of the steps:

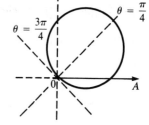

$r = 5(\cos \theta + \sin \theta)$

$\theta = \dfrac{\pi}{4}$

$\theta = \dfrac{3\pi}{4}$

   $(x^2 - 5x + \ ) + (y^2 - 5y + \ ) = 0$

   $(x^2 - 5x + 25/4) + (y^2 - 5y + 25/4) = 50/4$

   $(x - 5/2)^2 + (y - 5/2)^2 = (5/\sqrt{2})^2$

   So this is a circle centered at $(5/2, 5/2)$ with a radius of ≈ 3.54

13. Very similar to # 11.

   After completing the square the equation should look like this:

   $(x - 5/2)^2 + (y + 5/2)^2 = 25/2$

   which is a circle centered at $(5/2, -5/2)$ with a radius of $5/\sqrt{2}$

   (≈ 3.54 again).

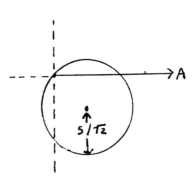

15. $r = 2 + 2 \sin \theta$ is the same as $r = 2 (1 + \sin \theta)$ which will be a cardioid similar to that in example 2 in the text but this one will be twice as large. It has symmetry about the line $\theta = \pi/2$ because

$$\sin (\pi - \theta) = \sin \pi \cos \theta - \cos \pi \sin \theta$$
$$= - (-1) \sin \theta = \sin \theta$$

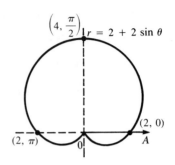

17. $r = 2 - 2 \sin \theta = 2 (1 - \sin \theta)$ (symmetric around $\theta = \pi/2$ again)

a table:

θ:	0	$\pi/6$	$\pi/4$	$\pi/3$	$\pi/2$	$3\pi/2$	$5\pi/3$	$7\pi/4$	$11\pi/6$
r:	2	1	$2 - \sqrt{2}$	$2 - \sqrt{3}$	0	4	$2 + \sqrt{3}$	$2 + \sqrt{2}$	3

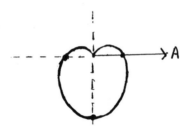

19. $r = -2 + 2 \sin \theta = -2 (1 - \sin \theta)$ and all the r values in the table for #17 will simply be negated. Plotting these new points should give you a graph that looks like the one at the right.

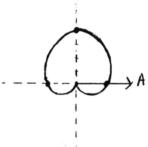

21. $r = -2 (1 + \sin \theta)$

Compare this to example 2 in the text and you should see that a table for this equation could be obtained from that table by multiplying the r values there by - 2.

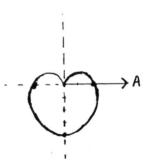

23. $r = 1 + 3 \sin \theta$

Since $\sin (\pi - \theta) = \sin \theta$ we have symmetry about $\theta = \pi/2$.

A table:

θ:	0	$\pi/6$	$\pi/4$	$\pi/3$	$\pi/2$	$3\pi/2$	$5\pi/3$	$7\pi/4$	$11\pi/6$
r:	1	5/2	$1 + 3\sqrt{2}/2$	$1 + 3\sqrt{3}/2$	4	-2	$1 - 3\sqrt{3}/2$	$1 - 3\sqrt{2}/2$	-1/2

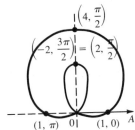

25. $r = -2 + 4 \cos \theta$ is symmetric about the polar axis

because $\cos (-\theta) = \cos \theta$. Angles from 0 to π should do.
Note: 4 being greater than 2 will cause an inward loop.

θ:	0	$\pi/6$	$\pi/4$	$\pi/3$	$\pi/2$
r:	2	$-2 + 2\sqrt{3}$	$-2 + 2\sqrt{2}$	0	-2

θ:	$2\pi/3$	$3\pi/4$	$5\pi/6$	π
r:	-4	$-2 - 2\sqrt{2}$	$-2 - 2\sqrt{3}$	-6

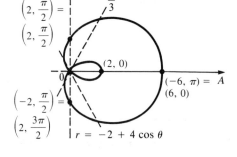

Note: when you plot the first five points above, you're
getting the top half of the "inner" loop. You can draw the
bottom half immediately from symmetry. When you plot the
next four points you should get the bottom part of the
"outer" loop. It's top half can be drawn from symmetry.

27. $r = -3 - 4 \cos \theta$

This is another limacon with a loop. It is symmetric about the polar axis. You can use the same
choices for θ as in #25 for your table. Good luck!

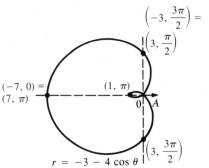

29. r = -3 - 4 sin θ

   Another limacon with a loop only this one will be symmetric

   about the line θ = π/2.

   θ: 0   π/6   π/4   π/3   π/2

   r: -3   -5   -3 - 2√2   -3 - 2√3   -7

   θ: 3π/2   5π/3   7π/4   11π/6

   r: 1   -3 + 2√3   -3 + 2√2   -1

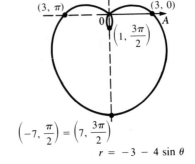

$r = -3 - 4 \sin \theta$

31. r = 4 - 3 cos θ

   (no inner loop here as 3 is less than 4)

   symmetry about the polar axis.

   Use the same θ-values as in #25 to calculate

   r-values from.

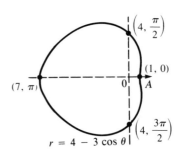

$r = 4 - 3 \cos \theta$

33. r = 4 + 3 sinθ

   limacon without loop (3 < 4) which will

   be symmetric to the line θ = π/2. Use values

   for θ as in # 29 and utilize symmetry.

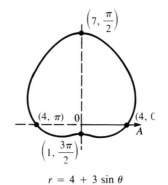

$r = 4 + 3 \sin \theta$

35. r = 3 sin 2θ is symmetric to the pole because

   if θ is replaced by θ + π the equation obtained,

   r = 3 sin (2θ + 2π) is equivalent to the original.

   In making a table, we'll halve our normal choices for θ:

   θ: 0   π/12   π/8   π/6   π/4

   r: 0   3/2   3√2/2   3√3/2   3

   θ: π/3   3π/8   5π/12   π/2

   r: 3√3/2   3√2/2   3/2   0

   Plotting just these points will give one leaf of the rose.

   As θ goes from π/2 to π, another leaf will form, etc.

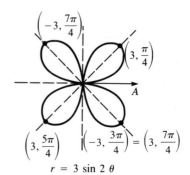

$r = 3 \sin 2\theta$

37. $r = 3 \cos 2\theta$ is also symmetric to the pole as well as

the polar axis [$\cos 2 \cdot (-\theta) = \cos 2\theta$].

Once again, halve the usual values:

θ: 0 $\pi/12$ $\pi/8$ $\pi/6$ $\pi/4$

r: 3 $3\sqrt{3}/2$ $3\sqrt{2}/2$ 3/2   0

   (above points constitute top half of one leaf)

θ: $\pi/3$ $3\pi/8$ $5\pi/12$ $\pi/2$

r: - 3/2 $-3\sqrt{2}/2$ $- 3\sqrt{3}/2$ -3

   (these points comprise left half of lower leaf)

symmetry considerations enable the filling in of the rest of the graph.

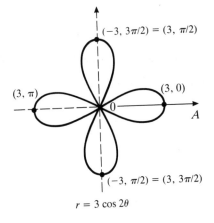

$r = 3 \cos 2\theta$

39. $r = 5 \sin 3\theta$ will be symmetric to the line $\theta = \pi/2$

choose values for θ that are a third of the usual fare.

θ: 0 $\pi/18$ $\pi/12$ $\pi/9$ $\pi/6$

r: 0   5/2 $5\sqrt{2}/2$ $5\sqrt{3}/2$ 5

    (half a leaf)

θ: $2\pi/9$ $\pi/4$ $5\pi/18$ $\pi/3$

r: $5\sqrt{3}/2$ $5\sqrt{2}/2$ 5/2   0

    (other half)

continuing in this fashion, the rest of the graph can be determined.

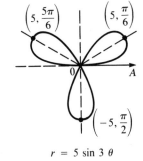

$r = 5 \sin 3\theta$

41. $r = - 5 \sin 3\theta$ (same table as #39 only with negative r's)

Also, the same symmetries.

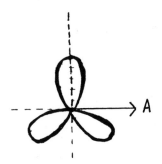

43. $r = 2 \cos 4\theta$ has all three symmetries

θ: 0 $\pi/24$ $\pi/16$ $\pi/12$ $\pi/8$

r: 2 $\sqrt{3}$ $\sqrt{2}$ 1   0

   (top half of one leaf)

by the time θ gets to $\pi/2$ in your table, another

leaf and a half will appear. Symmetry considerations

will then allow the drawing of an eight leafed rose.

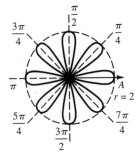

$r = 2 \cos 4\theta$

- 413 -

45. $r = -2 \cos 4\theta$ has as its graph a reflection of

the graph of #43 about the pole. The graphs end

up being identical.

47. $r = -5\theta$ has no symmetries, no repetitions of r values

θ: 0 $\pi/4$ $\pi/2$ $3\pi/4$ π $3\pi/2$ 2π

r: 0 $-5\pi/4$ $-5\pi/2$ $-15\pi/4$ -5π $-15\pi/2$ -10π

(a spiral which initially starts downward)

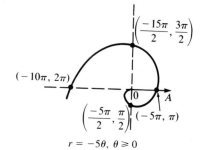

49. $r = e^{\theta}$

θ: 0 $\pi/4$ $\pi/2$ π $3\pi/2$

r: 1   2.19   4.81   23.1   111

(a really big spiral)

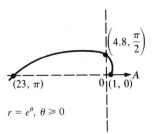

51. $r = e^{3\theta}$

θ: 0 $\pi/3$ $\pi/2$ π

r: 1   23.1   111   12392

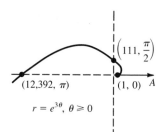

53. $r^2 = \sin 2\theta$ is symmetric about the pole because

$-r$ could be substituted for r with no change in the

equation. As in example 5, $\sin 2\theta$ must be ≥ 0 which

will occur if θ is in $[\, 0, \, \pi/2 \,]$ or $[\pi, \, 3\pi/2]$

θ: 0 $\pi/12$ $\pi/8$ $\pi/6$ $\pi/4$

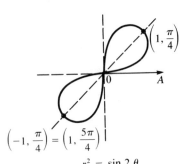

r: 0 $\pm 1/\sqrt{2}$ $\pm 2^{-.25}$ $\pm 3^{1/4} 2^{-1/2}$ ± 1

(estimate these using a calculator)

(by the time θ gets to $\pi/2$, the entire graph should have appeared)

55. $r^2 = -\sin 2\theta$

Because $\sin 2(\theta - \pi/2) = -\sin 2\theta$, a quarter turn

rotation of the graph of #53 counterclockwise will

produce the graph of this one.

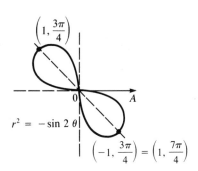

57. $r^2 = -25 \cos 2\theta$ requires that $\cos 2\theta$ be negative; this

occurs for θ from $[\pi/4, 3\pi/4]$ or $[5\pi/4, 7\pi/4]$

beginnings of a table:

θ: $\pi/4$ $\pi/3$ $\pi/2$ $2\pi/3$ $3\pi/4$

r: 0 $\pm\sqrt{12.5}$ ± 5 $\pm\sqrt{12.5}$ 0

(these points give the top loop - bottom one is

deduced by polar axis symmetry which is present

because $\cos 2(-\theta) = \cos 2\theta$

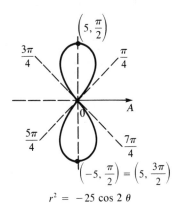

59. $r = \sin \theta \tan \theta$ and if both θ's are replaced by $-\theta$'s

the equation is unchanged $(-\sin \theta)(-\tan \theta) = $ original

and so the graph is symmetric about the polar axis.

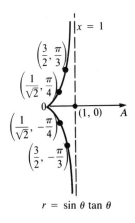

61. $r = 4 + 3 \csc \theta$ will be symmetric about $\theta = \pi/2$ because $\csc \theta = 1/\sin \theta$ and since

$\sin (\pi - \theta) = \sin \pi$, $\csc (\pi - \theta)$ will equal $\csc \theta$. θ cannot be 0, right?

θ: $\pi/6$ $\pi/2$ $7\pi/6$ $3\pi/2$ Note: the equation could be written as

r: 10   7   -2   1 $r \sin\theta = 4 \sin \theta + 3$ or $y = 4 \sin \theta + 3$ so as θ

approaches 0 we would have the graph approaching $y = 3$

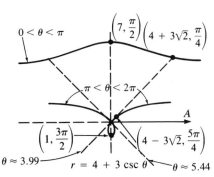

- 415-

63. $(r + 1)^2 = 3\theta$ requires that θ be greater than or equal to 0 and gives two r values for each positive θ value:

θ:	0	$\pi/2$	π	$3\pi/2$	2π	$5\pi/2$
r:	-1	1.2 or -3.2	-4.1 or 2.1	2.8 or -4.8	3.3	3.9

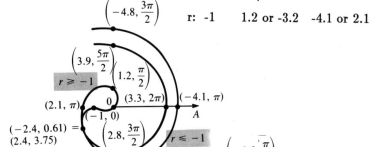

65. $r = |\cos\theta|$ is the same as $r^2 = r|\cos\theta|$ which is the same as $x^2 + y^2 = \pm x$ which is the same as $x^2 \pm x + y^2 = 0$ which, if you complete the square in x, gives $(x \pm 1/2)^2 + y^2 = 1/4$ which is a pair of circles, one centered at (-1/2, 0) the other at (1/2, 0) each with a radius of 1/2.

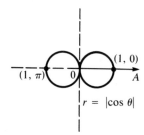

67. Multiply both sides by r to get $r^2 = b(r\sin\theta)$ which is the same as $x^2 + y^2 = by$.

Now complete the square: $x^2 + (y^2 - by + b^2/4) = 0 + b^2/4$

$$x^2 + (y - b/2)^2 = b^2/4 \quad \text{which is the equation of a circle}$$

centered at (0, b/2) with a radius of $\sqrt{b^2/4}$ or $|b/2|$

69. x-y version would be $(x + 2)^2 + (y - 3/2)^2 = 25/4$

Replacing x with r cos θ and y with r sin θ gives

$$(r\cos\theta + 2)^2 + (r\sin\theta - 3/2)^2 = 25/4$$

or $r^2\cos^2\theta + 4r\cos\theta + 4 + r^2\sin^2\theta - 3r\sin\theta + 9/4 = 25/4$

or $r^2(\sin^2\theta + \sin^2\theta) + 4r\cos\theta - 3r\sin\theta = 0$

or $r^2 = 3r\sin\theta - 4r\cos\theta$ and since r cannot be 0, divide both sides by r to get

$r = 3\sin\theta - 4\cos\theta$

71. Since $y = r\sin\theta$ the equation $r\sin\theta = a$ is perfectly equivalent to $y = a$ which is the equation of a horizontal line at a height of a.

73. (a) r sin θ = 3 is the same as y = 3, a horizontal line at a height of 3:

   (b) r cos θ = -2 is the same as x = -2, a vertical line with an x intercept of -2:

   (c) 3r cos θ = 8 is the same as r cos θ = 8/3 or x = 8/3, another vertical line:

   (d) r(2 sin θ - 3 cos θ) = 4 is the same as 2 r sin θ - 3 r cos θ = 4 which is the same

   as 2y - 3x = 4 which is the equation of a line with an x-intercept of -4/3 and a y-intercept

   of 2:

   (e) r(-5 sin θ + 10 cos θ) = 20 is equivalent to - 5y + 10x = 20, a straight line with a y-intercept

   of - 4 and an x-intercept of 2:

Problems 9.9

1. 5 i or 0 + 5 i has an absolute value of $\sqrt{0^2 + 5^2}$ = 5 ; it's argument, if not obvious from

   a quick sketch, is $\theta = \tan^{-1} 5/0 = \pi/2$ so in polar form, 5i = 5(cos $\pi/2$ + i sin $\pi/2$)

3. 8 or 8 + 0 i has an absolute value of $\sqrt{8^2 + 0^2}$ = 8 ; $\theta = \tan^{-1} 0/8 = 0$ and so the polar form

   of 8 is 8(cos 0 + i sin 0)

5. 1 - i or 1 + (-1) i has an absolute value of $\sqrt{1^2 + (-1)^2} = \sqrt{2}$

   On finding θ: a quick sketch of the complex number is invaluable for quadrant location and

   reasonability checks; the exact value for θ can always be found by evaluating $\tan^{-1}$ | b | / | a |

   and adjusting this answer for proper quadrant location (eg, if QII, subtract this answer from

   π; if QIII, add it to - π; if QIV, negate it, etc.)

   Here, we have QIV and $\tan^{-1}$ 1/1 to consider; $\tan^{-1} 1 = \pi/4$ which when negated will lie in QIV

   so θ = -$\pi/4$ and the desired polar form is $\sqrt{2}$ (cos (-$\pi/4$) + i sin (-$\pi/4$))

7. -1 - i or (-1) + (-1) i also has an absolute value of $\sqrt{2}$ and is clearly a resident of QIII. The in-

   verse tangent of 1 is $\pi/4$ again and we can add this to -π to get a QIII value for θ, namely -3$\pi/4$

   and so the polar form is $\sqrt{2}$ (cos (-3$\pi/4$) + i sin (-3$\pi/4$))

9. $-1 + \sqrt{3}$ i has an absolute value of $\sqrt{(-1)^2 + \sqrt{3}^2} = \sqrt{1 + 3} = 2$; the inverse tangent of $\sqrt{3}$ is $\pi/3$ and the number lies in QII so θ must be $\pi - \pi/3$ or $2\pi/3$. The polar form is then $2 (\cos 2\pi/3 + i \sin 2\pi/3)$

11. $-\sqrt{3} - i$ lies in the third quadrant; the inverse tangent of $1/\sqrt{3}$ is $\pi/6$ and the absolute value calculation (similar to that in #9) yields 2. Placing $\pi/6$ in QIII is done by adding it to $-\pi$ to obtain $-5\pi/6$ as θ. Polar form: $2 (\cos (-5\pi/6) + i \sin (-5\pi/6))$

13. $4 + 3i$ lies in QI and has an absolute value of 5 $\left(\text{from } \sqrt{3^2 + 4^2} \right)$ while $\tan^{-1} 3/4 \approx 0.6435$ so the polar form is $5 (\cos 0.6435 + i \sin 0.6435)$

15. $3 + 5i$ is in QI and has an absolute value of $\sqrt{3^2 + 5^2}$ or $\sqrt{34}$ and the inverse tangent of $5/3$ is approximately 1.0304 which can be used "as is" because of the QI location.
Polar form: $\sqrt{34} (\cos 1.0304 + i \sin 1.0304)$

17. $5 + 2$ i is also in quadrant I and its absolute value is $\sqrt{29}$ while its argument is the inverse tangent of $2/5$ which is approximately 0.3805 ; it's polar form is $\sqrt{29} (\cos 0.3805 + i \sin 0.3805)$

19. $-5 + 3$ i   is in QII and it's absolute value is $\sqrt{(-5)^2 + 3^2} = \sqrt{34}$; the argument in this case will be $\pi + \tan^{-1} 3/(-5) \approx 2.6012$. The polar form is $\sqrt{34} (\cos (2.6012) + i \sin (2.6012))$.

21. $-100 + 100$ i lies in QII ; it's absolute value is given by $\sqrt{(-100)^2 + 100^2} = \sqrt{20,000} = 100\sqrt{2}$ and the angle involved in finding θ is $\tan^{-1} 100/100 = \pi/4$. To account for the QII location, subtract $\pi/4$ from π to get $\theta = 3\pi/4$. Polar form: $100\sqrt{2} (\cos 3\pi/4 + i \sin 3\pi/4)$

23. $4 (\cos \pi/2 + i \sin \pi/2) = 4 [0 + i (1)] = 4$ i

25. $12 (\cos \pi + i \sin \pi) = 12 [(-1) + i (0)] = -12$

27. $8\sqrt{2} (\cos \pi/3 + i \sin \pi/3) = 8\sqrt{2} \left(1/2 + i (\sqrt{3}/2) \right) = 4\sqrt{2} + 4\sqrt{6}$ i

29. $4 (\cos 5\pi/6 + i \sin 5\pi/6) = 4 \left(- \sqrt{3}/2 + i (1/2) \right) = -2\sqrt{3} + 2$ i

31. $4 (\cos 2\pi/3 + i \sin 2\pi/3) = 4 \left(-1/2 + i (\sqrt{3}/2) \right) = -2 + 2\sqrt{3}$ i

33. $1/4 \left(\cos \pi/4 + i \sin \pi/4 \right) = 1/4 \left(\sqrt{2}/2 + i \sqrt{2}/2 \right) = \sqrt{2}/8 + \sqrt{2} \, i/8$

35. $1/4 \left(\cos 3\pi/4 + i \sin 3\pi/4 \right) = 1/4 \left((- \sqrt{2}/2) + i (\sqrt{2}/2) \right) = - \sqrt{2}/8 + \sqrt{2} \, i/8$

37. $5 \left(\cos 2 + i \sin 2 \right) \approx 5 (-0.4161 + i (0.9093)) = -2.0805 + 4.5465 \, i$ (note: the 2 was 2 radians)

39. $6 \left(\cos (- \pi/15) + i \sin (- \pi/15) \right) \approx 6 (0.9781 + (-0.2079) \, i) = 5.8686 - 1.2475 \, i$

41. $z_1 z_2 = 4 \cdot 3 \left(\cos (\pi/3 + \pi/6) + i \sin (\pi/3 + \pi/6) \right) = 12 (\cos \pi/2 + i \sin \pi/2)$

$z_1 / z_2 = 4/3 \left(\cos (\pi/3 - \pi/6) + i \sin (\pi/3 - \pi/6) \right) = 4/3 (\cos \pi/6 + i \sin \pi/6)$

43. $z_1 z_2 = 1 \cdot 2/3 \left(\cos (2 + (-1)) + i \sin (2 + (-1)) \right) = 2/3 (\cos 1 + i \sin 1)$

$z_1 / z_2 = 1 / (2/3) \left(\cos (2 - (-1)) + i \sin (2 - (-1)) \right) = 3/2 (\cos 3 + i \sin 3)$

45. $z_1 z_2 = 5 \cdot 1/7 \left(\cos (- 95° + 9°) + i \sin (- 95° + 9°) \right) = 5/7 \left(\cos (- 86°) + i \sin (- 86°) \right)$

$z_1 / z_2 = 5/ (1/7) \left(\cos (- 95° - 9°) + i \sin (- 95° - 9°) \right) = 35 \left(\cos (- 104°) + i \sin (-104°) \right)$

47. Here is the triangle inequality: $| z_1 + z_2 | \leq | z_1| + | z_2|$

Let $z_1 = a_1 + b_1 i = r_1 (\cos \theta_1 + i \sin \theta_1)$ and let $z_2 = a_2 + b_2 i = r_2 (\cos \theta_2 + i \sin \theta_2)$
Note: from these "lets" $| z_1| = r_1$ and $| z_2| = r_2$

We must show that $| z_1 + z_2 | = | z_1| + | z_2|$ if and only if $\theta_1 = \theta_2$.
THE IF PART:
assume $\theta_1 = \theta_2$ and start with the left side of what we must show:

$$| z_1 + z_2| = | r_1(\cos \theta_1 + i \sin \theta_1) + r_2(\cos \theta_1 + i \sin \theta_1) | \quad \text{(since } \theta_1 = \theta_2 \text{)}$$
$$= | (r_1 + r_2) (\cos \theta_1 + i \sin \theta_1) | \qquad \text{(by factoring)}$$
$$= r_1 + r_2 \qquad \text{(definition of absolute value)}$$
$$= | z_1| + | z_2| \qquad \text{(substitution -- see the Note above)}$$

47. (continued)

      THE ONLY IF PART:

      assume that $|z_1 + z_2| = |z_1| + |z_2|$

      We must show that this forces θ_1 to equal θ_2 !

      Our assumption in this part means that we are assuming that

$$|r_1(\cos\theta_1 + i\sin\theta_1) + r_2(\cos\theta_2 + i\sin\theta_2)| = r_1 + r_2$$

      grouping real and imaginary parts on the left gives us

$$|(r_1\cos\theta_1 + r_2\cos\theta_2) + i(r_1\sin\theta_1 + r_2\sin\theta_2)| \text{ which by definition is}$$

$$\sqrt{(r_1\cos\theta_1 + r_2\cos\theta_2)^2 + (r_1\sin\theta_1 + r_2\sin\theta_2)^2}$$

      if this is expanded and the $\sin^2 x + \cos^2 x = 1$ identity is used after some rearranging and factoring you should get

$$\sqrt{r_1^2 + 2r_1r_2(\cos\theta_1\cos\theta_2 + \sin\theta_1\sin\theta_2) + r_2^2} \text{ and this will equal the desired } r_1 + r_2$$

      only if $\cos\theta_1\cos\theta_2 + \sin\theta_1\sin\theta_2 = 1$ and this is the same as requiring that
$\cos(\theta_1 - \theta_2) = 1$ and this happens if $\theta_1 - \theta_2 = 0$ which implies that $\theta_1 = \theta_2$ which is
what we had to prove in this the ONLY IF part of the proof.

49.   Let $z = x + yi$ and let $z_0 = h + ki$

      The stipulation $|z - z_0| = a$ becomes $|(x + yi) - (h + ki)| = a$ which is equivalent to

      saying that $|(x - h) + (y - k)i| = a$ and this is the same as insisting that

$$\sqrt{(x - h)^2 + (y - k)^2} = a \text{ and if you square both sides you find yourself with the equation}$$

of a circle centered at (h, k), ie., at z_0 with a radius of a.

Problems 9.10

1. $(1+i)^{29} = \left(\sqrt{2}\left(\cos \pi/4 + i \sin \pi/4\right)\right)^{29} = \sqrt{2}^{29}\left(\cos 29\pi/4 + i \sin 29\pi/4\right)$

$= \sqrt{2}^{29}\left(\cos 5\pi/4 + i \sin 5\pi/4\right)$ Note: divided by 29 by 8 since $2\pi = \frac{8\pi}{4}$ to see the $\frac{5\pi}{4}$.

$= \sqrt{2}^{29}\left(-1/\sqrt{2} + i\left(-1/\sqrt{2}\right)\right) = -\sqrt{2}^{28} - \sqrt{2}^{28}\, i = -2^{14} - 2^{14}\, i$

3. $(1-i)^{15} = \left(\sqrt{2}\left(\cos(-\pi/4) + i \sin(-\pi/4)\right)\right)^{15} = \sqrt{2}^{15}\left(\cos(-15\pi/4) + i \sin(-15\pi/4)\right)$

$= 2^7 \sqrt{2}\left(\cos(-7\pi/4) + i \sin(-7\pi/4)\right) = 128\sqrt{2}\left(\cos \pi/4 + i \sin \pi/4\right) = 128\sqrt{2}\left(\frac{1}{\sqrt{2}} + i\frac{1}{\sqrt{2}}\right)$

$= 128 + 128\, i$

5. $(\sqrt{3} + i)^{15} = \left(2\left(\cos \pi/6 + i \sin \pi/6\right)\right)^{15} = 2^{15}\left(\cos 15\pi/6 + i \sin 15\pi/6\right)$

$= 2^{15}\left(\cos \pi/2 + i \sin \pi/2\right) = 2^{15}\left(0 + i\right) = 2^{15}i = 32{,}768\, i$

7. $(-1 - \sqrt{3}\,i)^{10} = \left(2\left(\cos(-2\pi/3) + i \sin(-2\pi/3)\right)\right)^{10} = 2^{10}\left(\cos(-20\pi/3) + i \sin(-20\pi/3)\right)$

$= 2^{10}\left(\cos(-2\pi/3) + i \sin(-2\pi/3)\right) = 2^{10}\left(-1/2 + (-\sqrt{3}/2)\,i\right) = -2^9 - 2^9\sqrt{3}\,i = -512 - 512\sqrt{3}\,i$

9. $(-4 + 3\,i)^5 = \left(5\left(\cos 2.4981 + i \sin 2.4981\right)\right)^5 = 5^5\left(\cos 12.4905 + i \sin 12.4905\right)$

$= 5^5\left(0.9971 + -0.0758\,i\right) = 3{,}116 - 237\, i$

11. $z^2 = 1$ can only have two solutions; $z = 1$ and $z = -1$ are they.

13. In solving $z^3 = 1$ the first step is to put 1 into the polar form $1\left(\cos 0 + i \sin 0\right)$

$z^3 = 1\left(\cos 0 + i \sin 0\right)$ has solutions $\sqrt[3]{1}\left(\cos \dfrac{0 + 2k\pi}{3} + i \sin \dfrac{0 + 2k\pi}{3}\right)$ for $k = 0,1,2$

13. (continued)

$$z_1 = 1 \ (\ \cos 0 + i \sin 0 \ ) = 1 \ (\ 1 + 0 \ ) = 1 \qquad \text{note: here k was 0}$$

$$z_2 = 1 \ (\ \cos 2\pi/3 + i \sin 2\pi/3 \ ) = 1 \ (\ - 1/2 + i \ (\sqrt{3}/2) \ ) = - 1/2 + \sqrt{3} \ i/2 \qquad \text{here, k was 1}$$

$$z_3 = 1 \ (\ \cos 4\pi/3 + i \sin 4\pi/3 \ ) = 1 \ (\ - 1/2 + (- \ \sqrt{3}/2) \ i \ ) = - 1/2 - \sqrt{3} \ i/2 \qquad \text{here, k was 2}$$

15. $z^2 = - i$ is the same as $z^2 = 1 \ (\ \cos \ (-\pi/2) + i \sin \ (-\pi/2) \ ) $ which will have as solutions

$$\sqrt{1} \ \left(\ \cos \frac{(-\pi/2) + 2k\pi}{2} + i \sin \frac{(-\pi/2) + 2k\pi}{2} \ \right) \text{ for k} = 0,1$$

if k = 0 we get $z_1 = 1 \ (\ \cos \ (-\pi/4) + i \sin \ (-\pi/4) \ ) = \ \sqrt{2}/2 - \sqrt{2} \ i/2$

if k = 1 we get $z_2 = 1 \ (\ \cos 3\pi/4 + i \sin 3\pi/4 \ ) = - \sqrt{2}/2 + \sqrt{2} \ i/2$

17. $z^2 = 2 - i$ is equivalent to $z^2 = \sqrt{5} \ (\ \cos \ (- \ 0.4636 \ ) + i \sin \ (- \ 0.4636 \ ))$ which has solutions

$$\sqrt{\sqrt{5}} \ \left(\ \cos \frac{- \ 0.4636 + 2k\pi}{2} + i \sin \frac{- \ 0.4636 + 2k\pi}{2} \ \right) \text{ for k} = 0, \ 1$$

which yields: $z_1 = 1.4953 \ (\ \cos \ (- \ 0.2318) + i \sin \ (- \ 0.2318) = 1.4553 - 0.3435 \ i$

$z_2 = 1.4953 \ (\ \cos \ 2.9098 + i \sin 2.9098 \ ) = - 1.4553 + 0.3435 \ i$

19. $z^3 = 2 + i = \sqrt{5} \ \left(\ \cos 0.4636 + i \sin 0.4636 \ \right)$ would have three solutions of the form:

$$\sqrt[3]{\sqrt{5}} \ \left(\ \cos \frac{0.4636 + 2k\pi}{3} + i \sin \frac{0.4636 + 2k\pi}{3} \ \right) \text{ for k} = 0, \ 1, \ 2$$

they are:

$$z_1 = 1.30766 \ (\ \cos 0.1545 + i \sin 0.1545 \ ) = 1.2921 + 0.2012 \ i$$
$$z_2 = 1.30766 \ (\ \cos 2.2489 + i \sin 2.2489 \ ) = - 0.8203 + 1.0184 \ i$$
$$z_3 = 1.30766 \ (\ \cos 4.3433 + i \sin 4.3433 \ ) = - 0.4718 - 1.2196 \ i$$

21. $z^4 = i = 1 (\cos \pi/2 + i \sin \pi/2)$ will have four solutions each of the form:

$$\sqrt[4]{1} \left(\cos \frac{\pi/2 + 2k\pi}{4} + i \sin \frac{\pi/2 + 2k\pi}{4} \right) \quad \text{for } k = 0, 1, 2, 3$$

Here they are:

$$z_1 = 1 (\cos \pi/8 + i \sin \pi/8) = 0.9239 + 0.3827 \, i$$
$$z_2 = 1 (\cos 5\pi/8 + i \sin 5\pi/8) = - 0.3827 + 0.9239 \, i$$
$$z_3 = 1 (\cos 9\pi/8 + i \sin 9\pi/8) = - 0.9239 - 0.3827 \, i$$
$$z_4 = 1 (\cos 13\pi/8 + i \sin 13\pi/8) = 0.3827 - 0.9239 \, i$$

23. $z^8 = i = 1 (\cos \pi/2 + i \sin \pi/2)$ will have 8 answers each of the form:

$$\sqrt[8]{1} \left(\cos \frac{\pi/2 + 2k\pi}{8} + i \sin \frac{\pi/2 + 2k\pi}{8} \right) \text{ for } k = 0, 1, 2, 3, 4, 5, 6, 7$$

and they are:

$$z_1 = \cos \pi/16 + i \sin \pi/16 = 0.9808 + 0.1951 \, i$$
$$z_2 = \cos 5\pi/16 + i \sin \pi/16 = 0.5556 + 0.8315 \, i$$
$$z_3 = \cos 9\pi/16 + i \sin 9\pi/16 = - 0.1951 + 0.9808 \, i$$
$$z_4 = \cos 13\pi/16 + i \sin 13\pi/16 = - 0.8315 + 0.5556 \, i$$
$$z_5 = \cos 17\pi/16 + i \sin 17\pi/16 = - 0.9808 - 0.1951 \, i$$
$$z_6 = \cos 21\pi/16 + i \sin 21\pi/16 = - 0.5556 - 0.8315 \, i$$
$$z_7 = \cos 25\pi/16 + i \sin 25\pi/16 = 0.1951 - 0.9808 \, i$$
$$z_8 = \cos 29\pi/16 + i \sin 29\pi/16 = 0.8315 - 0.5556 \, i$$

25. $z^4 = 1 - i = \sqrt{2} \left(\cos (- \pi/4) + i \sin (-\pi/4) \right)$ has four solutions gotten from:

$$\sqrt[4]{\sqrt{2}} \left(\cos \frac{(-\pi/4) + 2k\pi}{4} + i \sin \frac{(-\pi/4) + 2k\pi}{4} \right) \text{ for } k = 0, 1, 2, 3$$

$$z_1 = 1.0905 (\cos (-\pi/16) + i \sin (-\pi/16)) = 1.0695 - 0.2127 \, i$$
$$z_2 = 1.0905 (\cos 7\pi/16 + i \sin 7\pi/16) = 0.2127 + 1.0695 \, i$$
$$z_3 = 1.0905 (\cos 15\pi/16 + i \sin 15\pi/16 = - 1.0695 + 0.2127 \, i$$
$$z_4 = 1.0905 (\cos 23\pi/16 + i \sin 23\pi/16) = - 0.2127 - 1.0695 \, i$$

27. $z^5 = 32\,i = 32\ (\cos \pi/2 + i \sin \pi/2\ )$ has five solutions gotten from:

$$\sqrt[5]{32}\ \left(\ \cos \frac{\pi/2 + 2k\pi}{5} + i \sin \frac{\pi/2 + 2k\pi}{5}\ \right)\ \text{for k} = 0, 1, 2, 3, 4$$

so...

$z_1 = 2\ (\ \cos \pi/10 + i \sin \pi/10\ ) = 1.9021 + 0.6180\ i$

$z_2 = 2\ (\ \cos 5\pi/10 + i \sin 5\pi/10\ ) = 2i$

$z_3 = 2\ (\ \cos 9\pi/10 + i \sin 9\pi/10\ ) = -1.9021 + 0.6180\ i$

$z_4 = 2\ (\ \cos 13\pi/10 + i \sin 13\pi/10\ ) = -1.1756 - 1.6180\ i$

$z_5 = 2\ (\ \cos 17\pi/10 + i \sin 17\pi/10\ ) = 1.1756 - 1.6180\ i$

29. $z^6 = -64i = 64\ (\ \cos \frac{3\pi}{2} + i \sin \frac{3\pi}{2}\ )$ has six solutions obtained from:

$$\sqrt[6]{64}\ \left(\ \cos \frac{3\pi/2 + 2k\pi}{6} + i \sin \frac{3\pi/2 + 2k\pi}{6}\ \right)\ \text{for k} = 0, 1, 2, 3, 4, 5$$

$z_1 = 2\ (\ \cos \pi/4 + i \sin \pi/4\ ) = 1.4142 + 1.4142\ i$

$z_2 = 2\ (\ \cos 7\pi/12 + i \sin 7\pi/12\ ) = -0.5176 + 1.9319\ i$

$z_3 = 2\ (\ \cos 11\pi/12 + i \sin 11\pi/12\ ) = -1.9319 + 0.5176\ i$

$z_4 = 2\ (\ \cos 15\pi/12 + i \sin 15\pi/12\ ) = -1.4142 - 1.4142\ i$

$z_5 = 2\ (\ \cos 19\pi/12 + i \sin 19\pi/12\ ) = 0.5176 - 1.9319\ i$

$z_6 = 2\ (\ \cos 23\pi/12 + i \sin 23\pi/12\ ) = 1.9319 - 0.5176\ i$

31. $z^4 = -16 + 16\,i = \sqrt{512}\ (\ \cos 3\pi/4 + i \sin 3\pi/4\ )$ has four solutions obtained from:

$$\sqrt[4]{\sqrt{512}}\ \left(\ \cos \frac{3\pi/4 + 2k\pi}{4} + i \sin \frac{3\pi/4 + 2k\pi}{4}\ \right)\ \text{for k} = 0, 1, 2, 3$$

they are:

$z_1 = 2.181\ (\ \cos 3\pi/16 + i \sin 3\pi/16\ ) = 1.8134 + 1.2117\ i$

$z_2 = 2.181\ (\ \cos 11\pi/16 + i \sin 11\pi/16\ ) = -1.2117 + 1.8134\ i$

$z_3 = 2.181\ (\ \cos 19\pi/16 + i \sin 19\pi/16\ ) = -1.8134 - 1.2117\ i$

$z_4 = 2.181\ (\ \cos 27\pi/16 + i \sin 27\pi/16\ ) = 1.2117 - 1.8134\ i$

33. $z^4 = 16 - 16\,i = \sqrt{512}\,(\cos(-\pi/4) + i\sin(-\pi/4))$ will also have four roots:

$$2.181\left(\cos\frac{(-\pi/4) + 2k\pi}{4} + i\sin\frac{(-\pi/4) + 2k\pi}{4}\right) \text{ for } k = 0, 1, 2, 3$$

$$z_1 = 2.181\,(\cos(-\pi/16) + i\sin(-\pi/16)) = 2.1391 - 0.4255\,i$$
$$z_2 = 2.181\,(\cos 7\pi/16 + i\sin 7\pi/16) = 0.4255 + 2.1391\,i$$
$$z_3 = 2.181\,(\cos 15\pi/16 + i\sin 15\pi/16) = -2.1391 + 0.4255\,i$$
$$z_4 = 2.181\,(\cos 23\pi/16 + i\sin 23\pi/16) = -0.4255 - 2.1391\,i$$

35. H: z_k is an n th root of unity; that is, $(z_k)^n = 1$

$z_k \neq 1$

C: $1 + z_k + z_k^2 + \ldots + z_k^{n-1} = 0$

Proof: consider the polynomial $P(z) = z^n - 1$

since by our hypothesis $z_k^n = 1$ we must have z_k as a zero of this polynomial which means that $z - z_k$ is a factor of $P(z)$ and if $P(z)$ is divided by $z - z_k$ we can get at the other factor. We'll use synthetic division to do this:

$$z_k\,)\overline{\quad 1 \quad 0 \qquad 0 \,\ldots\, 0 \qquad -1 \quad}$$

with middle row $z_k \quad z_k^2 \ldots z_k^{n-1} \quad z_k^n$

and bottom row $1 \quad z_k \quad z_k^2 \ldots z_k^{n-1} \quad \boxed{z_k^n - 1}$ (this remainder is 0 by the way)

at this point we can piece together two factors of $P(z)$:

$$P(z) = z^n - 1 = (z - z_k)(1 \cdot z^{n-1} + z_k \cdot z^{n-2} + z_k^2 \cdot z^{n-3} + \ldots + z_k^{n-1})$$

we know that $P(1) = 0$ so substitute a 1 for each z in the above factorization:

$$(1 - z_k)(1 + z_k + z_k^2 + \ldots + z_k^{n-1}) = 0$$

and since z_k can't be 1 it cannot be the first factor that's 0 so it must be the second one and that is the desired conclusion.

37. H: $\operatorname{sgn} b = \dfrac{b}{|b|}$ and $z = a + b\,i$

   C: $\sqrt{z} = \pm\left(\sqrt{\dfrac{|z| + a}{2}} + \operatorname{sgn} b \sqrt{\dfrac{|z| - a}{2}}\,i \right)$

Proof: what we'll do is square the right side of the conclusion and see if we wind up with z

(the right side squared) $= \dfrac{|z| + a}{2} + 2\sqrt{\dfrac{|z| + a}{2}}\,\operatorname{sgn} b \sqrt{\dfrac{|z| - a}{2}}\,i + (\operatorname{sgn} b)^2 \cdot \dfrac{|z| - a}{2}\,i^2$

$= \dfrac{|z|}{2} + \dfrac{a}{2} + \operatorname{sgn} b\,\sqrt{|z|^2 - a^2}\,i + \dfrac{b^2}{|b|^2} \cdot \dfrac{|z| - a}{2} \cdot (-1)$

$= \dfrac{|z|}{2} + \dfrac{a}{2} + \dfrac{b}{|b|}\sqrt{(a^2 + b^2) - a^2}\,i - \dfrac{|z|}{2} + \dfrac{a}{2}$

$= a + \dfrac{b}{|b|}\sqrt{b^2}\,i$

$= a + \dfrac{b}{|b|}\cdot |b|\,i = a + b\,i = z$

39. When finding $\sqrt{2 + 3\,i}$ use the formula proven in # 37 with $a = 2$, $b = 3$, $|z| = \sqrt{13}$, $\operatorname{sgn} b = 1$

   to get: $\pm\left(\sqrt{\dfrac{\sqrt{13} + 2}{2}} + \sqrt{\dfrac{\sqrt{13} - 2}{2}}\,i \right) \approx \pm\,(\,1.674 + 0.896\,i\,)$

41. The square roots of $-4 - 5\,i$ will involve the formula in #37 with $a = -4$, $b = -5$, $|z| = \sqrt{41}$, and $\operatorname{sgn} b = -1$:

   the roots are $\pm\left(\sqrt{\dfrac{\sqrt{41} - 4}{2}} - \sqrt{\dfrac{\sqrt{41} + 4}{2}}\,i \right) \approx \pm\,(\,1.0962 - 2.2807\,i\,)$

Chapter Review for 9

1. Use law of sines after finding the third angle to be $(180 - 35 - 75)° = 70°$

$$\frac{a}{\sin A} = \frac{b}{\sin B} = \frac{c}{\sin C} \quad \text{which becomes} \quad \frac{a}{\sin 35°} = \frac{7}{\sin 70°} = \frac{c}{\sin 75°}$$

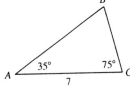

so $a = \dfrac{7 \sin 35°}{\sin 70°} = 4.273$

and $c = \dfrac{7 \sin 75°}{\sin 70°} = 7.195$

3. 2 sides & included angle $\Rightarrow$ law of cosines!

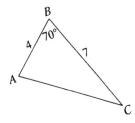

$b^2 = a^2 + c^2 - 2\,a\,c\,\cos B$

$\quad = 49 + 16 - 56 \cos 70°$

$\quad = 45.846 \quad$ so $\ b = 6.771$

Now switch to law of sines:

$$\frac{b}{\sin B} = \frac{a}{\sin A} \quad \text{becomes} \quad \frac{6.771}{\sin 70°} = \frac{7}{\sin A} \quad \text{so} \quad A = \sin^{-1}\frac{7 \sin 70°}{6.771} = 76.3°$$

$\angle\,C$ is found by subtracting $\angle\,A$ and $\angle\,B$ from $180°$; it's $33.7°$

5. $\angle\,C$ in the figure is $(180 - 20 - 120)° = 40°$

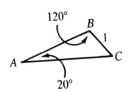

$$\frac{c}{\sin 40°} = \frac{1}{\sin 20°} = \frac{b}{\sin 120°}$$

so $c = \dfrac{\sin 40°}{\sin 20°} = 1.879 \quad$ and $b = \dfrac{\sin 120°}{\sin 20°} = 2.532$

7. 3 sides given $\Rightarrow$ use law of cosines reversed!

$$\cos C = \frac{a^2 + b^2 - c^2}{2\,a\,b} = \frac{25 + 16 - 64}{40} = -0.575$$

so $\angle C = \cos^{-1}(-0.575) = 125.1°$

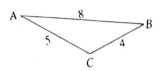

similarly you can find $\angle B = 30.8°$ and then

that leaves $24.1°$ for $\angle A$

9. $\angle A = (180 - 40 - 95)° = 45°$

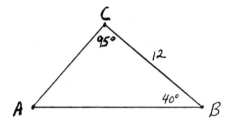

$$\frac{c}{\sin 95°} = \frac{12}{\sin 45°} = \frac{b}{\sin 40°} \quad (\text{law of sines})$$

solving the left proportion: the right proportion:

$$c = \frac{12 \sin 95°}{\sin 45°} = 16.906 \qquad b = \frac{12 \sin 40°}{\sin 45°} = 10.908$$

11. $c^2 = 16 + 25 - 2(4)(5)\cos 47°$ (law of cosines)

$\quad = 13.72$

$c \approx 3.704$ and now switch to law of sines:

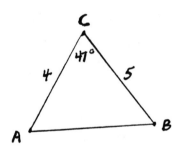

$$\frac{5}{\sin A} = \frac{3.704}{\sin 47°} \text{ so } \angle A = \sin^{-1}\left(\frac{5 \sin 47°}{3.704}\right) \approx 80.8°$$

so there are $(180 - 47 - 80.8)°$ or $52.2°$ left for $\angle B$

13. $\cos C = \dfrac{a^2 + b^2 - c^2}{2\,a\,b} = \dfrac{49 + 4 - 64}{2(7)(2)} = -0.3929$

so $\angle C = \cos^{-1}(-0.3929) = 113.1°$; now to law of sines:

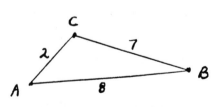

$$\frac{7}{\sin A} = \frac{8}{\sin 113.1°} \text{ so } \angle A = \sin^{-1}\left(\frac{7 \sin 113.1°}{8}\right) = 53.6°$$

which leaves $13.3°$ for $\angle B$.

15. The more accurate the sketch, the better (especially when there might be two triangles).

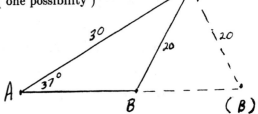

First use law of sines:

$$\frac{20}{\sin 37°} = \frac{30}{\sin B} \quad \text{so } \angle B = \sin^{-1}\left(\frac{30 \sin 37°}{20}\right) = 64.5° \text{ (one possibility)}$$

This forces $\angle C$ to be $78.5°$ and the law of sines
can be used to find c: $\quad \dfrac{c}{\sin 78.5°} = \dfrac{20}{\sin 37°}$
so here $c = \dfrac{20 \sin 78.5°}{\sin 37°} = 32.56$

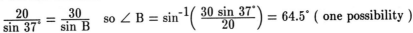

There is also a chance that $\angle B$ could be $115.5°$, the supplement of $64.5°$

This would force $\angle C$ to be only $27.5°$; now there's a new value of c to

be found from: $\qquad \dfrac{c}{\sin 27.5°} = \dfrac{20}{\sin 37°}$

From this c is 15.35

17. The law of sines applies: $\dfrac{30}{\sin 47°} = \dfrac{20}{\sin B}$

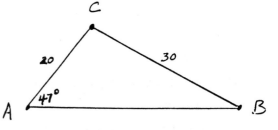

so $\angle B = \sin^{-1}\left(\dfrac{20 \sin 47°}{30}\right) = 29.2°$ which
forces $\angle C$ to be $103.8°$

Another application of the law of sines will give c:

$$\frac{30}{\sin 47°} = \frac{c}{\sin 103.8°} \qquad \text{so } c = 39.83$$

19. Trying the law of sines yields: $\dfrac{3}{\sin 107°} = \dfrac{5}{\sin B}$ from which $\angle B = \sin^{-1}\left(\dfrac{5 \sin 107°}{3}\right)$

$= \sin^{-1} 1.59$ and this is impossible!

21. Use Heron's formula: $\text{Area} = \sqrt{s(s - a)(s - b)(s - c)}$ where $s = \frac{1}{2}(a + b + c)$

Here $\quad s = \frac{1}{2}(5 + 6 + 7) = 9$; $\quad \text{Area} = \sqrt{9(9 - 5)(9 - 6)(9 - 7)} = \sqrt{216} \approx 14.697$ sq. units

23. 2 sides and an included angle given $\Rightarrow$ use law of cosines

$$AB^2 = 112^2 + 89^2 - 2\,(89)\,(112) \cos 23.33°$$
$$= 2159 \quad \text{so } AB \approx 46.46 \text{ ft}$$

25. We'll find EF and AF and then add them to the elevation

at C: $\sin 7.417° = \dfrac{BD \text{ (an auxilliary line)}}{516}$ so

BD = 66.61

We need BA in order to find AF:

law of sines: $\dfrac{BA}{\sin \angle ACB} = \dfrac{CB}{\sin \angle CAB}$ becomes

$$\dfrac{BA}{\sin 38.083} = \dfrac{516}{\sin 5.833°}$$

Note: think about those angles!

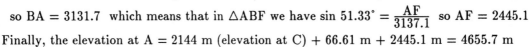

so BA = 3131.7 which means that in △ABF we have $\sin 51.33° = \dfrac{AF}{3137.1}$ so AF = 2445.1

Finally, the elevation at A = 2144 m (elevation at C) + 66.61 m + 2445.1 m = 4655.7 m

27. Let P represent the port, A represent the point at which
the ship changed course, and B represent the ship's final
location. We are asked to find BP and to do this with the
law of cosines we'll need ∠ PAB. ∠ PAS = 25° by alternate
interior angles and ∠ SAB = 82° because of the second
bearing given in the problem. Therefore, ∠ PAB = 107°
and $PB^2 = 102^2 + 88^2 - 2(102)(88) \cos 107°$

$$= 10404 + 7744 + 5248.66$$

and PB ≈ 153 km

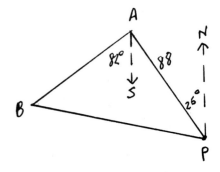

29. A = airplane; P = radar station; Q = 2nd radar station

The bearing asked for is ∠ NQA (same as ∠ QAS)

∠ QPA = ∠ NPQ (87.5°) - 16.25° = 71.25°

QA can be found by the law of cosines:

$QA^2 = 94^2 + 186^2 - 2(94)(186) \cos 71.25°$ so QA = 179.4

Now we can find ∠ PAQ (law of cosines--we know 3 sides)

and then subtract ∠ PAS (also 16.25°) in order to obtain ∠ QAS.

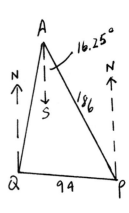

$\cos \angle PAQ = \dfrac{179.4^2 + 186^2 - 94^2}{2(179.4)(186)} = 0.86825$ so ∠ PAQ = 29.74° which means that

∠ QAS = 29.74° - 16.25° = 13.5° = ∠ NQA so the bearing is N 13°30' E

31. $v = -3i + 3j$ has magnitude $\sqrt{(-3)^2 + 3^2} = \sqrt{18} = 3\sqrt{2}$

v lies in quadrant II; therefore $\theta = \pi + \tan^{-1}\dfrac{3}{(-3)} = \pi + (-\pi/4) = 3\pi/4$

33. $v = (\sqrt{3}, 1)$ is in quadrant I and has magnitude $|v| = \sqrt{\sqrt{3}^2 + 1^2} = \sqrt{3+1} = 2$

$\theta = \tan^{-1}\dfrac{1}{\sqrt{3}} = \pi/6$

35. If $P = (1, -2)$ and $Q = (7, 12)$ the vector $PQ = (7-1, 12-(-2)) = (6, 14)$

37. (a) $-3v = -3(-3, -4) = (9, 12)$

    (b) $u + v = (-4 + -3, 1 + -4) = (-7, -3)$

    (c) $3u - 6v = 3(-4, 1) - 6(-3, -4) = (-12 + 18, 3 + 24) = (6, 27)$

39. $\dfrac{v}{|v|} = \dfrac{2i + 5j}{\sqrt{2^2 + 5^2}} = \dfrac{2}{\sqrt{29}}i + \dfrac{5}{\sqrt{29}}j$
            41. $\dfrac{v}{|v|} = \dfrac{3i + 4j}{5} = \dfrac{3}{5}i + \dfrac{4}{5}j$

43. $\dfrac{v}{|v|} = \dfrac{ai - aj}{\sqrt{a^2 + a^2}} = \dfrac{a}{|a|\sqrt{2}}i - \dfrac{a}{|a|\sqrt{2}}j$
      45. $-\dfrac{v}{|v|} = -\dfrac{5i + 2j}{\sqrt{29}} = -\dfrac{5}{\sqrt{29}}i - \dfrac{2}{\sqrt{29}}j$

47. $-\dfrac{v}{|v|} = -\dfrac{10i - 7j}{\sqrt{149}} = -\dfrac{10}{\sqrt{149}}i + \dfrac{7}{\sqrt{149}}j$

49. $v = |v|$ (unit vector in direction $\pi/2$) $= 1(\cos \pi/2\,i + \sin \pi/2\,j) = 1(0i + 1j) = j$.

51. $v = 7(\cos 5\pi/6\,i + \sin 5\pi/6\,j) = 7(-\sqrt{3}/2\,i + 1/2\,j) = -\dfrac{7\sqrt{3}}{2}i + \dfrac{7}{2}j$.

53. $u \cdot v = (-4)(0) + (0)(11) = 0$; $\cos \phi = \dfrac{u \cdot v}{|v|\,|u|} = \dfrac{0}{4(11)} = 0$.

55. $u \cdot v = (-1)(4) + (-2)(5) = -14$; $\cos \phi = \dfrac{-14}{(\sqrt{5})(\sqrt{41})} = -\dfrac{14}{\sqrt{205}}$

57. $u \cdot v = (4)(5) + (-5)(-4) = 40$ so $\cos \phi$ (which

equals $\dfrac{\text{dot product}}{\text{product of magnitudes}}$) will not be 0 (so

not orthogonal); also, $\cos \phi$ will not equal 1 because

the product of the magnitudes is 41 so these vectors are neither orthogonal nor parallel.

59. $\cos \phi = \dfrac{(-7)(1) + (-7)(1)}{\sqrt{98}\ \sqrt{2}} = \dfrac{-14}{\sqrt{196}} = \dfrac{-14}{14} = -1$

so these vectors are parallel ($\phi = \pi$).

61. $\cos \phi = \dfrac{(-7)(-1) + (-7)(-1)}{\sqrt{98}\ \sqrt{2}} = \dfrac{14}{14} = 1$

so these vectors are parallel ($\phi = 0$).

63. $\text{Proj}_\mathbf{v}\mathbf{u} = \dfrac{\mathbf{u} \cdot \mathbf{v}}{|\mathbf{v}|^2}\ \mathbf{v} = \dfrac{(14)(1) + (0)(1)}{(\sqrt{2})^2}\ (\mathbf{i} + \mathbf{j}) = 7\mathbf{i} + 7\mathbf{j}$

65. $\text{Proj}_\mathbf{v}\mathbf{u} = \dfrac{\mathbf{u} \cdot \mathbf{v}}{|\mathbf{v}|^2}\ \mathbf{v} = \dfrac{(3)(3) + (-2)(2)}{\sqrt{13}^2}\ (3\mathbf{i} + 2\mathbf{j}) = \dfrac{5}{13}(3\mathbf{i} + 2\mathbf{j}) = \dfrac{15}{13}\mathbf{i} + \dfrac{10}{13}\mathbf{j}$

67. $\text{Proj}_\mathbf{v}\mathbf{u} = \dfrac{(2)(-3) + (-5)(-7)}{(\sqrt{58})^2}\ (-3\mathbf{i} - 7\mathbf{j}) = \dfrac{1}{2}(-3\mathbf{i} - 7\mathbf{j}) = -\dfrac{3}{2}\mathbf{i} - \dfrac{7}{2}\mathbf{j}$

69. unit vector in direction of $\theta = \cos \pi/4\ \mathbf{i} + \sin \pi/4\ \mathbf{j} = \dfrac{\sqrt{2}}{2}\mathbf{i} + \dfrac{\sqrt{2}}{2}\mathbf{j}$
so the full fledged force vector (F's magnitude times the unit vector above) is $\mathbf{F} = 2(\dfrac{\sqrt{2}}{2}\mathbf{i} + \dfrac{\sqrt{2}}{2}\mathbf{j})$
$$= \sqrt{2}\ \mathbf{i} + \sqrt{2}\ \mathbf{j}.$$

The direction that the motion is taking is represented by the vector $\mathbf{D} = (2 - 1)\mathbf{i} + (4 - 6)\mathbf{j}$
and so $\mathbf{D} = \mathbf{i} - 2\mathbf{j}$.
The work done is the dot product $\mathbf{F} \cdot \mathbf{D} = (\sqrt{2})(1) + (\sqrt{2})(-2) = -\sqrt{2}$ J.

71. unit vector in direction of $\theta = \cos \pi/6\ \mathbf{i} + \sin \pi/6\ \mathbf{j} = \dfrac{\sqrt{3}}{2}\mathbf{i} + \dfrac{1}{2}\mathbf{j}$

so$|\ \mathbf{F}\ |$ times the vector above will give us F: $\mathbf{F} = \dfrac{11\sqrt{3}}{2}\mathbf{i} + \dfrac{11}{2}\mathbf{j}$

$\mathbf{D} = (-7 - (-1))\mathbf{i} + (-4 - (-2))\mathbf{j} = -6\mathbf{i} - 2\mathbf{j}$ and so work $= \mathbf{F} \cdot \mathbf{D} = \left(\dfrac{11\sqrt{3}}{2}\right)(-6) + \left(\dfrac{11}{2}\right)(-2)$

$$= (-33\sqrt{3} - 11)\text{ J}.$$

73. From the height of 12 and the length of 45 given, we can find the angle of elevation θ of the plane which will be the same as $\angle$ O in the force triangle (OW represents the weight and NW the component of the weight DOWN the slide (this has to be 80 to match the force pulling up).

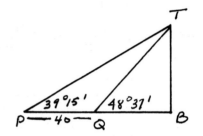

We seek the weight OW: $\sin \theta = \frac{12}{45} = 0.2667 \Rightarrow \theta = 15.466°$

In right$\triangle$ ONW we have the relationship $\sin O = \frac{\text{opp}}{\text{hyp}}$ which becomes $\sin 15.466° = \frac{80}{\text{OW}}$

so $OW = \frac{80}{\sin 15.466°} = 300$ pounds

75. P = first point on a road;　Q = closer point

B = bottom of pole　　T = top of pole

We seek TB. Plan: deduce some angles in $\triangle PQT$ and then find TQ by the law of sines. Then find TB in $\triangle QTB$ using ordinary right triangle trigonometry.

$\angle$ PQT $= 131.38°$ (the supplement of $48.62°$)

$\angle$ PTQ $= 9.37°$ ($180°$ cap on triangles)

the law of sines now says that $\frac{40}{\sin 9.37°} = \frac{TQ}{\sin 39.25°}$ so that TQ ≈ 155.5

in right $\triangle$ QBT $\sin \angle$ BQT $= \frac{\text{opp}}{\text{hyp}}$ so $\sin 48.62° = \frac{TB}{155.5}$ and TB ≈ 116.7 ft

77. From problem #76 we know that the boat is 318 km from X.

Main steps in arriving at that: 1) see that $\angle$ XAB is $63.433°$
2) find x to be 138 by law of cosines 3) find $\angle$ AXB to be $46.54°$ by using law of sines in $\triangle$AXB 4) see that $\angle$ NXB is $13.74°$ and $\angle$ XBF is $151.1°$ 6) use law of cosines in $\triangle$FBX to get 318.

Now, to find the bearing from the port X to the sub F we must determine $\angle$ NXF which can be seen to be an "east of north" bearing:

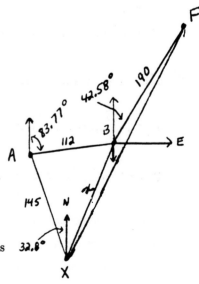

Using the law of sines in $\triangle$FBX: $\frac{190}{\sin \angle \text{ BXF}} = \frac{318}{\sin 151.1°}$

so $\sin \angle$ BXF ≈ 0.2888

so $\angle$ BXF $\approx 16.78°$ and adding this to $\angle$ BXN, $13.74°$, gives us $30.5°$ which would yield the bearing N $30.5°$ E

79. Wind is blowing from N 16°37' E at 60 mph

Want final bearing to be N 23°40'E

Want groundspeed to end up at 475 mph

Find take-off course and airspeed!

O = point of take off

OG = groundspeed   GW = opposite wind vector

OW = airspeed (what we're after)

Since airspeed + wind = groundspeed we could also write

airspeed = groundspeed + (opposite of wind)

Note: opposite wind is now IN the direction N 16°37' E

∠ OGW is findable: ∠ OGN is the supplement of

the 23.67° bearing which makes it 156.33°. ∠ NGW

meanwhile is the wind angle 16.62° so ∠ OGW is the

sum of these or 172.95°. Now we can use the law of cosines:

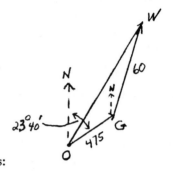

$$OW^2 = OG^2 + GW^2 - 2 \ (OG)(GW) \cos \angle OGW$$

$$= 475^2 + 60^2 - 2 \ (60)(475) \cos 172.95° = 285,797 \text{ and so OW, the airspeed, is 534.6 mph}$$

The bearing asked for is the angle between North and OW. We can find it by finding ∠ WOG

and subtracting it from 23.67°, the bearing for the groundspeed vector. Use law of sines:

$$\frac{534.6}{\sin 172.95°} = \frac{60}{\sin \angle WOG} \quad \text{which leads to } \sin \angle WOG = 0.01377 \text{ so } \angle WOG \approx 0.79°$$

which means the desired bearing is 23.67° - 0.79° = 22.88° or 22.9° east of north.

81. $r = 3$ and $\theta = \pi/6$ so $x = r \cos \theta = 3 \cos \pi/6 = 3 \ (\sqrt{3}/2) = \frac{3\sqrt{3}}{2}$

$$y = r \sin \theta = 3 \sin \pi/6 = 3 \ (1/2) = \tfrac{3}{2} \quad \Rightarrow \quad \left(\tfrac{3\sqrt{3}}{2}, \tfrac{3}{2}\right) \text{ is the}$$

solution.

83. $x = r\cos \theta = 4\cos \frac{2\pi}{3} = 4(-\tfrac{1}{2}) = -2; \ y = r\sin \theta = 4\sin \frac{2\pi}{3} = 4 \ (\frac{\sqrt{3}}{2}) = 2\sqrt{3}; \ \text{ans: } (-2, \ 2\sqrt{3})$

85. $r = \sqrt{x^2 + y^2} = \sqrt{4^2 + 0^2} = \sqrt{16} = 4 ; \ \theta = \tan^{-1} \tfrac{y}{x} = \tan^{-1} 0 = 0 ; \quad \text{answer: } (\ 4, \ 0 \ )$

87. $r = \sqrt{\sqrt{3}^2 + (-1)^2} = \sqrt{3 + 1} = 2; \ \theta \ (\text{QIV}) = 2\pi + \tan^{-1} \frac{-1}{\sqrt{3}} = 2\pi + (-\frac{\pi}{6}) = \frac{11\pi}{6}; \ \text{ans: } (2, \frac{11\pi}{6})$

89. $r = \sqrt{(-6)^2 + (-6)^2} = \sqrt{72} = 6\sqrt{2} ; \ \theta \ (\text{QIII}) = \pi + \tan^{-1} \frac{(-6)}{(-6)} = \pi + \pi/4 = 5\pi/4; \ \text{ans: } (6\sqrt{2}, \frac{5\pi}{4})$

91. $r = 8$ is equivalent to $\sqrt{x^2 + y^2} = 8$ or $x^2 + y^2 = 64$

which is a circle centered at the origin (pole) with a radius

of 8. The graph has all of the possible symmetries.

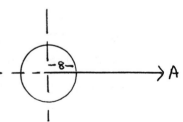

93. $r = 2 \cos \theta$ is equivalent to $r^2 = 2r \cos \theta$ which in turn is

equivalent to $x^2 + y^2 = 2x$ which needs its square completed: $\_ \_ \_ \_$

$x^2 - 2x + 1 + y^2 = 1$ or $(x - 1)^2 + y^2 = 1$ and this is

definitely a circle centered at $(1, 0)$ with a radius of 1.

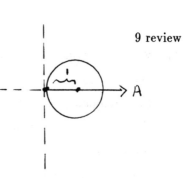

95. $r = 3 - 2 \sin \theta$ (a limacon without a loop since the coefficient of $\sin \theta$

isn't greater than the 3). Symmetry to the line $\theta = \pi/2$

so in making a table use values from 0 to $\pi/2$ and from π to $3\pi/2$

θ:	0	$\pi/6$	$\pi/4$	$\pi/3$	$\pi/2$
r:	3	2	$3 - \sqrt{2}$	$3 - \sqrt{3}$	1

θ:	π	$7\pi/6$	$5\pi/4$	$4\pi/3$	$3\pi/2$
r:	3	4	$3 + \sqrt{2}$	$3 + \sqrt{3}$	5

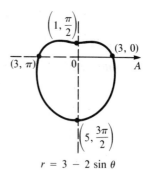

$r = 3 - 2 \sin \theta$

97. $r = 5 \cos 2\theta$ has symmetry to the polar axis (θ replacable by $- \theta$) so the

table need only go from 0 to π and when choosing values for θ

choose values that are half of familiar ones:

θ:	0	$\pi/12$	$\pi/8$	$\pi/6$	$\pi/4$
r:	5	$5\sqrt{3}/2$	$5\sqrt{2}/2$	$5/2$	0

Plotting these points gives the top half of one leaf of a rose;

By the time θ gets up to π in our table we will have four of

these "half leaves" or two leaves and the entire graph will

show four leaves.

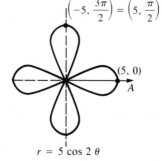

$r = 5 \cos 2\theta$

99. $r = 3 \sin 4\theta$ has all the symmetries. θ values from 0 to $\pi/2$

will do: (take 1/4 of the standard values for θ)

θ:	0	$\pi/24$	$\pi/16$	$\pi/12$	$\pi/8$
r:	0	$3/2$	$3\sqrt{2}/2$	$3\sqrt{3}/2$	3

we've gone 1/4 of the way to $\pi/2$ and we have 1/2 of a leaf.

Therefore, by the time we get to $\pi/2$ we'll have traced out

two leaves and the symmetry considerations will lead to the

final graph with 8 leaves.

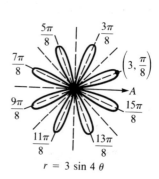

$r = 3 \sin 4\theta$

101. Multiply both sides by r to obtain $r^2 = 4 r \sin \theta - 6 r \cos \theta$ which is equivalent to
$x^2 + y^2 = 4y - 6x$ or $x^2 + y^2 + 6x - 4y = 0$ or $(x^2 + 6x + 9) + (y^2 - 4y + 4) = 9 + 4$
which is the same as $(x + 3)^2 + (y - 2)^2 = 13$.

103. $- 5 = - 5 + 0 i$ the absolute value, r, is $\sqrt{(-5)^2 + 0^2} = \sqrt{25} = 5$; the argument, θ, is given by
$\theta = \pi + \tan^{-1} 0/(-5) = \pi + 0 = \pi$; polar representation is r $(\cos \theta + i \sin \theta)$ which in this
case is $5 (\cos \pi + i \sin \pi)$.

105. $0 - \sqrt{2} i$ has $r = \sqrt{0^2 + (-\sqrt{2})^2} = \sqrt{2}$; $\theta = - \pi/2$ (sketch is easiest way to see this)
polar form: $\sqrt{2} [\cos (-\pi/2) + i \sin (-\pi/2)]$

107. $-2 + 2 i$ has $r = \sqrt{(-2)^2 + 2^2} = \sqrt{8} = 2\sqrt{2}$; a sketch reveals θ to be $3\pi/4$ and so the polar
form is $2\sqrt{2} (\cos 3\pi/4 + i \sin 3\pi/4)$

109. $-2 - 2 i$ has the same r-value as #73 ; a sketch here reveals θ to be $-3\pi/4$ which makes the
polar form $2\sqrt{2} (\cos -3\pi/4 + i \sin -3\pi/4)$

111. $4 - 7 i$ has $r = \sqrt{(4)^2 + (-7)^2} = \sqrt{65}$; θ is in QIV and is gotten from $\tan^{-1} (-7/4)$
$= - 1.052$ which makes the polar form $\sqrt{65} \left(\cos (- 1.052) + i \sin (- 1.052) \right)$

113. $- 3 - 5 i$ has $r = \sqrt{9 + 25} = \sqrt{34}$; θ is in QIII and is given by $- \pi + \tan^{-1} (-5)/(-3)$ or
$- \pi + 1.03 \approx - 2.111$ so the polar form is $\sqrt{34} [\cos (-2.111) + i \sin (-2.111)]$

115. $12 \left(\cos (-\pi/6) + i \sin (-\pi/6) \right) = 12 \left(\sqrt{3}/2 + i (- \frac{1}{2}) \right) = 6\sqrt{3} - 6 i$

117. $6 \left(\cos (-3\pi/4) + i \sin (-3\pi/4) \right) = 6 \left((-\sqrt{2}/2) + i (-\sqrt{2}/2) \right) = - 3\sqrt{2} - 3\sqrt{2} i$

119. $z_1 \cdot z_2 = 8 \cdot 4 [\cos (2 + 0.85) + i \sin (2 + 0.85)] = 16 (\cos 2.85 + i \sin 2.85)$

121. $(\sqrt{3} - i)^{12} = \left(2 [\cos (-\pi/6) + i \sin (-\pi/6)] \right)^{12} = 2^{12} (\cos (-2\pi) + i \sin (-2\pi)$
$= 4096(1 + 0) = 4096$

123. $(-2+7i)^6 = \left(\sqrt{53}\,(\cos 1.849 + i\sin 1.849\,)\right)^6 = 53^3\,(\cos 11.094 + i\sin 11.094\,)$

$$= 148{,}877\,[\,0.0983 + i\,(-0.995)\,] = 14{,}715 - 148{,}148\,i$$

Note: the angle of 1.849 was found by adding the inverse tangent of $7/(-2)$ to π

125. $z^3 = -1 + i = \sqrt{2}\left(\cos 3\pi/4 + i\sin 3\pi/4\right)$ will have three solutions each of the form

$$\sqrt[3]{\sqrt{2}}\left(\cos\frac{3\pi/4 + 2k\pi}{3} + i\sin\frac{3\pi/4 + 2k\pi}{3}\right)\text{ for } k = 0,\,1,\,2$$

$z_1 = \sqrt[6]{2}\,(\cos \pi/4 + i\sin \pi/4\,) = 0.7937 + 0.7937\,i$

$z_2 = \sqrt[6]{2}\,(\cos 11\pi/12 + i\sin 11\pi/12\,) = -1.0842 + 0.2905\,i$

$z_3 = \sqrt[6]{2}\,(\cos 19\pi/12 + i\sin 19\pi/12\,) = 0.2905 - 1.0842\,i$

127. $z^2 = -9i = 9\,(\cos(-\pi/2) + i\sin(-\pi/2))$ has 2 answers of the form

$$\sqrt{9}\left(\cos\frac{-\pi/2 + 2k\pi}{2} + i\sin\frac{-\pi/2 + 2k\pi}{2}\right)\text{ for } k = 0,\,1\ \text{ and they are:}$$

$z_1 = 3\,(\cos(-\pi/4) + i\sin(-\pi/4)\,) = 3\left(\frac{\sqrt{2}}{2} - \frac{\sqrt{2}}{2}i\right) = \frac{3\sqrt{2}}{2} - \frac{3\sqrt{2}}{2}i$

$z_2 = 3\,(\cos 3\pi/4 + i\sin 3\pi/4\,) = -3\left(-\frac{\sqrt{2}}{2} + \frac{\sqrt{2}}{2}i\right) = -\frac{3\sqrt{2}}{2} + \frac{3\sqrt{2}}{2}i$

129. $z^5 = -i = 1\,[\cos(-\pi/2) + i\sin(-\pi/2)\,]$ will have 5 roots of the form:

$$1\left(\cos\frac{(-\pi/2) + 2k\pi}{5} + i\sin\frac{(-\pi/2) + 2k\pi}{5}\right)\text{ for } k = 0,\,1,\,2,\,3,\,4$$

$z_1 = 1\,[\cos(-\pi/10) + i\sin(-\pi/10)\,] = 0.9511 - 0.3090\,i$

$z_2 = 1\,[\cos(3\pi/10) + i\sin(3\pi/10)\,] = 0.5878 + 0.8090\,i$

$z_3 = 1\,[\cos(7\pi/10) + i\sin(7\pi/10)\,] = -0.5878 + 0.8090\,i$

129. (continued)

$$z_4 = 1 \, [\, \cos \, (11\pi/10) + i \sin \, (11\pi/10) \,] = -0.9511 - 0.3090 \, i$$

$$z_5 = 1 \, [\, \cos \, (3\pi/2) + i \sin \, (3\pi/2) \,] = -i$$

CHAPTER TEN
Systems of Equations and Inequalities

Chapter Objectives.

In this chapter you learn about solving systems of equations and many other applications of matrices. In the first section you learn about solving small systems of linear equations. After you have become familiar with how to solve these simpler systems and some of the properties of these systems, you learn (in Section 2) a method called Gaussian elimination; this method gives you a way of more easily solving large systems of linear equations, and this method is well suited for computer programming. This method can also be used in the Section 4 on partial fraction decomposition, which is used to rewrite complicated rational expressions as a sum of simpler ones. It is a method which you will need if you take Calculus II, when techniques of integration are studied. Section 3 discusses solutions to systems of non-linear equations.

Next you will learn about systems of linear inequalities and their graphs, which are very important in the study of linear programming; this is a useful technique which is widely used to find the optimal solutions to a large class of problems. The last section of the text is a discussion of linear programming and some of its applications.

Chapter Summary

- **Gaussian elimination** is a procedure used for reducing a linear system of equations to a simpler form in which solutions are readily obtainable.

- An m x n **matrix** is a rectangular array of numbers with m rows and n columns.

- **Elementary row operations**
  i. Replace a row with a nonzero multiple of that row, denoted by $R_i \rightarrow cR_i$.
  ii. Replace a row with the sum of the row and a multiple of some other row, denoted by $R_j \rightarrow R_j + cR_i$.
  iii. Interchange two rows, denoted by $R_i \longleftrightarrow R_j$

- An **inconsistent system** is a system with no solution.

- **A matrix is in row echelon form** if (i) all rows (if any) consisiting entirely of zeroes appear at the bottom of the matrix, (ii) the first nonzero number in any row not consisting entirely of zeroes is 1, (iii) two successive rows do not consist entirely of zeroes, then the first 1 in the lower row occurs farther to the right than the first one in the higher row.

- A linear inequality in two varibles takes one of the following forms:

  $ax + by > c$ $ax + by < c$

  $ax + by \geq c$ $ax + by \leq c$

- The set of points in $\Re^2$ that satisfy a linear inequality in two variables is a **half-plane.**

- **Procedure for Obtaining the Partial Fraction Decompostion of** $\dfrac{p(x)}{q(x)}$

  Step 1. If degree of $p(x) \geq$ degree of $q(x)$, first divide to obtain $\dfrac{p(x)}{q(x)} = d(x) + \dfrac{r(x)}{q(x)}$
  where degree of $r(x) <$ degree of $q(x)$. Then find the decomposition of $\dfrac{r(x)}{q(x)}$

  Step 2. If the leading coefficient of $q(x)$ is $a_n \neq 1$, then factor a_n out so that the denominator function does have leading coefficient 1.

  Step 3. Factor the new $q(x)$ into linear and/or irreducible (or unfactorable) quadratic factors. This is the hardest step. Since the leading coefficient of $q(x)$ is 1, these factors will all have the form $x - c$ or $x^2 + ax + b$.

  Step 4. For each factor of the form $x - c$, the partial fraction decomposition of $\dfrac{r(x)}{q(x)}$ contains a term of the form $\dfrac{A}{x - c}$.

  Step 5. For each factor of the form $(x - c)^k$, $k > 1$, the partial fraction decomposition contains a sum of k terms: $\dfrac{A_1}{x - c} + \dfrac{A_2}{(x - c)^2} + \cdots \dfrac{A_k}{(x - c)^k}$.

  Step 6. For each factor of the form $x^2 + ax + b$, the partial fraction decomposition contains a term of the form $\dfrac{Ax + B}{x^2 + ax + b}$.

Step 7. For each factor of the form $(x^2 + ax + b)^k$, $k > 1$, the partial fraction decomposition contains a sum of k terms:

$$\frac{A_1x + B_1}{x^2 + ax + b} + \frac{A_2x + B_2}{(x^2 + ax + b)^2} + \cdots + \frac{A_kx + B_k}{(x^2 + ax + b)^k}$$

Step 8. Add the terms obtained in Steps 4, 5, 6, and 7 to obtain a rational function with denominator equal to $q(x)$.

Step 9. Equate the numerator in Step 8 to $p(x)$ or $r(x)$ to determine the coefficients (the A_k's and the B_k's).

- **A standard linear programming problem** in two variables takes the form

    Maximize
    $$f = ax + by \qquad\qquad (*)$$

    Subject to
    $$a_{11}x + a_{12}y \leq b_1$$
    $$a_{21}x + a_{22}y \leq b_2$$
    $$\cdot \qquad \cdot \qquad \cdot$$
    $$\cdot \qquad \cdot \qquad \cdot$$
    $$\cdot \qquad \cdot \qquad \cdot$$
    $$a_{m1}x + a_{m2}y \leq b_m$$

    $$x \geq 0, \ y \geq 0$$

    where $b_1, b_2, \cdots, b_m \geq 0$

- The set of all points that satisfy all the linear inequlities above is called the **constraint set** of the problem

- The function f in $(*)$ is called the **objective function**

- Any point in the constraint set is called a **feasible solution**

- The maximum and minimum values of the objective function are always taken at the corner points.

- The Corner-Point Method

    Step 1. Find all the possible corner points. These are points of intersection of two of the lines making up the constraint set.
    Step 2. Determine the actual corner points. These are the possible corner points that satisfy all the remaining constraints.
    Step 3. Compute the value of f, the objective function, at each actual corner point.
    Step 4. The maximum value of f is the largest number obtained in Step 3. The minimum value of f is the smallest number obtained in Step 3.

- A linear programming problem for which the objective function can take arbitrarily large values is called unbounded.

- A linear programming problem that has no feasible solutions is called **infeasible**.

SOLUTIONS TO CHAPTER TEN PROBLEMS

Problems 10.1

1. Multiply the second equation by 2 to get $-2x + 6y = 10$.

    Add the first equation to this to obtain $7y = 14$ or $y = 2$

    Plug this into one of the original equations and solve for x to obtain $x = 1$.

    If plugged into first equation, $2x + 2 = 4$, or $2x = 2$, so $x = 1$. Solution is $(1, 2)$

3. Multiply the first equation by $-\frac{3}{2}$ to obtain $-3x + \frac{3}{2}y = 6$.

    Add second equation to this, obtain $\frac{7}{2}y = 0$ so $y = 0$. Substitute into an original to obtain $x = -2$.

5. Multiply first equation by 4 to obtain $4x - 4y = 8$. Add to second equation, obtain $6x = 3$ or $x = \frac{1}{2}$. Substitute in to obtain $y = -\frac{3}{2}$.

7. Multiply the second equation by 2 and obtain $-4x + 6y = 0$.

Adding the two equations gives $0 = 0$, so there are an infinite number of solutions. Solving either equation for y, $y = \frac{2}{3}x$ so the solutions are $x = $ any real number and $y = \frac{2}{3}x$.

9. Multiply the first equation by 3 to obtain $6x + 9y = 12$.

Multiply the second equation by -2 to obtain $-6x - 8y = -10$

Add the two equations to obtain $y = 2$. Plug into either equation, obtain $x = -1$.

11. Multiply the first equation by b, get $abx + b^2y = bc$

Multiply the second equation by -a, get $-abx - a^2y = -ac$

Add and obtain $(b^2 - a^2)y = bc - ac$

So $y = \dfrac{c(b - a)}{(b - a)(b + a)} = \dfrac{c}{b + a}$ (if $b \neq \pm a$)

Multiply the first equation by a to get $a^2x + aby = ac$

Multiply the first equation by -b to get $-b^2x - aby = -bc$

Add the two and obtain $(a^2 - b^2)x = ac - bc$. As above, $x = \dfrac{c}{a + b}$ ($a \neq \pm b$)

(You could have substituted the y value into any of the equations to get x).

13. If the two equations are added, obtain $2ax = 2c$, or $x = \frac{c}{a}$. Therefore, $a \neq 0$.

If the second is subtracted from the first, obtain $2by = 0$; then $b = 0$ or $y = 0$.

But if $b = 0$, then y can be anything and the solution would not be unique. Therefore, $b \neq 0$.

So if $a \neq 0$ and $b \neq 0$, the unique solution is $x = \frac{c}{a}$, $y = 0$.

15. Multiply first equation by a, obtain $a^2x - bay = ac$

Multiply the second equation by b, obtain $b^2x + bay = bd$.

Add the two equations, and obtain $(a^2 + b^2)x = ac + bd$

Therefore, $x = \dfrac{ac + bd}{a^2 + b^2}$

Multiply the original first equation by b, obtain $bax - b^2y = bc$

Multiply the second equation by a, obtain $abx + a^2y = ad$

Subtract the first from the second and obtain $(a^2 + b^2)y = ad - bc$

Therefore, $y = \dfrac{ad - bc}{a^2 + b^2}$. For there to be no solutions, a and b both must be zero and either c or d must be nonzero.

17. Multiply the second equation by 2, add to the first equation and obtain $13x = 13$, so $x = 1$. Solve for y and obtain $y = -3$. Point of intersection is $(1, -3)$.

19. If the first equation is multiplied by -2 and added to the second, $0 = -4$ is obtained. No point of intersection.

21. Multiply the second equation by $-\frac{3}{2}$ and add to the first to obtain $-\frac{25}{2}y = \frac{7}{2}$, so $y = -\frac{7}{25}$
    $3x + 2(-\frac{7}{25}) = 4$ so $3x - \frac{14}{25} = \frac{100}{25}$ or $3x = \frac{114}{25}$ so $x = \frac{114}{75} = \frac{38}{25}$

23. Let A = the number of millions of dollars invested in A
    Let B = the number of dollars invested in B.
    Obtain the equations $.80A + .40B = 3$ and $.20A + .60B = 1$.
    Multiply the second equation by -4 and add to the first equatin to obtain
    $B = 0.5$.   Plug this into either equation and obtain $A = 3.5$.
    So one-half million dollars was invested in B; 3.5 million in A.

25. Let p = the number of pliers, s = the number of scissors.
    The two equations obtained are $2p + s = 140$ and $4p + 2s = 290$. If the first
    equation is multiplied by -2 and added to the second equation, $0 = 10$ is
    obtained. So there is no answer that will use up all of the materials.

27. Let c = the number of acres used for corn, s = the number used for soybeans.
    The equations obtained are $c + s = 500$ and $6c + 2s = 1200$
    Multiply the first equation by -2 and add to the second, giving $4c = 200$.
    Therefore $c = 50$ and since the total is 500, $s = 450$.
    Therefore, 50 acres are planted with corn and 450 acres with soybeans.

Problems 10.2

For the problems involving 3 equations in 3 variables, try to get the 3 x 3 identity on the left of the dotted line by first getting a 1 in position (1,1) and 0's underneath; next, get a 1 in the (2, 2) position and a 0 underneath; next a 1 in the (3, 3) position and 0's above; finally, a 0 in the (1, 2) position.

1. Start with
$$\begin{pmatrix} 1 & 1 & -1 & \vdots & 1 \\ 2 & -2 & 4 & \vdots & 0 \\ -1 & 2 & 1 & \vdots & -3 \end{pmatrix}$$

$R_2 \rightarrow R_2 + -2R_1$
$$\begin{pmatrix} 1 & 1 & -1 & \vdots & 1 \\ 0 & -4 & 6 & \vdots & -2 \\ -1 & 2 & 1 & \vdots & -3 \end{pmatrix}$$

$R_3 \rightarrow R_3 + R_1$
$$\begin{pmatrix} 1 & 1 & -1 & \vdots & 1 \\ 0 & -4 & 6 & \vdots & -2 \\ 0 & 3 & 0 & \vdots & -2 \end{pmatrix}$$

$R_2 \rightarrow -\frac{1}{4}R_2$
$$\begin{pmatrix} 1 & 1 & -1 & \vdots & 1 \\ 0 & 1 & -3/2 & \vdots & 1/2 \\ 0 & 3 & 0 & \vdots & -2 \end{pmatrix}$$

$R_3 \rightarrow R_3 + -3R_2$
$$\begin{pmatrix} 1 & 1 & -1 & \vdots & 1 \\ 0 & 1 & -3/2 & \vdots & 1/2 \\ 0 & 0 & 9/2 & \vdots & -7/2 \end{pmatrix}$$

$R_1 \rightarrow R_1 + -R_2$
$$\begin{pmatrix} 1 & 0 & 1/2 & \vdots & 1/2 \\ 0 & 1 & -3/2 & \vdots & 1/2 \\ 0 & 0 & 9/2 & \vdots & -7/2 \end{pmatrix}$$

$R_3 \rightarrow \frac{2}{9}R_3$

$$\begin{pmatrix} 1 & 0 & 1/2 & \vdots & 1/2 \\ 0 & 1 & -3/2 & \vdots & 1/2 \\ 0 & 0 & 1 & \vdots & -7/9 \end{pmatrix}$$

$R_2 \rightarrow R_2 + \frac{3}{2}R_3$

$$\begin{pmatrix} 1 & 0 & 1/2 & \vdots & 1/2 \\ 0 & 1 & 0 & \vdots & -2/3 \\ 0 & 0 & 1 & \vdots & -7/9 \end{pmatrix}$$

$R_1 \rightarrow R_1 + -\frac{1}{2}R_3$

$$\begin{pmatrix} 1 & 0 & 0 & \vdots & 8/9 \\ 0 & 1 & 0 & \vdots & -\frac{2}{3} \\ 0 & 0 & 1 & \vdots & -\frac{7}{9} \end{pmatrix}$$

So the solution is $x = \frac{8}{9}$, $y = -\frac{2}{3}$, $z = -\frac{7}{9}$

3. Start with

$$\begin{pmatrix} 2 & 4 & -6 & \vdots & 8 \\ -1 & 3 & 2 & \vdots & 6 \\ 1 & -1 & -1 & \vdots & -2 \end{pmatrix}$$

$R_1 \rightarrow \frac{1}{2}R_1$

$$\begin{pmatrix} 1 & 2 & -3 & \vdots & 4 \\ -1 & 3 & 2 & \vdots & 6 \\ 1 & -1 & -1 & \vdots & -2 \end{pmatrix}$$

$R_2 \rightarrow R_2 + R_1$
$R_3 \rightarrow R_3 + -R_1$

$$\begin{pmatrix} 1 & 2 & -3 & \vdots & 4 \\ 0 & 5 & -1 & \vdots & 10 \\ 0 & -3 & 2 & \vdots & -6 \end{pmatrix}$$

$R_2 \rightarrow \frac{1}{5}R_2$

$$\begin{pmatrix} 1 & 2 & -3 & \vdots & 4 \\ 0 & 1 & -1/5 & \vdots & 2 \\ 0 & -3 & 2 & \vdots & -6 \end{pmatrix}$$

$R_3 \rightarrow R_3 + 3R_2$
$R_1 \rightarrow R_1 + -2R_2$

$$\left(\begin{array}{ccc|c} 1 & 0 & -13/5 & 0 \\ 0 & 1 & -1/5 & 2 \\ 0 & 0 & 7/5 & 0 \end{array} \right)$$

$R_3 \rightarrow \frac{5}{7}R_3$

$$\left(\begin{array}{ccc|c} 1 & 0 & -13/5 & 0 \\ 0 & 1 & -1/5 & 2 \\ 0 & 0 & 1 & 0 \end{array} \right)$$

$R_1 \rightarrow R_1 + \frac{13}{5}R_2$
$R_2 \rightarrow R_2 + \frac{1}{5}R_3$

$$\left(\begin{array}{ccc|c} 1 & 0 & 0 & 0 \\ 0 & 1 & 0 & 2 \\ 0 & 0 & 1 & 0 \end{array} \right)$$

So the solution is x = 0, y = 2, z = 0.

5. Start with

$$\left(\begin{array}{ccc|c} 1 & 2 & 3 & 1 \\ 2 & -1 & 4 & 2 \\ 3 & 1 & 1 & 5 \end{array} \right)$$

$R_2 \rightarrow R_2 + -2R_1$
$R_3 \rightarrow R_3 + -3R_1$

$$\left(\begin{array}{ccc|c} 1 & 2 & 3 & 1 \\ 0 & -5 & -2 & 0 \\ 0 & -5 & -8 & 2 \end{array} \right)$$

$R_2 \rightarrow -\frac{1}{5}R_2$

$$\left(\begin{array}{ccc|c} 1 & 2 & 3 & 1 \\ 0 & 1 & 2/5 & 0 \\ 0 & -5 & -8 & 2 \end{array} \right)$$

$R_3 \rightarrow R_3 + 5R_2$

$$\begin{pmatrix} 1 & 2 & 3 & \vdots & 1 \\ 0 & 1 & 2/5 & \vdots & 0 \\ 0 & 0 & -6 & \vdots & 2 \end{pmatrix}$$

$R_1 \rightarrow R_1 + -2R_2$

$$\begin{pmatrix} 1 & 0 & 11/5 & \vdots & 1 \\ 0 & 1 & 2/5 & \vdots & 0 \\ 0 & 0 & -6 & \vdots & 2 \end{pmatrix}$$

$R_3 \rightarrow -\frac{1}{6}R_3$

$$\begin{pmatrix} 1 & 0 & 11/5 & \vdots & 1 \\ 0 & 1 & 2/5 & \vdots & 0 \\ 0 & 0 & 1 & \vdots & -1/3 \end{pmatrix}$$

$R_1 \rightarrow R_1 + -\frac{11}{5}R_3$
$R_2 \rightarrow R_2 + -\frac{2}{5}R_3$

$$\begin{pmatrix} 1 & 0 & 0 & \vdots & 26/15 \\ 0 & 1 & 0 & \vdots & 2/15 \\ 0 & 0 & 1 & \vdots & -1/3 \end{pmatrix}$$

So the solutions are $x = \frac{26}{15}$, $y = \frac{2}{15}$, $z = -\frac{1}{3}$

7. Start with

$$\begin{pmatrix} 1 & 1 & 1 & \vdots & -2 \\ -2 & 2 & 1 & \vdots & -18 \\ 3 & 2 & 2 & \vdots & 6 \end{pmatrix}$$

$R_2 \rightarrow R_2 + 2R_1$
$R_3 \rightarrow R_3 + -3R_1$

$$\begin{pmatrix} 1 & 1 & 1 & \vdots & -2 \\ 0 & 4 & 3 & \vdots & -22 \\ 0 & -1 & -1 & \vdots & 12 \end{pmatrix}$$

$R_2 \leftarrow \rightarrow R_3$
$R_2 \rightarrow -R_2$

$$\begin{pmatrix} 1 & 1 & 1 & \vdots & -2 \\ 0 & 1 & 1 & \vdots & -12 \\ 0 & 4 & 3 & \vdots & -22 \end{pmatrix}$$

$R_3 \rightarrow R_3 + -4R_2$

$R_1 \rightarrow R_1 + -R_2$

$$\begin{pmatrix} 1 & 0 & 0 & \vdots & 10 \\ 0 & 1 & 1 & \vdots & -12 \\ 0 & 0 & -1 & \vdots & 26 \end{pmatrix}$$

$R_2 \rightarrow R_2 + R_3$

$R_3 \rightarrow -R_3$

$$\begin{pmatrix} 1 & 0 & 0 & \vdots & 10 \\ 0 & 1 & 0 & \vdots & 14 \\ 0 & 0 & 1 & \vdots & -26 \end{pmatrix}$$

Therefore, the solutions are $x = 10$, $y = 14$, and $z = -26$.

9.
$$\begin{pmatrix} 1 & 1 & -1 & \vdots & 0 \\ 4 & -1 & 5 & \vdots & 0 \\ 6 & 1 & 3 & \vdots & 0 \end{pmatrix}$$

$R_2 \rightarrow R_2 + -4R_1$

$R_3 \rightarrow R_3 + -6R_1$

$$\begin{pmatrix} 1 & 1 & -1 & \vdots & 0 \\ 0 & -5 & 9 & \vdots & 0 \\ 0 & -5 & 9 & \vdots & 0 \end{pmatrix}$$

$R_2 \rightarrow -\frac{1}{5}R_2$

$R_3 \rightarrow R_3 + 5R_2$ (the new R_2)

$$\begin{pmatrix} 1 & 1 & -1 & \vdots & 0 \\ 0 & 1 & -9/5 & \vdots & 0 \\ 0 & 0 & 0 & \vdots & 0 \end{pmatrix}$$

For the last step, do $R_1 \rightarrow R_1 + -R_2$. The top row then is $\quad 1 \quad 0 \quad 4/5 \quad 0$.

There is no leading 1 under the z column, meaning that z can take on any value.

Let z be any real number. Then $y - \frac{9}{5}z = 0$, or $y = \frac{9}{5}z$ and $x = -\frac{4}{5}z$. $x + \frac{4}{5}c = 0$, or $x = -\frac{4}{5}c$

11. Start with
$$\begin{pmatrix} 1 & -2 & 3 & \vdots & 4 \\ -2 & 4 & -6 & \vdots & 12 \end{pmatrix}$$

$R_2 \rightarrow R_2 + 2R_1$

$$\begin{pmatrix} 1 & -2 & 3 & \vdots & 4 \\ 0 & 0 & 0 & \vdots & 20 \end{pmatrix}$$

Therefore, there are no solutions because the last row is all 0's left of dotted line, and there is a nonzero number to the right of the dotted line.

13. Start with
$$\begin{pmatrix} 1 & 2 & 1 & \vdots & 4 \\ 2 & 5 & 3 & \vdots & 2 \end{pmatrix}$$

$R_2 \rightarrow R_2 + -2R_1$

$$\begin{pmatrix} 1 & 2 & 1 & \vdots & 4 \\ 0 & 1 & 1 & \vdots & -6 \end{pmatrix}$$

$R_1 \rightarrow R_1 + -2R_2$

$$\begin{pmatrix} 1 & 0 & -1 & \vdots & 16 \\ 0 & 1 & 1 & \vdots & -6 \end{pmatrix}$$

Solutions (where z can be any value), $y = -6 - z$, $x = 16 + z$

15. The initial matrix is
$$\begin{pmatrix} 2 & 6 & -4 & 2 & \vdots & 4 \\ 1 & 0 & -1 & 1 & \vdots & 5 \\ -3 & 2 & -2 & 0 & \vdots & -2 \end{pmatrix}$$

We will do $R_1 \rightarrow \frac{1}{2}R_1$, although $R_1 \leftarrow \rightarrow R_2$ would also work.

$$\begin{pmatrix} 1 & 3 & -2 & 1 & \vdots & 2 \\ 1 & 0 & -1 & 1 & \vdots & 5 \\ -3 & 2 & -2 & 0 & \vdots & -2 \end{pmatrix}$$

$R_2 \rightarrow R_2 + -R_1$
$R_3 \rightarrow R_3 + 3R_1$

$$\begin{pmatrix} 1 & 3 & -2 & 1 & \vdots & 2 \\ 0 & -3 & 1 & 0 & \vdots & 3 \\ 0 & 11 & -8 & 3 & \vdots & 4 \end{pmatrix}$$

$R_2 \to -\frac{1}{3}R_2$ will change the
second row to $\qquad (\; 0 \quad\quad 1 \quad\quad -1/3 \quad\quad 0 \; \vdots \; -1 \;)$

Now do $R_3 \to R_3 + -11R_2$ will change
the third row to $\qquad (\; 0 \quad\quad 0 \quad\quad -13/3 \quad\quad 3 \; \vdots \; 15 \;)$

Now do $R_3 \to -\frac{3}{13}R_3$ to
obtain the entire matrix

$$\begin{pmatrix} 1 & 3 & -2 & 1 & \vdots & 2 \\ 0 & 1 & -1/3 & 0 & \vdots & -1 \\ 0 & 0 & 1 & -9/13 & \vdots & -45/13 \end{pmatrix}$$

$R_2 \to R_2 + \frac{1}{3}R_3$
$R_1 \to R_1 + 2R_3$

$$\begin{pmatrix} 1 & 3 & 0 & -5/13 & \vdots & -64/13 \\ 0 & 1 & 0 & -3/13 & \vdots & -28/13 \\ 0 & 0 & 1 & -9/13 & \vdots & -45/13 \end{pmatrix}$$

Taking $R_1 \to R_1 + -3R_2$ gives
a new first row of $\qquad (\; 1 \quad\quad 0 \quad\quad 0 \quad\quad 4/13 \; \vdots \; 20/13 \;)$

Therefore, the solutions are w = any real number, $x = \frac{20}{13} - \frac{4}{13}w$, $y = -\frac{28}{13} + \frac{3}{13}w$, $z = -\frac{45}{13} + \frac{9}{13}w$

17. Start with
$$\begin{pmatrix} 1 & -2 & 1 & 1 & \vdots & 2 \\ 3 & 0 & 2 & -2 & \vdots & -8 \\ 0 & 4 & -1 & -1 & \vdots & 1 \\ 5 & 0 & 3 & -1 & \vdots & -3 \end{pmatrix}$$

$R_2 \to R_2 + -3R_1$
$R_4 \to R_4 + -5R_4$

$$\begin{pmatrix} 1 & -2 & 1 & 1 & \vdots & 2 \\ 0 & 6 & -1 & -5 & \vdots & -14 \\ 0 & 4 & -1 & -1 & \vdots & 1 \\ 0 & 10 & -2 & -6 & \vdots & -13 \end{pmatrix}$$

The operation $R_2 \to \frac{1}{6}R_2$ would make the second row
$$(\; 0 \quad\quad 1 \quad\quad -1/6 \quad\quad -5/6 \; \vdots \; -14/6 \;)$$

Now do the operations $R_3 \rightarrow R_3 + -4R_2$

and $\hspace{5em} R_4 \rightarrow R_4 + -10R_2$

$$\begin{pmatrix} 1 & -2 & 1 & 1 & \vdots & 2 \\ 0 & 1 & -1/6 & -5/6 & \vdots & -14/6 \\ 0 & 0 & -2/6 & 14/6 & \vdots & 62/6 \\ 0 & 0 & -2/6 & 14/6 & \vdots & 62/6 \end{pmatrix}$$

$R_3 \rightarrow -3R_3$

$R_4 \rightarrow R_4 + \frac{1}{2}R_3$ (new R_3)

$$\begin{pmatrix} 1 & -2 & 1 & 1 & \vdots & 2 \\ 0 & 1 & -1/6 & -5/6 & \vdots & -14/6 \\ 0 & 0 & 1 & -7 & \vdots & -31 \\ 0 & 0 & 0 & 0 & \vdots & 0 \end{pmatrix}$$

$R_2 \rightarrow R_2 + \frac{1}{6}R_3$

$R_1 \rightarrow R_1 + -R_3$

$$\begin{pmatrix} 1 & -2 & 0 & 8 & \vdots & 33 \\ 0 & 1 & 0 & -2 & \vdots & -15/2 \\ 0 & 0 & 1 & -7 & \vdots & -31 \\ 0 & 0 & 0 & 0 & \vdots & 0 \end{pmatrix}$$

$R_1 \rightarrow R_1 + 2R_2$ gives

a new top row of $\quad (\, 1 \quad\quad 0 \quad\quad 0 \quad\quad 4 \quad \vdots \quad 18 \,)$

So the solutions are $w =$ any real number, $x = 18 - 4w$, $y = -\frac{15}{2} + 2w$, $z = -31 + 7w$

19. Start with $\begin{pmatrix} 1 & 2 & \vdots & -1 \\ 3 & 1 & \vdots & 7 \\ 4 & 3 & \vdots & 6 \end{pmatrix}$

$R_2 \rightarrow R_2 + -3R_1$

$R_3 \rightarrow R_3 + -4R_3$

$$\begin{pmatrix} 1 & 2 & \vdots & -1 \\ 0 & -5 & \vdots & 10 \\ 0 & -5 & \vdots & 10 \end{pmatrix}$$

$R_3 \rightarrow R_3 + -R_2$

$R_2 \rightarrow -\frac{1}{5}R_2$

$$\begin{pmatrix} 1 & 2 & \vdots & -1 \\ 0 & 1 & \vdots & -2 \\ 0 & 0 & \vdots & 0 \end{pmatrix}$$

$R_1 \rightarrow R_1 + -2R_2$

$$\begin{pmatrix} 1 & 0 & \vdots & 3 \\ 0 & 1 & \vdots & -2 \\ 0 & 0 & \vdots & 0 \end{pmatrix}$$

Solution: $x = 3$, $y = -2$.

21. Yes, but not reduced.   23. Yes, and reduced.

25. No; the leading 1 in the second row is to the left of the leading 1 in the first row.

27. Yes, but not reduced.

29. No; the leading 1 in row three is directly beneath the leading 1 in row two.

31. Let $x =$ the number of units of product 1

   $y =$ the number of units of product 2

   $z =$ the number of units of product 3

   Obtain the equations $.2x + .5y + .4z = 1281$

   $.1x + .4y + .3z = 942$

   $.3x + .3y + .5z = 1185$

   Solve, the solutions are $x = 1100$, $y = 1450$, $z = 840$.

   So 1100 units of product 1, 1450 units of product 2, 840 units of product 3.

33. Let x = the number of days spent in England   y = the number of days spent in France

    z = the number of days spent in Spain

    Obtain the following equations:

    $30x + 20y + 20z = 340$ (housing)

    $20x + 30y + 20z = 320$ (food)

    $10x + 10y + 10z = 140$ (incidentals)

    The solution is x = 6, y = 4, z = 4

    So 6 days were spent in England, 4 days in France, and 4 days in Spain.

35. Let x = the number of shares of Eastern stock, y = the number of shares of Hilton stock

    z = the number of shares of McDonald's stock

    Equations: $-1x - 1.50y + .50z = -350$ (2 days ago)

    $1.50x - .50y + 1z = 600$ (yesterday)

    Two equations in 3 variables cannot have a unique solution (evident from Gaussian elimination) but

    if z = 200 (the total number of shares of McDonald's stock), then 2 equations in 2 variables are

    obtained, and the solutions are y = 100 (the number of shares of Hilton stock is 100)

    and x = 300 (meaning 300 shares of Eastern stock).

37. Let x = the number of hours spent grazing, y = the number of hours spent moving,

    z = the number of hours spent resting.

    The equations are $200x = 150y + 50z$ (moving + resting = grazing)

    $x + y + z = 24$ (hours in a day)

    If $z \geq 6$ (the number of hours spent resting, then grazing will be $\frac{72}{7} - \frac{2}{7}z$ hours

    (that is x) and moving will be $\frac{96}{7} - \frac{5}{7}z$ (that is y). We must chose z so that

    neither moving nor grazing are negative, so $z \leq \frac{96}{5}$ hours.

39. Let x = the number of units of I

    y = the number of units of II

    z = the number of units of III

    Equations: $10x + 20y + 50z = 1600$

    $30x + 30y \qquad = 1200$

    $60x + 50y + 50z = 3200$

    The solution is x = 20, y = 20, z = 20. So 20 units of each should be produced.

41. Let x = the number of batches of love potion to be made

    y = the number of batches of cold remedy to be made

    Equations: $3\frac{1}{13}x + 5\frac{5}{13}x = 10$ (ground 4 leaf clovers)

    $2\frac{2}{13}x + 10\frac{10}{13}y = 14$ (mandrake)

    To start, multiply each equation by 13 to obtain $40x + 70y = 130$

    and $28x + 140y = 182$.   Solve and obtain

    x = 1.5 batches of love potion,   y = 1 batches of cold remedy

43. Let x = the number of sodas   y = the number of shakes

    Equations: $1x + 1y = 160$ (syrup; 5 qt. = 160 oz.)

    $4x + 3y = 512$ (ice cream; 4 gal. = 512 oz.)

    Solution is x = 32 sodas,   y = 128 shakes.

45. Let x = the number of units of type A feed to be given

    y = the number of units of type B feed

    Equations: $10x + 12y = 90$ (protein)

    $15x + 8y = 110$ (carbohydrates)

    Solution: x = 6 units of A,   y = 2.5 units of B

47. Let x = the number of attendants. Then 8x = the total number of hours

    worked by the attendants.

    Let y = the number of mechanics. Then 8y = the total number of hours

    worked by the mechanics.

    Equations: $\frac{3}{4}(8y) = 24$ or $6y = 24$ or $y = 4$

    $\frac{1}{4}(8y) + 8x = 32$. Since y = 4, plug in and obtain x = 3.

    So should hire 4 mechanics and 3 attendants.

49. Start with $\begin{pmatrix} 1 & -1 & 3 & \vdots & b \\ 2 & 3 & -1 & \vdots & a \\ 3 & 7 & -5 & \vdots & c \end{pmatrix}$

$R_2 \rightarrow R_2 + -2R_1$
$R_3 \rightarrow R_3 + -3R_1$

$$\begin{pmatrix} 1 & -1 & 3 & \vdots & b \\ 0 & 5 & -7 & \vdots & a-2b \\ 0 & 10 & -14 & \vdots & c-3b \end{pmatrix}$$

If the second row is multiplied by -2 and added to the third row,

(0   0   0   ┊   -2a+4b+c-3b)   is obtained.

The system will be inconsistent if that last expression is not zero.

So must have -2a + 4b + c - 3b = 0 ,   which reduces to b = 2a - c.

51. The solution obtained is x = 1.90081   y = 4.19411   z = -11.34852

53. Start with
$$\begin{pmatrix} 1 & 2 & \vdots & 0 \\ 2 & 4 & \vdots & 0 \end{pmatrix}$$

$R_2 \rightarrow R_2 + -2R_1$

$$\begin{pmatrix} 1 & 2 & \vdots & 0 \\ 0 & 0 & \vdots & 0 \end{pmatrix}$$

So y = any real number, and x = -2y

55. Use Gauss or notice that the second equation is obtained by multiplying the first one by -1. Therefore, the solution can be expressed as

y = any real number, x = 5y.

57. Start with
$$\begin{pmatrix} 1 & 1 & -1 & \vdots & 0 \\ 2 & -4 & 3 & \vdots & 0 \\ -1 & -7 & 6 & \vdots & 0 \end{pmatrix}$$

$R_2 \rightarrow R_2 + -2R_1$
$R_3 \rightarrow R_3 + R_1$

$$\begin{pmatrix} 1 & 1 & -1 & \vdots & 0 \\ 0 & -6 & 5 & \vdots & 0 \\ 0 & -6 & 5 & \vdots & 0 \end{pmatrix}$$

$R_2 \rightarrow -\frac{1}{6}R_2$
$R_3 \rightarrow R_3 + 6R_2$ (the new R_2)

- 456 -

$$\begin{pmatrix} 1 & 1 & -1 & \vdots & 0 \\ 0 & 1 & -5/6 & \vdots & 0 \\ 0 & 0 & 0 & \vdots & 0 \end{pmatrix}$$

$R_1 \to R_1 + -R_2$ makes

the first row $\quad (1 \quad 0 \quad -1/6 \;\vdots\; 0)$

So the solutions are z = any real number, $x = \frac{1}{6}z$, $y = \frac{5}{6}z$

59. Start with $\begin{pmatrix} 2 & 3 & -1 & \vdots & 0 \\ 6 & -5 & 7 & \vdots & 0 \end{pmatrix}$

$R_1 \to \frac{1}{2}R_1$

$R_2 \to R_2 + -6R_1$ (the new R_1)

$$\begin{pmatrix} 1 & 3/2 & -1/2 & \vdots & 0 \\ 0 & -14 & 10 & \vdots & 0 \end{pmatrix}$$

$R_2 \to -\frac{1}{14}R_2$

$R_1 \to R_1 + -\frac{3}{2}R_2$

$$\begin{pmatrix} 1 & 0 & 4/7 & \vdots & 0 \\ 0 & 1 & -5/7 & \vdots & 0 \end{pmatrix}$$

So the solution can be written as z = any real number, $x = -\frac{4}{7}z$, $y = \frac{5}{7}z$

61. Start with $\begin{pmatrix} 1 & -1 & 7 & -1 & \vdots & 0 \\ 2 & 3 & -8 & 1 & \vdots & 0 \end{pmatrix}$

$R_2 \to R_2 + -2R_1$

$$\begin{pmatrix} 1 & -1 & 7 & -1 & \vdots & 0 \\ 0 & 5 & -22 & 3 & \vdots & 0 \end{pmatrix}$$

$R_2 \to \frac{1}{5}R_2$

$R_1 \to R_1 + R_2$ (the new R_2)

$$\begin{pmatrix} 1 & 0 & 13/5 & -2/5 & \vdots & 0 \\ 0 & 1 & -22/5 & 3/5 & \vdots & 0 \end{pmatrix}$$

Solution is z = any real number, w = any real number, $x = -\frac{13}{5}z + \frac{2}{5}w$, $y = \frac{22}{5}z - \frac{3}{5}w$

63. We interchange the first 2 rows first. So the initial

matrix will be
$$\begin{pmatrix} 1 & 2 & -1 & 4 & \vdots & 0 \\ -2 & 0 & 0 & 7 & \vdots & 0 \\ 3 & 0 & -1 & 5 & \vdots & 0 \\ 4 & 2 & 3 & 0 & \vdots & 0 \end{pmatrix}$$

$R_2 \rightarrow R_2 + 2R_1$
$R_3 \rightarrow R_3 + \text{-}3R_1$
$R_4 \rightarrow R_4 + \text{-}4R_1$

$$\begin{pmatrix} 1 & 2 & -1 & 4 & \vdots & 0 \\ 0 & 4 & -2 & 15 & \vdots & 0 \\ 0 & -6 & 2 & -7 & \vdots & 0 \\ 0 & -6 & 7 & -16 & \vdots & 0 \end{pmatrix}$$

$R_2 \rightarrow \frac{1}{4}R_2$
$R_3 \rightarrow R_3 + 6R_2$
$R_4 \rightarrow R_4 + 6R_2$

$$\begin{pmatrix} 1 & 2 & -1 & 4 & \vdots & 0 \\ 0 & 1 & -1/2 & 15/4 & \vdots & 0 \\ 0 & 0 & -1 & 31/2 & \vdots & 0 \\ 0 & 0 & 4 & 13/2 & \vdots & 0 \end{pmatrix}$$

Continuing, you'll eventually obtain

$$\begin{pmatrix} 1 & 0 & 0 & 0 & \vdots & 0 \\ 0 & 1 & 0 & 0 & \vdots & 0 \\ 0 & 0 & 1 & 0 & \vdots & 0 \\ 0 & 0 & 0 & 1 & \vdots & 0 \end{pmatrix}$$

Therefore, the solution is x = y = z = w = 0.

65. All three of the equations are really the same; the second is -2 times the first and the third is 4 times the first. The solution to any of them can be written x = 3y, y = any real number.

67. Two equations in two variables have an inifinite number of solutions if and only if one equation is a multiple of the other. In this case, if we multiply the first equation by the correct constant, we should get the second equation. If the first equation is multiplied by a_{21}/a_{11} we get $a_{11} \cdot a_{21}/a_{11} = a_{21}$ for the coefficient of x, which agrees with the coefficient of x in the second equation; looking at the y terms, we must have that $a_{12} \cdot a_{21}/a_{11} = a_{22}$. Multiply both sides of the last relationship by a_{11} and obtain $a_{12}a_{21} = a_{11}a_{22}$ or $a_{11}a_{22} - a_{12}a_{21} = 0$. All of these pass through the origin. If $a_{11}a_{22} = a_{12}a_{21}$, then they are both the same line and therefore have an infinite number of solutions.

69. Start with
$$\begin{pmatrix} 1 & -7 & 1 & \vdots & 0 \\ 2 & -3 & 5 & \vdots & 0 \\ 4 & 11 & k & \vdots & 0 \end{pmatrix}$$

$R_2 \rightarrow R_2 + -2R_1$
$R_3 \rightarrow R_3 + -4R_1$

$$\begin{pmatrix} 1 & -7 & 1 & \vdots & 0 \\ 0 & 11 & 3 & \vdots & 0 \\ 0 & 39 & -4+k & \vdots & 0 \end{pmatrix}$$

$R_2 \rightarrow \frac{1}{11}R_2$ gives new second row
$$(\; 0 \quad 1 \quad 3/11 \; \vdots \; 0 \;)$$

$R_3 \rightarrow R_3 + -39R_2$ gives new third row
$$0 \quad 0 \quad -117/11 \; - 4 + k \quad 0$$

To have solutions, $-\frac{117}{11} - 4 + k = 0$ or $-117 - 44 + 11k = 0$
$$11k = 161 \text{ or } k = \frac{161}{11}$$

Problems 10.3

1. Square both sides of the second equation to obtain $y^2 = x^2$.

Substitute into first equation and
obtain $x^2 + x^2 = 4$
$$2x^2 = 4$$
$$x^2 = 2$$
$x = \pm\sqrt{2}$. It follows from equation two that
the two solutions are $(\sqrt{2}, \sqrt{2})$ and $(-\sqrt{2}, -\sqrt{2})$

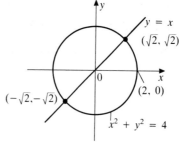

- 459 -

3. Substitute the second equation into the first equation and obtain

$$x^2 + (x - 1)^2 = 25$$

$$x^2 + x^2 - 2x + 1 = 25$$

$$2x^2 - 2x - 24 = 0$$

$$2(x^2 - x - 12) = 0$$

$$2(x - 4)(x + 3) = 0$$

$$x = 4 \text{ or } x = -3.$$

Substitute these values into the second equation to

obtain (4, 3) and (-3, -4)

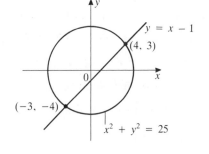

5. $x^2 = 2x$ so $x^2 - 2x = 0$ or $x(x - 2) = 0$

which has solutions $x = 0$ and $x = 2$.

Solutions are (0, 0) and (2, 4)

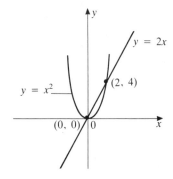

7. Multiply the first equation by -2,

add to the second equation and obtain

$$-3y^2 = -12$$

$$y^2 = 4$$

$y = \pm 2$. Substitute this into either equation

to obtain solutions

(3, 2) (3, -2) (-3, 2) (-3, -2)

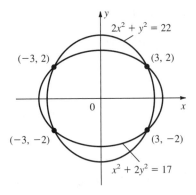

9. Add the two equations, obtain $2x^2 = 8$

$$x^2 = 4$$

This has solutions 2 and -2.

Plug into either equation,

obtain solutions (2, 0) and (-2, 0)

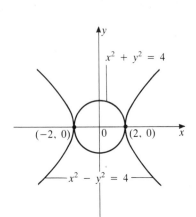

11. Add the two equations and obtain $-2y + y^2 = 4$

$$y^2 - 2y - 4 = 0$$

Therefore, $y = \dfrac{2 \pm \sqrt{20}}{2} = 1 \pm \sqrt{5}$

Plug into equation 2, obtain solutions

$(2\sqrt{5}, 1 + \sqrt{5})$ and $(-2\sqrt{5}, 1 - \sqrt{5})$

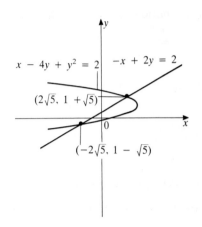

13. Multiply the first equation by 4,

add to the second and obtain $23y^2 = 0$.

This means $y = 0$,

and the solution is $(0, 0)$.

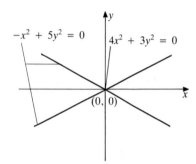

15. Solve the second equation for x and obtain

$x = 1 - \frac{4}{3}y$. Subsitute this into the first and obtain

$-(1 - \frac{4}{3}y)y = 15$

$\frac{4}{3}y^2 - y - 15 = 0$

$4y^2 - 3y - 45 = 0$

$(4y - 15)(y + 3) = 0$

$y = \frac{15}{4}$ or $y = -3$. Plug into second line above.

Solutions: $(-4, \frac{15}{4})$ and $(5, -3)$.

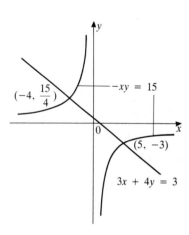

17. The first equation give $\frac{y}{2} = x^2$. Substitute into second equation to obtain $\frac{y}{2} + y^2 = 4$.

This gives $2y^2 + y - 8 = 0$. Solution is

$$y = \frac{-1 \pm \sqrt{65}}{4}.$$ Now divide by 2 and take $\pm$ the

square root to obtain x. Since what's under the $\sqrt{}$ cannot be negative, obtain the solutions

$$x = \pm\sqrt{\frac{-1 + \sqrt{65}}{8}} \approx \pm 0.94 \quad y = \frac{-1 + \sqrt{65}}{4} \approx 1.77$$

19. First equation gives $y = \frac{4}{x}$. Substitute into second equation to obtain $x + \frac{4}{x} = 6$.

Now subtract 6 from both sides, multiply through by x to obtain $x^2 - 6x + 4 = 0$.

Use the quadratic formula to obtain $x = 3 \pm \sqrt{5}$. Now $\frac{4}{3 + \sqrt{5}} = 3 - \sqrt{5}$ and $\frac{4}{3 - \sqrt{5}} = 3 + \sqrt{5}$

so solutions are $(3 + \sqrt{5}, 3 - \sqrt{5})$ and $(3 - \sqrt{5}, 3 + \sqrt{5})$.

For problems 21 - 32 (except 29), use substitution $u = \frac{1}{x}$ and $v = \frac{1}{y}$

21. The equations become $u - 3v = 6$ and $-4u + 2v = 6$. Multiply the first by 4, add to the second and obtain $-10v = 30$, or $v = -3$. Then $u = -3$ also, and the solution is $\left(-\frac{1}{3}, -\frac{1}{3}\right)$.

23. The equations become $2u - 8v = 5$ and $-3u + 12v = 8$. Multiply the first by 3, the second by 2, then add the equations. Obtain $0 = 31$, which means there are no solutions.

25. The equations become $3u + v = 0$ and $2u - 3v = 0$. This system of equations has the solution $u = 0$ and $v = 0$. But $\frac{1}{x}$ cannot equal 0, so there are no solutions.

27. The equations become $u + v = 4$ and $u - v = 4$. Add and obtain $2u = 8$, which has solutions $u = \pm 4$ and $v = 0$. Again, $\frac{1}{y}$ cannot $= 0$, so there are no solutions.

29. Solve each for y, set $=$ and get $x^2 - 9 = -x^2 - 2x + 3$ or $2x^2 + 2x - 12 = 0$.

Now divide each side by 2, factor, obtain $(x - 2)(x + 3) = 0$. Solutions are $(2, -5)$ and $(-3, 0)$.

31. Obtain $u - 2v + 3w = 11$, $4u + v - w = 4$, and $2u - v + 3w = 10$. Use Gaussian elimination to obtain solution of $w = 1$, $v = -3$ and $u = 2$, which translates to $\left(\frac{1}{2}, -\frac{1}{3}, 1\right)$

33. Set equations =, getting $e^x = e^{2x}$ and take ln of both sides to obtain $x = 2x$. This has solution $x = 0$. Since $e^0 = 1$, $y = 1$. Solution is $(0, 1)$

35. Set the two equations =, getting $e^x + 1 = e^{2x} - 11$. Now subtract left side from both sides and obtain $0 = e^{2x} - e^x - 12$. Now use the substitution $u = e^x$, and the equation becomes $0 = u^2 - u - 12$. That equation factors and has solutions $u = 4$ and $u = -3$, which means $e^x = 4$ or $e^x = -3$ (the latter is impossible).
Therefore, $x = \ln 4$ and $y = e^{\ln 4} + 1 = 4 + 1 = 5$. Solutions is $(\ln 4, 5)$

37. Set =, subtact to obtain $2 = \ln x$, or $e^2 = x$. Since $\ln e^2 = 2$, $y = 2 + 3 = 5$.
Solution is $(e^2, 5)$.

39. Solve first for y and obtain $y = 3 - x$. Substitue into second equation to obtain $3^x 2^{3-x} = 8$.
Take ln of both sides. Apply $\ln(ab) = \ln a + \ln b$ and $\ln a^r = r \ln a$ to obtain
$x \ln 3 + (3 - x)\ln 2 = \ln 8$
$x \ln 3 + 3 \ln 2 - x \ln 2 = \ln 8$
$x(\ln 3 - \ln 2) = \ln 8 - 3\ln 2 = \ln 8 - \ln 8 = 0$
Therefore $x = 0$ and so $y = 3$. Solution is $(0, 3)$.

41. Set equations = obtain $2x^2 = -2$,
which has solutions $x = \pm i$.
Solutions are $(i, 2)$ and $(-i, 2)$.

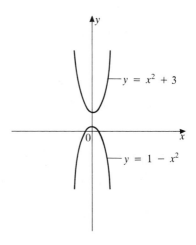

43. Add the equations and obtain

$2x^2 = 5$, or $x^2 = \frac{5}{2}$. This gives

$$x = \pm\sqrt{\frac{5}{2}} = \pm\frac{\sqrt{10}}{2}$$

and $y = \pm\frac{\sqrt{3}}{2}i$. The solutions are

$$\left(\frac{\sqrt{10}}{2}, \frac{\sqrt{6}}{2}i\right), \left(\frac{\sqrt{10}}{2}, -\frac{\sqrt{6}}{2}i\right),$$

$$\left(-\frac{\sqrt{10}}{2}, \frac{\sqrt{6}}{2}i\right), \left(-\frac{\sqrt{10}}{2}, -\frac{\sqrt{6}}{2}i\right).$$

$$\left(\text{note } \frac{\sqrt{10}}{2} = \sqrt{\frac{5}{2}} \text{ and } \frac{\sqrt{6}}{2} = \sqrt{\frac{3}{2}}\right)$$

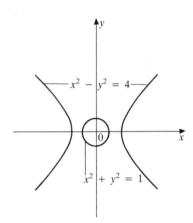

45. Subtract the second equation from the first
and obtain $7y^2 = -5$. This has solutions

$$y = \pm\sqrt{\frac{5}{7}}\, i = \pm\frac{\sqrt{35}}{7}i. \text{ The value of } x$$

can $= \pm\sqrt{\frac{336}{49}} = \pm\frac{4\sqrt{21}}{7}.$

Take the 4 possible + and - cases

to obtain
$$\left(\frac{4\sqrt{21}}{7}i, \frac{\sqrt{35}}{7}\right), \left(\frac{4\sqrt{21}}{7}i, -\frac{\sqrt{35}}{7}\right),$$

$$\left(-\frac{4\sqrt{21}}{7}i, \frac{\sqrt{35}}{7}\right), \left(-\frac{4\sqrt{21}}{7}i, -\frac{\sqrt{35}}{7}\right).$$

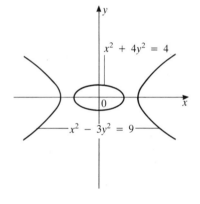

47. Let the numbers be x and y. Obtain the equations $x + y = 20$ and $xy = 96$. Solve the first
equation for y, substitute into the second and obtain $x(20 - x) = 96$.
Multiply out and factor to obtain $0 = (x - 8)(x - 12)$. The possible x values are 8 and 12 and
the corresponding possible y values are 12 and 8. So the two numbers are 8 and 12.

49. Let x and y be the two numbers. Then $xy = 48$ and $\frac{x}{y} = 3$. Rewrite the second equation as
$x = 3y$, and substitute it into the second equation. Obtain $3y^2 = 48$, which has solutions
$y = \pm 4$, but we are to take the positive value 4. The value of x is 12, so numbers are 4 and 12.

51. Since the sum of the radii is 4, obtain $r_1 + r_2 = 4$, or $r_2 = 4 - r_1$. From the area info,

    obtain the equation $\pi r_1^2 + \pi r_2^2 = \frac{25}{2}\pi$. Dividing both sides by π and substituting in the first

    equation gives $r_1^2 + (4 - r_1)^2 = \frac{25}{2}$. Multiply out, subtract $\frac{25}{2}$ from both sides, then multiply the

    resulting equation by 2 giving $4r_1^2 - 16r_1 + 7 = 0$. Factor and obtain $(2r_1 - 1)(2r_1 - 7) = 0$.

    So the radii of the two circles are $\frac{1}{2}$ cm. and $\frac{7}{2}$cm.

53. Let x = number of people, y = cost per person. Then obtain equations $xy = 36,000$ and

    $(x + 6)(y - 10) = 36,000$. Multiply out the second equation to obtain

    $xy + 6y - 10x - 60 = 36,000$. Solve the first equation for x, plug this into the second equation,

    multiply by x on both sides and obtain $0 = 10x^2 + 60x - 216,000$.

    Now divide both sides by 10, factor and obtain $(x - 144)(x + 150) = 0$.

    The only positive solution is = 144, which means there were 144 people. The corresponding y

    value is $250 per person.

Problems 10.4

1. $\frac{1}{x^2 - 1} = \frac{A}{x - 1} + \frac{B}{x + 1}$. Multiply both sides by $(x - 1)(x + 1)$ to obtain

   $1 = A(x + 1) + B(x - 1)$. Let $x = 1$ and obtain $1 = 2A$ or $A = \frac{1}{2}$

   Let $x = -1$ and obtain $-2B = 1$ or $B = -\frac{1}{2}$.

   Therefore, get $\frac{1/2}{x - 1} + \frac{-1/2}{x + 1}$

3. $\frac{7x - 1}{(x - 3)(x + 1)} = \frac{A}{x - 3} + \frac{B}{x + 1}$. Multiply both sides by $(x - 3)(x + 1)$ to obtain

   $7x - 1 = A(x + 1) + B(x - 3)$ Let $x = 3$ and obtain $21 - 1 = A(4)$ or $A = 5$

   Let $x = -1$ and obtain $-7 - 1 = -4B$ or $B = 2$

   Therefore, you get $\frac{2}{x + 1} + \frac{5}{x - 3}$

5. First, you must divide $2x^2 - 3x - 25$ by $x^2 - 4x + 3$ (since degree numerator = degree denom).

You get a quotient of 2 and a remainder of $5x - 31$

Find the partial fraction decomposition of $\dfrac{5x - 31}{x^2 - 4x + 3}$

$\dfrac{5x - 31}{(x - 3)(x - 1)} = \dfrac{A}{x - 3} + \dfrac{B}{x - 1}$. Multiply both sides by $(x - 3)(x - 1)$.

Obtain $A(x - 1) + B(x - 3) = 5x - 31$ Let $x = 1$. Get $-2B = -26$ or $B = 13$

Let $x = 3$. Get $2A = -16$ or $A = -8$.

So the decomposition of the original problem is $2 + \dfrac{-8}{x - 3} + \dfrac{13}{x - 1}$

7. Using the hint, find the partial fraction decomposition of $\dfrac{x + 2}{x^2 - 4x - 5}$

$\dfrac{x + 2}{(x - 5)(x + 1)} = \dfrac{A}{x - 5} + \dfrac{B}{x + 1}$. Multiply both sides by $(x - 5)(x + 1)$

and obtain $x + 2 = A(x + 1) + B(x - 5)$ Let $x = 5$ and obtain $7 = 6A$ or $A = \dfrac{7}{6}$

Let $x = -1$ and obtain $1 = -6B$ or $B = -\dfrac{1}{6}$

To obtain the partial fraction decomposition of the original problem,

multiply these constants by $\dfrac{1}{3}$ to obtain the decomposition
$\dfrac{7/18}{x - 5} + \dfrac{-1/18}{x + 1}$.

9. Factor denom. $\dfrac{1}{(x + 2)(x - 2)(x - 1)} = \dfrac{A}{x + 2} + \dfrac{B}{x - 2} + \dfrac{C}{x - 1}$. Multiply by $(x + 2)(x - 2)(x - 1)$

obtain $1 = A(x - 2)(x - 1) + B(x + 2)(x - 1) + C(x + 2)(x - 2)$

Let $x = 2$ and obtain $1 = A \cdot 0 + B(4)(1) + C \cdot 0$ so $1 = 4B$ and $B = \dfrac{1}{4}$

Let $x = -2$ and obtain $1 = 12A$ or $A = \dfrac{1}{12}$

Let $x = 1$ and obtain $1 = -3C$ or $C = -\dfrac{1}{3}$

Therefore, the decomposition is $\dfrac{1/12}{x + 2} + \dfrac{1/4}{x - 2} + \dfrac{-1/3}{x - 1}$

11. $\dfrac{-3x^2 - 5x + 4}{(x + 1)(x + 2)(x + 3)} = \dfrac{A}{x + 1} + \dfrac{B}{x + 2} + \dfrac{C}{x + 3}$. Multiply $(x + 1)(x + 2)(x + 3)$

obtain $-3x^2 - 5x + 4 = A(x + 2)(x + 3) + B(x + 1)(x + 3) + C(x + 1)(x + 2)$.

Let $x = -2$ and obtain $-3(-2)^2 - 5(-2) + 4 = A\cdot 0 + B(-1)(1) + C\cdot 0$

$$2 = -B \text{ or } B = -2$$

Let $x = -1$ and obtain $6 = 2A$ or $A = 3$ Let $x = -3$ and obtain $-8 = 2C$ or $C = -4$

Decomposition is $\dfrac{3}{x + 1} + \dfrac{-2}{x + 2} + \dfrac{-4}{x + 3}$

13. $\dfrac{x + 2}{x^3 - x} = \dfrac{x + 2}{x(x - 1)(x + 1)} = \dfrac{A}{x} + \dfrac{B}{x + 1} + \dfrac{C}{x - 1}$. So

$x + 2 = A(x + 1)(x - 1) + B(x)(x - 1) + Cx(x + 1)$

Let $x = 0$ Obtain $2 = -A$ or $A = -2$

Let $x = 1$ obtain $3 = 2C$ or $C = \dfrac{3}{2}$

Let $x = -1$ obtain $1 = 2B$ or $B = \dfrac{1}{2}$

The decomposition is $\dfrac{-2}{x} + \dfrac{1/2}{x + 1} + \dfrac{3/2}{x - 1}$

15. $\dfrac{3x^2 - 8x + 6}{(x - 2)^3} = \dfrac{A}{x - 2} + \dfrac{B}{(x - 2)^2} + \dfrac{C}{(x - 2)^3}$ or $3x^2 - 8x + 6 = A(x - 2)^2 + B(x - 2) + C$

Let $x = 2$ and obtain $12 - 16 + 6 = C$ or $C = 2$. Multiply out the above and obtain

$3x^2 - 8x + 6 = Ax^2 - 4Ax + 4A + Bx - 2B + 2$

Compare the coefficients of x^2; $3 = A$

Compare the coefficients of x; $-8 = -12 + B$ so $B = 4$

Decomposition is $\dfrac{3}{x - 2} + \dfrac{4}{(x - 2)^2} + \dfrac{2}{(x - 2)^3}$

17. Divide out the fraction (degree numerator is greater than degree of denominator) and obtain a quotient of $x^2 + 4$ and a remainder of 2; therefore, the decomposition is $x^2 + 4 + \dfrac{2}{x^2 + 1}$.

19. $\dfrac{2x + 5}{(x + 1)^2} = \dfrac{A}{x + 1} + \dfrac{B}{(x + 1)^2}$ or $2x + 5 = A(x + 1) + B$ (both sides multiplied by $(x + 1)^2$).

Let $x = -1$ and obtain $3 = B$

Comparing the coefficients of x, one obtains $A = 2$.   Decomposition is $\dfrac{2}{x + 1} + \dfrac{3}{(x + 1)^2}$

21. $\dfrac{1-x}{(1+x)^3} = \dfrac{A}{1+x} + \dfrac{B}{(1+x)^2} + \dfrac{C}{(1+x)^3}$ or $1-x = A(1+x)^2 + B(1+x) + C$

Let $x = -1$, obtain $2 = C$

Multiply out and get $1 - x = A + 2Ax + Ax^2 + B + Bx + 2$

Comparing the coefficients of x^2; $A = 0$

Comparing the coefficients of x; $-1 = B$

Decomposition is $\dfrac{-1}{(1+x)^2} + \dfrac{2}{(1+x)^3}$

23. $\dfrac{x+4}{(x+3)(x-2)(x^2+x+1)} = \dfrac{A}{x+3} + \dfrac{B}{x-2} + \dfrac{Cx+D}{x^2+x+1}$

$x + 4 = A(x-2)(x^2+x+1) + B(x+3)(x^2+x+1) + (Cx+D)(x+3)(x-2)$

Let $x = 2$ obtain $6 = 35B$ or $B = \dfrac{6}{35}$

Let $x = -3$ obtain $1 = -35A$ or $A = \dfrac{-1}{35}$

Compare coefficents of x^3; $0 = A + B + C = -\dfrac{1}{35} + \dfrac{6}{35} + C$ so $C = -\dfrac{5}{35} = -\dfrac{1}{7}$

Comparing constants; $4 = -2A + 3B - 6D = \dfrac{2}{35} + \dfrac{18}{35} - 6D$

then get $6D = \dfrac{2}{35} + \dfrac{18}{35} - \dfrac{140}{35} = \dfrac{-120}{35}$ or $D = \dfrac{-20}{35} = -\dfrac{4}{7}$

The decomposition is $\dfrac{-1/35}{x+3} + \dfrac{6/35}{x-2} + \dfrac{-1/7x - 4/7}{x^2+x+1}$

25. First factor $x^4 - 5x^2 + 4 = (x^2 - 4)(x^2 - 1) = (x-2)(x+2)(x-1)(x+1)$

$\dfrac{1}{(x+2)(x-2)(x+1)(x-1)} = \dfrac{A}{x+2} + \dfrac{B}{x-2} + \dfrac{C}{x+1} + \dfrac{D}{x-1}$

$1 = A(x-2)(x+1)(x-1) + B(x+2)(x+1)(x-1) + C(x+2)(x-2)(x-1)$
$\qquad + D(x+2)(x-2)(x+1)$

Let $x = 2$ get $1 = 12B$ or $B = \dfrac{1}{12}$ Let $x = -2$ get $1 = -12A$ or $A = -\dfrac{1}{12}$

Let $x = 1$ get $1 = -6D$ or $D = -\dfrac{1}{6}$ Let $x = -1$ get $1 = 6C$ or $C = \dfrac{1}{6}$

Decomposition $\dfrac{-1/12}{x+2} + \dfrac{1/12}{x-2} + \dfrac{1/6}{x+1} + \dfrac{-1/6}{x-1}$

27. The denominator factors as $(x^2 - 9)^2 = (x + 3)^2(x - 3)^2$. So

$$\frac{x}{(x - 3)^2(x + 3)^2} = \frac{A}{x + 3} + \frac{B}{(x + 3)^2} + \frac{C}{x - 3} + \frac{D}{(x - 3)^2}$$

$x = A(x + 3)(x - 3)^2 + B(x - 3)^2 + C(x - 3)(x + 3)^2 + D(x + 3)^2$

Let $x = 3$ then $3 = 36D$ or $D = \frac{1}{12}$

Let $x = -3$ then $-3 = 36B$ r $B = -\frac{1}{12}$

Compare the coefficients of x^3; $0 = A + C$ or $A = -C$

Compare the constant terms; $0 = 27A + 9B - 27C + 9D$

$$0 = 27A + 9(-\frac{1}{12}) - 27C + 9(\frac{1}{12})$$

$$0 = A - C = -C - C = -2C$$

Therefore, $C = 0$ and $A = 0$

Decomposition is $\dfrac{-1/12}{(x + 3)^2} + \dfrac{1/12}{(x - 3)^2}$

29. First divide out x from the numerator and denominator to give $\dfrac{x^3}{16 - x^2}$.

Next, divide (degree numerator is greater than degree denominator) and obtain a quotient of $-x$ and remainder of $+16x$. Find the partial fraction decomposition of $\dfrac{16x}{16 - x^2}$

$\dfrac{16x}{(4 - x)(4 + x)} = \dfrac{A}{4 - x} + \dfrac{B}{4 + x}$. Multiply by $(4 - x)(4 + x)$, obtain $16x = A(4 + x) + B(4 - x)$

Let $x = 4$, obtain $64 = 8A$, or $A = 8$. Let $x = -4$, get $-64 = 8B$, or $B = -8$.

Therefore, decomposition $= -x + \dfrac{8}{4 - x} + \dfrac{-8}{4 + x}$

31. $\dfrac{x^2 + 2}{x(x - 1)^2(x + 1)} = \dfrac{A}{x} + \dfrac{B}{x - 1} + \dfrac{C}{(x - 1)^2} + \dfrac{D}{x + 1}$

$x^2 + 2 = A(x - 1)^2(x + 1) + Bx(x - 1)(x + 1) + Cx(x + 1) + Dx(x - 1)^2$

Let $x = 0$ get $2 = A$

Let $x = 1$ get $3 = 2C$ or $C = \frac{3}{2}$

Let $x = -1$ get $3 = -4D$ or $D = -\frac{3}{4}$

Compare coefficients of x^3; $A + B + D = 0$ so $2 + B - \frac{3}{4} = 0$ or $B = -\frac{5}{4}$

Decomposition is $\dfrac{2}{x} + \dfrac{-5/4}{x - 1} + \dfrac{3/2}{(x - 1)^2} + \dfrac{-3/4}{x + 1}$

33. $\dfrac{-x^3 + x^2 + 1}{(x^2 + 3)(x^2 + 2)} = \dfrac{Ax + B}{x^2 + 3} + \dfrac{Cx + D}{x^2 + 2}$ Multiply by $(x^2 + 3)(x^2 + 2)$ and obtain

$-x^3 + x^2 + 1 = Ax^3 + 2Ax + Bx^2 + 2B + Cx^3 + 3Cx + Dx^2 + 3D$

Compare the coefficeints of x^3; $-1 = A + C$. Compare the coefficients of x^2 ; $1 = D + B$

Compare the coefficients of x; $0 = 2A + 3C$. Subtract twice the first equation

from this one and obtain $2 = C$; therefore, $A = -3$

Compare the constant terms; $1 = 2B + 3D$. Subtract twice the x^2 equation from this one and

obtain $-1 = D$; therefore $B = 2$. Decomposition:

$$\dfrac{-3x + 2}{x^2 + 3} + \dfrac{2x - 1}{x^2 + 2}$$

35. $\dfrac{x^4 + 1}{x^3(x^2 + 1)} = \dfrac{A}{x} + \dfrac{B}{x^2} + \dfrac{C}{x^3} + \dfrac{Dx + E}{x^2 + 1}$

$x^4 + 1 = Ax^2(x^2 + 1) + Bx(x^2 + 1) + C(x^2 + 1) + (Dx + E)x^3$

  Let $x = 0$ get $1 = C$

$x^4 + 1 = Ax^4 + Ax^2 + Bx^3 + Bx + x^2 + 1 + Dx^4 + Ex^3$

Compare coefficients of x^4; $A + D = 1$. Compare coefficients of x^3; $B + E = 0$

Compare coefficients of x^2; $A + 1 = 0$ so $A = -1$ and $D = 2$

Compare coefficients of x; $B = 0$ so $E = 0$

Decomposition is $\dfrac{-1}{x} + \dfrac{1}{x^3} + \dfrac{2x}{x^2 + 1}$

37. $\dfrac{5x^3 + 11x^2 - 18x + 17}{(x^2 + x + 1)(x^2 - 4x - 9)} = \dfrac{Ax + B}{x^2 + x + 1} + \dfrac{Cx + D}{x^2 - 4x - 9}$ (no more factoring is

possible)

Multiply through by $(x^2 + x + 1)(x^2 - 4x - 9)$ to obtain

$5x^3 + 11x^2 - 18x + 17 = (Ax + B)(x^2 - 4x - 9) + (Cx + D)(x^2 + x + 1)$

$= Ax^3 + Bx^2 - 4Ax^2 - 4Bx + -9Ax - 9B + Cx^3 + Dx^2 + Cx^2 + Cx + Dx + D$

To find the constants, compare coefficients:

x^3: $5 = A + C$ x^2: $11 = B - 4A + D + C$ x: $-18 = -9A - 4B + C + D$

constant: $17 = -9B + D$

Solve these four equations in four variables simultaneously (using Gaussian elimination if you like)

and obtain $A = \dfrac{23}{5}$, $B = \dfrac{6}{5}$, $C = \dfrac{2}{5}$, $D = \dfrac{139}{5}$

Therefore, the decomposition is $\dfrac{\frac{23}{5}x + \frac{6}{5}}{x^2 + x + 1} + \dfrac{\frac{2}{5}x + \frac{139}{5}}{x^2 - 4x - 9} = \dfrac{23x + 6}{5(x^2 + x + 1)} + \dfrac{2x + 139}{5(x^2 - 4x - 9)}$

39. $\dfrac{x^3 + 1}{(x^2 + 4x + 13)^2} = \dfrac{Ax + B}{x^2 + 4x + 13} + \dfrac{Cx + D}{(x^2 + 4x + 13)^2}$

$x^3 + 1 = (Ax + B)(x^2 + 4x + 13) + Cx + D$

$x^3 + 1 = Ax^3 + 4Ax^2 + 13Ax + Bx^2 + 4Bx + 13B + Cx + D$

Compare x^3; $A = 1$

Compare x^2; $4A + B = 0$.   Since $A = 1$, $B = -4$

Compare x; $13A + 4B + C = 0$. From above, $13 - 16 + C = 0$ or $C = 3$

Compare constant; $-52 + D = 1$ so $D = 53$

Decomposition is $\dfrac{x - 4}{x^2 + 4x + 13} + \dfrac{3x + 53}{(x^2 + 4x + 13)^2}$

Problems 10.5

1.

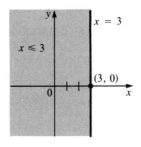

3.

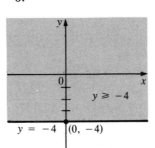

5.

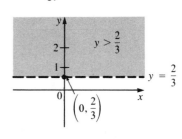

7.

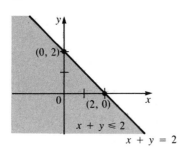

9.

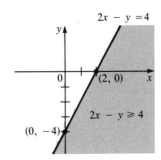

11.

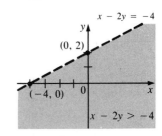

13.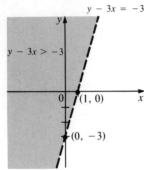

$y - 3x = -3$

$y - 3x > -3$

$(1, 0)$

$(0, -3)$

15.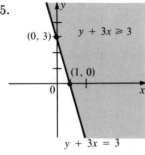

$(0, 3)$

$y + 3x \geq 3$

$(1, 0)$

$y + 3x = 3$

17.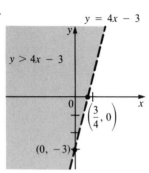

$y = 4x - 3$

$y > 4x - 3$

$\left(\frac{3}{4}, 0\right)$

$(0, -3)$

19.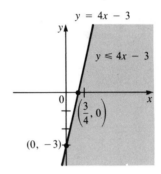

$y = 4x - 3$

$y \leq 4x - 3$

$\left(\frac{3}{4}, 0\right)$

$(0, -3)$

21.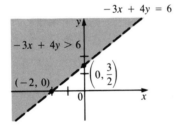

$-3x + 4y = 6$

$-3x + 4y > 6$

$\left(0, \frac{3}{2}\right)$

$(-2, 0)$

23.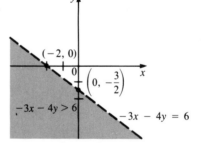

$(-2, 0)$

$\left(0, -\frac{3}{2}\right)$

$-3x - 4y > 6$

$-3x - 4y = 6$

25.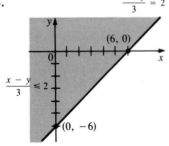

$\frac{x - y}{3} = 2$

$(6, 0)$

$\frac{x - y}{3} \leq 2$

$(0, -6)$

27.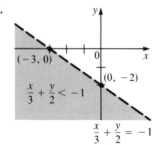

$(-3, 0)$

$(0, -2)$

$\frac{x}{3} + \frac{y}{2} < -1$

$\frac{x}{3} + \frac{y}{2} = -1$

29.

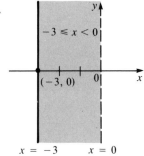

$-3 \le x < 0$

$(-3, 0)$

0

$x = -3$ $x = 0$

31.

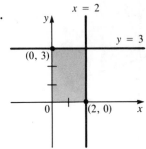

$x = 2$

$y = 3$

$(0, 3)$

0 $(2, 0)$

33.

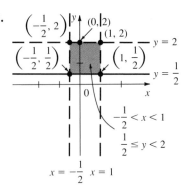

$\left(-\frac{1}{2}, 2\right)$ $(0, 2)$ $(1, 2)$

$y = 2$

$\left(-\frac{1}{2}, \frac{1}{2}\right)$ $\left(1, \frac{1}{2}\right)$

$y = \frac{1}{2}$

$-\frac{1}{2} < x < 1$

$\frac{1}{2} \le y < 2$

$x = -\frac{1}{2}$ $x = 1$

35.

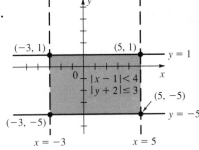

$(-3, 1)$ $(5, 1)$

$y = 1$

$|x - 1| < 4$

$|y + 2| \le 3$

$(5, -5)$

$y = -5$

$(-3, -5)$

$x = -3$ $x = 5$

37.

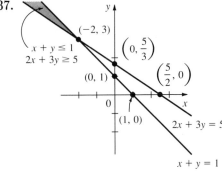

$(-2, 3)$

$\left(0, \frac{5}{3}\right)$

$x + y \le 1$

$2x + 3y \ge 5$

$(0, 1)$ $\left(\frac{5}{2}, 0\right)$

$(1, 0)$

$2x + 3y = 5$

$x + y = 1$

39.

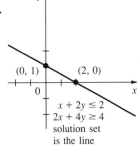

$(0, 1)$ $(2, 0)$

0

$x + 2y \le 2$

$2x + 4y \ge 4$

solution set
is the line

41.

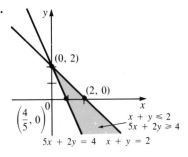

$(0, 2)$

$(2, 0)$

$\left(\frac{4}{5}, 0\right)$ 0

$x + y \le 2$

$5x + 2y \ge 4$

$5x + 2y = 4$ $x + y = 2$

43.

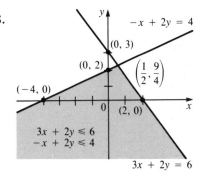

$-x + 2y = 4$

$(0, 3)$

$(0, 2)$ $\left(\frac{1}{2}, \frac{9}{4}\right)$

$(-4, 0)$

0 $(2, 0)$

$3x + 2y \le 6$

$-x + 2y \le 4$

$3x + 2y = 6$

45.

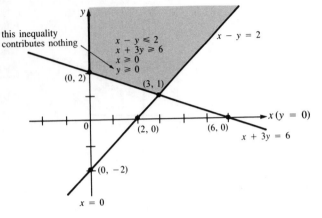

this inequality contributes nothing

$x - y \leqslant 2$
$x + 3y \geqslant 6$
$x \geqslant 0$
$y \geqslant 0$

$x - y = 2$

$(0, 2)$

$(3, 1)$

$(2, 0)$ $(6, 0)$ $x(y = 0)$

$x + 3y = 6$

$(0, -2)$

$x = 0$

Problems 10.6

1. Using the corner points in the example and that $P = 5x + 5y$:

    $(65, 40) \rightarrow 5(65) + 5(40) = 325 + 200 = 525.$ $(0, 0) \rightarrow 0$

    $(95, 0) \rightarrow 475$ $(0, 66) \rightarrow 330$

    So, the maximum profit is $525, and it occurs when there are 65 chairs and 40 tables are made.

3. Now the constraints are changed. We now have : $30x + 40y \leq 11{,}400$

    $$4x + 6y \leq 1650$$

    $$x \geq 0 \quad y \geq 0$$

    To plot the constraints: For the first one, let $x = 0$ and obtain $y = 285$, then let $y = 0$ so $x = 380$.

    The intercepts of the second constraint are $(0, 275)$ and $(412.5, 0)$.

    To find the points of intersection, multiply the second by $-\frac{15}{2}$ and add to the first constraint.

    You obtain $5y = 975$ or $y = 195$. Plug this into either of the others, obtain $x = 120$.

    Now we evaluate the expression $P = 5x + 6y$ at the four corner points.

    $(0, 0) \rightarrow 0$ $(0, 275) \rightarrow 1650$

    $(380, 0) \rightarrow 1900$ $(120, 195) \rightarrow 1770$

    Therefore, 380 chairs and 0 tables should be made; profit will be $1900.

5. Use the same corner points as #3, but use $P = 8x + 2y$.

    $(0, 0) \rightarrow 0$ $(0, 275) \rightarrow 550$

    $(380, 0) \rightarrow 3040$ $(120, 195) \rightarrow 1350$

    Again, make 380 chairs and 0 tables but this time with a profit of $3040

7. If that constraint is eliminated, you end up with this region:

   There are only 3 corner points.

   $(0, 10) \rightarrow 5000$

   $(5, 10) \rightarrow 6500$

   $(5, 5) \rightarrow 4000$ (this is the minimum)

9. The intercepts of the first constraint are $(0, 4)$ and $(4, 0)$.

   The intercepts of the second constraint are $(0, 5)$ and $(\frac{5}{2}, 0)$.

   Subtracting constraint 1 from constraint 2 gives $x = 1$ and therefore $y = 3$ (constraint 1).

   Evaluate at the 4 corner points.

   $(0, 0) \rightarrow 0$ $(1, 3) \rightarrow 13$

   $(0, 4) \rightarrow 12$ $(\frac{5}{2}, 0) \rightarrow 10.$

   Maximum f value is 13 when $x = 1$ and $y = 3$.

11. Intercepts of constraint 1 are $(0, 4)$ and $(4, 0)$, of constraint 2 are $(5, 0)$ and $(0, \frac{10}{3})$, and

    of constraint 3 are $(3, 0)$ and $(0, 6)$. With three constraints, you need to find where each pair

    intersect. First, take -2 times constraint 1 and add to constraint 2. You end up with $y = 2$ and

    therfore $x = 2$ by constraint 1. Now take -2 times constraint 2 and add to constraint 3. You

    get $-4y = -8$ or $y = 2$ and so $x = 2$ again. If you look at where 1 and 3 intersect, you get

    $(2, 2)$ again. All 3 constraints intersect at the same spot.

    Evaluating: $(0, 0) \rightarrow 0$ $(3, 0) \rightarrow 6$

    $(2, 2) \rightarrow 10$ $(0, \frac{10}{3}) \rightarrow 10$

    The maximum value is 10, and it occurs all along the line $2x + 3y = 10$ between $(0, \frac{10}{3})$ and $(2, 2)$.

13. Intercepts of constraints are: 1: $(0, 10)$ and $(1, 0)$; 2:$(10, 0)$ and $(0, 1)$ 3: $(3, 0)$ and $(0, 2)$.

    See graph for coordinates of intersection points.

    $(0, 0) \rightarrow 0$ $(1, 0) \rightarrow 3$

    $(0, 1) \rightarrow 5$ $(\frac{10}{11}, \frac{10}{11}) \rightarrow \frac{80}{11} \approx 7.27$

    Maximum value is $\frac{80}{11}$ which occurs at $x = \frac{10}{11}$, $y = \frac{10}{11}$

15. This one has the same feasible set as #13.

$(0, 1) \rightarrow 12$ $(0, 0) \rightarrow 0$

$(1, 0) \rightarrow 1$ $(\frac{10}{11}, \frac{10}{11}) \rightarrow \frac{130}{11} \approx 11.82$ Maximum is 12, occurs when x = 0 and y = 1.

17. Intercepts of constraint 1 are $(4, 0)$ and $(0, 2)$;

of constraint 2 are $(3, 0)$ and $(0, 3)$.

Subtract constrant 2 from constraint 1

to obtain y = 1 and so x = 2.

$(0, 3) \rightarrow 15$ $(4, 0) \rightarrow 16$ $(2, 1) \rightarrow 13$

Minimum value is 13 and occurs when x = 2 and y = 1.

19.

$(0, 1) \rightarrow 7$ $(1, 0) \rightarrow 3$

$(\frac{1}{13}, \frac{8}{13}) \rightarrow \frac{59}{13} \approx 4.54.$

So the minimum value is 3 and

occurs when x = 1 and y = 0.

21. Let x = number of pounds of Food I,

y = number of pounds of Food II.

We want to minimize cost C = 50x + 100y

(in cents)

subject to: (carbs) $0.90x + 0.60y \geq 2$ or $9x + 6y \geq 20$

               (prots) $0.10x + 0.40y \geq 1$ or $1x + 4y \geq 10$

                $x \geq 0, y \geq 0$

(Change percentages to decimals; the unknown percentages are found by subtracting from 1 or 100%).

Intercepts of constraint 1 are $(\frac{20}{9}, 0)$ and $(0, \frac{10}{3})$ and of constraint 2 are $(0, 2.5)$ and $(10, 0)$.

Intersection is $(\frac{2}{3}, \frac{7}{3})$.

Evaluate: $(10, 0) \rightarrow 500$ $(0, \frac{10}{3}) \rightarrow \frac{1000}{3}$ $(\frac{2}{3}, \frac{7}{3}) \rightarrow \frac{800}{3}$

Minimum is $\frac{800}{3}$ cents or $\approx \$2.67$ which occurs when $x = \frac{2}{3}$ and $y = \frac{7}{3}$

The preice per pound: $\frac{2}{3}$lb + $\frac{7}{3}$lb. = 3 lb. $\frac{\$2.67}{3\text{lb.}} \approx \$0.89$ per pound.

23. Let x = number of species 1   y = number of species 2.

   The constraints are as follows: $5x + y \geq 10$

$$2x + 2y \geq 12$$

$$x \geq 0, y \geq 0$$

   The graph at right shows the feasible region.

   $(0, 10) \rightarrow 20$ $(1, 5) \rightarrow 13$ $(4, 2) \rightarrow 16$ $(12, 0) \rightarrow 36$

   Minimum value is 13 and occurs when x = 1 and y = 5.

Review Exercises for Chapter Ten

1. Multiply the second equation by -2 and obtain $4x - 6y = -8$. Add to   this the first equation
   $3x + 6y = 9$ and obtain $7x = 1$ or $x = \frac{1}{7}$
   Substitute this into either of the original equations to get $y = \frac{10}{7}$.

3. Multiply the first equation by $\frac{2}{3}$ to obtain $2x - 4y = 6$. Add to the second
   equation $-2x + 4y = 9$ and obtain $0 = 15$ which means there are no solutions.

5. Start with
$$\begin{pmatrix} 1 & 1 & 1 & \vdots & 0 \\ 2 & -1 & 2 & \vdots & 0 \\ -3 & 2 & 3 & \vdots & 0 \end{pmatrix}$$

$R_2 \rightarrow R_2 + -2R_1$
$R_1 \rightarrow R_2 + 3R_3, R_3 + 3R_1$
$$\begin{pmatrix} 1 & 1 & 1 & \vdots & 0 \\ 0 & -3 & 0 & \vdots & 0 \\ 0 & 5 & 6 & \vdots & 0 \end{pmatrix}$$

$R_2 \rightarrow -\frac{1}{3}R_2$
$R_3 \rightarrow R_3 + -5R_2$
$R_3 \rightarrow \frac{1}{6}R_3, R_1 \rightarrow R_1 + -R_3$
$R_1 \rightarrow R_1 + -R_2$
$$\begin{pmatrix} 1 & 0 & 0 & \vdots & 0 \\ 0 & 1 & 0 & \vdots & 0 \\ 0 & 0 & 1 & \vdots & 0 \end{pmatrix}$$

Therefore, x = 0, y = 0, z = 0

7. Start with

$$\begin{pmatrix} 1 & 1 & 1 & \vdots & 4 \\ 1 & -2 & 1 & \vdots & 7 \\ -1 & 1 & 3 & \vdots & 6 \end{pmatrix}$$

$R_2 \rightarrow R_2 + -R_1$
$R_3 \rightarrow R_3 + R_1$

$$\begin{pmatrix} 1 & 1 & 1 & \vdots & 4 \\ 0 & -3 & 0 & \vdots & 3 \\ 0 & 2 & 4 & \vdots & 10 \end{pmatrix}$$

$R_2 \rightarrow -\frac{1}{3}R_2$

$$\begin{pmatrix} 1 & 1 & 1 & \vdots & 4 \\ 0 & 1 & 0 & \vdots & -1 \\ 0 & 2 & 4 & \vdots & 10 \end{pmatrix}$$

$R_3 \rightarrow R_3 + -2R_2$
$R_1 \rightarrow R_1 + -R_2$

$$\begin{pmatrix} 1 & 0 & 1 & \vdots & 5 \\ 0 & 1 & 0 & \vdots & -1 \\ 0 & 0 & 4 & \vdots & 12 \end{pmatrix}$$

$R_3 \rightarrow \frac{1}{4}R_3$
$R_1 \rightarrow R_1 + -R_3$

$$\begin{pmatrix} 1 & 0 & 0 & \vdots & 2 \\ 0 & 1 & 0 & \vdots & -1 \\ 0 & 0 & 1 & \vdots & 3 \end{pmatrix}$$

So the solutions are x = 2   y = -1   z = 3

9. Start with

$$\begin{pmatrix} 2 & 1 & -3 & \vdots & 0 \\ 4 & -1 & 1 & \vdots & 0 \end{pmatrix}$$

$R_1 \rightarrow \frac{1}{2}R_1, R_2 \rightarrow R_2 + -4R_1$ (new R_1)

$$\begin{pmatrix} 1 & 1/2 & -3/2 & \vdots & 0 \\ 0 & -3 & 7 & \vdots & 0 \end{pmatrix}$$

$R_2 \rightarrow -\frac{1}{3}R_2, R_1 \rightarrow R_1 + -\frac{1}{2}R_2$

$$\begin{pmatrix} 1 & 0 & -1/3 & \vdots & 0 \\ 0 & 1 & -7/3 & \vdots & 0 \end{pmatrix}$$

The solutions are z any real number, $x = \frac{1}{3}z$, $y = \frac{7}{3}z$

11. Start with
$$\begin{pmatrix} 1 & 1 & \vdots & 1 \\ 2 & 1 & \vdots & 3 \\ 3 & 1 & \vdots & 4 \end{pmatrix}$$

$R_2 \rightarrow R_2 + -2R_1$

$R_3 \rightarrow R_3 + -3R_3$

$$\begin{pmatrix} 1 & 1 & \vdots & 1 \\ 0 & -1 & \vdots & 1 \\ 0 & -2 & \vdots & 1 \end{pmatrix}$$

$R_2 \rightarrow -R_2$ and $R_3 \rightarrow R_3 + 2R_2$ make the last row $\begin{pmatrix} 0 & 0 & \vdots & -1 \end{pmatrix}$ so there are no solutions.

13. The matrix is
$$\begin{pmatrix} 1 & 1 & 1 & 1 & \vdots & 0 \\ 2 & -3 & -1 & 4 & \vdots & 0 \\ -2 & 4 & 1 & -2 & \vdots & 0 \\ 5 & -1 & 2 & 1 & \vdots & 0 \end{pmatrix}$$

Following the same steps as in #12, reduce the left part of the matrix to the 4 x 4 identity and get that the solution is x = y = z = w = 0

15. The matrix for this system is already in reduced row echelon form. The solution to this system can immediately be read as w = any real number, z = any real number, x = -z, y = w + 4

17. Yes. The first non-zero element in each row is 1, each such 1 lies further to the right than the 1 in the row before it, and all numbers in the column under such 1's are zeroes.

19. No. The first nonzero element in row 3 lies to the left of the first nonzero element in row 2.

21. $R_1 \rightarrow \frac{1}{2}R_2$, $R_2 \rightarrow R_2 + -R_1$ yields $\begin{pmatrix} 1 & 4 & \vdots & -1 \\ 0 & -4 & \vdots & -5 \end{pmatrix}$ Now $R_1 \rightarrow R_1 + R_2$ $\begin{pmatrix} 1 & 0 & \vdots & -6 \\ 0 & -4 & \vdots & -5 \end{pmatrix}$

If reduced row echelon form is desired, divide each entry of Row 2 by -4.

23. $\dfrac{1}{(x - 2)(x + 3)} = \dfrac{A}{x - 2} + \dfrac{B}{x + 3}$. Multiply through by $(x - 2)(x + 3)$ to obtain

$1 = A(x + 3) + B(x - 2)$. When $x = 2$, obtain $1 = 5A$ or $A = \dfrac{1}{5}$

When $x = -3$, obtain $1 = -5B$ or $B = -\dfrac{1}{5}$.

Decomposition is $\dfrac{1/5}{x - 2} + \dfrac{-1/5}{x + 3}$

25. $\dfrac{x^2 + 1}{(x + 1)(x + 2)(x - 2)} = \dfrac{A}{x + 1} + \dfrac{B}{x + 2} + \dfrac{C}{x - 2}$. Multiply throught by $(x + 1)(x + 2)(x - 2)$

to obtain $x^2 + 1 = A(x - 2)(x + 2) + B(x + 1)(x - 2) + C(x + 1)(x + 2)$.

Let $x = -1$, obtain $2 = A(-3)(1)$ or $A = -\dfrac{2}{3}$.   Let $x = 2$, obtain $5 = C(3)(4)$, or $C = \dfrac{5}{12}$.

Let $x = -2$, obtain $5 = 4B$, or $B = \dfrac{5}{4}$.

Decomposition is $\dfrac{-2/3}{x + 1} + \dfrac{5/4}{x + 2} + \dfrac{5/12}{x - 2}$

27. $\dfrac{x^2 - 2x + 4}{(x^2 + 2)(x^2 - 2x + 3)} = \dfrac{Ax + B}{x^2 + 2} + \dfrac{Cx + D}{x^2 - 2x + 3}$

$x^2 - 2x + 4 = (Ax + B)(x^2 - 2x + 3) + (Cx + D)(x^2 + 2)$

$= Ax^3 + Bx^2 - 2Ax^2 - 2Bx + 3Ax + 3B + Cx^3 + Dx^2 + 2Cx + 2D$

Compare the coefficients of x^3; $0 = A + C$ or $A = -C$

Compare the coefficients of x^2; $1 = B - 2A + D$

Compare the coefficients of x; $-2 = -2B + 3A + 2C$

Compare the constant terms; $4 = 3B + 2D$

Solve the above system, obtain $A = \dfrac{2}{9}$, $B = \dfrac{10}{9}$, $C = -\dfrac{2}{9}$, $D = \dfrac{1}{3}$

The solution is $\dfrac{(2/9)x + (10/9)}{x^2 + 2} + \dfrac{(-2/9)x + (1/3)}{x^2 - 2x + 3}$

29. Substitute second equation into first, obtain $x^2 + x^2 = 9$, or $2x^2 = 9$.   Therefore, $x^2 = \dfrac{9}{2}$

and so $x = \pm\dfrac{3}{\sqrt{2}}$.   Since $y = -x$, the ordered pairs are

$$\left(\dfrac{3}{\sqrt{2}}, -\dfrac{3}{\sqrt{2}}\right) \text{ and } \left(-\dfrac{3}{\sqrt{2}}, \dfrac{3}{\sqrt{2}}\right)$$

31. Subtract the second equation from the first to obtain $3y^2 = 12$, or $y = \pm 2$

Using the second equation, obtain $x^2 = 2 + 4 = 6$, or $x = \pm\sqrt{6}$.

Solutions are $(\sqrt{6}, 2)$, $(\sqrt{6}, -2)$, $(-\sqrt{6}, 2)$, $(-\sqrt{6}, -2)$.

33. From first equation, $y = \frac{2}{x}$. Substitute this into the second equation and obtain

$4x - \frac{4}{x} = 8$. Subtract 8 from both sides, multiply both sides by x and obtain

$4x^2 - 8x - 4 = 0$, or $4(x^2 - 2x - 1) = 0$. Use the quadratic formula to find the solutions

$$x = \frac{2 \pm \sqrt{4 - 4(1)(-1)}}{2(1)} = \frac{2 \pm \sqrt{8}}{2} = \frac{2 \pm 2\sqrt{2}}{2} = 1 \pm \sqrt{2}$$

When $x = 1 + \sqrt{2}$, $y = \frac{2}{1 + \sqrt{2}}$ (you can rationalize denom and obtain $-2 + 2\sqrt{2}$)

Similar for $x = 1 - \sqrt{2}$. Solutions are $(1 + \sqrt{2}, -2 + 2\sqrt{2})$, $(1 - \sqrt{2}, -2 - 2\sqrt{2})$

35. Use substitution $u = \frac{1}{x}$ and $v = \frac{1}{y}$. Equations become $7u + 5v = 4$ and $-u + 2v = -3$.

Multiply the second equation by 7, add to the first and obtain $19v = -17$, or $v = -\frac{17}{19}$.

From the second equation, $-u = -3 - 2(-\frac{17}{19})$, or $u = \frac{23}{19}$.

Take reciprocals to obtain $x = \frac{19}{23}$ and $y = -\frac{19}{17}$.

37. Square the first equation to obtain $y^2 = e^{4x}$. From equations 1 and 2, obtain $y = y^2$, which

is true when $y = 0$ or $y = 1$. Now y cannot $= 0$, since $y = e^{2x}$ and e^{2x} is always positive.

When $y = 1$, then the power of e must be 0; that is, $x = 0$. So $(0, 1)$ is the only solution.

39.
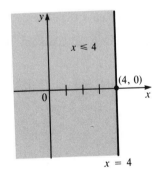

41.
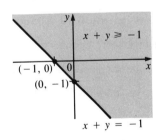

43.
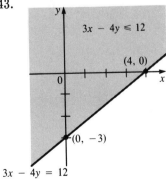

45.
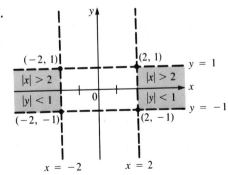

47.

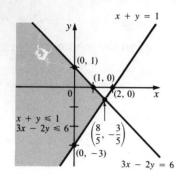

49. Intercepts of first constraint are $(0, 4)$ and $(2, 0)$; of second constraint are $(0, \frac{8}{3})$ and $(0, 8)$.

to find the point of intersection multiply the second equation by -2 and add to the first to obtain

$-5y = -12$, or $y = \frac{12}{5}$. The corresponding x is $\frac{4}{5}$

Evaluating the objective function: $(0, 0) \to 0$ $\qquad$ $(\frac{4}{5}, \frac{12}{5}) \to \frac{68}{5} = 13\frac{3}{5}$

$(0, \frac{8}{3}) \to \frac{40}{3} = 13\frac{1}{3}$ $\qquad$ $(2, 0) \to 4$

The maximum value is $\frac{68}{5}$ and occurs when x = $\frac{4}{5}$ and y = $\frac{12}{5}$.

51. The first constraint has intercepts $(0, 2)$ and $(4, 0)$; the second one has intercepts $(3, 0)$ and $(0, \frac{3}{2})$. They do not intersect, so the second one is superfluous. The third constraint has intercepts at $(0, 10)$ and $(2, 0)$. It intersects the first constraint at $(\frac{16}{9}, \frac{10}{9})$.

Evaluating the objective function: $(0, 10) \to 20$ $\qquad$ $(4, 0) \to 12$ $\qquad$ $(\frac{16}{9}, \frac{10}{9}) \to \frac{68}{9}$ (the min).

53. The feasible set is unbounded and therefore there is no solution to the problem.

CHAPTER ELEVEN
Matrices and Determinants

Chapter Objectives

In this chapter is a discussion of matrices; what matrices are, arithmetic operations involving matrices, inverses of matrices, determinants, and applications of all of these concepts. You will want to know the methods described and their applications. Emphasized are the applications to solving systems of equations, including Cramer's Rule.

Chapter Summary

- An m x n matrix is a rectangular array of numbers with m rows and n columns.

- A **row vector** is a matrix with one row.
- A **column vector** is a matrix with one column.

- **Addition of Matrices:** If A and B are m x n matrices then A + B is the m x n matrix obtained by adding the corresponding components of A and B.

- **Dot Product:** If $\mathbf{a} = \begin{pmatrix} a_1 \\ a_2 \\ : \\ a_n \end{pmatrix}$ and $\mathbf{b} = (b_1, b_2, \dots , b_n)$ then $\mathbf{a} \cdot \mathbf{b} = a_1 b_1 + a_2 b_2 + \cdots + a_n b_n$

- **Matrix Product:** If A is an m x n matrix and B is an n x p matrix, then AB is an m x p matrix whose i,j^{th} component is the dot product of the ith row of A and the jth column of B.

- **Identity Matrix:** An n x n matrix, denoted by I, with 1's down the main diagonal and 0's everywhere else. For every n x n matrix A, AI = IA = A.

- **Matrix Inverse:** If A is an n x n matrix, then the inverse of A, if it exists, is denoted by A^{-1} and is the matrix where $AA^{-1} = A^{-1}A = I$, the identity matrix.

- If $A = \begin{pmatrix} a & b \\ c & d \end{pmatrix}$ and $D = ad - bc \neq 0$, then A is invertible and

  $A^{-1} = \begin{pmatrix} d/D & -b/D \\ -c/D & a/D \end{pmatrix}$

- The **determinant** of the 2 x 2 matrix $A = \begin{pmatrix} a & b \\ c & d \end{pmatrix}$, denoted by det A,

  is defined by det $A = ad - bc$

- The i,j^{th} **minor** of an n x n matrix is the (n - 1) x (n - 1) matrix obtained by deleting the ith row and j^{th} column of A.
- The i,j^{th} **cofactor** of A is the determinant of the i,j^{th} minor of A multiplied by $(-1)^{i+j}$
- det $A = a_{11}A_{11} + a_{12}A_{12} + ... + a_{1n}$, where A_{ij} is the i,j^{th} cofactor of A.

- Properties of Determinants

  1. If any row (column) is a multiple of another row (column), then det $A = 0$.
  2. Multiplying a row(column) of A by a constant has the effect of multiplying det A by that constant.
  3. Interchanging any two rows (columns) of A has the effect of multiplying det A by -1.
  4. If any row (column) of A is multiplied by a constant and added to a defferent row (column) of A, then det A is unchanged.

- If A is a square (n x n) matrix, then A is invertible if and only if det $A \neq 0$.

Problems 11.1

1. 2 x 2; since the two dimensions are the same, the matrix is square.

3. 2 x 2; square   5. 3 x 2; dimensions are not the same, so not square

7. 2 x 4; not square   9. 3 x 1; not square   11. 3 x 3; square   13. 1 x 4; not square

15. no; none of the corresponding elements are equal   17. No; they are not of the same dimension

19. Yes   21. Multiply every entry by 4 to obtain $\begin{pmatrix} 8 & 20 \\ 12 & -4 \end{pmatrix}$

23. Taking negatives of everything, obtain $\begin{pmatrix} -1 & -2 & -5 \\ -7 & 2 & 0 \end{pmatrix}$

25. not of same dimension, cannot be added.

27. $\begin{pmatrix} 0 & 12 & 8 \\ 20 & -36 & 0 \\ -32 & 20 & 12 \end{pmatrix} + \begin{pmatrix} -12 & 18 & 15 \\ 15 & 6 & 3 \\ 3 & 15 & -6 \end{pmatrix} = \begin{pmatrix} -12 & 30 & 23 \\ 35 & -30 & 3 \\ -29 & 35 & 6 \end{pmatrix}$

29. $\begin{pmatrix} 5 & 6 & 2 \\ 3 & 4 & 1 \\ 0 & -7 & 2 \end{pmatrix}$

31. $\begin{pmatrix} 2 & 6 \\ 4 & 10 \\ -2 & 4 \end{pmatrix}$

33. $\begin{pmatrix} 0 & 4 \\ 6 & 11 \\ -8 & 5 \end{pmatrix}$ 35. $\begin{pmatrix} 0 & 0 \\ 0 & 0 \\ 0 & 0 \end{pmatrix}$

37. $\begin{pmatrix} -2 & 4 \\ 7 & 15 \\ -15 & 10 \end{pmatrix}$

39. $\begin{pmatrix} 2 & 6 \\ 4 & 10 \\ -2 & 4 \end{pmatrix} + \begin{pmatrix} 6 & 0 \\ -3 & -12 \\ 21 & -15 \end{pmatrix} + \begin{pmatrix} -4 & 4 \\ 16 & 24 \\ -28 & 12 \end{pmatrix} = \begin{pmatrix} 4 & 10 \\ 17 & 22 \\ -9 & 1 \end{pmatrix}$

41. D must equal 2A + B

$$\begin{pmatrix} 2 & 6 \\ 4 & 10 \\ -2 & 4 \end{pmatrix} + \begin{pmatrix} -2 & 0 \\ 1 & 4 \\ -7 & 5 \end{pmatrix} = \begin{pmatrix} 0 & 6 \\ 5 & 14 \\ -9 & 9 \end{pmatrix}$$

43.

$$\begin{pmatrix} 0 & 0 & 2 \\ 3 & 1 & 0 \\ 0 & -2 & 4 \end{pmatrix} - \begin{pmatrix} 2 & -2 & 4 \\ 4 & 8 & 10 \\ 0 & 2 & -2 \end{pmatrix} = \begin{pmatrix} -2 & 2 & -2 \\ -1 & -7 & -10 \\ 0 & -4 & 6 \end{pmatrix}$$

45.

$$\begin{pmatrix} 1 & 1 & 5 \\ 8 & 5 & 10 \\ 7 & -7 & 3 \end{pmatrix}$$

47.

$$\begin{pmatrix} 0 & 0 & 2 \\ 3 & 1 & 0 \\ 0 & -2 & 4 \end{pmatrix} + \begin{pmatrix} -1 & 1 & -2 \\ -2 & -4 & -5 \\ 0 & -1 & 1 \end{pmatrix} + \begin{pmatrix} 0 & -2 & -1 \\ -3 & 0 & -5 \\ -7 & 6 & 0 \end{pmatrix} = \begin{pmatrix} -1 & -1 & -1 \\ -2 & -3 & -10 \\ -7 & 3 & 5 \end{pmatrix}$$

49. D = -A - B - C $= \begin{pmatrix} -1 & -1 & -5 \\ -8 & -5 & -10 \\ -7 & 7 & -3 \end{pmatrix}$ (Just the negative of the matrix in #45)

51.

$$1.25 \begin{pmatrix} 6 & 19 & 14 & 46 \\ 8 & 28 & 12 & 40 \\ 4 & 26 & 17 & 55 \end{pmatrix} = \begin{pmatrix} 7.5 & 23.75 & 17.5 & 57.5 \\ 10 & 35 & 15 & 50 \\ 5 & 32.5 & 21.25 & 68.75 \end{pmatrix}$$

53.

$$1.30 \begin{pmatrix} 7 & 12 & 2 & 28 \\ 5 & 14 & 8 & 17 \\ 6 & 9 & 5 & 33 \end{pmatrix} + .8 \begin{pmatrix} 6 & 21 & 8 & 41 \\ 10 & 19 & 14 & 33 \\ 2 & 26 & 5 & 28 \end{pmatrix}$$

$$= \begin{pmatrix} 13.9 & 32.4 & 9 & 69.2 \\ 14.5 & 33.4 & 21.6 & 48.5 \\ 9.4 & 32.5 & 10.5 & 65.3 \end{pmatrix}$$

Problems 11.2

1. $2 \cdot 4 + 3 \cdot (-2) = 8 - 6 = 2$

3. $1 \cdot 2 + 7 \cdot (-3) + 5 \cdot 4 = 2 - 21 + 20 = 1$

5. $6 + 0 + 2 + 24 = 32$

7. Let $\mathbf{a} = (a_1, a_2, \ldots, a_n)$. Then $\mathbf{a} \cdot \mathbf{a} = a_1^2 + a_2^2 + \ldots + a_n^2$. Since the square of every real number is non-negative, the sum is non-negative and so $\mathbf{a} \cdot \mathbf{a} \geq 0$.

9. Take the following dot product

$$(25 \quad 45 \quad 20 \quad 30) \begin{pmatrix} 8 \\ 15 \\ 12 \\ 14 \end{pmatrix}$$

Obtain $200 + 675 + 240 + 420 = \$1535$

What to keep in mind for multiplication is;

If A is m x n and B is p x q, AB is defined only if n = p and AB will have dimension m x q. The (i, j) element of AB comes from taking the dot product of row i of A with column j of B.

11. A is 2 x 2, B is 2 x 2, so multiplication can be done, and AB is 2 x 2.

Row 1 of A with column 1 of j = $3 \cdot (-2) + 4 \cdot 0 = -6 + 0 = -6$

Row 1 of A with column 2 of j = $3 \cdot 1 + 4 \cdot 3 = 3 + 12 = 15$

Row 2 of A with column 1 of j = $4 \cdot (-2) + -1 \cdot 0 = -8 + 0 = -8$

Row 2 of A with column 2 of j = $4 \cdot 1 + -1 \cdot 3 = 4 - 3 = 1.$

$$\begin{pmatrix} -6 & 15 \\ -8 & 1 \end{pmatrix}$$

13. Following the pattern in number 13, $-1 - 2 = -3$ $0 + -3 = -3$

$-1 + 2 = 1$ $0 + 3 = 3$

$$\begin{pmatrix} -3 & -3 \\ 1 & 3 \end{pmatrix}$$

15. A is 2 x 3, B is 3 x 2, so AB is defined and has dimension 2 x 2

Row 1 with column 1 is $2 \cdot 1 + 4 \cdot 4 + 3 \cdot (-3) = 2 + 16 - 9 = 9$

Row 1 with column 2 is $2 \cdot (-2) + 4 \cdot 3 + 3 \cdot 0 = -4 + 12 + 0 = 8$

Row 2 with column 1 is $1 \cdot 1 + 0 \cdot 4 + -2 \cdot (-3) = 1 + 0 + 6 = 7$

Row 2 with column 2 is $1 \cdot (-2) + 0 \cdot 3 + (-2) \cdot 0 = -2 + 0 + 0 = -2$

$$\begin{pmatrix} 9 & 8 \\ 7 & -2 \end{pmatrix}$$

17. A 2 x 3 multiplied by a 2 x 2 is not defined.

19. 3 x 3 times 3 x 3 yields 3 x 3.

$3 + 4 - 8 = -1$ $0 + 2 + 0 = 2$ $1 - 6 - 4 = -9$

$-9 + 0 + 8 = -1$ $0 + 0 + 0 = 0$ $-3 + 0 + 4 = 1$

$6 + 6 + 4 = 16$ $0 + 3 + 0 = 3$ $2 - 9 + 2 = -5$

$$\begin{pmatrix} -1 & 2 & -9 \\ -1 & 0 & 1 \\ 16 & 3 & -5 \end{pmatrix}$$

21. A 1 x 4 times a 4 x 2 gives a 1 x 2.

$3 + 8 + 0 - 4 = 7$ $-6 + 16 + 0 + 6 = 16$ Ans: $(7 \quad 16)$

23. As in #19;
$3 + 0 + 0 = 3$ $0 - 2 + 0 = \text{-}2$ $0 + 0 + 1 = 1$

$4 + 0 + 0 = 4$ $0 + 0 + 0 = 0$ $0 + 0 + 6 = 6$

$5 + 0 + 0 = 5$ $0 + 1 + 0 = 1$ $0 + 0 + 9 = 9$

$$\begin{pmatrix} 3 & \text{-}2 & 1 \\ 4 & 0 & 6 \\ 5 & 1 & 9 \end{pmatrix}$$

25. Multiplying by the identity gives the same matrix:
$$\begin{pmatrix} a & b & c \\ d & e & f \\ g & h & j \end{pmatrix}$$

27. a) 5, since there are 5 columns in A and each column represents one person.

 b) 4, since B has 4 columns.

 c) multiply A by B to give
$$\begin{pmatrix} 1 & 3 & 1 & 2 \\ 0 & 1 & 2 & 2 \end{pmatrix}$$

29. a)   100   200   400   100   b)
$$\begin{pmatrix} 46 \\ 34 \\ 15 \\ 10 \end{pmatrix}$$

 c) Take the dot product of the answers for a) and b) and obtain

 $4600 + 6800 + 6000 + 1000 = \$18,400$

31. a)
$$\begin{pmatrix} 80,000 & 45,000 & 40,000 \\ 50 & 20 & 10 \end{pmatrix}$$
    b)
$$\begin{pmatrix} 1 \\ 3 \\ 1 \end{pmatrix}$$

 c) The product of these matrices is
$$\begin{pmatrix} 255,000 \\ 120 \end{pmatrix}$$
    so answers are \$255,000 and 120 shares.

33. a)

	A	T	C
Br	1/5	2/5	1/5
milk	0	1	1/2
cof	1/10	0	0
ch	1/8	1/8	1/16

b)

A	2
T	1
C	3

c) The product yields

Br	7/5
mi	5/2
cof	1/5
ch	9/16

35.
$$\begin{pmatrix} 2 & -1 \\ 4 & 5 \end{pmatrix} \begin{pmatrix} x \\ y \end{pmatrix} = \begin{pmatrix} 3 \\ 7 \end{pmatrix}$$

37.
$$\begin{pmatrix} 3 & 6 & -7 \\ 2 & -1 & 3 \end{pmatrix} \begin{pmatrix} x \\ y \\ z \end{pmatrix} = \begin{pmatrix} 0 \\ 1 \end{pmatrix}$$

39.
$$\begin{pmatrix} 2 & 0 & 4 \\ 3 & 2 & 0 \\ 0 & -4 & -9 \end{pmatrix} \begin{pmatrix} x \\ y \\ z \end{pmatrix} = \begin{pmatrix} 9 \\ 6 \\ 3 \end{pmatrix}$$

41. Let $A = \begin{pmatrix} a & b \\ c & d \end{pmatrix}$ $\qquad B = \begin{pmatrix} k & l \\ m & n \end{pmatrix}$ $\qquad C = \begin{pmatrix} x & y \\ z & w \end{pmatrix}$

$$(BC) = \begin{matrix} kx + lz & ky + lw \\ mx + nz & my + nw \end{matrix}$$

$$A(BC) = \begin{matrix} a(kx + lz) + b(mx + nz) & a(ky + lw) + b(my + nw) \\ c(kx + lz) + d(mx + nz) & c(ky + lw) + d(my + nw) \end{matrix}$$

$$= \begin{matrix} akx + alz + bmx + bnz & aky + alw + bmy + bnw \\ ckx + clz + dmx + dnz & cky + clw + dmy + dnw \end{matrix}$$

$$(AB) = \begin{matrix} ak + bm & al + bn \\ ck + dm & cl + dn \end{matrix}$$

$$(AB)C = \begin{matrix} (ak + bm)x + (al + bn)\,z & (ak + bm)y + (al + bn)z \\ (ck + dm)\,x + (cl + dn)z & (ck + dm)y + (cl + dn)z \end{matrix}$$

$$= \begin{matrix} akx + bmx + alz + bnz & aky + bmy + alw + bnw \\ ckx + amx + clz + dnz & cky + dmy + clw + dnw \end{matrix}$$

$$= A(BC)$$

Problems 11.3

To find the inverse of a matrix, we create a new matrix by placing the matrix given to the left of the identity matrix of the same size. If, when the orignial matrix has been transformed into reduced row -echelon form, the original matrix has become the identity matrix, the identity matrix has been transformed into the inverse. If this is not possible, the original matrix has no inverse.

1. Start with $\begin{pmatrix} 1 & 3 & \vdots & 1 & 0 \\ 2 & 5 & \vdots & 0 & 1 \end{pmatrix}$

$R_2 \rightarrow R_2 + -2R_1$

$\begin{pmatrix} 1 & 3 & \vdots & 1 & 0 \\ 0 & -1 & \vdots & -2 & 1 \end{pmatrix}$

$R_2 \rightarrow -R_2$

$\begin{pmatrix} 1 & 3 & \vdots & 1 & 0 \\ 0 & 1 & \vdots & 2 & -1 \end{pmatrix}$

$R_1 \rightarrow R_1 + -3R_2$

$\begin{pmatrix} 1 & 0 & \vdots & -5 & 3 \\ 0 & 1 & \vdots & 2 & -1 \end{pmatrix}$

Therefore, the inverse is $\begin{pmatrix} -5 & 3 \\ 2 & -1 \end{pmatrix}$

3. Start with $\begin{pmatrix} 0 & 1 & \vdots & 1 & 0 \\ 1 & 0 & \vdots & 0 & 1 \end{pmatrix}$

$R_1 \longleftrightarrow R_2$

$\begin{pmatrix} 1 & 0 & \vdots & 0 & 1 \\ 0 & 1 & \vdots & 1 & 0 \end{pmatrix}$

Therefore, the inverse is $\begin{pmatrix} 0 & 1 \\ 1 & 0 \end{pmatrix}$

5. Start with $\begin{pmatrix} a & a & \vdots & 1 & 0 \\ b & b & \vdots & 0 & 1 \end{pmatrix}$

$R_1 \rightarrow \frac{1}{a}R_1$

$R_2 \rightarrow R_2 + -bR_1$

$\begin{pmatrix} 1 & 1 & \vdots & 1/a & 0 \\ 0 & 0 & \vdots & -b/a & 0 \end{pmatrix}$

The matrix at left has all zeroes in the bottom row, so it cannot be transformed into the identity matrix. Therfore, the given matrix has no inverse.

7. Start with
$$\begin{pmatrix} 1 & 2 & 3 & \vdots & 1 & 0 & 0 \\ 2 & 1 & -1 & \vdots & 0 & 1 & 0 \\ 3 & 2 & 3 & \vdots & 0 & 0 & 1 \end{pmatrix}$$

$R_2 \rightarrow R_2 + -2R_1$
$R_3 \rightarrow R_3 + -3R_1$

$$\begin{pmatrix} 1 & 2 & 3 & \vdots & 1 & 0 & 0 \\ 0 & -3 & -7 & \vdots & -2 & 1 & 0 \\ 0 & -4 & -6 & \vdots & -3 & 0 & 1 \end{pmatrix}$$

$R_2 \rightarrow -\frac{1}{3}R_2$

$$\begin{pmatrix} 1 & 2 & 3 & \vdots & 1 & 0 & 0 \\ 0 & 1 & 7/3 & \vdots & 2/3 & -1/3 & 0 \\ 0 & -4 & -6 & \vdots & -3 & 0 & 1 \end{pmatrix}$$

$R_3 \rightarrow R_3 + 4R_2$
$R_1 \rightarrow R_1 + -2R_2$

$$\begin{pmatrix} 1 & 0 & -5/3 & \vdots & -1/3 & 2/3 & 0 \\ 0 & 1 & 7/3 & \vdots & 2/3 & -1/3 & 0 \\ 0 & 0 & 10/3 & \vdots & -1/3 & -4/3 & 1 \end{pmatrix}$$

$R_3 \rightarrow \frac{3}{10}R_3$

$$\begin{pmatrix} 1 & 0 & -5/3 & \vdots & -1/3 & 2/3 & 0 \\ 0 & 1 & 7/3 & \vdots & 2/3 & -1/3 & 0 \\ 0 & 0 & 1 & \vdots & -1/10 & -2/5 & 3/10 \end{pmatrix}$$

$R_2 \rightarrow R_2 + -\frac{7}{3}R_3$
$R_1 \rightarrow R_1 + \frac{5}{3}R_3$

$$\begin{pmatrix} 1 & 0 & 0 & \vdots & -1/2 & 0 & 1/2 \\ 0 & 1 & 0 & \vdots & 9/10 & 3/5 & -7/10 \\ 0 & 0 & 1 & \vdots & -1/10 & -2/5 & 3/10 \end{pmatrix}$$

Therefore, the inverse is
$$\begin{pmatrix} -1/2 & 0 & 1/2 \\ 9/10 & 3/5 & -7/10 \\ -1/10 & -2/5 & 3/10 \end{pmatrix}$$

9. **Start with**

$$\left(\begin{array}{ccc|ccc} 1 & 3 & 0 & 1 & 0 & 0 \\ 4 & 2 & 3 & 0 & 1 & 0 \\ 6 & 8 & 3 & 0 & 0 & 1 \end{array}\right)$$

$R_2 \rightarrow R_2 + -4R_1$
$R_3 \rightarrow R_3 + -6R_1$

$$\left(\begin{array}{ccc|ccc} 1 & 3 & 0 & 1 & 0 & 0 \\ 0 & -10 & 3 & -4 & 1 & 0 \\ 0 & -10 & 3 & -6 & 0 & 1 \end{array}\right)$$

Now note that the third row of the matrix to the left of the dotted line equals the second row.

Further operations will make the third row to the left all 0's, meaning the matrix has no inverse.

11. **Start with**

$$\left(\begin{array}{ccc|ccc} 2 & 1 & 0 & 1 & 0 & 0 \\ 0 & 1 & 3 & 0 & 1 & 0 \\ -2 & 3 & 2 & 0 & 0 & 1 \end{array}\right)$$

$R_1 \rightarrow \frac{1}{2}R_1$

$$\left(\begin{array}{ccc|ccc} 1 & 1/2 & 0 & 1/2 & 0 & 0 \\ 0 & 1 & 3 & 0 & 1 & 0 \\ -2 & 3 & 2 & 0 & 0 & 1 \end{array}\right)$$

$R_3 \rightarrow R_3 + 2R_1$

$$\left(\begin{array}{ccc|ccc} 1 & 1/2 & 0 & 1/2 & 0 & 0 \\ 0 & 1 & 3 & 0 & 1 & 0 \\ 0 & 4 & 2 & 1 & 0 & 1 \end{array}\right)$$

$R_3 \rightarrow R_3 + -4R_2$
$R_1 \rightarrow R_1 + -\frac{1}{2}R_2$

$$\left(\begin{array}{ccc|ccc} 1 & 0 & -3/2 & 1/2 & -1/2 & 0 \\ 0 & 1 & 3 & 0 & 1 & 0 \\ 0 & 0 & -10 & 1 & -4 & 1 \end{array}\right)$$

$R_3 \rightarrow -\frac{1}{10}R_3$

$$\left(\begin{array}{ccc|ccc} 1 & 0 & -3/2 & 1/2 & -1/2 & 0 \\ 0 & 1 & 3 & 0 & 1 & 0 \\ 0 & 0 & 1 & -1/10 & 2/5 & -1/10 \end{array}\right)$$

$R_2 \rightarrow R_2 + -3R_3$
$R_1 \rightarrow R_1 + \frac{3}{2}R_3$

$$\begin{pmatrix} 1 & 0 & 0 & \vdots & 7/20 & 1/10 & -3/20 \\ 0 & 1 & 0 & \vdots & 3/10 & -1/5 & 3/10 \\ 0 & 0 & 1 & \vdots & -1/10 & 2/5 & -1/10 \end{pmatrix}$$

Therefore, the inverse is

$$\begin{pmatrix} 7/20 & 1/10 & -3/20 \\ 3/10 & -1/5 & 3/10 \\ -1/10 & 2/5 & -1/10 \end{pmatrix}$$

13. Begin with

$$\begin{pmatrix} 1 & 0 & 1 & 2 & \vdots & 1 & 0 & 0 & 0 \\ 0 & 1 & -1 & 0 & \vdots & 0 & 1 & 0 & 0 \\ 2 & 1 & 2 & 1 & \vdots & 0 & 0 & 1 & 0 \\ 1 & 2 & 1 & 0 & \vdots & 0 & 0 & 0 & 1 \end{pmatrix}$$

$R_3 \to R_3 + -2R_1$
$R_4 \to R_4 + -R_1$

$$\begin{pmatrix} 1 & 0 & 1 & 2 & \vdots & 1 & 0 & 0 & 0 \\ 0 & 1 & -1 & 0 & \vdots & 0 & 1 & 0 & 0 \\ 0 & 1 & 0 & -3 & \vdots & -2 & 0 & 1 & 0 \\ 0 & 2 & 0 & -2 & \vdots & -1 & 0 & 0 & 1 \end{pmatrix}$$

$R_3 \to R_3 + -R_2$
$R_4 \to R_4 + -2R_2$

$$\begin{pmatrix} 1 & 0 & 1 & 2 & \vdots & 1 & 0 & 0 & 0 \\ 0 & 1 & -1 & 0 & \vdots & 0 & 1 & 0 & 0 \\ 0 & 0 & 1 & -3 & \vdots & -2 & -1 & 1 & 0 \\ 0 & 0 & 2 & -2 & \vdots & -1 & -2 & 0 & 1 \end{pmatrix}$$

$R_4 \to R_4 + -2R_3$
$R_2 \to R_2 + R_3$
$R_1 \to R_1 + 2R_3$

$$\begin{pmatrix} 1 & 0 & 0 & 5 & \vdots & 3 & 1 & -1 & 0 \\ 0 & 1 & 0 & -3 & \vdots & -2 & 0 & 1 & 0 \\ 0 & 0 & 1 & -3 & \vdots & -2 & -1 & 1 & 0 \\ 0 & 0 & 0 & 4 & \vdots & 3 & 0 & -2 & 1 \end{pmatrix}$$

$R_4 \rightarrow \frac{1}{4}R_4$ yields a new Row 4

$$(0 \qquad 0 \qquad 0 \; \vdots \; 1 \qquad 3/4 \qquad 0 \qquad -1/2 \quad 1/4)$$

$R_3 \rightarrow R_3 + 3R_4$

$R_2 \rightarrow R_2 + 3R_4$

$R_1 \rightarrow R_1 + -5R_4$

$$\begin{pmatrix} 1 & 0 & 0 & 0 & \vdots & -3/4 & 1 & 3/2 & -5/4 \\ 0 & 1 & 0 & 0 & \vdots & 1/4 & 0 & -1/2 & 3/4 \\ 0 & 0 & 1 & 0 & \vdots & 1/4 & -1 & -1/2 & 3/4 \\ 0 & 0 & 0 & 1 & \vdots & 3/4 & 0 & -1/2 & 1/4 \end{pmatrix}$$

The inverse is the matrix to the right of the dotted line.

15. Start with

$$\begin{pmatrix} -1 & 1 & 2 & 3 & \vdots & 1 & 0 & 0 & 0 \\ 3 & 2 & 1 & 0 & \vdots & 0 & 1 & 0 & 0 \\ 1 & 1 & -1 & 1 & \vdots & 0 & 0 & 1 & 0 \\ 3 & 4 & 2 & 4 & \vdots & 0 & 0 & 0 & 1 \end{pmatrix}$$

$R_2 \rightarrow R_2 + 3R_1$

$R_3 \rightarrow R_3 + R_1$ Doing these steps yields the following to the left of the dotted line:

$R_4 \rightarrow R_4 + 3R_1$

$$\begin{pmatrix} -1 & 1 & 2 & 3 \\ 0 & 5 & 7 & 9 \\ 0 & 2 & 1 & 4 \\ 0 & 7 & 8 & 13 \end{pmatrix}$$

You can continue further, but if you'll notice, Row 4 is the sum of Rows 2 and 3. This means if you continue with the row operations, eventually you'll get the last row is all 0's which means there is no inverse.

17. All you have to show is that when you multiply $\begin{pmatrix} 3 & 4 \\ -2 & 3 \end{pmatrix}$

by itself (the multiplication is defined) you get $\begin{pmatrix} 1 & 0 \\ 0 & 1 \end{pmatrix}$

19. Start with $\begin{pmatrix} a & b \\ c & d \end{pmatrix}$

If $a \neq 0$ then do the following:

$R_1 \rightarrow \frac{1}{a} R_1$

$R_2 \rightarrow R_2 + -cR_1$

$$\begin{pmatrix} 1 & b/a \\ 0 & -cb/a + d \end{pmatrix}$$

But ad - bc = 0. Divide both sides of the equation by a and obtain d - cb/a = 0, which would mean the row-echelon form of A has the second row all zeroes and hence A has no inverse.

If a = 0, then for ad - bc = 0, b or c must equal 0.

If b = 0, the original matrix has 0's across the top, no inverse.

If c = 0, the original matrix is $\begin{pmatrix} 0 & b \\ 0 & d \end{pmatrix}$

and $R_2 \rightarrow R_2 + -\frac{d}{b} R_1$ would make the second row all zeroes.

Problems 11.4

The determinant of a 2 x 2 matrix $\begin{pmatrix} a & b \\ c & d \end{pmatrix}$ is found by taking ad - bc.

1. $1 \cdot 4 - 3 \cdot 2 = 4 - 6 = -2$ 3. $3 \cdot 2 - 4 \cdot 8 = 6 - 32 = -26$

5. $6 - 6 = 0$

The determinant of any triangular matrix (a square matrix where all of the elements above, or, all of the elements below the diagonal are zero) is computed by taking the product of the diagonal elements. This comes from using the first column of the matrix for expanding the determinant, then using the second column of the submatrix remaining, etc.

7. From the above comments, the determinant is $(3)(2)(-6) = -36$

From the definition $(3)(2)(-6) + (0)(5)(0) + (0)(0)(0) = -36 + 0 + 0 = -36$

$(0)(2)(0) + (0)(5)(3) + (-6)(0)(0) = 0 + 0 + 0$

-36 - 0 = -36

9. $-1 \cdot 5 \cdot 0 = 0$ (another triangular matrix) **11.** abc (triangular again!)

13. Exchanging rows 1 and 4, and then exchanging rows 2 and 3 gives a diagonal matrix with the elements d, c, b, a on the diagonal. This matrix   has determinant abcd. Therefore, the original matrix has determinant $(-1)^2$abcd = abcd (since there were 2 switches of rows).

We can use the simpler method described in the text for finding the determinant of a 3 x 3 matrix.

15. $(2) \cdot (4) \cdot (-2) + 3 \cdot 2 \cdot (-1) + (-1) \cdot 3 \cdot 2 = -16 - 6 - 6 = -28$

$(-1)(4)(-1) + (3)(2)(2) + (-2)(3)(2) = 4 + 12 - 12 = 4$

$-28 - 4 = -32$

17. $(2)(-1)(7) + (3)(5)(6) + (2)(4)(2) = -14 + 90 + 16 = 92$

$(6)(-1)(2) + (2)(5)(2) + (7)(4)(3) = -12 + 20 + 84 = 92$

$92 - 92 = 0$

19.

Let D = determinant of coefficient matrix = $\begin{pmatrix} 1 & 1 & 1 \\ 3 & 0 & 5 \\ -2 & 3 & -1 \end{pmatrix}$

This determinant is found as follows:

$(1)(0)(-1) + (1)(5)(-2) + (1)(3)(3) = 0 - 10 + 9 = -1$

$(-2)(0)(1) + (3)(5)(1) + (-1)(3)(1) = 0 + 15 - 3 = 12$

$-1 - 12 = -13.$

To find the values of the variables, first replace the first column of the coefficient matrix by the constants and find the determinant of this matrix. Divide the result by D = -13 to get x.

$\begin{pmatrix} 5 & 1 & 1 \\ -3 & 0 & 5 \\ 2 & 3 & -1 \end{pmatrix} = -77$

$x = \frac{-77}{-13} = \frac{77}{13}$

To find y, replace the second row of the coeficient matrix by the constants and find the determinant of the resulting matrix

$\begin{pmatrix} 1 & 5 & 1 \\ 3 & -3 & 5 \\ -2 & 2 & -1 \end{pmatrix} = -42$

So $y = \frac{-42}{-13} = \frac{42}{13}$

Use the same technique for z: need the determinant of
$$\begin{pmatrix} 1 & 1 & 5 \\ 3 & 0 & -3 \\ -2 & 3 & 2 \end{pmatrix} = 54$$

So $z = \dfrac{54}{-13} = -\dfrac{54}{13}$

21. This problem is done identically to #19. The determinant of the coefficient matrix is

$$\begin{array}{ccc} -3 & 2 & 5 \\ 4 & -1 & 2 \\ -2 & 5 & -3 \end{array} = 127.$$ Replacing the first, second, and third columns, respectively,

by the constants, yields matrices with determinants of -113, 111, and 91 respectively.

Therefore, the solution is $x = -\dfrac{113}{127}$ $\quad y = \dfrac{111}{127}$ $\quad z = \dfrac{91}{127}$

23. Lower triangular, the determinant is the product of the diagonal elements. Anyway, the determinant is 0 and therefore the matrix is not invertible.

25. To find the determinant: $2 + -15 + 2 = -11$

$3 + 5 + -4 = 4$

$-11 - 4 = 15$. The matrix is invertible since its determinant is -15.

27. $288 + 80 - 40 = 328.$ $\qquad 120 + 240 - 32 = 328 .$ $\qquad 328 - 328 = 0.$

Since the determinant is 0, the matrix is not invertible.

Review Exercises for Chapter Eleven

1. $\begin{pmatrix} -6 & 3 \\ 0 & 12 \\ 6 & 9 \end{pmatrix}$

3. $\begin{pmatrix} -12 & -18 & 6 \\ -6 & -3 & 0 \\ 3 & -15 & -6 \end{pmatrix} + \begin{pmatrix} 10 & 2 & 0 \\ -6 & 8 & 4 \\ 8 & -4 & 2 \end{pmatrix} = \begin{pmatrix} -2 & -16 & 6 \\ -12 & 5 & 4 \\ 11 & -19 & -4 \end{pmatrix}$

5. 2 x 4 times 4 x 3 gives 2 x 3.

$10 + 6 + 1 + 0 = 17$ $14 + 0 + 0 + 25 = 39$ $2 + 9 + 0 + 30 = 41$

$0 + 12 + 2 + 0 = 14$ $0 + 0 + 0 + 20 = 20$ $0 + 18 + 0 \ 24 = 42$

17	39	41
14	20	42

7. 2 x 5 times 5 x 2 gives 2 x 2

$7 + 0 - 3 - 5 + 10 = 9$ $1 + 0 + 0 - 6 + 15 = 10$

$14 + 2 - 6 + 10 + 10 = 30$ $2 + 3 + 0 + 12 + 15 = 32$

9	10
30	32

9. -4   11. $(1)(4)(6) = 24$ 13. $16 + 4 + 18 + 16 - 6 + 12 = 60$

15. Write the coefficient matrix
$$\begin{pmatrix} 1 & -1 & 1 \\ 2 & 0 & -5 \\ 0 & 3 & -1 \end{pmatrix}$$

The determinant is $6 + 15 - 2 = 19$

Replace the first column of the coefficient matrix by the constants and obtain a matrix with determinant 123; therefore, $x = \frac{123}{19}$. Replace the second column and obtain a matrix with determinant 24, so $y = \frac{24}{19}$. Do the same with column 3 and obtain $z = \frac{34}{19}$

17. Start with
$$\begin{pmatrix} 4 & 7 & \vdots & 1 & 0 \\ 2 & 5 & \vdots & 0 & 1 \end{pmatrix}$$

$R_1 \rightarrow \frac{1}{4}R_1$
$$\begin{pmatrix} 1 & 7/4 & \vdots & 1/4 & 0 \\ 2 & 5 & \vdots & 0 & 1 \end{pmatrix}$$

$R_2 \rightarrow R_2 + -2R_1$
$$\begin{pmatrix} 1 & 7/4 & \vdots & 1/4 & 0 \\ 0 & 3/2 & \vdots & -1/2 & 1 \end{pmatrix}$$

$R_2 \rightarrow \frac{2}{3}R_2, R_1 \rightarrow R_1 + -\frac{7}{4}R_2$
$$\begin{pmatrix} 1 & 0 & \vdots & 5/6 & -7/6 \\ 0 & 1 & \vdots & -1/3 & 2/3 \end{pmatrix}$$
The inverse is to the right of the dotted line.

- 500 -

19. Start with

$$\left(\begin{array}{ccc|ccc} 1 & 2 & 0 & 1 & 0 & 0 \\ 2 & 1 & -1 & 0 & 1 & 0 \\ 3 & 1 & 1 & 0 & 0 & 1 \end{array}\right)$$

$R_2 \to R_2 + -2R_1$
$R_3 \to R_3 + -3R_3$

$$\left(\begin{array}{ccc|ccc} 1 & 2 & 0 & 1 & 0 & 0 \\ 0 & -3 & -1 & -2 & 1 & 0 \\ 0 & -5 & 1 & -3 & 0 & 1 \end{array}\right)$$

$R_2 \to -\frac{1}{3}R_2$
$R_3 \to R_3 + 5R_2$ (new R_2)

$$\left(\begin{array}{ccc|ccc} 1 & 2 & 0 & 1 & 0 & 0 \\ 0 & 1 & 1/3 & 2/3 & -1/3 & 0 \\ 0 & 0 & 8/3 & 1/3 & -5/3 & 1 \end{array}\right)$$

$R_3 \to \frac{3}{8}R_3$
$R_2 \to R_2 + -\frac{1}{3}R_3$
$R_1 \to R_1 + -2R_2$, $R_1 \to R_1 + \frac{2}{3}R_3$

$$\left(\begin{array}{ccc|ccc} 1 & 0 & 0 & -1/4 & 1/4 & 1/4 \\ 0 & 1 & 0 & 5/8 & -1/8 & -1/8 \\ 0 & 0 & 1 & 1/8 & -5/8 & 3/8 \end{array}\right)$$

The matrix on the right is the inverse.

21. Start with

$$\left(\begin{array}{ccc|ccc} 2 & 1 & 0 & 1 & 0 & 0 \\ -1 & 2 & 3 & 0 & 1 & 0 \\ 0 & 2 & 1 & 0 & 0 & 1 \end{array}\right)$$

$R_2 \leftarrow \to R_1$
$R_1 \to -R_1$

$$\left(\begin{array}{ccc|ccc} 1 & 2 & -3 & 0 & -1 & 0 \\ 2 & 1 & 0 & 1 & 0 & 0 \\ 0 & 2 & 1 & 0 & 0 & 1 \end{array}\right)$$

$R_2 \to R_2 + -2R_1$

$$\left(\begin{array}{ccc|ccc} 1 & -2 & -3 & 0 & -1 & 0 \\ 0 & 5 & 6 & 1 & 2 & 0 \\ 0 & 2 & 1 & 0 & 0 & 1 \end{array}\right)$$

$R_2 \rightarrow \frac{1}{5}R_2$ changes row 2 to

$$(0 \quad 1 \quad 6/5 \ \vdots \ 1/5 \quad 2/5 \quad 0)$$

$R_1 \rightarrow R_1 + 2R_2$

$R_3 \rightarrow R_3 + -2R_2$

$$\begin{pmatrix} 1 & 0 & -3/5 & 2/5 & -1/5 & 0 \\ 0 & 1 & 6/5 & 1/5 & 2/5 & 0 \\ 0 & 0 & -7/5 & -2/5 & -4/5 & 1 \end{pmatrix}$$

$R_3 \rightarrow -\frac{5}{7}R_3$ changes row 3 to

$$(0 \quad 0 \quad 1 \quad 2/7 \ \vdots \ 4/7 \quad -5/7)$$

$R_2 \rightarrow R_2 + -\frac{6}{5}R_3$

$R_1 \rightarrow R_1 + \frac{3}{5}R_3$

$$\begin{pmatrix} 1 & 0 & 0 & 4/7 & 1/7 & -3/7 \\ 0 & 1 & 0 & -1/7 & -2/7 & 6/7 \\ 0 & 0 & 1 & 2/7 & 4/7 & -5/7 \end{pmatrix}$$

The inverse is the matrix to the right of the dotted line.

23.
$$\begin{pmatrix} 1 & 2 & 0 \\ 2 & 1 & -1 \\ 3 & 1 & 1 \end{pmatrix} \begin{pmatrix} x \\ y \\ z \end{pmatrix} = \begin{pmatrix} 3 \\ -1 \\ 7 \end{pmatrix}$$

The coefficient matrix is the one from # 19. Therefore, just multiply the answer in 19 by vector on the right side of the equal sign above. You will obtain $x = \frac{3}{4}$, $y = \frac{9}{8}$, $z = \frac{29}{8}$

CHAPTER TWELVE
Introduction to Discrete Mathematics

Chapter Objectives

Mathematics, computer science, and other disciplines have discovered the importance of discrete mathematics. Some of the important topics in the area of discrete mathematics are introduced here; if you are going to study any of the sciences you are likely to encounter these topics again. Mathematical induction is discussed first. Many important theorems can be proven using the Principle of Mathematical Induction, and you will learn how to use this Principle to do some proofs.

Next up are sequences and the summation (or $\sum$) notation, which are also important. You will again encounter these concepts near the end of Calculus I. The topics in Sections 2, 3 and 4, which include recursion, geometric series and geometric progressions, have many applications, especially in mathematics and computer science.

You then come to counting the number of objects in a set; important methods here involve the Fundamental Principle of Counting, permutations, and combinations. These topics are further discussed in conjunction with some beginning probability theory, which you will see again if you study statistics.

As you should know by now, $(x + y)^2 \neq x^2 + y^2$. In Chapter 1 you saw formulas for $(x + y)^2$ and $(x + y)^3$. A general theorem, called the Binomial Theorem, tells you how to find $(x + y)^n$ for any n. You will learn how to prove the theorem using mathematical induction, and how to apply it in a number of areas.

Chapter Summary

- **Mathematical Induction:** To prove that something is true for each positive integer, prove it for the first integer. Then assume it true for the integer k and prove it true for the integer k + 1.

- **The Sigma Notation:**

$$\sum_{k=1}^{n} a_k = a_1 + a_2 + a_3 + \cdots + a_n$$

- The sequence $a_1, a_2, a_3, \cdots$ defined by $a_{n+1} = a_n + d$ is called an **arithmetic sequence**. For an arithmetic sequence, $a_n = a_1 + (n - 1)d$

- 503 -

- The sequence defined by $a_{n+1} = ra_n$, $r \neq 0, 1$, is called a **geometric sequence**. For a geometric sequence, $a_n = a_0 r^n$.

- The sum of a geometric progression is a sum of the form

$$S_n = 1 + r + r^2 + r^3 + \cdots + r^n = \sum_{k=0}^{n} r^k = \frac{1 - r^{n+1}}{1 - r} \quad \text{for } r \neq 1$$

- A geometric series is an infinite sum of the form

$$S = 1 + r + r^2 + r^3 + \cdots + = \sum_{k=0}^{\infty} r^k = \frac{1}{1 - r} \quad \text{if } -1 < r < 1.$$

- A **permutation of n objects** is a rearrangement of the objects.

- **Factorial:** n factorial, denoted by n!, is the product of the first n integers.
  $$n! = n(n - 1)(n - 2)(n - 3)\cdots(3)(2)(1).$$

- The number of permuatations of n objects is n!

- The number of **permutations** of n objects taken k at a time, denoted by $P_{n,k}$, is given by
  $$P_{n,k} = n(n - 1)(n - 2)\cdots(n - k + 1).$$

- A **combination** of n objects taken k at a time is any selection of the n objects without regard to order.

- The number of combinations of n objects taken k at a time, denoted by $\binom{n}{k}$,
  is given by $\binom{n}{k} = \dfrac{n!}{(n - k)!k!}$

- The **sample space S of an experiment** is the set of all possible outcomes.

- An **event E** is a subset of the sample space. One outcome is called a **simple event**.

- A **sample space** is an equiprobable space if all simple events are equally likely.

- **Probablility in an equiprobable space:**

$$P(E) = \frac{\text{number of outcomes in E}}{\text{number of outcomes in S}}$$

- **The Binomial Theorem:**
  $$(x + y)^n = x^n + \binom{n}{1}x^{n-1}y + \binom{n}{2}x^{n-2}y^2 + \cdots + \binom{n}{j}x^{n-j}y^j + \cdots + \binom{n}{n-1}xy^{n-1} + y^n$$

SOLUTIONS TO CHAPTER TWELVE PROBLEMS

Problems 12.1

1. First, is it true for n = 1?   2 = 1(1 + 1) so it is.

   Now assume it is true for n = k. Then $2 + 4 + 6 + \cdots + 2k = k(k + 1)$.

   We must now show that it is true for n = k + 1; that is, we must show that

   $2 + 4 + 6 + \cdots + 2k + 2(k + 1) = (k + 1)[(k + 1) + 1]$. We know that

   $2 + 4 + 6 + \cdots + 2k + 2(k + 1) = k(k + 1) + 2(k + 1)$ [Induction Hypothesis]

   $$= (k + 2)(k + 1) = (k + 1)[(k + 1) + 1] \quad \text{QED}$$

3. First, is it true for n = 1? $2 = \dfrac{1(3 \cdot 1 + 1)}{2}$ so it is.

   Now assume it is true for n = k. Then $2 + 5 + 8 + \cdots + (3k - 1) = \dfrac{k(3k + 1)}{2}$.

   We must now show it is true for n = k + 1; that is, we must show that

   $$2 + 5 + 8 + \cdots (3k - 1) + (3k + 2) = \frac{(k + 1)[3(k +1) + 1]}{2} = \frac{(k + 1)(3k + 4)}{2} = \frac{3k^2 + 7k + 4}{2}$$

   Add 3k + 2 to both sides of the equation in the induction hypothesis and get

   $$2 + 5 + 8 + \cdots + (3k - 1) + (3k + 2) = \frac{k(3k + 1)}{2} + (3k + 2) = \frac{3k^2 + k}{2} + \frac{6k + 4}{2} =$$

   $\dfrac{3k^2 + 7k + 4}{2}$. QED

5. Is it true for n = 1? Yes, $2^1 > 1$.

   Now assume true for n = k; that is, $2^k > k$.

   We must now prove it true for n = k + 1, or that $2^{k+1} > k + 1$

   Multiply both sides of the induction hypothesis by 2 to obtain $2^{k+1} > 2k$.

   Now $2k = k + k > k + 1$ if k > 1, which it is. Therefore, $2^{k+1} > k + 1$. QED

7. Is it true for n = 0? Yes, since $1 = 2^1 - 1$.

   Now assume true for n = k; that is, $1 + 2 + 4 + \cdots + 2^k = 2^{k+1} - 1$.

   We must now prove that it is true for n = k + 1, or that $1 + 2 + 4 + \cdots + 2^k + 2^{k+1} = 2^{k+2} - 1$

   Add 2^{k+1} to both sides of the induction hypothesis and obtain

   $1 + 2 + 4 + \cdots + 2^k + 2^{k+1} = 2^{k+1} - 1 + 2^{k+1} = 2 \cdot 2^{k+1} - 1 = 2^{k+2} - 1$ QED

9. Is it true for $n = 0$? Yes, since $1 = 2 - \dfrac{1}{2^0} = 2 - 1$.

Now assume true for $n = k$; that is, $1 + \dfrac{1}{2} + \dfrac{1}{4} + \cdots + \dfrac{1}{2^k} = 2 - \dfrac{1}{2^k}$.

We must now prove for $n = k + 1$; that is, $1 + \dfrac{1}{2} + \dfrac{1}{4} + \cdots + \dfrac{1}{2^k} + \dfrac{1}{2^{k+1}} = 2 - \dfrac{1}{2^{k+1}}$.

Add $\dfrac{1}{2^{k+1}}$ to both sides of the induction hypothesis and obtain

$$1 + \dfrac{1}{2} + \dfrac{1}{4} + \cdots + \dfrac{1}{2^k} + \dfrac{1}{2^{k+1}} = 2 - \dfrac{1}{2^k} + \dfrac{1}{2^{k+1}} = 2 - \dfrac{2}{2^{k+1}} + \dfrac{1}{2^{k+1}} = 2 - \dfrac{1}{2^{k+1}} \quad \text{QED}$$

11. Is it true for $n = 1$? Yes, $1^3 = \dfrac{1^2(1+1)^2}{4}$.

Now assume that it is true for $n = k$; that is,

$$1^3 + 2^3 + 3^3 + \cdots + k^3 = \dfrac{k^2(k+1)^2}{4}.$$

We must prove that it is true for $n = k + 1$; that is,

$$1^3 + 2^3 + 3^3 + \cdots + k^3 + (k+1)^3 = \dfrac{(k+1)^2[(k+1)+1]^2}{4}$$

$$= \dfrac{k^4 + 6k^3 + 13k^2 + 12k + 4}{4}$$

Add $(k+1)^3$ to both sides of the induction hypothesis.
$$1^3 + 2^3 + \cdots + k^3 + (k+1)^3 = \dfrac{k^2(k+1)^2}{4} + (k+1)^3$$

$$= \dfrac{k^4 + 2k^3 + k^2}{4} + k^3 + 3k^2 + 3k + 1 = \dfrac{k^4 + 2k^3 + k^2}{4} + \dfrac{4k^3 + 12k^2 + 12k + 4}{4}$$

$$= \dfrac{k^4 + 6k^3 + 13k^2 + 12k + 4}{4} \qquad \text{QED}$$

13. Is it true for n = 1? Yes, $1 \cdot 2 = \dfrac{1 \cdot (1 + 1) \cdot (4(1) - 1)}{3}$

Assume true for n = k, that is $1 \cdot 2 + 3 \cdot 4 + \cdots + (2k - 1)(2k) = \dfrac{k(k + 1)(4k - 1)}{3}$

Now prove for n = k + 1; that is, $1 \cdot 2 + 3 \cdot 4 + \cdots + (2k - 1)(2k) + (2k + 1)(2k + 2)$

$= \dfrac{(k + 1)(k + 2)(4k + 3)}{3} = \dfrac{4k^3 + 15k^2 + 17k + 6}{3}.$

Add $[2(k + 1) - 1][2(k + 1] = (2k + 1)(2k + 2)$ to both sides of the Ind. Hyp.

Obtain $1 \cdot 2 + 3 \cdot 4 + \cdots (2k - 1)(2k) + (2k + 1)(2k + 2)$
$= \dfrac{k(k + 1)(4k - 1)}{3} + (2k + 1)(2k + 2) = \dfrac{4k^3 + 3k^2 - k}{3} + 4k^2 + 6k + 2$

$= \dfrac{4k^3 + 3k^2 - k}{3} + \dfrac{12k^2 + 18k + 6}{3} = \dfrac{4k^3 + 15k^2 + 17k + 6}{3}$ QED

Several of the following use the fact that if an integer m divides evenly into an integer a and divides evenly into an integer b, then m divides evenly into a + b.

15. Is it true for n = 1? Yes, since $1 + 1^2 = 2$ is even.

Assume that $k^2 + k$ is even. Now prove true for k; that is,

we now must prove that $(k + 1)^2 + (k + 1)$ is even.

But $(k + 1)^2 + (k + 1) = k^2 + 2k + 1 + k + 1 = (k^2 + k) + (2k + 2)$

Now 2 divides evenly into $k^2 + k$ by the induction hypothesis.

It is evident that 2 divides evenly into 2k and that 2 divides evenly into 2.

Therefore, 2 divides evenly into $(k^2 + k) + (2k + 2)$, meaning the number is even. QED

17. Is it true for n = 1? Yes, because $1(1^2 + 5) = 6$ is divisible by 6.

Now assume true for k; that is, $k(k^2 + 5)$ is divisible by 6.

We now must prove that $(k + 1)[(k + 1)^2 + 5]$ is divisible by 6.

$(k + 1)[(k + 1)^2 + 5] = (k + 1)(k^2 + 2k + 6) = (k + 1)[(k^2 + 5) + (2k + 1)]$

$= k(k^2 + 5) + (k^2 + 5) + k(2k + 1) + (2k + 1) = k(k^2 + 5) + 3(k^2 + k) + 6.$

Now $k(k^2 + 5)$ is divisible by 6 by the Ind. Hyp., $3(k^2 + k)$ is clearly

divisible by 3 and is even (divisible by 2) by problem #15, so is divisible by 6, and certainly

6 is divisible by 6, so the expression given is divisible by 6.

19. The problem is true if n = 1, since x^1 - 1 is divisible by x - 1.

Now assume that x^k - 1 is divisible by x - 1.

We have to prove that x^{k+1} - 1 is divisible by x - 1.

Now $x^{k+1} - 1 = x^k x - 1 = x^k x - x + x - 1 = x(x^k - 1) + (x - 1)$.

The first term is divisible by x - 1 by the induction hypothesis, and the

second term in the sum is divisible by x - 1, so the expression is. QED

21. k = 21 is the smallest value that works, since $20 \cdot 20 = 20^2$, but $20 \cdot 21 < 21^2$.

Assume true for n = k, that is, $20k < k^2$. (Ind. Hyp)

Prove true for n = k + 1; that is, prove that $20(k + 1) < (k + 1)^2$

Start with induction hypothesis and add 20 to both sides:

$20k + 20 < k^2 + 20 < k^2 + 2k \ (k \geq 21) \ < k^2 + 2k + 1 = (k + 1)^2$ QED

23. Using trial and error and a calculator, obtain the value k = 238.

We assume 100 log k < k for k > 238. We must show that 100 log (k + 1) < k + 1.

$100 \log (k + 1) = 100(\log(k + 1) - \log k + \log k) = 100 \left(\log \left(\frac{k + 1}{k} \right) + \log k \right)$

$= 100 \left(\log \left(1 + \frac{1}{k} \right) + \log k \right) < 100 \left(\log \left(1 + \frac{1}{238} \right) + \log k \right) \approx 100(0.00182 + \log k)$

$< 100(0.01 + \log k) = 100 \log k + 1 < k + 1$

25. For k = 5, $5^3 < e^5$ since 125 < 148.4. Try k = 4 and you'll see that's too small.

Now we assume that $k^3 < e^k$ for k > 5. Must show that $(k + 1)^3 < e^{k+1}$.

$(k + 1)^3 = k^3 \left(1 + \frac{1}{k} \right)^3 < e^k \left(1 + \frac{1}{k} \right)^3 \leq e^k \left(1 + \frac{1}{6} \right)^3 \approx 1.59 e^k < e \cdot e^k = e^{k+1}$.

27. k = 6, since $e^6 < 6!$ but $e^5 > 5!$

We assume that $e^k < k!$ for k > 6. Must show that $e^{k+1} < (k + 1)!$

Multiply both sides of the induction hypothesis by e to obtain $e \cdot e^k < e \cdot k!$

and so $e^{k+1} < e \cdot k! < (k + 1)k! = (k + 1)!$

29. It is true for n = 1; $(ab)^1 = ab = a^1 b^1$. Now assume true for n = k; $(ab)^k = a^k b^k$.

Must prove that $(ab)^{k+1} = a^{k+1} b^{k+1}$.

$(ab)^{k+1} = (ab)^k (ab) = a^k b^k ab$ (Ind. Hyp) $= a^k a b^k b$ (commutative property for multiplication)

$= a^{k+1} b^{k+1}$.

31. Assume (2k - 1) is even for some integer k. Then 2 divides 2k - 1.

Then 2 also divides 2k - 1 + 2 = 2k + 1.

There is nothing wrong with the proof. We have not proven an initial case,

namely n = 1. The statement is not true for n = 1, so you cannot attempt to

prove this statement, (which is false), by induction.

Problems 12.2

For problems 1-25, just plug in n = 1, n = 2, n = 3, n = 4, and n = 5.

1. 1, 2, 3, 4, 5   3. 2, 6, 10, 14, 18   5. $\frac{1}{3}, \frac{1}{4}, \frac{1}{5}, \frac{1}{6}, \frac{1}{7}$ 7. 2, 6, 12, 20, 30

9. $2, \frac{3}{2}, \frac{4}{3}, \frac{5}{4}, \frac{6}{5}$ 11. $\frac{3}{2}, \frac{7}{4}, \frac{15}{8}, \frac{31}{16}, \frac{63}{32}$ 13. 1, 16, 81, 256, 625   15. $-\frac{1}{4}, \frac{1}{5}, -\frac{1}{6}, \frac{1}{7}, -\frac{1}{8}$

17. e, e^2, e^3, e^4, e^5 19. $e, e^{1/2}, e^{1/3}, e^{1/4}, e^{1/5}$ 21. 1, 1, 1, 1, 1

23. $-\frac{2}{3}, \frac{4}{9}, -\frac{8}{27}, \frac{16}{81}, -\frac{32}{243}$ 25. -1, -2, -3, -4, -5

Any sequence that alternates will have $(-1)^n$ or $(-1)^{n+1}$. Recall that you are to start with n = 1.

27. $(-1)^{n+1}n$ 29. n^3 31. $\frac{(-1)^{n+1}}{(n+1)^2}$ 33. $(-1)^{n+1}\frac{2^{n+1}+1}{2^n} = (-1)^{n+1}(2 + \frac{1}{2^n})$

35. $\frac{n}{2n+1}$ 37. $\frac{(-1)^{n+1}}{3^n}$ 39. $\frac{n^2}{n+1}$

41. 3(1) + 3(2) + 3(3) + 3(4) +3(5) = 3 + 6 + 9 + 12 + 15 = 45

43. 1 + 1 + 1 + 1 + 1 + 1 + 1 = 7

45. $(-2)^2 + (-1)^2 + 0^2 + 1^2 + 2^2 + 3^2 + 4^2 = 4 + 1 + 0 + 1 + 4 + 9 + 16 = 35$

47. $\sum_{k=0}^{4} 2^k$ 49. $\sum_{k=2}^{n} \frac{k}{k+1} = \sum_{k=1}^{n-1} \frac{k+1}{k+2}$

51. $\displaystyle\sum_{k=1}^{n} k^{1/k}$ 53. $\displaystyle\sum_{k=1}^{5} x^{4k+2}$ 55. $\displaystyle\sum_{k=1}^{8} (2k-1)(2k+1) = \sum_{k=0}^{7} (2k+1)(2k+3)$

57. $(32)^{-3}\displaystyle\sum_{k=1}^{32} k^2$ 59. $(64)^{-4}\displaystyle\sum_{k=1}^{64} k^3$ 61. $\dfrac{1}{50}\displaystyle\sum_{k=1}^{50} \sqrt{\dfrac{k}{50}} = (50)^{-3/2}\sum_{k=1}^{50} \sqrt{k}$

63. $\displaystyle\sum_{1=1}^{7} (7-i)^2$. **This one adds squares of the numbers from 0 to 6; the others add the squares of the numbers from 1 to 7.**

65. a) $\displaystyle\sum_{k=1}^{n} [g(k) - g(k-1)] = [g(1) - g(0)] + [g(2) - g(1)] - [g(3) - g(2)] + \cdots$
$$+ [g(n-1) - g(n-2)] + [g(n) - g(n-1)] = g(n) - g(0)$$

b) Let $g(k) = \frac{1}{2}k\cdot(k+1)$. Then
$$g(k) - g(k-1) = \tfrac{1}{2}k\cdot(k+1) - \tfrac{1}{2}(k-1)k = \tfrac{1}{2}k^2 + \tfrac{1}{2}k - \tfrac{1}{2}k^2 + \tfrac{1}{2}k = k.$$

From part a), $\displaystyle\sum_{k=1}^{n} \left(\tfrac{1}{2}k\cdot(k+1)\right) - \left(\tfrac{1}{2}(k-1)k\right) = \sum_{k=1}^{n} g(k) - g(k-1)$

$= g(n) - g(0) = \frac{1}{2}n(n+1) - \frac{1}{2}(0)(0+1) = \frac{1}{2}n(n+1)$.

If $G(k) = \frac{1}{3}k\cdot(k+1)\cdot(k+2)$, then

$G(k) - G(k-1) = \left(\tfrac{1}{3}k(k+1)(k+2)\right) - \left(\tfrac{1}{3}(k-1)(k)(k+1)\right)$
$= \tfrac{1}{3}k(k+1)\big((k+2) - (k-1)\big) = \tfrac{1}{3}k(k+1)(3) = k(k+1)$

67. Let $g(k) = \frac{1}{4}k(k+1)(k+2)(k+3)$; then $g(k-1) = \frac{1}{4}(k-1)(k)(k+1)(k+2)$
$g(k) - g(k-1) = \frac{1}{4}k(k+1)(k+2)(k+3) - \frac{1}{4}(k-1)(k)(k+1)(k+2)$
$= \frac{1}{4}k(k+1)(k+2)\big((k+3) - (k-1)\big) = \frac{1}{4}k(k+1)(k+2)4 = k(k+1)(k+2)$

$\displaystyle\sum_{k=1}^{n} k(k+1)(k+2) = g(n) - g(0) = \frac{1}{4}n(n+1)(n+2)(n+3) - 0$

$\displaystyle\sum_{k=1}^{n} (k^3 + 3k^2 + 2k) = \frac{1}{4}n(n+1)(n+2)(n+3)$ (continued)

$$\sum_{k=1}^{n} k^3 + 3\sum_{k=1}^{n} k^2 + 2\sum_{k=1}^{n} k = \frac{1}{4}n(n+1)(n+2)(n+3)$$

$$\sum_{k=1}^{n} k^3 = \frac{1}{4}n(n+1)(n+2)(n+3) - 3\sum_{k=1}^{n} k^2 - 2\sum_{k=1}^{n} k$$

$$= \frac{1}{4}n(n+1)(n+2)(n+3) - \frac{3}{6}n(n+1)(2n+1) - \frac{2}{2}n(n+1)$$

$$= \frac{1}{4}n^4 + \frac{3}{2}n^3 + \frac{11}{4}n^2 + \frac{3}{2}n - n^3 - \frac{3}{2}n^2 - \frac{1}{2}n - n^2 - n$$

$$= \frac{1}{4}n^4 + \frac{1}{2}n^3 + \frac{1}{4}n^2 = \frac{n^2(n+1)^2}{4}$$

69. $\dfrac{1}{n^2 + 2n} = \dfrac{A}{n} + \dfrac{B}{n+2}$ by partial fraction decompostion, since $n^2 + 2n = n(n+2)$

$1 = A(n+2) + Bn$

Let $n = 0$ and obtain $1 = 2A$ or $A = \frac{1}{2}$. Let $n = -2$ and obtain $-2B = 1$ or $B = -\frac{1}{2}$

$$\sum_{k=1}^{n} \frac{1}{k^2 + 2k} = \sum_{k=1}^{n} \left(\frac{1/2}{k} - \frac{1/2}{k+2}\right) = \left(\frac{1}{2} - \frac{1}{6}\right) + \left(\frac{1}{4} - \frac{1}{8}\right) + \left(\frac{1}{6} - \frac{1}{10}\right) + \cdots + \left(\frac{1/2}{n} - \frac{1/2}{n+2}\right)$$

$$= \frac{3}{4} - \frac{1/2}{n+1} - \frac{1/2}{n+2} = \frac{1}{2}\left(\frac{3}{2} - \frac{1}{n+1} - \frac{1}{n+2}\right)$$

71. $\displaystyle\sum_{k=1}^{N} (a_k + b_k) = (a_1 + b_1) + (a_2 + b_2) + (a_3 + b_3) + \cdots + (a_N + b_N)$

$$= (a_1 + a_2 + a_3 + \cdots + a_N) + (b_1 + b_2 + b_3 + \cdots + b_N)$$

$$= \sum_{k=1}^{N} a_k + \sum_{k=1}^{N} b_k$$

Problems 12.3

1. $a_n = 1 + (n-1)3 = 1 + 3n - 3 = 3n - 2$ 3. $a_n = 4 + (n-1)(-3) = 4 - 3n + 3 = 7 - 3n$

5. $a_n = -12 + (n - 1)\, 6 = -12 + 6n - 6 = -18 + 6n$

7. 5, 6, 7, 8, 9   9. $\frac{17}{3}, \frac{16}{3}, 5, \frac{14}{3}, \frac{13}{3}$ 11. $\frac{13}{2}, \frac{25}{2}, \frac{37}{2}, \frac{49}{2}, \frac{61}{2}$ 13. $a_n = 2^n$

15. $a_n = 6(-1)^n$ 17. $a_n = 7\left(\frac{1}{3}\right)^n$ 19. 5, 10, 20, 40, 80, 160   21. 3, -3, 3, -3, 3, -3

23. 1, 4, 16, 64, 256, 1024

25. arithmetic since no exponents: each term differs by 6.

27. geometric: each term is 6 times greater than the previous.

29. write as $\frac{4}{5} + \frac{3}{5}n$ to see it is arithmetic: each term differs by $\frac{3}{5}$.

31. combining + and exponents gives neither.

33. geometric; the common ratio is -1

35. arithmetic; the common difference is 10

37. geometric; the common ratio is -1 (the sequence is same as $\{(-1)^n\}$).

39. same as $\{3 \cdot (-2)^n\}$ so geometic with common ratio -2.

41. The sequence is $\{3 + 4(n - 1)\}$, so when n = 10, get 39.

43. The sequence is $\{12 - 3(n - 1)\}$ so when n = 100, get -285.

45. The sequence is $\{4 \cdot \left(\frac{1}{2}\right)^n\}$. To get the eight term, plug in n = 7 (since a geometric sequence begins with n = 0) and obtain $\frac{1}{32}$.

47. $120 = a_6 = a_1 + 5(6 - 1) = a_1 + 25$. Therefore, $a_1 = 95$.

49. $\frac{26 - 10}{8 - 4} = \frac{16}{4} = 4$ 51. $\frac{4 - 8}{24 - 8} = \frac{-4}{16} = -\frac{1}{4}$ 53. $80 = a_4 = a_0 \cdot 2^4 = 16a_0$ so $a_0 = 5$

55. $r^{7-2} = \frac{972}{4} = 243$. If $r^5 = 243$, then $r = 3$ 57. $r^{7-3} = \frac{15}{4}$ so $r = \sqrt[4]{\frac{15}{4}} \approx 1.3916$

59. $5000(1.08)^5 = \$7346.65$ 61. $10,000(1.065)^{10} = \$18,771.37$

63. If r is the interest rate for the year and there are m periods per year, the

rate per period is $\frac{r}{m}$. The interest after 1 period (m periods per year) is $P_0 \cdot \frac{r}{m}$, so

the total value of the investment is $P_0 + P_0 \cdot \frac{r}{m}$ or $P_1 = P_0\left(1 + \frac{r}{m}\right)^1$.

To get P_2, add the amount after 1 period to the interest accumulated on this

amount to obtain $P_0\left(1 + \frac{r}{m}\right) + P_0\left(1 + \frac{r}{m}\right)\frac{r}{m} = P_0\left(1 + \frac{r}{m}\right)^2$. To get each succeeding

value, just multiply by $\left(1 + \frac{r}{m}\right)$. That proves the problem.

65. $5000(1.07)^8 = \$8590.93$

67. annual compounding gives $(1.06)^{10} = 1.79085$ which means you'll have 79.085% more than

what you started with.

Quarterly compounding gives $\left(1 + \frac{.06}{4}\right)^{40} = 1.81402$, or an increase of 81.402%.

69. In problem #69, you get the 5% compounded daily is better.

5% compounded daily: $\left(1 + \frac{.05}{365}\right)^{365} = 1.05126$

$5\frac{1}{8}$% compounded semiannually: $\left(1 + \frac{.05125}{2}\right)^2 = 1.0519066$, a better return.

71. P_n is the probability that a cumulative score of n is attained. The only way

a score of n + 1 can be attained is if a "1" is obtained on the next flip. Since

there is a $\frac{1}{2}$ probability of that happening, $P_{n+1} = \frac{1}{2}P_n$ or, equivalently,

$P_n = \frac{1}{2}P_{n-1}$, Since $P_0 = 1$, each $P_1 = \frac{1}{2}$, $P_2 = \frac{1}{2} \cdot \frac{1}{2} = \frac{1}{4}$, or $P_n = \frac{1}{2^n}$

73. We are given that $a_0 = 4$. Plug this into the equation given and obtain $a_1 = -1$.

Because of the $\frac{1}{2}$ in the equation we know that $a_n = b + c\left(\frac{1}{2}\right)^n$ for some

constants b and c. Now $4 = a_0 = b + c\left(\frac{1}{2}\right)^0 = b + c$ and

$-1 = b + c\left(\frac{1}{2}\right)^1 = b + \frac{1}{2}c$. Solve these two equations simultaneously to obtain

$c = 10$ and $b = -6$; therefore, $a_n = -6 + 10\left(\frac{1}{2}\right)^n$.

75. As in #73, $a_n = b + c(-2)^n$. Now $5 = a_0 = b + c$ and $a_1 = 0 = b + c(-2)^1$ or $b - 2c$. Solve simultaneously to obtain $b = \frac{10}{3}$ and $c = \frac{5}{3}$, or $a_n = \frac{10}{3} + \frac{5}{3}(-2)^n$.

77. a) $P_{n+1} = P_n(1.08) - 300$, because you add on the 8% interest before subtracting the $300.

b) $P_n = b + c(1.08)^n$. Therefore, since $10{,}000 = P_0$, $10{,}000 = b + c$ and $10{,}500 = P_1 = b + 1.08c$
Solve simultaneously to obtain $b = 375$ and $c = 6250$, so $P_n = 3750 + 6250(1.08)^n$

c) Plug in $n = 5$ into the equation in b) and obtain $12{,}933.30$. Plug in $n = 10$ and
obtain $17{,}243.28$.

79. $a_{n+1} = \frac{n+2}{n+1}a_n$. Since $a_1 = 2$, $a_2 = \frac{3}{2}(2) = 3$ $a_3 = \frac{4}{3}(3) = 4$

$a_n = 2 + (n-1)\cdot 1 = 2 + n - 1 = 1 + n$

81. $a_1 = 2$ $\qquad a_2 = 2 + 1^2 = 3$ $\qquad a_3 = 3 + 2^2 = 3 + 4 = 7$

$a_4 = 7 + 3^2 = 16$ $\qquad a_5 = 16 + 4^2 = 32$ $\qquad a_6 = 32 + 5^2 = 57$

83. $a_1 = 4$ $\qquad a_2 = \frac{4}{2+4} = \frac{2}{3}$ $\qquad a_3 = \frac{2/3}{2 + 2/3} = \frac{1}{4}$

$a_4 = \frac{1/4}{2 + 1/4} = \frac{1}{9}$ $\qquad a_5 = \frac{1/9}{2 + 1/9} = \frac{1}{19}$ $\qquad a_6 = \frac{1/19}{2 + 1/19} = \frac{1}{39}$

85. By the hint, the result is true for $k = 0$. Now we assume it is true for $n = k$;
that is, the solution to $a_{k+1} = ra_k + b$ is

$a_k = \frac{b}{1-r} + \left(a_0 - \frac{b}{1-r}\right)r^k$. We must prove that it is true for $n = k + 1$; that
is, prove that the solution to $a_{k+2} = ra_{k+1} + b$ is $\qquad a_{k+1} = \frac{b}{1-r} + \left(a_0 - \frac{b}{1-r}\right)r^{k+1}$.

$r(a_{k+1}) + b = \frac{rb}{1-r} + \left(a_0 - \frac{b}{1-r}\right)r^{k+2} + b = \frac{rb}{1-r} + \left(a_0 - \frac{b}{1-r}\right)r^{k+2} + \frac{b - rb}{1-r}$
$= \frac{b}{1-r} + \left(a_0 - \frac{b}{1-r}\right)r^{k+2}$

Problems 12.4

In problems 1-11, use $1 + r + r^2 + \cdots + r^n = \dfrac{1 - r^{n+1}}{1 - r}$

1. $\dfrac{1 - 4^5}{1 - 4} = \dfrac{1 - 1024}{-3} = 341$

3. $\dfrac{1 - (-5)^6}{1 - (-5)} = \dfrac{-15,624}{6} = -2604$

5. $1 - 0.3 + 0.3^2 - \cdots + 0.3^8 = \dfrac{1 - (-0.3)^9}{1 - (-.3)} = \dfrac{1 + (.3)^9}{1.3}$. This includes two terms

that are not given in the original problem, so they have to be subtracted off:

the sum is $\dfrac{1 + (.3)^9}{1.3} - 1 + 0.3 \approx 0.0692459.$

Another way to write the series is $(0.3)^2 \, [1 - 0.3 + (0.3)^2 - (0.3)^3 + \cdots + (0.3^6]$

$$= (0.3)^2 \left(\dfrac{1 + (0.3)^7}{1 - (-0.3)} \right) \approx 0.0692459$$

7. $\dfrac{1 - (-1/b^2)^8}{1 - (-1/b^2)} = \dfrac{1 - \dfrac{1}{b^{16}}}{1 + \dfrac{1}{b^2}} = \dfrac{b^{16} - 1}{b^{16} + b^{14}}$

9. $\dfrac{1 - (\sqrt{2})^9}{1 - \sqrt{2}} = \dfrac{1 - 16\sqrt{2}}{1 - \sqrt{2}} \cdot \dfrac{1 + \sqrt{2}}{1 + \sqrt{2}} = \dfrac{1 + \sqrt{2} - 16\sqrt{2} - 32}{1 - 2} = 15\sqrt{2} + 31$

11. Can use a calculator or the fact that $1 - 4 + 16 - 64 + 256 - 1024 + 4096$

$$= \dfrac{1 - (-4)^7}{1 - (-4)} = \dfrac{-16,385}{5} = 3277 \quad \text{so } -1 + 4 - 16 + 64 - 256 + 1024 - 4096 = -3277.$$

To get the expression in the book, add 1 and subtract 4 to get -3280.

For 13-21, use $S = \dfrac{1}{1 - r}$

13. $\dfrac{1}{1 - 1/3} = \dfrac{1}{2/3} = \dfrac{3}{2}$

15. $\dfrac{1}{1 - 1/10} = \dfrac{1}{9/10} = \dfrac{10}{9}$

17. $\dfrac{1}{1 - 1/e} = \dfrac{e}{e - 1}$

19. $\dfrac{1}{1 - (-0.31)} = \dfrac{1}{1.31} = \dfrac{1}{131/100} = \dfrac{100}{131} \approx 0.763359$

21. $\dfrac{3}{5}\left(1 - \dfrac{1}{5} + \dfrac{1}{25} - \cdots \right) = \dfrac{3}{5} \cdot \dfrac{1}{1 - (-1/5)} = \dfrac{3}{5} \cdot \dfrac{1}{6/5} = \dfrac{3}{5} \cdot \dfrac{5}{6} = \dfrac{1}{2}$

23. $n \ln .5 < \ln .01$. Divide both sides by $\ln .5$ (this reverses inequality since

$$n > \frac{\ln .01}{\ln .5} \qquad\qquad \ln .5 \text{ is negative)}$$

$n > 6.6438561$ So if n is a positive integer, $n \geq 7$

25. As in #23, take ln of both sides and obtain $n \ln .99 < \ln .01$

$$n > \frac{\ln .01}{\ln .99} \quad \text{or} \quad n > 458.21057 \text{ or } n \geq 459 \text{ if n is a positive integer}$$

27. a) It is true that when Achilles catches up to where the tortoise was, the tortoise will have moved on. However, the time intervals that it takes Achilles to reach where the tortoise was become so small that he eventually will catch up. See part b.

   b) Use $d = rt$. If the tortoise has a 40-km head start, he has a 40 hour head start. The distances covered by each are equal.

   $1(40 + t) = 201t$

   $40 + t = 201t$

   $40 = 200t$

   $t = \frac{40}{200} = \frac{1}{5}$ hour or 12 minutes.

In 29-37, just use the formula $B\left(\frac{(1 + r)^n - 1}{r}\right)$

29. Let $B = 500$, $r = .002$, $n = 10$, and obtain $5045.24

31. Let $B = 500$, $r = .10$, and $n = 10$, and obtain $7968.71

33. Let $B = 4000$, $r = .08$, $n = 60$, and obtain $5,012,853.18

35. Let $B = 200$, $r = .064$, $n = 16$, and obtain $5306.72

37. Let $B = 10$, $r = \frac{.08}{52}$, $n = 52 \cdot 10 = 520$ and obtain $7957.13

39. There are two ways to do this problem.

    One way is to consider the infinite series $0.42(1 + .01 + .0001 + \cdots)$

    and obtaining the sum $0.42 \cdot \dfrac{1}{1 - .01} = \dfrac{42}{100} \cdot \dfrac{100}{99} = \dfrac{42}{99} = \dfrac{14}{33}$.

    Another way is to consider

    $x = 0.4242424242\ldots$

    $100x = 42.4242424242\ldots$

    Subtract the first line from the second and obtain $99x = 42$ or $x = \dfrac{42}{99} = \dfrac{14}{33}$

41. As in #39, consider $.71(1 + .01 + .0001 + \cdots)$ and obtain $\dfrac{71}{99}$.

    Can also do by setting $x = .71717171\ldots$

43. Now consider the series $0.312(1 + 0.001 + .000001 + \cdots)$

    This has sum $0.312 \cdot \dfrac{1}{1 - .001} = \dfrac{312}{1000} \cdot \dfrac{1000}{999} = \dfrac{312}{999} = \dfrac{104}{333}$.

    Can also do by setting $x = 0.312312312\ldots$ and $1000x = 312.312312312\ldots$

45. Can write this as $0.11 + .00362(1 + .001 + .000001 + \cdots)$

    This has sum $0.11 + .00362 \cdot \dfrac{1}{1 - .001} = \dfrac{11}{100} + \dfrac{362}{100,000} \cdot \dfrac{1000}{999} = \dfrac{10989}{99,900} + \dfrac{362}{99,900} = \dfrac{11,351}{99.900}$

47. This is $19(1 + .01 + .0001 + \cdots) = 19 \cdot \dfrac{1}{1 - .01} = 19 \cdot \dfrac{100}{99} = \dfrac{1900}{99}$

49. Let x = the number of minutes past the hour when the minute hand is exactly

    over the hour hand.

    In 1 minute, the minute hand moves $\dfrac{1}{60}$ of the way around the clock, or

    $\dfrac{1}{60}$ of 360°, which equals 6°. The hour hand moves $\dfrac{1}{12} \cdot \dfrac{1}{60} = \dfrac{1}{720}$ of the way around,

    or 0.5°. The minute hand starts at the top, which we will use as 0° for

    reference. The hour hand starts at 7; in degrees, $\dfrac{7}{12}$ of 360°, which equals

    210°. Set the hour hand's position x minutes after the hour equal to the

    minute hand's position to obtain $210 + .5x = 6x$.

    $$210 = 5.5x$$

    $$x \approx 38.1818 \text{ minutes.}$$

So it is 38.1818 minutes after the hour when they are on top of each other.

Problems 12.5

1.

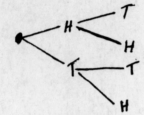

Two possibilities (H or T) for first flip, two for the second; $2\cdot2 = 4$.

3.

Six possibilities on each of the two rolls; $6\cdot6 = 36$.

5.

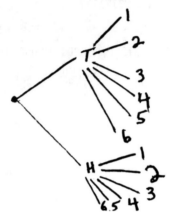

$2\cdot6 = 12$

7.

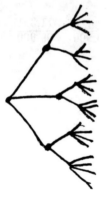

$3\cdot2\cdot4 = 24$

9.

$7\cdot6\cdot5 = 210$

11. 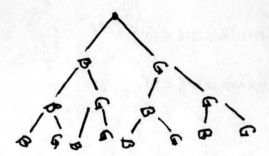 $2 \cdot 2 \cdot 2 = 8$

13. If three letters are used, the number of species that can be classified is $26 \cdot 26 \cdot 26 = 17,576$, which is not enough. Using 4 letters would permit the classification of 456,976 species, which is more than enough. The answer is 4.

15. Since there are 4 choices for each of the three letters, take $4 \cdot 4 \cdot 4 = 64$.

17. Take $6 \cdot 6 \cdot 6 \cdot 8 \cdot 8 = 13,824$ 19. $26 \cdot 9 \cdot 9 \cdot 9 \cdot 9 = 170,586$ 21. $8 \cdot 7 \cdot 6 = 336$

23. $47 \cdot 51 \cdot 54 \cdot 55 = 7,119,090$

Problems 12.6

1. $7! = 7 \cdot 6 \cdot 5 \cdot 4 \cdot 3 \cdot 2 \cdot 1 = 5040$

3. There are 6 distinct letters, so the answer is $6! = 720$

The next group of problems uses $P_{n,k} = \dfrac{n!}{(n-k)!}$

5. $\dfrac{6!}{3!} = 6 \cdot 5 \cdot 4 = 120$ 7. $\dfrac{8!}{4!} = 8 \cdot 7 \cdot 6 \cdot 5 = 1680$ 9. $\dfrac{12!}{4!} = 12 \cdot 11 \cdot 10 \cdot 9 \cdot 8 \cdot 7 \cdot 6 \cdot 5 = 19,958,400$

11. $\dfrac{9!}{1!} = 362,880$ 13. $\dfrac{n!}{n!} = 1$ 15. $\dfrac{n!}{0!} = n!$ (since 0! is defined to be 1)

17. $\dfrac{n!}{2!} = n(n-1)(n-2)\cdots 4 \cdot 3$ 19. $P_{7,4} = \dfrac{7!}{3!} = 7 \cdot 6 \cdot 5 \cdot 4 = 840$

21. $50!$, which is approximately equal to 3.0414×10^{64}

23. $P_{50,10} = \frac{50!}{40!} = 50 \cdot 49 \cdot 48 \cdot 47 \cdot 46 \cdot 45 \cdot 44 \cdot 43 \cdot 42 \cdot 41 \approx 3.7276 \times 10^{16}$

25. There are 5 letters, but 3 are identical, so the answer is $\frac{5!}{3!} = 20$

For combination problems, use $\binom{n}{k} = \frac{n!}{k!(n-k)!}$

27. $\frac{6!}{2!4!} = \frac{6 \cdot 5 \cdot 4!}{2 \cdot 4!} = \frac{30}{2} = 15$ 29. $\frac{5!}{2!3!} = \frac{120}{12} = 10$ 31. $\frac{11!}{3!8!} = \frac{11 \cdot 10 \cdot 9 \cdot 8!}{6 \cdot 8!} = \frac{990}{6} = 165$

33. $\frac{9!}{4!5!} = \frac{9 \cdot 8 \cdot 7 \cdot 6}{24} = 126$ 35. $\frac{11!}{7!4!} = 330$ 37. $\frac{13!}{1!12!} = 13$ 39. $\frac{8!}{1!7!} = 8$

41. $\frac{5!}{0!5!} = 1$ since $0! = 1$ (recall also that $\binom{n}{n} = 1$ for any n)

43. 1   45. $\frac{n!}{1!(n-1)!} = \frac{n(n-1)!}{(n-1)!} = n$ 47. $\frac{n!}{(n-2)!2!} = \frac{n(n-1)(n-2)!}{(n-2)!2} = \frac{n(n-1)}{2}$

49. $\binom{10}{4}$, since you are choosing 4 people from 10. $\binom{10}{4} = \frac{10!}{6!4!} = 210$

51. Choose 13 from 52, or $\binom{52}{13} = \frac{52!}{13!39!} = 635{,}013{,}559{,}680$

53. Of the 9, choose a group of 6 who vote yes. Then the answer is $\binom{9}{6} = \frac{9!}{3!6!} = 84$

55. The number of ways to choose the questions on probability is $\binom{10}{3}$, and the number of ways to choose the questions on matrix theory is $\binom{8}{2}$. Multiply the two together and obtain $120 \cdot 28 = 3360$.

57. There are 13 ways to choose the denomination making up the three of a kind. Once that is done, there are $\binom{4}{3}$ ways to choose the cards from that denomination (which of the three suits are used). Once the three of a kind denomination is chosen, there are 12 remaining denominations which could be chosen for the pair. Once the denomination of the pair is chosen, there are $\binom{4}{2}$ ways to choose the cards from that denomination. So, multiply $13 \cdot \binom{4}{3} \cdot 12 \cdot \binom{4}{2} = 13 \cdot 4 \cdot 12 \cdot 6 = 3744$

59. $\binom{10}{4} = 210$

61. The only one-letter word fitting the bill is X.

    There are two two-letter words: XX and XY (since YX and XY are the same).

    The three letter words are XXX, XXY, and XYY. Total is 6.

63. A straight can start at A, 2, 3, 4, 5, 6, 7, 8, 9, or 10, which is 10 possible denominations.

    For each card of the five card straight, there are 4 possible choices for the card making up the denomination; therefore, there are $10 \cdot 4^5 = 10,240$ possible straights (this includes straight flushes).

65. There are $\binom{13}{2}$ ways to choose the denominations of the 2 pair.

    Within each denomination, there are $\binom{4}{2}$ ways to choose the cards from the particular denomination. Then there are 44 cards left from which to choose the fifth card. The answer is $\binom{13}{2} \cdot \binom{4}{2} \cdot \binom{4}{2} \cdot 44 = 78 \cdot 6 \cdot 6 \cdot 44 = 123,552$

67. $5 \cdot 3 = 15$

69. You are choosing 5 from a group of 8; $\binom{8}{5} = \frac{8!}{5!3!} = \frac{8 \cdot 7 \cdot 6}{6} = 56$

71. $P_{20,2} = 20 \cdot 19 = 380$ (there are 20 choices for first and then 19 for last)

73. $4^{20} \approx 1.0995 \text{ X } 10^{12}$

75. First select the 37 of 100 that advanced; then, from the remaining 63, select the 51 that declined to obtain $\binom{100}{37} \cdot \binom{63}{51}$. Or, select the decliners first and then the ones that advanced to get $\binom{100}{51} \cdot \binom{49}{37}$. These expressions give the same answer, approximately $9.1261 \text{ X } 10^{39}$.

77. $P_{9,2} \cdot P_{7,2} = (9 \cdot 8)(7 \cdot 6) = 3024$

79. Label the teams A, B, C, D respectively; then there are $\binom{15}{4} \cdot \binom{11}{3} \cdot \binom{8}{5} \cdot \binom{3}{3}$ ways of arranging the children, because any 4 of the 15 could be placed on team A, then any three of the remaining 11 on team B, then any 5 of the remaining 8 on team C, and all three of the remaining people on team D. Multiply out these numbers and obtain 12,612,600.

81. $\binom{12}{8} + \binom{12}{9} + \binom{12}{10} + \binom{12}{11} + \binom{12}{12} = 794$

83. Consider the Smiths as a block and each of the four other people as separate
blocks. The number of ways to arrange 5 blocks is $5 \cdot 4 \cdot 3 \cdot 2 \cdot 1 = 120$. Since
the Smiths can be arranged in two different ways within this block, multiply
120 by 2 which equals 240 total arrangements where they are together.

85. First, calculate how many ways to get a number with exactly one digit besides
0 repeating. Remember a 0 cannot be in the first three digits.

A) First, we will count up the number of ways that a number can have
a repeated digit if the repeated digit is not a 0.

i) The number of ways that both occurences of the doubled digit occur in the three digit
prefix is computed as follows: first, select the two places in the prefix where the digit occurs;
next, select the other digit in the prefix; third, select the remaining 4 digits. There are
$\binom{3}{2} \cdot 8 \cdot 8 \cdot 7 \cdot 6 \cdot 5$ to do this (the first 8 occurs because the other digit in the prefix cannot be a 0
nor the repeated digit; since you have then used up two of
the digits, you now have a choice of 8 digits for the next number, etc.).

ii) Now count the number of ways that one occurence of the doubled digit is in the prefix
and one in the last 4 digits. First, select the postion of the doubled digit in the prefix;
second, select the other two digits in the prefix (neither of which can be 0); third, select the
position of the repeated digit in the last four digits; fourth, select the other three
digits in the last four. There are $\binom{3}{1} \cdot 8 \cdot 7 \cdot \binom{4}{1} \cdot 7 \cdot 6 \cdot 5$ ways to do this.

iii) Now we have to count up the number of ways the repeated digit can occur in the last
four digits. First, select the first three digits (they cannot be 0 nor the repeated digit); next,
select the positions of the repeated digit; third, select the other two digits of the last four.
There are $8 \cdot 7 \cdot 6 \cdot \binom{4}{2} \cdot 6 \cdot 5$ ways to do this.

(continued)

Add the expressions found in parts i), ii), and iii) to obtain the number of ways a specific digit other than 0 could be repeated exactly once. Multiply this number by 9 to get the total number of numbers containing a digit other than 0 which is repeated exactly once.

B) Now calculate how many numbers have exactly two occurences of 0.
First, select the first 3 digits (they can be any numbers except 0);
second, select the position of the two 0's (they must occur in the last
four digits); third, select the last two digits. There are $9 \cdot 8 \cdot 7 \cdot \binom{4}{2} \cdot 6 \cdot 5$ ways to do this.

To get the final answer, add the answers in parts A and B to get 2,268,000!!!!

87. There are 9 choices for the third digit (0 is not permitted), 10 for the fourth, 10 for the fifth, 10 for the sixth, and 10 for the seventh, so the answer is $9 \cdot 10 \cdot 10 \cdot 10 \cdot 10 = 90,000$.

89. a) $4^5 = 1024$, since there are 4 choices for each of the 5.
   b) First, select which blood type is repeated. There are 4 ways to do that.
   Now select which two of the five people are the ones with the same
   type. Then select how to assign the remaining three types to the
   three people that are left. There are $4 \cdot \binom{5}{2} \cdot 3 \cdot 2 \cdot 1 = 240$ ways to do this.

91. a) $2^5 = 32$, since 2 possibilities for each of the 5 children.
   b) $2^3 = 8$, since it is only the 3 middle children where there is a choice.
   c) GBBBG, BGBBG, BBGBG, BBBGG are the possibilities, so there are 4.
   A G must go at the end, and there are 4 choices for where the other G may go.

Problems 12.7

1. a) The sample space consists of the set containing each of the 52 cards.
   b) $\frac{1}{52}$

3. a) Each of the $\binom{10}{5} = 252$ possible choice of 5 people from the 10 person group.
   b) $\frac{1}{252}$

5. a) There are $P_{6,2} = 30$ different pairings of 2 people of the six.

b) $\frac{1}{30}$

7. There are 12! possible arrangements of the gradess so a) $\frac{1}{12!} \approx 2.0877 \times 10^{-9}$

b) $1 - \frac{1}{12!}$

c) 0. How can all but one be correct? This is not possible.

9. a) $\dfrac{\binom{20}{6}}{\binom{28}{6}} = \dfrac{10,336}{376,740} \approx 0.1029$ b) $\dfrac{\binom{20}{3} \cdot \binom{8}{3}}{\binom{28}{6}} = \dfrac{63,840}{376,740} \approx 0.1695$

11. There are 13 possibilities for the three of a kind, and three of the four cards from that denomination must be chosen. There are then 12 possibilities for the denomination of the pair, and two of the four cards must be chosen. Therefore, the answer is

$$\frac{13 \cdot \binom{4}{3} \cdot 12 \cdot \binom{4}{2}}{\binom{52}{5}} = \frac{3744}{2,598,960} \approx 0.00144 \approx \frac{1}{694}$$

13. First, choose the denomination of the three of a kind. There are 13 ways to do that. Then there are $\binom{4}{3}$ ways to choose the three cards from that denomination to form the three of a kind. Then there are $\binom{12}{2}$ ways to choose the denominations of the two cards that don't match, and 4 ways to choose the card from that denomination in each case. The answer is, then,

$$\frac{13 \cdot \binom{4}{3} \cdot \binom{12}{2} \cdot 4^2}{\binom{52}{5}} = \frac{54912}{2,598,960}.$$ Another way to think of choosing the cards that

don't match the three of a kind is to pick any two of the remaining 48 cards, but then subtract off the number of pairs. This gives the answer

$$\frac{13 \cdot \binom{4}{3}\left(\binom{48}{2} - 12 \cdot \binom{4}{2} \right)}{\binom{52}{5}}$$ which gives the same number.

15. First select the denomination of the pair. There are 13 ways to do that. Then select the two cards making up the pair. There are $\binom{4}{2}$ ways to do that. Now select three of the remaining 12 denominations for the cards that do not match; for each denomination, there are 4 possibilities for the card to be chosen from that denomination. Therefore, the answer is

$$\frac{13 \cdot \binom{4}{2} \cdot \binom{12}{2} \cdot 4^3}{\binom{52}{5}} = \frac{1,098,240}{2,589,960} \approx 0.4227$$

17. $\frac{1}{4^5} = \frac{1}{1024}$

19. There are $3 + 8 + 3 + 2 = 16$ people. The number of ways to choose 3 people is $\binom{16}{3} = 560$.

We must find the number of ways of choosing exactly 2 people having the same blood type.

The number of ways of choosing 2 people with A, 1 not A is $\binom{3}{2} \cdot 13 = 3 \cdot 13 = 39$

The number for choosing 2 people with O, one not O is $\binom{8}{2} \cdot 8 = 28 \cdot 8 = 224$

The number of ways of choosing 2 with B, 1 not B is $\binom{3}{2} \cdot 13 = 39$

The number of ways of choosing 2 with AB, 1 not AB is $\binom{2}{2} \cdot 14 = 14$.

Add these to get 316 ways of getting exactly 2 people with the same type.

The desired probability is then $\frac{316}{560} = \frac{79}{140} \approx 0.5643$.

21. Since there are 2 possibilities for each of the 4 children, there are $2^4 = 16$ possible arrangements of the sexes of the four children. a) The ones where all of the children are of the same sex are BBBB and GGGG, so $\frac{2}{16} = \frac{1}{8}$.

b) Possibilities are BBBB, BBBG, BBGG, BGGG, GGGG so the answer is $\frac{5}{16}$.

23. a) You can choose any 4 of the 15 people, so the sample space is all four-person subsets of the 15 people.

b) $\frac{1}{1365}$ since $\binom{15}{4} = 1365$

c) $\frac{1}{1365}$ since there is only 1 way to choose 4 left-handers from 4 lefties.

Problems 12.8

1. $\frac{5!}{3!2!} = \frac{5 \cdot 4}{2} = 10$

3. $\frac{7!}{3!4!} = \frac{7 \cdot 6 \cdot 5}{3 \cdot 2 \cdot 1} = 35$

5. $\dfrac{11!}{8!3!} = \dfrac{11\cdot10\cdot9}{6} = 165$ 7. 1, since $\binom{n}{0} = 1$ for any n.

9. $x^5 + \binom{5}{1}x^4(-y)^1 + \binom{5}{2}x^3(-y)^2 + \binom{5}{3}x^2(-y)^3 + \binom{5}{4}x^1(-y)^4 + (-y)^5$

$\quad x^5 - 5x^4y + 10x^3y^2 - 10x^2y^3 + 5xy^4 - y^5$

11. $(x)^4 + \binom{4}{1}(x)^3(-2y) + \binom{4}{2}(x)^2(-2y)^2 + \binom{4}{3}x(-2y)^3 + (-2y)^4 = x^4 - 8x^3y + 24x^2y^2 - 32xy^3 + 16y^4$

13. $a^8 + \binom{8}{1}a^7b^1 + \binom{8}{2}a^6b^2 + \binom{8}{3}a^5b^3 + \binom{8}{4}a^4b^4 + \binom{8}{5}a^3b^5 + \binom{8}{6}a^2b^6 + \binom{8}{7}a^1b^7 + b^8$

$\quad = a^8 + 8a^7b + 28a^6b^2 + 56a^5b^3 + 70a^4b^4 + 56a^3b^5 + 28a^2b^6 + 8ab^7 + b^8$

15. $(2a)^5 + \binom{5}{1}(2a)^4(-3b) + \binom{5}{2}(2a)^3(-3b)^2 + \binom{5}{3}(2a)^2(-3b)^3 + \binom{5}{4}(2a)(-3b)^4 + (-3b)^5$

$\quad = 32a^5 - 240a^4b + 720a^3b^2 - 1080a^2b^3 + 810ab^4 - 243b^5$

17. $\left(\dfrac{u}{3}\right)^3 + \binom{3}{1}\left(\dfrac{u}{3}\right)^2\left(-\dfrac{v}{4}\right) + \binom{3}{2}\left(\dfrac{u}{3}\right)\left(-\dfrac{v}{4}\right)^2 + \left(-\dfrac{v}{4}\right)^3 = \dfrac{u^3}{27} - \dfrac{u^2v}{12} + \dfrac{uv^2}{16} - \dfrac{v^3}{64}$

19. $(x^2)^4 + \binom{4}{1}(x^2)^3(2y) + \binom{4}{2}(x^2)^2(2y)^2 + \binom{4}{3}2x^2(2y)^3 + (2y)^4$

$\quad\quad\quad = x^8 + 8x^6y + 24x^4y^2 + 32x^2y^3 + 16y^4$

21. $(a^2)^4 + \binom{4}{1}(a^2)^3(b^3) + \binom{4}{2}(a^2)^2(b^3)^2 + \binom{4}{3}(a^2)(b^3)^3 + (b^3)^4$

$\quad = a^8 + 4a^6b^3 + 6a^4b^6 + 4a^2b^9 + b^{12}$

-437-

23. $(\sqrt{x})^6 + \binom{6}{1}(\sqrt{x})^5\sqrt{y} + \binom{6}{2}(\sqrt{x})^4(\sqrt{y})^2 + \binom{6}{3}(\sqrt{x})^3(\sqrt{y})^3 + \binom{6}{4}(\sqrt{x})^2(\sqrt{y})^4$

$\quad + \binom{6}{5}\sqrt{x}(\sqrt{y})^5 + (\sqrt{y})^6$

$\quad = x^3 + 6x^2\sqrt{xy} + 15x^2y + 20xy\sqrt{xy} + 15xy^2 + 6y^2\sqrt{xy} + y^3$
$\quad = x^3 + 6x^{5/2}y^{1/2} + 15x^2y + 20x^{3/2}y^{3/2} + 15xy^2 + 6x^{1/2}y^{5/2} + y^3$

25. $(3\sqrt{x} + 3\sqrt{y})^4 = [3(\sqrt{x} + \sqrt{y})]^4 = 3^4(\sqrt{x} + \sqrt{y})^4$

$$= 81\left((\sqrt{x})^4 + \binom{4}{1}(\sqrt{x})^3\sqrt{y} + \binom{4}{2}(\sqrt{x})^2(\sqrt{y})^2 + \binom{4}{3}\sqrt{x}(\sqrt{y})^3 + (\sqrt{y})^4\right)$$

$$= 81x^2 + 324x\sqrt{xy} + 486xy + 342y\sqrt{xy} + 81y^2$$

$$= 81x^2 + 324x^{3/2}y^{1/2} + 486xy + 342x^{1/2}y^{3/2} + 81y^2$$

27. $(ab)^3 + \binom{3}{1}(ab)^2(-cd) + \binom{3}{2}ab(-cd)^2 + (-cd)^3 = a^3b^3 - 3a^2b^2cd + 3abc^2d^2 - c^3d^3$

29. $w^4 + \binom{4}{1}w^3\left(-\frac{u}{v}\right) + \binom{4}{2}w^2\left(-\frac{u}{v}\right)^2 + \binom{4}{3}w\left(-\frac{u}{v}\right)^3 + \left(-\frac{u}{v}\right)^4$

$$= w^4 - 4w^3\cdot\frac{u}{v} + 6w^2\cdot\frac{u^2}{v^2} - 4w\cdot\frac{u^3}{v^3} + \frac{u^4}{v^4}$$

31. $1 - 10a + 45a^2 - 120a^3 + 210a^4 - 252a^5 + 210a^6 - 120a^7 + 45a^8 - 10a^9 + a^{10}$

33. $1 + 3\sqrt{x} + 3(\sqrt{x})^2 + (\sqrt{x})^3 = 1 + 3\sqrt{x} + 3x + x\sqrt{x} = 1 + 3x^{1/2} + 3x + x^{3/2}$

35. $(\sqrt{5})^4 + \binom{4}{1}(\sqrt{5})^3\sqrt{7} + \binom{4}{2}(\sqrt{5})^2(\sqrt{7})^2 + \binom{4}{3}\sqrt{5}(\sqrt{7})^3 + (\sqrt{7})^4$

$$= 25 + 20\sqrt{35} + 210 + 28\sqrt{35} + 49 = 284 + 48\sqrt{35}$$

37. The coefficients will just be 1, $\binom{4}{1}$, $\binom{4}{2}$, $\binom{4}{3}$, and 1 so obtain $1 + 4z^6 + 6z^{12} + 4z^{18} + z^{24}$

39. Put $(u + v)$ in one group and -2 in another. Apply the Binomial Theorem.
$$(u + v)^4 + \binom{4}{1}(u + v)^3(-2) + \binom{4}{2}(u + v)^2(-2)^2 + \binom{4}{3}(u + v)(-2)^3 + (-2)^4$$

$$= u^4 + 4u^3v + 6u^2v^2 + 4uv^3 + v^4 - 8(u^3 + 3u^2v + 3uv^2 + v^3)$$
$$+ 24(u^2 + 2uv + v^2) - 32(u + v) + 16$$
$$= u^4 + 4u^3v + 6u^2v^2 + 4uv^3 + v^4 - 8u^3 - 24u^2v - 24uv^2 - 8v^3 + 24u^2 + 48uv$$
$$+ 24v^2 - 32u - 32v + 16$$

41. This is the same as #39 except x, y and z take the roles of u, v and -2 above.

Obtain $(x + y)^4 + 4(x + y)^3 z + 6(x + y)^2 z^2 + 4(x + y)z^3 + z^4$

$= x^4 + 4x^3 y + 6x^2 y^2 + 4xy^3 + y^4 + 4x^3 z + 12x^2 yz + 12xy^2 z + 4y^3 z$

$\quad + 6x^2 z^2 + 12xyz^2 + 6y^2 z^2 + 4xz^3 + 4yz^3 + z^4$

43. $x^{5n} + 5x^{4n} y^n + 10x^{3n} y^{2n} + 10x^{2n} y^{3n} + 5x^n y^{4n} + y^{5n}$

45. It would be $-\binom{10}{3} = -120$ 47. $\binom{8}{4} = 70$

49. $\binom{n}{k} = \dfrac{n!}{(n - k)!k!}$ $\binom{n}{n - k} = \dfrac{n!}{[(n - (n - k)]!(n - k)!} = \dfrac{n!}{k!(n - k)!}.$

They should be equal since $(a + b)^n = (b + a)^n$ and so the kth coefficient of $(a + b)^n$ is equal to the $(n - k)^{th}$ coefficient of $(b + a)^n$.

51. $(1 + 1)^n = \binom{n}{0}1^n + \binom{n}{1}1^{n-1}1^1 + \binom{n}{2}1^{n-2}1^2 + \cdots + \binom{n}{n-1}1^1 1^{n-1} + \binom{n}{n}1^n$

$2^n = \binom{n}{0} + \binom{n}{1} + \binom{n}{2} + \cdots + \binom{n}{n-1} + \binom{n}{n}$

53. $100! \approx 9.3248476 \times 10^{157}$ $200! \approx 7.883293288 \times 10^{374}$

55.   1   7   21   35   35   21   7   1   (obtained by adding adjacent elements in the previous row)

Review Exercises for Chapter Twelve

1. Is it true for n = 1? Does $1 = \dfrac{3^1 - 1}{2}$? Yes, the right side = 1.

Now assume it is true for n = k; that is, $1 + 3 + 3^2 + \cdots + 3^{k-1} = \dfrac{3^k - 1}{2}$. (Ind Hyp).

Is it true for n = k + 1 ? That is, does $1 + 3 + 3^2 + \cdots + 3^{k-1} + 3^k = \dfrac{3^{k+1} - 1}{2}$?

$1 + 3 + 3^2 + \cdots + 3^{k-1} + 3^k = \dfrac{3^k - 1}{2} + 3^k$ by the Induction Hypothesis.

$$= \dfrac{3^k - 1}{2} + \dfrac{2 \cdot 3^k}{2} = \dfrac{3 \cdot 3^k - 1}{2} = \dfrac{3^{k+1} - 1}{2} \quad \text{QED}$$

3. Is it true for k = 1? Does $1 = 1(2 \cdot 1 - 1)$? Yes.

Now assume it is true for $n = k$, that is assume $1 + 5 + 9 + \cdots + (4k - 3) = k(2k - 1)$.

Now prove that it is true for $n = k + 1$; that is, we want to prove that

$1 + 5 + 9 + \cdots (4k - 3) + (4(k + 1) - 3) = (k + 1)(2(k + 1) - 1) = (k + 1)(2k + 1) = 2k^2 + 3k + 1$

The left side of the above equals $1 + 5 + 9 + (4k - 3) + (4k + 1)$

$$= k(2k - 1) + (4k + 1) \text{ by Ind. Hyp.}$$

$$= 2k^2 - k + 4k + 1 = 2k^2 + 3k + 1 \quad \text{QED}$$

5. 10, 13, 16, 19, 22

7. $4 + 3(n - 1) = 3n + 1$

9. $3 + 5 + 7 + 9 + 11 + 13 = 48$

11. $\displaystyle\sum_{k=0}^{m} 3^k$

13. $\displaystyle\sum_{k=1}^{6} (-1)^{k+1} x^{4k-1}$ (are several equivalent answers)

15. $6 - \frac{1}{2}(n - 1) = -\frac{1}{2}n + \frac{13}{2}$, starting with $n = 1$

17. $-27 \cdot \left(-\frac{1}{3}\right)^n = (-3)^{3-n}$, starting with $n = 0$

19. 2, 6, 18, 54, 162, 486

21. $1, \frac{5}{3}, \frac{17}{9}, \frac{53}{27}, \frac{161}{81}, \frac{485}{243}$

23. $4^9 = 262{,}144$

25. Because of the -3, $a_n = a + b(-3)^n$. Now $1 = a_0 = a + b(-3)^0 = a + b$.

Plug $n = 0$ into the formula given in the problem to obtain $-1 = a_1 = a - 3b$

Solve the two equations simultaneously to obtain $a = 2$ and $b = -1$;

therefore, $a_n = 2 - 3^n$.

27. $\dfrac{1 - 3^9}{1 - 3} = \dfrac{1 - 19{,}683}{-2} = 9840$

29. $\dfrac{1}{1 - (-1/2)} = \dfrac{1}{3/2} = \dfrac{2}{3}$

31. $2^4 = 16$

33. $5 \cdot 6 \cdot 8 \cdot = 240$

35. $9 \cdot 8 \cdot 7 = 504$

37. $P_{7,5} = \dfrac{7!}{2!} = 7 \cdot 6 \cdot 5 \cdot 4 \cdot 3 = 2520$

39. a) $\dfrac{9!}{4!5!} = \dfrac{9 \cdot 8 \cdot 7 \cdot 6 \cdot 5!}{4 \cdot 3 \cdot 2 \cdot 5!} = 126$

b) $\dfrac{7!}{2!5!} = \dfrac{7 \cdot 6}{2} = 21$

c) $\dfrac{9!}{8!1!} = 9$

d) 9

e) $\dfrac{10!}{6!4!} = \dfrac{10 \cdot 9 \cdot 8 \cdot 7}{4 \cdot 3 \cdot 2} = 210$

41. $11! = 39,916,800$

43. There are 7 letters, but one is repeated twice and another three times, so the answer is $\frac{7!}{2!3!} = \frac{7 \cdot 6 \cdot 5 \cdot 4 \cdot 3 \cdot 2 \cdot 1}{2 \cdot 6} = 420$

45. There are 4 choices for the suit, and then $\binom{13}{5}$ choices for the 5 cards in that suit which make up the flush, so obtain $4 \cdot \binom{13}{5} = 5148.$ 40 of these are straight flushes.

47. A straight can start at any of the denominations A, 2, 3 ,... up to 10.

Each of the five cards can be any of the four suits. The answer is

$$\frac{10 \cdot 4^5}{\binom{52}{5}} = \frac{10,240}{2,598,960} \approx 0.00394 \approx \frac{1}{253.8} \text{ (including straight flushes)}$$

49. $\frac{1}{P_{4,2}} = \frac{1}{12}$

51. Either select the advancing stocks and then the declining ones, or vice-versa; the answer can be written as $\binom{50}{18}\binom{32}{13}$ or $\binom{50}{13}\binom{37}{18} \approx 6.27 \times 10^{21}.$

53. a) $\frac{1}{2^{10}} = \frac{1}{1024}$ b) $\frac{\binom{10}{3}}{2^{10}} = \frac{120}{1024} = \frac{15}{128}$

55. a) $4^4 = 256$ b) $4 \cdot 3 \cdot 2 \cdot 1 = 24$

57. $a^6 + \binom{6}{1}a^5 b + \binom{6}{2}a^4 b^2 + \binom{6}{3}a^3 b^3 + \binom{6}{4}a^2 b^4 + \binom{6}{5}ab^5 + b^6$

$= a^6 + 6a^5 b + 15a^4 b^2 + 20a^3 b^3 + 15a^2 b^4 + 6ab^5 + b^6$

59. $(3x)^5 + \binom{5}{1}(3x)^4(-5y) + \binom{5}{2}(3x)^3(-5y)^2 + \binom{5}{3}(3x)^2(-5y)^3 + \binom{5}{4}(3x)(-5y)^4 + (-5y)^5$

$= 243x^5 - 2025x^4 y + 6750x^3 y^2 - 11250x^2 y^3 + 9375xy^4 - 3125y^5$